*Routledge Revivals*

# Monitoring Active Volcanoes

Originally published in 1995, *Monitoring Active Volcanoes* is a comprehensive text which addresses the importance of volcano surveillance in the context of forecasting eruptive activity and mitigating its effects. The traditional core of seismic and ground deformation monitoring is discussed, along with more innovative techniques involving the recording of microgravity and micro-magnetic variations, and the changing compositions of volcanic gases and liquids. The role of satellites is stressed, particularly with regard to the capabilities for measuring surface deformation, recognizing thermal anomalies and monitoring gas and ash plumes from space platforms. This book provides an invaluable insight into how and why volcanoes are monitored. It will be of interest to volcanologists, geophysicists and earth scientists.

# Monitoring Active Volcanoes

## Strategies, Procedures and Techniques

Edited by Bill McGuire, Christopher R.J. Kilburn and John Murray

Routledge
Taylor & Francis Group

First published in 1995 by UCL Press

This edition first published in 2022 by Routledge
4 Park Square, Milton Park, Abingdon, Oxon, OX14 4RN
and by Routledge
605 Third Avenue, New York, NY 10158

*Routledge is an imprint of the Taylor & Francis Group, an informa business*

© 1995 W. J. McGuire, C. Kilburn, J. Murray and contributors

The right of Bill McGuire, Christopher Kilburn and John Murray to be identified as the authors of this work has been asserted by them in accordance with sections 77 and 78 of the Copyright, Designs and Patents Act 1988.

Publisher's Note
The publisher has gone to great lengths to ensure the quality of this reprint but points out that some imperfections in the original copies may be apparent.

ISBN 13: 978-1-032-35481-1 (hbk)
ISBN 13: 978-1-003-32708-0 (ebk)
ISBN 13: 978-1-032-35484-2 (pbk)
Book DOI 10.4324/9781032354811

# Monitoring active volcanoes

Strategies, procedures and techniques

Edited by

Bill McGuire
*Cheltenham & Gloucester College of Higher Education*
*and University College London*

Christopher Kilburn
*University College London*

John Murray
*Open University*

UCL
PRESS

First published in 1995 by UCL Press

UCL Press Limited
University College London
Gower Street
London WC1E 6BT

The name of University College London (UCL) is a registered trade mark
used by UCL Press with the consent of the owner.

ISBN: 1-85728-036-9

**Library of Congress Cataloging-in-Publication Data**

Monitoring active volcanoes / edited by Bill McGuire.
     p.    cm.
   Includes bibliographical references and index.
   ISBN 1-85728-036-9
   1. Volcanism—Research—Methodology. I. McGuire, Bill. 1954–
QE522.M66 1995
551.2'1'072—dc8820                  94-43478
                                      CIP

Typeset in Times and Optima.
Printed and bound by
Biddles Ltd., Guildford and King's Lynn, England.

# Contents

# Preface

This book seeks to chart recent innovation and progress in the methods and strategies developed to monitor active volcanoes and to forecast their activity. The climactic eruption of Mount St Helens in 1980 heralded a decade rich both in the number and variety of volcanic eruptions, and in ideas about how to monitor them. Over this period, the more traditional "core" techniques of seismic and ground deformation monitoring have been augmented by innovative methods involving, for example, the measurement of microgravity and micromagnetic variations, and the changing compositions of volcanic gases and ground waters. At the same time all techniques have benefited enormously from the plummeting costs of computer, electronic, and communications hardware, from the development of more portable instrumentation, and from greatly increased computing power. The role of satellite platforms in volcano monitoring has become more clearly defined, with capabilities for the detection of thermal anomalies, the remote measurement of ground deformation, and the observation and tracking of ash plumes already demonstrated.

Following an introductory overview and consideration of the crucial role played by data acquisition and telemetry in contemporary volcano monitoring, all the major surveillance techniques in current operational use are discussed. A case-study approach is used throughout to illustrate how the various methods have been fruitfully utilized at volcanoes around the world. The closing chapters are devoted to discussion of forecasting, both of the behaviour of lava flows, and of eruptions in general, and of future prospects for the science of volcanology in this International Decade for Natural Disaster Reduction.

In addition to the contributors and editors, numerous people have provided invaluable help and support in the preparation of this book. In particular those colleagues who agreed to review chapters; Tom Casadevall, Gennaro Corrado, Paul Cross, John Ewert, Peter Francis, Lori Glaze, Michele Halbwachs, Duncan Hawksbee, John Horsfall, Dallas Jackson, Steve McNutt, Tom Murray, Domenico Patella, Don Peterson, Harry Pinkerton, Dave Rothery, Freystein Sigmundsson, and Don Swanson. The editors would also like to thank Roger Jones, Kate Williams, and all at UCL Press for their patience, advice, and encouragement – particularly when the going got rough!

BILL McGUIRE
DECEMBER 1994

This book is dedicated to the memory of Geoff Brown and to all those who have lost their lives while striving to increase our understanding of volcanoes and their activity.

# List of contributors

*R. J. Andres* Department of Earth Sciences, Tennessee Technological University, Cookeville, TN 38505, USA.

*P. Bachelery* Département des Sciences de la Terre, Université de la Réunion, 15 Av. René Cassin, 97489 Saint Denis, Réunion Island, France.

*P. A. Blum* Institut de Physique du Globe de Paris, 4 Place Jussieu, 75252 Paris Cedex 05, France.

*A. Bonneville* Université Française du Pacifique, Centre Universitaire de Polynésie Français, BP 6570 Faaa - Aéroport, Tahiti.

*H. Delorme* Institut de Physique du Globe de Paris, 4 Place Jussieu, 75252 Paris Cedex 05, France.

*F. Ferrucci* Dipartimento di Scienze della Terra, Universitá della Calabria, I-87063 Arcavacata di Rende (Cosenza), Italy. (also at - Gruppo Nazionale per la Vulcanologia, Centro Operativo, Piazza Roma 2, I-95123 Catania, Italy.)

*M. Halbwachs* Laboratoire d'Instrumentation Géophysique, Université de Savoie, BP 1104-73011, Chambéry Cedex, France.

*C. R. J. Kilburn* Department of Geological Sciences, University College London, Gower Street, London, WC1E 6BT, UK.

*P. Kowalski* Observatoire Volcanologique du Piton de la Fournaise, 14 RN3 27ème km, 97418 La Plaine des Cafres, Isle de Réunion, France.

*J. F. Lénat* Centre de Recherches Volcanologiques, Observatoire de Physique du Globe de Clermont, Université Blaise Pascal, 5, Rue Kessler, 63038 Clermont-Ferrand Cedex, France.

*P. Lesage* Laboratoire d'Instrumentation Géophysique, Université de Savoie, BP 1104-73011, Chambéry Cedex, France.

*W. J. McGuire* Centre for Volcanic Research, Department of Geography and Geology, Cheltenham & Gloucester College of Higher Education, Francis Close Hall, Swindon Road, Cheltenham GL50 4AZ, UK. (Also at Department of Geological Sciences, University College London, Gower Street, London WC1E 6BT, UK.)

*J. B. Murray* Department of Earth Sciences, The Open University, Walton Hall, Milton Keynes MK7 6AA, UK.

*G. Nunnari* Dipartimento Elettrico, Elettronico e Sistemistico, Universita di Catania, Viale Andrea Doria 9, I-95125 Catania, Italy.

*C. Oppenheimer* Department of Geography, University of Cambridge, Downing Place, Cambridge CB2 3EN, UK.

*H. Pinkerton* Environmental Sciences Division, IEBS, Lancaster University, Lancaster LA1 4YQ, UK.

*G. Puglisi* Istituto Internazionale di Vulcanologia, Piazza Roma 2, 95123 Catania, Sicilia, Italy.

*A. D. Pullen* Department of Civil Engineering, Imperial College, London SW7 2BU, UK.

*W. I. Rose* Department of Geological Engineering, Geology, and Geophysics, Michigan Technological University, Houghton, MI 49931, USA.

D. A. Rothery Department of Earth Sciences, The Open University, Walton Hall, Milton Keynes MK7 6AA, UK.

*H. Rymer* Department of Earth Sciences, The Open University, Walton Hall, Milton Keynes MK7 6AA, UK.

*S. J. Saunders* Neotectonics Research Unit, Brunel University College, Borough Road, Isleworth, Middlesex TW7 5DU, UK.

*D. Tedesco* Institute for study of the Earth's interior, Okatama University, Misasa, Tottori-ken 682-01, Japan.

*R. I. Tilling* U. S. Geological Survey, Branch of Volcanic and Geothermal Processes, 345 Middlefield Road, MS-910, Menlo Park, CA 94025, USA.

*J. P. Toutain* Laboratoire de Géochimie, Université Paul Sabatier, 38, rue des 36 ponts, 31400 Toulouse, France.

*J. Vademeulebrouk* Laboratoire d'Instrumentation Géophsique, Université de Savoie, BP 1104-73011, Chambéry Cedex, France.

*L. Wilson* Department of Earth Sciences, IEBS, Lancaster University, Lancaster LA1 4YQ, UK.

*J. Zlotnicki* Laboratoire de Géophysique d'Orléans (Département des Obervatoires IPGP) 3D av de la recherche scientifique, 45071 Orléans cedex 02, France.

# 1 Monitoring active volcanoes – an introduction

W. J. McGuire

## 1.1 Why monitor volcanoes?

Active volcanoes attract both pure and applied scientists, offering the former a tantalizing, if often obscure, "window on the Earth's interior", and the latter the opportunity to pursue the grail of successful eruption prediction. Volcanoes provide clues to processes operating deep beneath the surface, in the mantle or lower crust. Volcano-related seismicity and eruption periodicity can provide information on the nature and rate of magma ascent from the source region, while monitoring the changing composition of magma and volcanic gases can offer an insight into the fractionation processes operating during magma ascent and storage. Similarly, analysis of mantle or deep-crustal xenoliths ejected during eruptions supplies clues to the composition of the magma source region and the melting processes operating therein. At shallower depths, the changing shape of the ground surface provides information on magma accumulation and transport within the volcanic pile itself, the nature and changing form of this high-level volcanic "plumbing system" being further constrained by the monitoring of variations in the local gravity, magnetic and electrical fields, changes in the compositions of volcanic gases and the strength and location of sources of thermal radiance.

This chapter is designed to provide an introduction to some of the contemporary issues associated with volcano monitoring. The development of sensible monitoring strategies is discussed, and the role of monitoring in eruption forecasting and prediction, and hazard mitigation is highlighted. In addition, recent advances and developments in monitoring techniques are briefly introduced to provide a "taster" for the succeeding specialist chapters. In view of the recent tragic deaths of a number of scientists while undertaking volcano monitoring, the chapter is concluded with a number of safety guidelines which, if followed, should help to minimize the risks involved in studying active volcanoes

The incentive to understand how volcanoes function has never been driven purely by scientific curiosity. Ever since systematic monitoring began with the opening in 1847 of the Osservatorio Vesuviano on the flanks of Vesuvius, a primary concern has been to develop a greater understanding of how and why volcanoes erupt, in order to predict, assess, and mitigate their effects on nearby communities. It is no coincidence that the first observatory was established on a volcano located in a densely populated region of Italy, nor that the next observatories to be established, at Mount Asama (Japan), and Kilauea (Hawaii), became operational within a decade of the devastating eruption of Mont Pelée on the island of Martinique (Caribbean), which claimed 29000 lives in 1902.

As illustrated by the 1991 eruption of Pinatubo in the Philippines (PVOT 1991), volcanic eruptions can be enormously destructive, not only in terms of injury, loss of life, and damage to property, but also in their ability to seriously disrupt the social and economic infrastructure of a society for many years following the eruption. Although, due to accurate eruption forecasting and implementation of an effective evacuation policy, less than 500 lives were lost during the Philippine event, over 200000 of the local inhabitants required transport to safe locations. Many of these displaced persons remain without permanent homes, their towns and villages destroyed by extensive volcanic mudflows (lahars) associated with the eruption. Mudflow production is likely to continue for at least 10 years; a legacy of the heavy seasonal rains falling on the extensive ash deposits that mantle the flanks of Pinatubo, disrupting the regional economy into the next century.

The Pinatubo eruption, although one of the largest this century, simply constituted the latest in a sequence of disastrous volcanic events which have claimed over 28000 lives since 1980 (Tilling 1989). Following the climactic eruption of Mount St Helens in Washington State (USA) on 18 May of that year (Lipman & Mullineaux 1981), the trend continued with lethal eruptions at El Chichón (Mexico) in 1982 (Luhr & Varekamp 1984), and Nevado del Ruiz (Columbia) in 1985 (Voight 1990, Hall 1992)), and with the unexpected release of deadly, volcanic-derived $CO_2$ from the waters of Lakes Nyos and Monoun in Cameroon in 1986 (Kling et al. 1987, Baxter & Kapila 1989, Sigvaldason 1989). Looking back further in time, erupting volcanoes have claimed around 80000 lives since the turn of the century and over 260000 since AD 1700 (IAVCEI IDNDR Task Group 1990) (Table 1.1), constituting an average annual death rate of nearly 1000 lives. Over the period 1947–1981, the average number of deaths per volcanic disaster totalled 525; this compares with 190 in landslides, 856 due to tsunamis, and 2652 during earthquakes (Thompson 1982). Between 1980 and 1990, around 620000 people were affected by volcanic activity compared, for example, with over 28 million affected by earthquakes, and 524 million by floods (UNESCO 1993) (Table 1.2).

In the same way that the establishment of new volcano observatories represented a response to the Mont Pelée and other volcanic disasters in the early years of this century, so increased governmental and public awareness, and significant advances in both volcano monitoring strategies and techniques have been driven by the high-profile eruptions of the 1980s. Even so, both the authorities and the inhabitants of most volcanically active regions often remain poorly informed with regard to the threat posed by eruption-related hazards. At the same time, some form of surveillance is still only undertaken on about 150 of the world's 550 or so historically active volcanoes, with adequate, multiparameter monitoring being confined to only around a dozen of these, mainly in the developed world. A major problem, requiring solution, lies in the fact that most unmonitored or poorly monitored volcanoes are located in developing countries where population numbers are rapidly increasing. Competition for the most fertile agricultural land is seeing the slopes of potentially destructive volcanoes becoming increasingly crowded, particularly in southeast Asia and in Central and South America. To intensify the problem, most such countries lack both the motivation and organization to establish and maintain even rudimentary volcano monitoring programmes which, although simple, have

**Table 1.1** Volcanic disasters since AD 1700 involving over 1000 fatalities (simplified after IAVCEI IDNDR Task Group, 1990).

| Volcano | Country | Year | No. of fatalities |
|---|---|---|---|
| Awu | Indonesia | 1701 | 3000 |
| Oshima-Oshima | Japan | 1741 | 1475 |
| Cotopaxi | Ecuador | 1741 | 2000 |
| Makian | Indonesia | 1760 | 2000 |
| Papandayan | Indonesia | 1772 | 2957 |
| Laki | Iceland | 1783 | 9336 |
| Asama | Japan | 1783 | 1151 |
| Unzen | Japan | 1792 | 15188 |
| Mayon | Philippines | 1814 | 1200 |
| Tambora | Indonesia | 1815 | 92000 |
| Galunggung | Indonesia | 1822 | 4000 |
| Mayon | Philippines | 1825 | 1500 |
| Awu | Indonesia | 1856 | 3000 |
| Cotopaxi | Ecuador | 1877 | 1000 |
| Krakatau | Indonesia | 1883 | 36417 |
| Awu | Indonesia | 1892 | 1532 |
| Soufrière | St Vincent | 1902 | 1565 |
| Mt. Pelée | Martinique | 1902 | 29000 |
| Santa Maria | Guatemala | 1902 | 6000 |
| Ta al | Philippines | 1911 | 1332 |
| Kelut | Indonesia | 1919 | 5110 |
| Merapi | Indonesia | 1930 | 1300 |
| Lamington | Papua New Guinea | 1951 | 2942 |
| Agung | Indonesia | 1963 | 1900 |
| El Chichón | Mexico | 1982 | 1700 |
| Nevado del Ruiz | Colombia | 1985 | 25000 |
| Nyos | Cameroon | 1986 | 1746 |
| Total fatalities over the period: | | | 255 351* |

* Smaller, more frequent tolls add at least 10 000 to this figure.

**Table 1.2** Estimated number of people affected by natural disasters between 1980 and 1990 (adapted from UNESCO 1993).

| Disaster type | Numbers of people affected* ('000s) |
|---|---|
| Droughts | 952223 |
| Floods | 524638 |
| Windstorms | 150336 |
| Earthquakes | 28410 |
| Landslides | 3154 |
| *Volcanoes* | *620* |
| Wildfires | 612 |
| Tsunamis | 1 |
| Total | 1 660 044 |

* Figures exclude fatalities

the potential to save thousands or tens of thousands of lives. Where poverty, disease and civil strife are endemic and omnipresent, volcano monitoring comes near the bottom of the list of government priorities. This results in a lack of political will to develop volcano surveillance programmes; a circumstance which is compounded by limited financial resources and insufficient scientific and technical expertise. In order to improve this situation, the United Nations (UN) and the European Union have both recently initiated research programmes which will see international teams of scientists establishing surveillance programmes on some of the potentially more destructive volcanoes which are currently poorly studied (see Ch. 15).

If the detrimental impact of erupting volcanoes on society is to be reduced in the coming decades, increased levels of monitoring at larger numbers of volcanoes must be accompanied by an education campaign, aimed at tutoring both the authorities and the public in how to respond to a volcanic crisis. Such a two-pronged policy is strongly advocated by the International Association of Volcanology and the Chemistry of the Earth's Interior (IAVCEI) Task Group for the UN International Decade for Natural Disaster Reduction (IDNDR) in their 1990 report. Implementation of such a "volcano awareness" programme is vital if a second catastrophe of the Nevado del Ruiz type is to be avoided. During the November 1985 eruption the inability of scientists to convince the Colombian authorities of the seriousness of the lahar threat led to over 22000 needless deaths in the town of Armero and nearby villages (Hall 1992). Many, if not most, volcanic eruptions do not pose a threat to local inhabitants. The primary goal of any surveillance programme must therefore be to determine and predict those eruptions which are potentially hazardous. These need not necessarily be the largest eruptions, and there is no simple correlation between the magnitude of an eruption and the hazard it presents. This is graphically illustrated by the death toll resulting from the relatively minor eruption at Nevado del Ruiz.

## 1.2 Recent advances in ground-based volcano surveillance

In the decade since Mount St Helens erupted, the procedures and techniques used to monitor active volcanoes have become increasingly sophisticated and wide ranging. Consequently, the nature and operation of volcanic "plumbing systems" are better understood, resulting, at least for some volcanoes, in an ability to make more tightly constrained predictions about the form and timing of future eruptive behaviour. The increasing use of satellite remote sensing in recent years is worthy of special mention, and is discussed in more detail in the next section and in a number of the specialist chapters.

Volcano monitoring is defined by Banks et al. (1989) as referring "collectively to scientific studies that systematically observe, record, and analyze the changes – visible and invisible – in a volcano and its surroundings". Despite the increasing use of satellites for volcano surveillance, ground-based techniques continue to form the basis of most monitoring networks. The tried and tested "core" methods of seismic and ground-deformation monitoring remain of paramount importance, the former now having been successfully used to monitor magma transport into and within active volcanoes for nearly 150 years. Both techniques have, if anything, become more powerful, due to more sophisticated instrumentation and advances in data

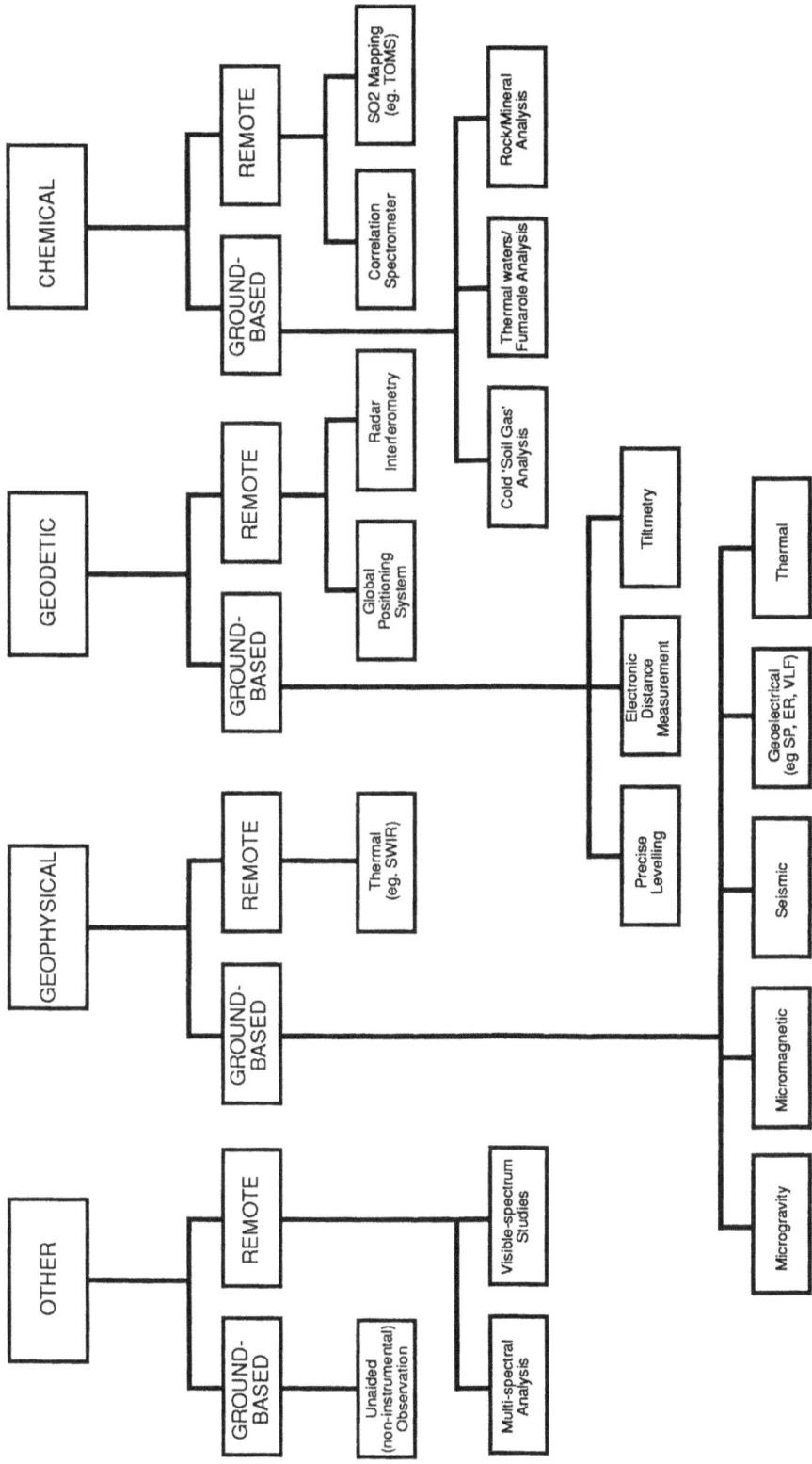

**Figure 1.1** Ground-based and remote-sensing methods and techniques currently available for volcano monitoring (after McGuire 1992). SWIR, short wave infrared; TOMS, Total Ozone Mapping Spectrometer; SP, self-potential; ER, electrical resistivity; VLF, very low frequency.

acquisition and interpretation, and are now supported by an impressive array of additional geophysical, geodetic, and geochemical techniques (Fig. 1.1).

A major advance in seismic monitoring over the last few years has resulted from the development of portable, personal computer (PC) based systems designed to measure the total seismic energy release at a volcano. Such a Real-time Seismic Amplitude Measurement (RSAM) system, developed by the United States Geological Survey Cascades Volcano Observatory (Endo & Murray 1991, Murray & Endo 1992), has been successfully used to monitor recent explosive activity at Mount St Helens and at Redoubt volcano (Alaska). Furthermore, following the onset of activity at Pinatubo, a radio-telemetred RSAM array comprising seven portable recorders operated by the USGS and Philippine Institute of Volcanology and Seismology (PHIVOLCS), proved invaluable in providing a "near-time" picture of developing seismic activity within the volcanic pile (PVOT 1991). As an inexpensive, portable, and user-friendly surveillance tool, RSAM systems are likely to become increasingly important in the rapid-response monitoring of reactivated, previously unmonitored volcanoes.

The deformation of the ground surface in response to the tapping and replenishment of a high-level magma reservoir was first recognized in Japan during the early part of this century by Omori (1914) who deduced its occurrence from the changing heights of benchmarks accompanying eruptions of Sakura-jima volcano. Since then, increasingly sophisticated techniques have been used on many volcanoes, notably Kilauea in Hawaii (e.g. Eaton 1959, Fiske & Kinoshita 1969, Dzurisin et al. 1984; Delaney et al. 1990), to monitor the operation of high-level magmatic "plumbing" systems and their relationship to eruptive events. Ground-deformation monitoring remains a vital tool in recording not only changes in magma flux associated with persistent, established reservoirs, but also in determining the location, form, and growth of newly intruded magma bodies, such as the cryptodome responsible for the climactic eruption of Mount St Helens (Lipman et al. 1981). In addition to elastic movements, permanent deformation related to dyke formation has also been recognized and successfully quantified, providing invaluable information on rift mechanisms, for example at Kilauea (Swanson et al. 1976, Hoffmann et al. 1990), Krafla (Ewart et al. 1991) and Etna (McGuire et al. 1990 1991), and the dimensional characteristics and propagation rates of dykes (Dieterich & Decker 1975, Murray & Pullen 1984, Toutain et al. 1992a; (see also Chs 4 and 5).

Contemporary ground deformation measurements are recorded using a range of methods. "Dry" tilt or single-setup surveys, using precise levelling techniques (Yamashita 1981, 1992, Ewert 1992) are still frequently used, offering a low-cost but effective way of monitoring relatively large ground tilts. A high-tech alternative is the electronic tiltmeter which, although not without drawbacks (Banks et al. 1989, Dzurisin 1992, see also Ch. 4), is now widely and increasingly used to measure radial and tangential (relative to the volcano's summit region) ground tilt (e.g. Delorme & Blum 1991). Laser or infrared electronic distance measurement (EDM) techniques are now routinely utilized to record the centimetre- to metre-scale extension and contraction of the ground that occurs in response to subsurface processes such as magma intrusion or withdrawal (e.g. Okamura et al. 1988, McGuire et al. 1990, 1991, Ewart et al. 1991). A recent innovation is the "total station" (Fig. 1.2), a combination of EDM and electronic theodolite which enables both horizontal and vertical changes of the ground surface to be recorded using a single instrument.

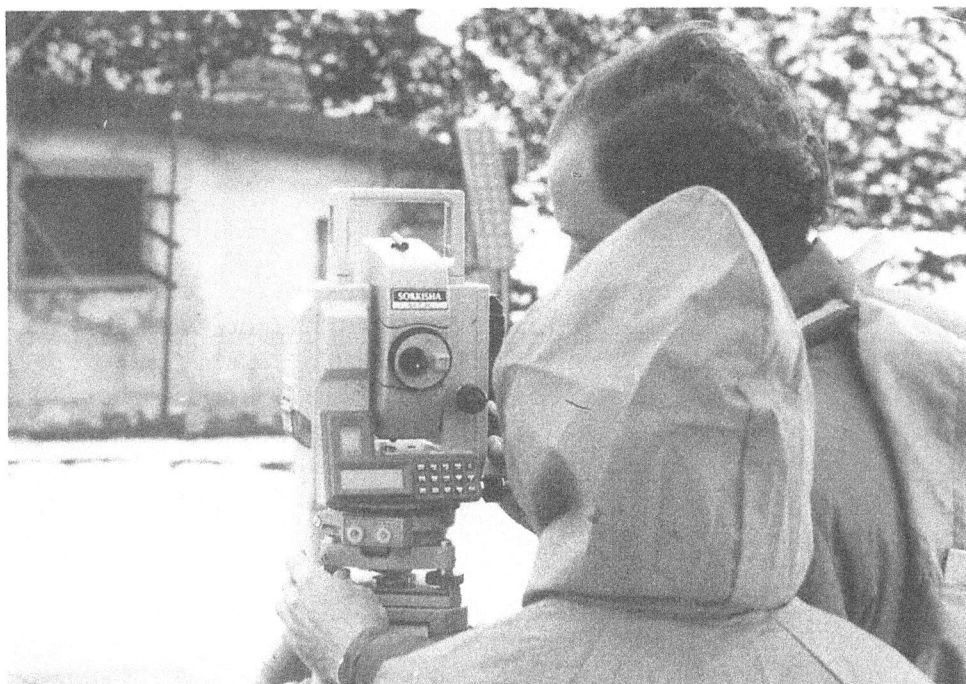

**Figure 1.2** Total stations combine an electronic theodolite with an electronic distance meter, making ground deformation monitoring at active volcanoes easier and quicker to accomplish.

Vertical ground movements can also be monitored independently, and more precisely, using levelling techniques, whereby the changing heights of benchmarks forming a traverse or network, are compared with the altitude of a stable base station. Although such precise levelling is a well established surveying tool, it has, over the past decade benefited considerably from new computerized instrumentation and the availability of levelling staffs which can be read automatically by laser. Some idea of the precisions made possible using near-contemporary ground deformation monitoring technology can be gained from Ewart et al. (1991), who undertook an intensive monitoring campaign at Krafla (Iceland) over the period 1975–1985, using a range of techniques. The authors report standard measurement errors for the Krafla campaigns of $1\,\mu$rad for a water-tube tiltmeter, $9$–$12\,\mu$rad for single-setup networks, 11 mm for distance measurements (along lines of the order of 1–2 km length), and $<24$ mm for elevation changes. As pointed out by Tilling in Chapter 14, there is also a useful role for low-tech ground deformation surveying, particularly in the developing world where limited funding will not permit the acquisition of expensive equipment or the training of local personnel to operate and maintain such high-tech instrumentation. A typical low-tech approach to ground deformation monitoring was very successfully adopted by Swanson et al. (1983) who used steel tapes to record episodes of accelerating growth of the Mount St Helens dome between 1980 and 1982. In combination with data gathered by other means, the results led to the prediction of 13 eruptions over the period. A recent and very comprehensive series of reviews covering the range of high- and low-tech ground deformation monitoring techniques used to monitor the Cascades volcanoes can be found in Ewert & Swanson (1992).

For a number of well-studied volcanoes, data gathered by means of ground deformation and seismic monitoring, are increasingly supported and better constrained by information acquired by more innovative surveillance techniques which, in isolation, are rarely sufficient to provide a clear picture of the behaviour of the subsurface "plumbing" system. Such techniques are primarily concerned with monitoring small variations in the local gravity and magnetic fields, or with changes in the temperature or electrical properties of near-surface materials.

Over the past decade, gravity surveys (Fig. 1.3) have been undertaken on a number of volcanoes including Krafla in Iceland (Johnson et al. 1980), Pacaya in Guatemala (Eggers 1983), and Poás in Costa Rica (Rymer & Brown 1989), and have been successful in recognizing sub- or intra-volcanic mass and/or density changes caused by the intrusion or withdrawal of fresh magma. At Poás, Rymer & Brown (1989) interpreted an overall subsurface mass increase in terms of magma intrusion beneath the summit crater lake. The cooling effect of the lake temporarily staved off an eruption by freezing the rising magma. However, once the lake had dried out due to heat generated by the intrusion, an eruption was soon initiated (Brown et al. 1991).

Microgravity variations associated with mass and density changes within and beneath active volcanoes are small, typically of the order of $100\,\mu\text{Gal}$ (where the average value of the Earth's gravity field is equal to around $1000\,\text{Gal}$). Because gravity decreases with increasing altitude (and vice versa), gravity surveys must be undertaken in tandem with precise levelling. This is necessary in order to distinguish the small variations in gravity due to subsurface mass or density changes, from those resulting from elevation changes caused, for example,

**Figure 1.3** Microgravity surveys can provide an invaluable volcano monitoring service, and can be undertaken even during infavourable meteorological conditions.

by magma intrusion or withdrawal. Gravity surveys undertaken in isolation have, therefore, only limited use in volcano surveillance.

During the 1980s, a number of workers (e.g. Johnston & Stacey 1969, Christoffel 1989, Zlotnicki & Le Mouel 1988, 1990) (see also Ch. 10) claimed some success in recognizing and monitoring the small temporal changes that occur in the intensity of the local magnetic field when fresh magma is intruded into the high levels of a volcanic pile prior to eruption. Recorded changes are very small (usually of the order of a few to 10 nT where the total magnetic field strength of the Earth commonly ranges between 30000 and 65000 nT), and the volcanic "signal" must be unravelled from diurnal and other variations in field strength. Nevertheless, a correlation between eruptive activity and magnetic variation has now been recorded for a number of volcanoes including Kilauea (Davis et al. 1979), Ruapehu (Johnston & Stacey 1969) and White Island (Christoffel 1989) in New Zealand, Soufriére in Guadeloupe (Zlotnicki 1986), and Piton de la Fournaise on Réunion island (Zlotnicki & Le Mouel 1988, 1990). At the latter volcano, a permanent array of high-sensitivity (0.25 nT), continuously-recording magnetometers was used to detect micromagnetic variations that preceded eruptions in 1985, 1986, and 1987, and which could be correlated with shallow seismicity and ground deformation caused by subsurface, pre-eruption magma emplacement (Zlotnicki & Le Mouel 1988). The manner in which intruded magma changes the local magnetic field is not yet clearly established, and more than one mechanism may be involved. A mechanism considered by Zlotnicki & Le Mouel (1990) for Piton de la Fournaise involves micromagnetic variations being representative of an electrokinetic effect due to fresh magma modifying the pattern of subsurface water circulation, and this is discussed further in Chapter 10. Shallow magma bodies, at temperatures above the Curie Point of iron-rich magnetic crystals such as magnetite, may also be detectable due to their low level of magnetization relative to surrounding rocks. This property enabled Dzurisin et al. (1990) to identify the magnetic anomaly associated with the dacitic dome at Mount St Helens that developed within the crater formed during the climactic eruption in May 1980, and to monitor changes in its thermal structure and cooling rate over time.

Several geoelectrical methods have also been used in recent years to monitor different aspects of activity at a number of volcanoes including Izu-Oshima (Japan) (Yukutake et al. 1987), Kilauea (Jackson et al. 1985), Etna (see Ch. 9), and Piton de la Fournaise (Lénat et al. 1989) (see also Ch. 9). The various techniques available are all primarily concerned with constructing a resistivity map of the internal structure of the volcano, or part of the volcano, under observation, and with demonstrating how this changes in response to periodic variations in the "plumbing" system such as the intrusion of fresh magma prior to eruption. The resistivity of volcanic materials relies heavily on their porosity and the resistivity of the fluids contained within the pore spaces. Changes in resistivity of active volcanoes result from a variety of mechanisms, many of which affect the rocks in the vicinity of a newly intruded body of magma, and include variations in porosity and pore-fluid composition and changes in temperature. Subsurface resistivity variations can be measured in two principal ways: (a) by generating artificial electromagnetic fields in the ground, or (b) by observing the distribution of naturally occurring, subsurface, static electric-field potentials (self potentials (SP)). At Kilauea, variants of the former technique have been successfully used both to detect the subsurface intrusion of dykes (Jackson et al. 1985) and to monitor the cooling of lava lakes and the positions of active lava tubes (Zablocki 1978, Jackson et al. 1987) (see also Ch. 9). SP changes have been detected at a number of volcanoes, and attributed, amongst other things, to the effect of fresh magma intrusions on the temperature, composition, and flow of intravolcanic hydrothermal systems. This mechanism has been proposed to explain SP changes associated with eruptive activity at both Piton de la Fournaise and Etna (see Ch. 9).

As might be expected, variations in the total thermal output of a volcano are closely related to changes in the "plumbing" system, and as such can be expected to be sensitive to the intrusions of fresh magma which often precede eruptive activity. Brown et al. (1991) drew attention to the fact that volcanoes represent efficient cooling systems which typically liberate between $10^8$ W and $10^{10}$ W continuously and that shed heat by a variety of mechanisms including evaporation from crater lakes, fumarolic discharges, and straightforward radiation. Monitoring the total thermal budget of a volcano might be expected to provide a clue to the timing of forthcoming activity. Unfortunately, calculating the value of this budget depends on the accurate assessment of both the absolute and relative contributions of the various mechanisms of both heat input and output, parameters which are almost impossible to measure with a precision which would enable small, but significant changes in the total thermal budget to be recognized and monitored. However, thermal budget monitoring, on a less ambitious level has been used successfully to examine changes in the thermal output of crater lakes. At Soufriére (St Vincent), for example, Shepherd & Sigurdsson (1978) recorded rising lake temperatures and evaporation rates prior to the onset of eruptive activity in 1971. At Poás (Costa Rica), the effects of rising magma on the thermal behaviour of the crater lake prior to the April 1989 eruption were more spectacular, with the lake completely drying up to reveal deposits of liquid sulphur (Brown et al. 1987, Brown et al. 1991, Oppenheimer & Stevenson 1989). In contrast to the situations at Soufriére and Poás, Hurst & Dibble (1981) reported a fall in crater lake thermal output prior to eruptive activity at Ruapehu (New Zealand), explaining this somewhat surprising phenomenon in terms of pre-eruption vent blockage hindering the entry of high-temperature volcanic gases into the lake. In the future, thermal monitoring of active volcanoes is likely to become increasingly important. It is also likely that the current trend towards the use of aircraft and satellite instrument platforms will continue (Glaze et al. 1989) offering the advantages of better spatial coverage and more rapid surveying and data gathering from previously inaccessible or poorly accessible volcanoes (see Ch. 7).

In recent years, geochemical monitoring has played an increasingly important role in complementing geophysical and geodetic surveillance, and in providing supporting premonitory evidence for impending eruption. In particular, observations of changes in the $SO_2$ concentration of the summit plumes of open vent volcanoes such as Etna, have enabled the level of magma to be estimated; increasing concentrations of the gas normally being the result of magma rising towards shallower depths within the volcanic pile. Significant increases in $SO_2$ concentration have recently been recorded, using a correlation spectrometer (COSPEC), at Etna (Caltabiano & Romano 1990) and Pinatubo (PVOT 1991). It is worth pointing out that, in the latter case, sudden falls in $SO_2$ concentrations also characterized the reactivation of the volcano in 1991, and were interpreted as resulting from temporary vent blockage as magma neared the surface. Changes in $SO_2$ emissions may also provide evidence for the high-level, internal movement of magma. At Kilauea, for example, Malinconico (1987) recognized increasing

$SO_2$ concentrations accompanying a dyke emplacement event along the southwest rift zone during 1979, an event which was independently verified by associated seismicity. However, the use of COSPEC to monitor $SO_2$ flux in gas plumes is rarely straightforward, and problems of estimating wind parameters and other factors have resulted in large discrepancies in published data. Casadevall et al. (1987) concluded, in fact, that for the amount of time and effort required, the continuous monitoring of $SO_2$ and $CO_2$ emission rates is simply not practical using currently available techniques.

As discussed in Chapter 12, the changing geochemistry of an active volcano can also be recorded and monitored by analyzing gases and fluids escaping in the form of fumaroles either at the summit or on the flanks. At Solfatara crater (Fig. 1.4), in the Campi Flegrei caldera (Naples), for example, fumarolic gas analysis during the bradyseismic (increasing seismicity accompanying ground deformation) crisis of 1982–84 revealed that fresh magmatic material was not approaching the surface. Here, increases in $H_2O$, $H_2S$ and the C/S ratio were interpreted (Tedesco et al. 1988) as resulting from an input of deeper fluids into a shallower hydrothermal system, a model supported by the lack of change in the concentration of "magmatic" gases such as CO. At Mount St Helens, Gerlach & Casadevall (1986) identified changes in fumarolic gas ratios before and during eruptive activity, noting in particular that eruptions were preceded by significant reductions in $CO_2$ emission rates, and that $H_2S/SO_2$ ratios during eruptive activity were characteristically high. These observations provide an additional useful tool in predicting further eruptions of the volcano.

**Figure 1.4** Geochemical monitoring of fumaroles at Solfatara (Campi Flegrei) played an important role in constraining the nature of the bradyseismic crises of the 1980s.

As well as hot fumarolic gases, low temperature gaseous emanations also escape through the flanks of active volcanoes. Increasing levels of magmatic gases such as $CO_2$, $H_2$, CO, He, and the radon isotope $^{222}$Rn, due to the ascent of magma to shallower depths, can be detected by sampling soils and analyzing the gases trapped in pore spaces. Due to its very low solubility in magmas, $CO_2$ is more abundant in volcanic gases than $SO_2$, and is always present in both summit gas plumes and in soil gases. At Etna, Allard et al. (1991) have recently shown that over half the daily production of $CO_2$ ($100 \times 10^9$ g) takes the form of soil-gas emanations. At Vulcano (Aeolian Islands), a feature of recent increasing unrest has been a rise in the levels of soil $CO_2$ (Barberi et al. 1991), and Baubron et al. (1990) reported degassing sites on the flanks where near-pure $CO_2$ is released into the atmosphere, enriched with both He and $^{222}$Rn. Due to considerable variability with time, the absolute abundances of $CO_2$, He and $^{222}$Rn are of limited use as a predictive tool at the La Fossa volcano on Vulcano. Toutain et al. (1992b) have recently suggested, however, that increasing ratios of He/$CO_2$ and $^{222}$Rn/$CO_2$ might indicate the rise of magma to shallower levels resulting in larger concentrations of both He and $^{222}$Rn.

## 1.3 Airborne and satellite monitoring

Over the past decade, the remote sensing of active volcanoes by airborne and satellite-based instruments has become increasingly feasible (Rothery 1992) (see also Chs 6, 7 and 11), and this trend is likely to continue into the next century. Such remote surveillance offers a number of advantages over ground-based monitoring, and Mouginis-Mark et al. (1989) highlight, in particular, the ability to monitor all volcanoes using the same techniques, thereby generating a globally-consistent dataset. They also stress the importance and usefulness of multispectral surveillance, and point to the absence of accessibility problems that may prevent or hinder ground-based monitoring. Satellite and airborne monitoring are complementary in that they offer different services to the volcanologist (Rothery & Pieri 1993). Airborne surveys are able to concentrate on smaller areas, offer better spatial resolution (typically down to 5 m), and can be rapidly repeated (minutes to a few hours). On the other hand, satellite-based instruments have the potential to monitor any volcano anywhere on the planet unimpeded by political constraints. In addition, they are able to provide a broader picture which allows better quality surveillance of large-scale features such as eruptive plumes (Fig 1.5).

Satellite remote sensing currently has application to three main areas of volcano surveillance, the observation and tracking of eruptive plumes (see Ch. 11), the recognition and interpretation of thermal anomalies (see Ch. 7), and the monitoring of ground surface deformation (see Ch. 6). The formation, growth and development of eruptive plumes and clouds can now be routinely monitored (Matson 1989, Sawada 1989) using the global system of geostationary meteorological satellites (GOES, GMS, METEOSAT and others), and by lower altitude, polar-orbiting weather satellites such as the US National Oceanic and Atmospheric Agency (NOAA) Tiros-N series, although this facility is not yet fully utilized, at least on a global basis. The latter vehicles carry an instrument known as the "Advanced Very High Resolution Radiom-

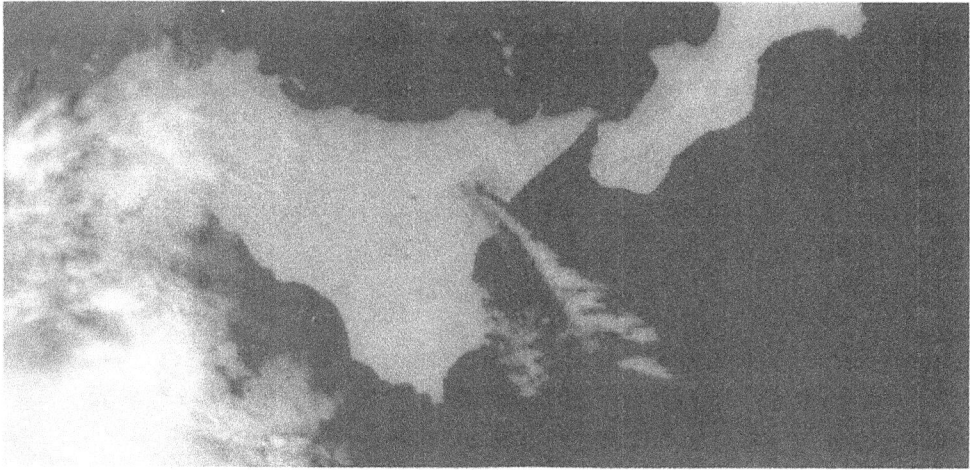

**Figure 1.5** Advanced Very High Resolution Radar (AVHRR) image of eruption plume at Mount Etna taken from the NOAA 7 satellite in May 1983.

eter" (AVHRR), which provides image data in both the visible and thermal infrared parts of the spectrum with spatial resolution of either 8 or 1 km per pixel (Holasek & Rose 1991). Low resolution pictures are automatically transmitted as a television (TV) type signal, and the lower resolution images can be picked up in real time by fairly low-tech, inexpensive receiving equipment linked to a PC (Rothery 1992). The rapid transmission time is important, particularly where the plume is the first evidence of eruptive activity, so that appropriate rapid-response action can be taken by both volcanologists and governmental authorities. In terms of hazard mitigation, plume monitoring has a number of important applications, notably in determining areas on the ground which are likely to experience heavy ash fall, and in providing an early warning to airlines of potentially dangerous ash clouds in the atmosphere. There have been a number of near-disastrous cases in recent years, of aircraft flying unwittingly into eruptive plumes at Galunggung (Indonesia) in 1982, and Redoubt (Alaska) in 1989 and losing power due to the clogging effects of melted ash in the engines (Fox 1988, Kienle et al. 1990).

As well as debris, volcanic plumes and clouds also act as effective mechanisms for carrying large volumes of volcanic gas, particularly $SO_2$ into the stratosphere. Because of the climatic implications of increased concentrations of sulphurous aerosols in the atmosphere, considerable interest has been shown in using satellites to map the stratospheric distribution of eruption-derived $SO_2$. This was successfully accomplished for the 1982 eruption of El Chichón (Mexico) (Krueger 1983), using the Total Ozone Mapping Spectrometer (TOMS) mounted on a NIMBUS 7 satellite, although the instrument was designed, as the name suggests, to look at atmospheric ozone concentrations (see Ch. 11).

Ash plumes indicate that an eruption is in progress, but they do not provide warning of a forthcoming eruption. This is a service which may increasingly be provided in the future by remote thermal monitoring of active volcanic terrains (Oppenheimer & Rothery 1991) (see also Ch. 7). Although a relatively new technique with a number of teething problems, satellite and airborne surveillance has already proved itself useful in detecting and interpreting thermal anomalies at volcanoes for which ground-based thermal data were not available. Notably, Francis & Rothery (1987) and Glaze et al. (1989) used high-resolution (30 m) images recorded in the short-wave infrared (SWIR) by the Thematic Mapper instrument on Landsat to identify "hot spots" in the summit crater of Láscar (Chile) 18 months prior to the 1986 eruption. Two drawbacks currently limit the potential of satellite thermal monitoring: its inability to penetrate cloud cover, and its failure to resolve low temperature thermal anomalies (Bonneville & Kerr 1987, Bonneville & Gouze 1992) caused by, for example, small changes (a few degrees celsius) in ground surface temperature above a replenishing magma reservoir. Although the latter capability may eventually become available, ground-based thermal monitoring will continue to play an important role.

Volcanic eruptions inevitably involve terrain changes. These may range in scale from the formation of a small lava flow field or cinder cone, to the destruction of a large part of the edifice or the formation of thick and extensive deposits of volcanic debris. Although such changes can easily be qualitatively assessed by means of airborne or satellite-based optical or infrared images, such data do not easily provide volumetric information on the amount of material either added or removed during the eruptive process. Such information can be derived from radar surveillance, a technique which has the added benefit of being equally effective if used at night or when the target is cloud covered. The future potential of spaceborne synthetic aperture radar (SAR) interferometry to monitor changes in topography at active volcanoes has recently been investigated using an airborne SAR instrument known as TOPSAR (Evans et al. 1992). In 1991 this was used to construct a digital terrain model (DTM) across Hekla volcano (Iceland), from which metre-scale information on slopes, flow-surface textures and thickness were derived. Similar satellite-based systems should be capable of monitoring the growth and development of lava flow fields during an eruption, allowing surface features to be analyzed and changing volume estimates to be calculated. Radar interferometry also has application both in recording topographic changes, and in the observation of ground deformation such as that which preceded the climactic 1980 eruption of Mount St Helens (Lipman & Mullineaux 1981). More recently, Massonet et al. (1993) used SAR images taken by the European ERS-1 satellite to map the small topographic changes caused by the 1992 Landers earthquake (California) allowing detailed reconstruction of the deformation field. The centimetric resolution of this technique bodes well for future application in monitoring ground-surface deformation in active volcanic terrains.

Remote monitoring of small-scale (centimetre to decimetre) ground surface deformation at active volcanoes remains difficult to accomplish, but is likely to become increasingly important over the next few years. Using the satellite-based Global Positioning System (GPS), a number of workers have already detected and monitored, with varying degrees of success, small displacements of the surface which are attributed to subsurface magma activity in volcanic terrains (e.g. Shimada et al. 1990) (see also Ch. 6). The GPS was initiated by the US Department of Defense in 1976 in order to provide a mechanism for accurately and precisely establishing the positions of objects on the Earth's surface. Only within the last few years, however, has it become more accessible as a civilian tool, utilized both for pinpointing locations and for determining distances between objects. The GPS is effectively a radio communications system (currently consisting of 21 satellites) which transmits fixed-frequency signals to receivers on the planet's surface. Establishing receivers above benchmarks in a distance-measurement network allows interbenchmark distances to be determined with a precision of < 1 ppm. At Etna, Briole et al. (1988) and Nunnari & Puglisi (1991) have demonstrated that the accuracy of GPS compares well with that of more traditional, ground-based EDM techniques, at least for distances of less than 25 km. This implies that the GPS should be able to detect the small horizontal displacements associated with magma emplacement prior to eruption. Recently, Shimada et al. (1990) used the GPS to monitor ground displacements associated with seismicity at the Teisha volcano (Japan) during 1989. Their results showed that the distance between two benchmarks 10 km apart increased by 14.5 cm during the seismic activity, and they attributed this to the opening of a magma-filled fracture which 10 days later fed a new eruption. There are still problems with the GPS, notably it lacks the precision to measure the small vertical displacements associated with the replenishment and tapping of magma reservoirs or the intrusion of new magma bodies. It is likely, however, that this capability will be achieved in the near future due largely to better atmospheric corrections, and increasing numbers of available satellites, including those making up the equivalent Russian Global Navigation System (GLONASS).

The application of future satellite platforms and instruments to volcano monitoring, including those which are expected to form part of theNational Aeronautics and Space Administration (NASA) Earth Observation System (EOS), are discussed in Chapter 15.

## 1.4 Monitoring strategies

The development of an efficient and workable strategy for volcano surveillance is desirable both at the level of the individual volcano and globally. In the former case, the primary role of monitoring should be to understand how a volcano functions in order, ultimately, to ensure that the threat imposed on the local population is minimized. In the latter, logical reasoning dictates that, given the constraints imposed by lack of political will, funding, and expertise, monitoring programmes should concentrate on those of the world's 550 or so historically active volcanoes that pose the greatest threat to those living in the vicinity, always bearing in mind that the most devastating future eruptions may originate at volcanoes with no record of historical activity. The current economic and political climate ensures that it is simply not feasible at present to establish and operate a monitoring programme incorporating all potentially dangerous volcanoes, and in consequence a policy of selectivity is required. This approach has been adopted by the United Nations (IAVCEI IDNDR Task Group 1990) and the European Union (EU), both of which have supported the initiation of monitoring programmes on restricted numbers of high-risk and "test-bed" volcanoes in both developed and developing countries (see Ch. 15). In addition to these 18 or so volcanoes selected for special study, many more require some form of continuous surveillance if the toll of volcanic disasters is not to rise in the future. This need not involve great expense nor complex technologies, in fact where expertise is unavailable and the infrastructure to support the utilization of elaborate equipment does not exist, a low-tech approach can be a positive advantage. Techniques involving, for example, the use of tape-measures to record changes in ground deformation or thermometers to monitor variations in fumarole or crater-lake temperatures, recording the frequency of felt-earth tremors, and compiling visual records of summit activity, may all provide sufficient warning of impending eruption, allowing time for more sophisticated monitoring equipment to be installed. Such a basic surveillance scheme has been established in the Philippines (Delos Reyes 1992), making use of volunteer observers living on or near active volcanoes, whose role is to report on any unusual events, such as sulphurous smells, dying vegetation or ground fissuring, which may presage an eruption (Table 1.3, Fig. 1.6)

The great weight of accumulated evidence gathered from over a century of volcano surveillance indicates that eruptions rarely occur without premonitory warning signs. In order

**Table 1.3** Example of the type of questionnaire distributed to residents in the vicinity of the Taal volcano in the Philippines (after Delos Reyes, 1992).

---

| Date | Time | [ ] Morning |
| | | [ ] Afternoon |
| | | [ ] Evening |

Place of observation _____

1. Steaming activity
   Intensity:     [ ] Strong     [ ] Moderate     [ ] Weak
   Height:        [ ] High       [ ] Medium       [ ] Low
   Colour:        [ ] Grey       [ ] Brown        [ ] White
   Direction:     [ ] North      [ ] South        [ ] East          [ ] West

2. Unusual changes in the ground
   Earthquake:                        [ ] Yes          [ ] No
   Rumbling sounds:                   [ ] Yes          [ ] No
   Ground:        [ ] Subsidence      [ ] Liquefaction [ ] Fissuring
   Sulphur smell:                     [ ] Yes          [ ] No
   Drying of vegetation:              [ ] Yes          [ ] No
   Unusual animal behaviour:          [ ] Yes          [ ] No
                  Type of animal _____

3. Unusual changes in lake water
   Taal Lake  [ ]   Main crater lake  [ ]
   Bubbling:                          [ ] Yes          [ ] No
   Increase in temperature:           [ ] Yes          [ ] No
   Increase in water level:           [ ] Yes          [ ] No
   Change in colour:                  [ ] Yes          [ ] No
   Unusual fish behaviour:            [ ] Yes          [ ] No
   Fish deaths:                       [ ] Yes          [ ] No
   Sulphur smell:                     [ ] Yes          [ ] No

4. Other observations/remarks _____
   _____
   _____

   Name of observer  _____
   Name of barangay  _____

---

GITAS–ON (HEIGHT)

TAAS (HIGH)

MEDYO TAAS (MEDIUM)

MUBO (LOW)

GUIDAGHANON (VOLUME)

KUSOG (STRONG)

MEDYO KUSOG (MODERATE)

HINAY (WEAK)

KOLOR (COLOUR)

KOLOR ABO (GREY)

KOLOR TSOKOLATE (BROWN)

PUTI (WHITE)

PAGDAKO SA BUNTOD (DOME GROWTH)

PAG ULAN SA ABO (ASHFALL)

PAGKAUGA SA TANOM (VEGETATION DRYING)

PAGLIKI SA YUTA (FISSURE)

PAGKATUMPAG (LANDSLIDE)

ULAN (RAIN)

PAGKAUNLOD SA YUTA (SUBSIDENCE)

NAGBAHA (MUDFLOW)

NAGLINOG (EARTHQUAKE)

PAGKAUGA SA ATABAY (WELL DRYING)

KATINGALAHANG LIHOK SA MGA HAYOP (ABNORMAL ANIMAL BEHAVIOUR)

DAGOOK GIKAN SA BUKID (RUMBLING SOUND)

BAHO SA ASOPRE (SULFUR SMELL)

$O_2$   $H_2S$

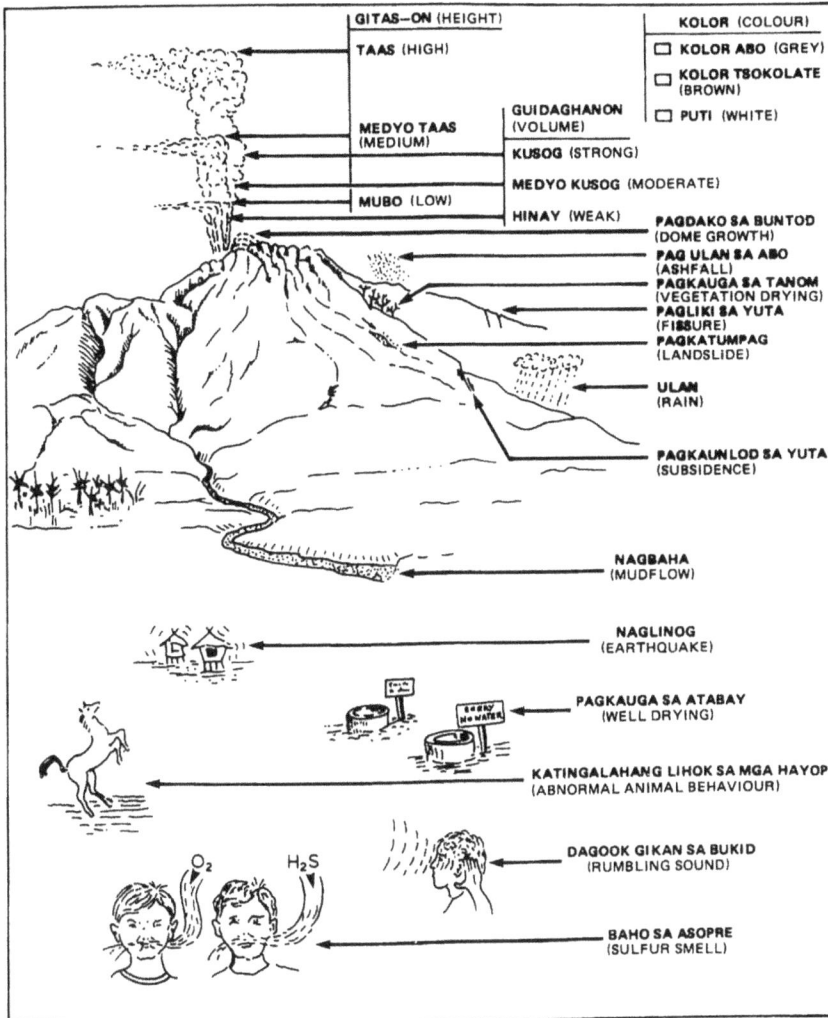

**Figure 1.6** A leaflet issued to the local residents around Hibok-Hibok volcano in the Philippines illustrates the precursive phenomena which may be observed prior to an eruption (from Delos Reyes 1992).

to recognize the anomalous behaviour that might presage an impending eruption, however, the normal behaviour pattern of the volcano under observation must be recognized and understood. A primary role for any volcano monitoring programme involves, therefore, establishing the level of this background or *baseline* activity, against which anomalous, pre-eruption behaviour can then be measured. Both the nature and scale of baseline activity are unique to each volcano. In order, therefore, to develop an accurate picture of how this varies over time, a volcano should be monitored over as long a period of inactivity as possible. Under ideal circumstances this should be a minimum of several years or even decades. Currently, baseline data are often of an incomplete nature, and only available for about a quarter of the world's active volcanoes, making the recognition of premonitory behaviour difficult, and holding open the real possibilities, on the one hand, of inadequate mitigation procedures leading to a disaster, and on the other, of expensive and unnecessary evacuation. A major goal of a global volcano monitoring strategy must, therefore, involve gathering better quality baseline data for larger numbers of the potentially more destructive volcanoes.

In their 1990 report, the International Association of Volcanology and Chemistry of the Earth's Interior (IAVCEI) Task Group for the UN International Decade for Natural Disaster Reduction (IDNDR) lay down guidelines for a minimum baseline volcano monitoring programme (Table 1.4) which, they suggest, should consist of the following four components: visual observation, continuous seismic monitoring using at least three seismometers, ground deformation using available techniques including EDM, tiltmeters and ideally GPS, and geochemical and thermal monitoring of hot springs and volcanic gases. The task group further recommend that this minimum surveillance network should be rapidly supplemented and upgraded at the first signs of unrest (e.g. ground inflation or increased seismicity) which might presage a forthcoming eruption. Ideally, under these circumstances, the number of seismometers would be increased to five or more to better constrain the depths and locations of seismic events. In order to develop a better picture of ground-surface deformation, EDM networks should be enlarged, where required, and additional numbers of continuously recording tiltmeters installed in appropriate locations. The level of accompanying geochemical monitoring, using a COSPEC and other techniques, should also be stepped up, in order to look for changes in gas flux and ratios that might indicate magma moving to shallow depths.

Although a combination of seismic and ground deformation monitoring can provide important and useful information on the reactivation of a volcano prior to a new eruption, under ideal circumstances a wider combination of monitoring approaches would be preferable. In such an integrated, multi-parameter surveillance network, a broad range of complementary techniques would gather detailed information on not only seismic activity and ground-surface deformation, but also on magnetic, gravity, and electrical resistivity changes, variations in the compositions of fumaroles, hot springs, and gas plumes, and the existence

Table 1.4 Minimum monitoring programmes recommended by the International Association of Volcanology and Chemistry of the Earth's Interior (IAVCEI) Task Group for the International Decade for Natural Disaster Reduction (IDNDR).

*Baseline monitoring*
　Visual observation
　Seismic monitoring (minimum of 3 seismometers)
　Ground deformation monitoring (EDM, tiltmeters, and ideally GPS)
　Geochemical and thermal monitoring of hot springs and volcanic gases

*Upgraded monitoring following signs of unrest*
　Number of seismometers increased to a minimum of 5
　EDM network expanded
　Additional numbers of continuously recording tiltmeters
　Increased level of geochemical monitoring including use of COSPEC

COSPEC, correlation spectrometer; EDM, electronic distance measurement; GPS, Global Positioning System.

and nature of thermal anomalies. This type of data-gathering strategy is mutually supportive and is, therefore, much better suited to constraining the form and function of the magmatic plumbing system, and consequently more useful in reliable eruption prediction and hazard mitigation. Such well-coordinated surveillance networks do exist, notably at Kilauea, Mount St Helens, and Piton de la Fournaise, but unfortunately they are not located at the most dangerous volcanoes. Notwithstanding the initiatives launched recently by the UN and European Union (EU), it is unlikely that financial support will be forthcoming to allow such networks to be established at a significant number, if any, of the life-threatening volcanoes in the developing countries. The best approach in the circumstances is likely to involve, therefore, the establishment of baseline surveillance programmes on the highest-risk volcanoes, combined with the facility for rapid supplementaion and upgrading should the need arise.

## 1.5 Predicting volcanic eruptions

Forecasting or predicting volcanic eruptions is essentially an exercise in pattern recognition. The difference between a forecast and a prediction is defined in Swanson et al. (1985), and is discussed in some detail in Chapter 14. In broad terms, however, a prediction is generally more precise, both in terms of the possible nature of the eruptive event and the size of the predictive "window". Predictions are likely to be short term in the sense that they will probably be announced months to hours before the expected event. Forecasts may be short term, but may also involve estimates of the nature and timing of activity years for decades hence. As discussed in the previous section, it is vital that the "normal" or *baseline* pattern of activity be sufficiently recognizable and understood so that any departure from this can be rapidly identified as anomalous behaviour that might presage impending eruptive activity. On well-studied, regularly erupting volcanoes, experience gained over several decades provides sufficient data to distinguish effectively between baseline and pre-eruptive behaviour. At Kilauea, for example, ground-surface inflation in the summit region, accompanied by discrete, short-period seismic events, typically indicates that replenishment of the magma reservoir is underway (Tilling et al. 1987), at a later stage, experience has shown that sudden deflation accompanied by an increase in the frequency of long-period seismic events and volcanic tremor marks the lateral intrusion of magma into one of the rift zones. Decker (1986) reports how dyke-related, pre-eruption seismicity at Kilauea has been used to forecast the positions of flank eruptions, allowing the authorities to constrain the size of the potential evacuation area. There is as yet, however, no way of predicting whether or not such an intrusive event will actually intersect the ground surface to produce an eruption.

Notwithstanding successes at Kilauea and a small number of other sites, eruption prediction at most active volcanoes is fraught with difficulty. This may arise from one or more of a range of commonly encountered problems, notably an absence of sufficient baseline data, a high degree of variability in precursory activity, progressive but non-eruption related temporal changes in the behaviour of the volcano under observation, and difficulties in distinguishing between anomalous activity that presages an eruption, and that which culminates merely in an intrusive event. Faced with a period of increasing unrest, volcanologists may find themselves under pressure to make predictions on the basis of poor quality data, which, on the one hand may lead the authorities to organize unneccessary, expensive, and unpopular evacuation, and on the other to considerable loss of life due to implementation of a "stay put" policy. Repeatedly erring on the side of caution can lead to the development of "cry-wolf syndrome", making it more difficult to persuade both authorities and the local population of the need for evacuation in a future volcanic crisis, while the implications of the alternative scenario are obvious and highly detrimental to the professional reputation and integrity of volcanologists as a group. A prime example of this dilemma is illustrated by the 1976–7 eruption of Soufrière (Guadeloupe) where 72 000 inhabitants were evacuated for four months largely due to the misidentification of juvenile material in ejected tephra (see Ch. 14). A similar predicament faced volcanologists during the Rabaul (Papua New Guinea) volcanic crisis between 1983 and 1985 (McKee et al. 1985) (see also Ch. 14). Here, increasing shallow seismicity and accelerating uplift in the Rabaul caldera suggested that a potentially destructive eruption was at least possible, if not likely. Detailed contingency plans for evacuation were drawn up but in the end not required as activity abruptly returned to pre-crisis levels. This scenario is likely to be repeated many times over the next few decades, as surveillance levels are increased at previously unmonitored active volcanoes. It results, at least partly, from insufficient knowledge of *baseline* activity, which can only be enhanced by longer periods of monitoring.

To compound the problem of anomalous behaviour without eruption, there are occasions when eruptions occur without significant precursory activity. If, for example, an absence of structural integrity (e.g. the volcano is fractured) prevents the build up of stresses in the edifice, then both ground deformation and seismicity during pre-eruption, subsurface magma accumulation may be minimal. This situation prevailed prior to the 1991–3 eruption of Etna, one of the largest there this century, when magma appears to have passively filled a fracture system formed a few years earlier (Fig. 1.7), with little accompanying seismicity or significant ground deformation (Rymer et al. 1993). The worth of alternative volcano monitoring techniques was well demonstrated here, as only microgravity surveys were able to detect the mass changes associated with the emplacement of fresh magma at shallower levels.

It is extremely unlikely that any volcano can become reactivated without some form of precursive activity that can be detected provided that the appropriate monitoring equipment is in place. The major problem lies, however, in predicting the nature and timing of a potentially destructive event that may come at any time following reactivation. Many of the great volcanic disasters, including Krakatoa in 1883 and Mont Pelée in 1902, followed long episodes of relatively harmless activity. There is always a worry, therefore, that even a wealth of baseline data may not be sufficient to provide warning of some of the rarer and more destructive events. In other words, predicting small, inconsequential eruptions is likely to be much easier than predicting the "big one".

Despite the difficulties outlined above, there have been a number of very successful predictions made about impending activity at a number of recently active volcanoes. Worthy of particular note, and discussed in more detail in Chapter 14, are the very precise predictions made about the timing of dome-related eruptions at Mount St Helens since June 1980. Swanson

**Figure 1.7** Part of a 6 km long fracture system formed during 1989 at Mount Etna, and filled passively with magma during 1991, causing only minimal ground deformation and no seismicity. The eruption which followed in December 1991 lasted for nearly 15 months and was volumetrically the largest this century.

et al. (1985) were able to define successively smaller predictive windows on the basis of increasing rates of a range of precursory activity, including ground tilt, seismic-energy release, and dome expansion. Another success was achieved at Galunggung volcano (Indonesia), where rapid augmentation of a basic surveillance network at the start of the 1982–83 eruptive cycle enabled some of the later eruptive events to be predicted, thereby making an important contribution to hazard mitigation during the nine month long period of activity (Punongbayan & Tilling 1989). At Pinatubo, a system of alert levels was established following reactivation of the volcano in April 1991 (PVOT 1991) (see also Ch. 14). The successive declarations from alert level 1 (low-level unrest) to alert level 5 (eruption in progress), over the period April to mid-June, effectively constituted the defining of progressively smaller predictive windows. The evacuation of an increasingly larger area around the volcano, in response to the successively raised alert levels, undoubtedly saved many thousands of lives which otherwise would have been lost during the climactic eruption of 15 June.

## 1.6 Monitoring and volcanic hazard mitigation

Volcano monitoring has the stated goal of understanding better how volcanoes function, and why and when they erupt. This is not, however, a purely academic exercise. Apart from scientific curiosity, the primary impetus behind the "need to know" lies in a duty to save lives and minimize the destructive impact of erupting volcanoes on human society. Considerable energy is released during a volcanic eruption, much of which is expended in transporting large volumes of often hot material over considerable distances, sometimes at velocities of over 100 km/hr$^{-1}$. The effects of such behaviour are usually hazardous to the local population, and sometimes to those further afield. Several varieties of volcanic hazard are identified (Blong 1984, Crandell et al. 1984, Tilling 1989), some more threatening to life and property than others, and the principal types are listed in Table 1.5 (after Scott 1989a). While lives and property in the immediate vicinity of an erupting volcano are liable to be more at risk from lavas (Guest & Murray 1979, Williams & Moore 1983), pyroclastic- and debris-flows, floods, rock avalanches, and related phenomena (e.g. Miller & Smith 1977, Davies et al. 1978, Janda et al. 1981, Fisher & Heiken 1982, Ui 1983, Siebert 1984, Pierson 1985, Lowe et al. 1986, Brantley & Waitt 1988), the effects on settlements at greater distances from the volcano are likely to be limited to those caused by ash fall (e.g. Blong 1981). In the case of major events that involve the destruction of large parts of the volcanic edifice, blast effects, atmospheric shock waves, and large-scale rock avalanches or debris flows may, however, affect hundreds or thousands of square kilometres of the adjacent land area (e.g. Gorshkov 1963, Lipman & Mullineaux 1981, Bogoyavlenskaya et al. 1985). Where lakes fill dormant volcanic craters, gas releases caused by convective overturn can pose a serious threat (Kling et al. 1987, Sigurdsson et al. 1987, Baxter & Kapila 1989, Sigvaldason 1989), while eruptions of coastal or island volcanoes may generate tsunamis which are potentially far more destructive than the direct effects of the erupted products (Latter 1981, Francis 1985). Gigantic eruptions such as that of Toba (Indonesia) around 75 000 years BP may have a catastrophic effect on the global climate (Stothers et al. 1989) which cannot effectively be mitigated.

Successful volcanic hazard mitigation relies, firstly, upon a detailed understanding of a volcano's previous behaviour over as long a period as possible, and secondly, on effective monitoring. The former is largely reliant upon geological studies aimed at establishing the type, frequency, and extent of potentially hazardous activity occurring during both historic and prehistoric times. These data are then used to construct hazard zonation maps designed

Table 1.5 Principal types of volcanic hazard (adapted from Scott, 1989).

Lava flows and collapsing lava domes
Pyroclastic density currents:
   pyroclastic flows
   hot pyroclastic surges
   cold (or base) surges
Directed blasts
Lahars, debris flows, and floods
Slope-failure related debris avalanches and related phenomena
Tephra falls and ballistic projectiles
Volcanic gases
Volcanogenic earthquakes
Atmospheric shock waves
Tsunamis

to delineate areas at risk from the various types of volcanic hazard (Crandell et al. 1984, Scott 1989b). The role of monitoring is primarily one of establishing, as precisely as possible, the timing and nature of an impending eruption so that the inhabitants of at-risk areas identified on the hazard zonation maps can be warned or evacuated, as appropriate, before the onset of hazardous eruptive activity. There is, therefore, a close link between volcano surveillance and volcanic hazard mitigation. In recent years, there have been a number of occasions where well-conducted monitoring during periods of pre-eruptive and syn-eruptive unrest, together with effective hazard-zonation mapping, have led to the implementation of suitably constrained and timely evacuation policies and the consequent saving of many thousands of lives (Punongbayan & Tilling 1989) (see also Ch. 14).

At Mount St Helens, a surveillance network was already in place prior to its reactivation in 1980. On the basis of its previous activity, the volcano was recognized as being potentially dangerous and hazard zonation maps already produced (Crandell & Mullineaux 1978). Furthermore, Crandell et al. (1975) had forecast that Mount St Helens would "erupt again, perhaps before the end of this century". Consequently, when the first seismic events were recorded in March 1980, the monitoring network was rapidly augmented, and the hazard zonation maps made available to the authorities and the public. Ground deformation techniques were used to monitor the growth of the bulge on the northern side of the volcano, and recognition of the effects and extent of its detachment, based upon the hazard zonation maps, led to restricted access to the at-risk area, thereby limiting the death toll during the climactic eruption on 18 May to only 57. Although certainly a successful exercise in hazard mitigation, a number of lessons were learned from the Mount St Helens experience. Perhaps the most important involved the surprising extent of the area affected by blast damage, this proved to be up to 15 times greater than that predicted from studies of past eruptions (Miller et al. 1981).

More recently, volcano monitoring was able to play a more direct role in warning of volcanic hazards at Pinatubo in the Philippines. Here, lahars produced by heavy, seasonal rains on extensive, unconsolidated pyroclastic deposits ejected primarily during the eruption on 15–16 June 1991 (Pierson 1992), are likely to pose a threat to the local population into the next century (Fig. 1.8). In order to warn, and if neccessary evacuate, population centres in the river valleys which channel the lahars, they have to be detected high up on the flanks of the volcano. Traditionally, trip-wire systems have been adopted to warn of impending volcanic mudflows. At Pinatubo, however, seismometers are used to detect the ground motion caused by the movement of a water–debris mixture. The seismic signal is distinguishable from others common to a volaño during its active cycle, and the system is not liable to be destroyed by a lahar provided that the instruments are judiciously placed.

The Nevado del Ruiz disaster of 13 November 1985, (Herd et al. 1986, Voight 1988, 1990, Hall 1992) illustrated in the most devastating manner possible the ineffectiveness of hazard zonation and surveillance data for hazard mitigation, in the absence of an appropriate response by both authorities and public. As discussed by Tilling in Chapter 14, the mudflow threat at Ruiz was recognized and understood by volcanologists and this information imparted to the authorities in good time. A number of factors, however, including poor communications between government officials and the local population, indecisiveness, lack of understanding of the threat and appreciation of its seriousness, and even poor weather on the night of the

**Figure 1.8** Mudflows constituted a major problem to the local population following the 1991 eruption of Pinatubo. Heavy seasonal rains falling on the extensive tephra deposits produced during the eruption are likely to continue to generate mudflowws for several years to come.

eruption, all combined to set the scene for the deaths of over 22 000 inhabitants of the town of Armero and nearby settlements. The Ruiz catastrophe represents an important lesson for the future, namely that improvements in monitoring, particularly on volcanoes in the Third World and developing countries, will be insufficient to reduce the detrimental impact of volcanic eruptions without an accompanying programme of public awareness and improved relations and communications between scientists, authorities, and local inhabitants. An important step forward in this area lies in the training of local scientists who are better placed to understand the needs and concerns of both the authorities and inhabitants in their own countries. Such a training scheme is already underway at the Centre for the Study of Active Volcanoes (CSAV) at the University of Hawaii at Hilo (Anderson & Decker 1992), where scientists from developing countries are tutored in volcano monitoring, volcanic hazards, and how to deal with the authorities and the media during volcanic crises.

## 1.7 Safe surveillance of active volcanoes

Over the 3-year period during which this text was conceived and written, eleven volcanologists were killed and a number of others injured while monitoring active volcanoes. In June

1991, Harry Glicken (USA), Maurice Krafft (France), and Katja Krafft (France), were killed in a pyroclastic flow eruption at Unzen volcano (Japan), while carrying out scientific observations and filming this particularly dangerous volcanic hazard (see Ch. 14). More recently, on 14 January 1993, six more volcanologists (José Arlés Zapata, Fernando Cuenca, Néstor García and Carlos Trujillo of Colombia, Igor Menyailov of Russia, and Geoff Brown of the UK) were killed by ejecta during a short explosive eruption at Galeras volcano in Columbia (Smithsonian Institution 1992), while on 12 March 1993, Victor Hugo Pérez and Egdo Acuaro Sánchez died in a phreatic explosion at Guagua Pichincha (Equador). Active volcanoes undoubtedly represent high-risk environments, and there are few volcanologists who would argue that volcano surveillance does not involve certain elements of danger. There are simple precautions that can be taken to minimize the risk of death or injury in certain situations, but sadly there is also evidence that such precautions are not always taken. On the basis of discussions and consultations with volcanologists from around the world, the following guidelines are proposed to reduce the risk factor while working in the summit region of a volcano during its active cycle. Particular attention is drawn to the wearing of a protective helmet, a simple precaution which is a potential life saver.

(a) A protective helmet should be carried, or preferably worn, while in the vicinity of active vents, craters, or domes. This should be of the mountaineering/rock-climbing type and should be capable of being securely fastened with a chin strap.

(b) Where volcanic gases are a potential problem, a gas mask should be carried and worn when needed. Appropriate filters (usually of the acid-gas type) should be inserted, and a spare set carried.

(c) The number of observers in a high-risk area should be limited. Three is an optimum number, allowing one person to go for help, and another to offer assistance should someone become incapacitated. It is not recommended that an observer enter a high-risk area alone. Large, poorly equipped groups associated, for example, with conference visits, are to be discouraged.

(d) Ideally, the observing team should be in radio contact with a member stationed outside the probable high-risk area, or with a local authority/observatory base. If radios are not available, then the intention of the team to enter a potentially dangerous part of the volcano should be lodged with an appropriate authority (police, observatory staff, or even local hotel proprietor), together with an estimate of the duration of the visit.

(e) Surveillance, sampling, etc., should be accomplished as rapidly as is feasible, without affecting the quality of data gathered, so as to reduce the length of time in a high-risk area.

(f) A first-aid kit should be carried by one of the observing team who, preferably, should have some knowledge of basic first aid.

(g) Spare food, water, and, where appropriate, warm clothing, should be carried, to be used in the event of one or more members of the observing team being stranded in the summit region overnight and/or being incapacitated by accident or eruptive activity.

Monitoring effusive eruptions is also potentially dangerous, and here also certain simple but effective precautions should be taken. In this context, Tilling & Peterson (1993) present some basic guidelines for studying active lavas in safety, including the wearing of protective clothing, working upwind of flows, learning to recognize potentially unstable ground, and being constantly aware of escape routes. More detailed recommendations designed to ensure greater safety during volcano monitoring have recently been agreed by IAVCEI members and are included in the Appendix.

# References

Allard, P., J. Carbonelle, D. Dajlevic, J. Le Bronec, P. Morel, M. C. Robe, J. A. Maurenas, R. Faivre-Pierret, D. Martin, J-C. Sabroux, P. Zettwoog 1991. Eruptive and diffuse emissions of $CO_2$ from Mount Etna. *Nature* **351**, 387–91.

Anderson, J. L. & R. W. Decker 1992. Volcano risk mitigation through training. In *Geohazards: natural and man made*, G. J. H. McCall, D. J. C. Laming, S. C. Scott (eds), 7–12. London: Chapman & Hall.

Banks, N. G., R. I. Tilling, D. H. Harlow, J. W. Ewert 1989. Volcano monitoring and short-term forecasts. In *Short courses in geology, vol 1: Volcanic hazards*, R. I. Tilling (ed.), 51–80. Washington DC: American Geophysical Union.

Barberi, F., G. Neri, M. Valenza, L. Villari 1991. Volcanological research in Italy (1987–1990) report to IAVCEI. *Bollettino Geofisica Teorica Applicata Supplemento* **32**, 413–88.

Baubron, J. C., P. Allard, J. P. Toutain 1990. Diffuse volcanic emissions of carbon dioxide from Vulcano Island, Italy. *Nature* **344**, 51–53.

Baxter, P. J. & M. Kapila 1989. Acute health impact of the gas release at Lake Nyos, Cameroon 1986. *Journal of Volcanology and Geothermal Research* **39**, 265–75.

Blong, R. J. 1981. Some effects of tephra falls on buildings. In *Tephra Studies*, S. Self & R. S. J. Sparks (eds), 405–20. Boston: Reidel.

Blong, R. J. 1984. *Volcanic hazards: a sourcebook on the effects of eruptions*. Orlando, Fl: Academic Press.

Bogoyavlenskaya, G. E., O. A. Braitseva, I. V. Melekestsev, V. Yu. Kiriyanov, C. D. Miller 1985. Catastrophic eruptions of the directed-blast type at Mount St Helens, Bezymianny, and Shiveluch volcanoes. *Journal of Geodynamics* **3**, 189–218.

Bonneville, A. & P. Gouze 1992. Thermal survey of Mount Etna volcano from space. *Geophysical Research Letters* **19**, 725–8.

Bonneville, A. & Y. Kerr 1987. A thermal forerunner of the 28th March 1983 Mount Etna eruption from satellite thermal infrared data. *Journal of Geodynamics* **7**, 1–31.

Brantley, S. R. & R. B. Waitt 1988. Interrelations among pyroclastic surges, pyroclastic flows, and lahars in Smith Creek valley during the first minutes of 18 May 1980 eruption of Mount St Helens, USA. *Bulletin Volcanologique* **50**, 304–26.

Briole, P., G. Nunnari, G. Puglisi, J. C. Ruegg 1988. Misure GPS sul Monte Etna: resultati della prima campagna. *Atti. del 7° Cenvegno Nazionale di Geofisica*, Conference Proceedings 855–64.

Brown, G. C., H. Rymer, R. S. Thorpe 1987. Gravity fields and the interpretation of volcanic structures. *Earth and Planetary Science Letters* **82**, 323–34.

Brown, G. C., H. Rymer, D. Stevenson 1991. Volcano monitoring by microgravity and energy budget analysis. *Journal of the Geological Society, London* **148**, 585–93.

Caltabiano, T. & R. Romano 1990. COSPEC data. In *Mount Etna: 1989 eruption*, F. Barberi, A. Bertagnani, P. Landi (eds). Pisa: Giardini.

Casadevall, T. J., J. B. Stokes, L. P. Greenland, L. L. Malinconico, J. R. Casadevall, B. T. Furukawa 1987. $SO_2$ and $CO_2$ emission rates at Kilauea volcano 1979-1984. In *United States Geological Survey Professional Paper 1350, Volcanism in Hawaii*, R. W. Decker, T. L. Wright, P. H. Stauffer (eds), 771–80. Washington DC: United States Government Printing Office.

Crandell, D. R. & D. R. Mullineaux 1978. *United States Geological Survey Bulletin 1383-C, Potential hazards from future eruptions of Mount St Helens volcano, Washington*. Washington DC: United States Government Printing Office.

Crandell, D. R., D. R. Mullineaux, M. Rubin 1975. Mount St Helens volcano: recent and future behaviour. *Science* **187**, 438–41.

Crandell, D. R., B. Booth, K. Kusumadinata, D. Shimozuru, G. P. L. Walker, D. Westercamp 1984. *Source-book for volcanic-hazard zonation*. Paris: UNESCO.

Christoffel, D. A. 1989. Variations in the magnetic field intensity at White Island volcano related to the 1976–82 eruption sequence. *New Zealand Geological Survey Bulletin* **103**, 109–18.

Davis, P. M., F. D. Staley, C. J. Zablocki, J. V. Olson 1979. Improved signal discrimination in tectonomagnetism: discovery of a volcano-magnetic effect at Kilauea, Hawaii. *Physics of the Earth and Planets International* **19**, 331-6.

Davies, D. K., M. W. Quearry, S. B. Bonis 1978. Glowing avalanches from the 1974 eruption of volcano Fuego, Guatemala. *Geological Society of America Bulletin* **89**, 369–84.

Decker, R. W. 1986. Forecasting volcanic eruptions. *Annual Review of Earth and Planetary Sciences* **14**, 267–91.

Delaney, P. T., R. S. Fiske, A. Miklius, A. T. Okamura, M. K. Sako 1990. Deep magma body beneath the summit and rift zones of Kilauea volcano, Hawaii. *Science* **247**, 1311–16.

Delorme, H. & P. A. Blum 1991. Tiltmeter network observations at Piton de la Fournaise volcano (Réunion Island) 1985-1990. In *Proceedings of the workshop on geodynamical instrumentation applied to volcanic areas*, M. van Ruymbeke & N. d'Oreye (eds). Luxembourg: Council of Europe.

Delos Reyes, P. J. 1992. Volunteer observers program: a tool for monitoring volcanic and seismic events in the Philippines. In *Geohazards: natural and man made*, G. J. H. McCall, D. J. C. Laming, S. C. Scott (eds), 13–24. London: Chapman & Hall.

Dieterich, J. H. & R. W. Decker 1975. Finite element modelling of surface deformation associated with volcanism. *Journal of Geophysical Research* **80**, 4094–102.

Dzurisin, D. 1992. Electronic tiltmeters for volcano monitoring: lessons from Mount St Helens. In *United States Geological Survey Bulletin 1966, Monitoring volcanoes: techniques and strategies used by the staff of the Cascades Volcano Observatory 1980–90*, J. W. Ewert & D. A. Swanson (eds), 69–84. Washington DC: United States Government Printing Office.

Dzurisin, D., R. Y. Koyanagi, T. T. English 1984. Magma supply and storage at Kilauea volcano, Hawaii; 1956-1983. *Journal of Volcanology and Geothermal Research* **21**, 177–206.

Dzurisin, D., R. P. Denliger, J. G. Rosenbaum 1990. Cooling rate and thermal structure determined from progressive magnetization of the dacite dome at Mount St Helens, Washington. *Journal of Geophysical Research* **95**, 2763–80.

Eaton, J. P. 1959. A portable water-tube tiltmeter. *Bulletin of the Seismological Society of America* **46**, 301–16.

Eggers, A. A. 1983. Temporal gravity and elevation changes at Pacaya volcano, Guatemala. *Journal of Volcanology and Geothermal Research* **19**, 223–37.

Endo, E. T. & T. L. Murray 1991. Real-time seismic amplitude measurement (RSAM): a volcano monitoring and prediction tool. *Bulletin of Volcanology* **53**, 533–45.

Evans, D. L., T. G. Farr, H. A. Zebker, J. J. van Zyl, P. J. Mouginis-Mark 1992. Radar interferometry studies of the Earth's topography. *Eos* **73**, 553, 557, 558.

Ewert, J. A., B. Voight, A. Björnsson 1991. Elastic deformation models of Krafla volcano, Iceland, for the decade 1975 through 1985. *Bulletin of Volcanology* **53**, 436–59.

Ewert, J. W. 1992. A single-setup trigonometric levelling method for monitoring ground-tilt changes. In *United States Geological Survey Bulletin 1966, Monitoring volcanoes: techniques and strategies used by the staff of the Cascades Volcano Observatory 1980–90*, J. W. Ewert & D. A. Swanson (eds), 151–8. Washington DC: United States Government Printing Office.

Ewert, J. W. & D. A. Swanson (eds) 1992. *United States Geological Survey Bulletin 1966*, Monitoring volcanoes: techniques and strategies used by the staff of the Cascades Volcano Observatory 1980–90. Washington DC: United States Government Printing Office.

Fisher, R. V. & G. Heiken 1982. Mount Pelée, Martinique: May 8 and 20 1902, pyroclastic flows and surges. *Journal of Volcanology and Geothermal Research* **12**, 339–71.

Fiske, R. S. & W. R. Kinoshita 1969. Inflation of Kilauea volcano prior to its 1967–68 eruption. *Science* **165**, 341–49.

Fox, T. 1988. Global airways volcano watch is steadily expanding. *International Civil Aviation Organisation*

*Bulletin*, **April**, 21–23.

Francis, P. W. 1985. The origin of the 1883 Krakatau tsunamis. *Journal of Volcanology and Geothermal Research* **25**, 349–63.

Francis, P. W. & D. A. Rothery 1987. Using the Landsat Thematic Mapper to detect and monitor active volcanoes. *Geology* **15**, 614–17.

Gerlach, T. M. & T. J. Casadevall 1986. Fumarole emissions at Mount St Helens volcano, June 1980 to October 1981: degassing of a magma-hydrothermal system. *Journal of Volcanology and Geothermal Research* **28**, 141–60.

Glaze, L. S., P. W. Francis, D. A. Rothery 1989. Measuring thermal budgets of active volcanoes by satellite remote sensing. *Nature* **338**, 144–6.

Gorshkov, G. S. 1963. Directed volcanic blasts. *Bulletin Volcanologique* **26**, 83–88.

Guest, J. E. & J. B. Murray 1979. An analysis of hazard from Mount Etna volcano. *Journal of the Geological Society of London* **136**, 347–54.

Hall, M. L. 1992. The 1985 Nevado del Ruiz eruption: scientific, social, and governmental response and interaction before the event. In *Geohazards, natural and man-made*, G. J. H. McCall, D. J. C. Lamming, S. C. Scott (eds), 43-52. London: Chapman & Hall.

Herd, D. G. & the Comité de Estudios Vulcanológicos 1986. The 1985 Ruiz Volcano disaster. *Eos* **67**, 457–60.

Hoffmann, J. P., G. E. Ulrich, M. O. Garcia 1990. Horizontal ground deformation patterns and magma storage during the Pu'u 'O'o eruption of Kilauea volcano, Hawaii: episodes 22–42. *Bulletin of Volcanology* **52**, 522–31.

Holasek, R. E. & W. I. Rose 1991. Anatomy of 1986 Augustine volcano eruptions as recorded by multispectral image processing of digital AVHRR weather satellite data. *Bulletin of Volcanology* **53**, 420–35.

Hurst, A. W. & R. R. Dibble 1981. Bathymetry, heat output, and convection in Ruapehu crater lake, New Zealand. *Journal of Volcanology and Geothermal Research* **9**, 215–36.

IAVCEI IDNDR Task Group 1990. Reducing volcanic disasters in the 1990s. *Bulletin of the Volcanological Society of Japan* **35**, 80–95.

Jackson, D. B., J. Kauahikaua, C. J. Zablocki 1985. Resistivity monitoring of an active volcano using the controlled-source electromagnetic technique: Kilauea volcano, Hawaii. *Journal of Geophysical Research* **90**, 12545–55.

Jackson, D. B., M. K. Hort, K. Hon, J. Kauahikaua 1987. Detection and mapping of active lava tubes using the VLF induction techniques, Kilauea volcano, Hawaii. *Eos* **68**, 1543.

Janda, R. J., K. M. Scott, K. M. Nolan, H. A. Martinson 1981. Lahar movements, effects, and deposits. In *United States Geological Survey Professional Paper 1250, The 1980 eruptions of Mount St Helens, Washington*, P. W. Lipman & D. R. Mullineaux (eds), 461–78. Washington DC: United States Government Printing Office.

Johnson, G. V., A. Björnsson, S. Sigurdsson 1980. Gravity and elevation changes caused by magma movement beneath the Krafla caldera, Northeast Iceland. *Journal of Geophysical Research* **47**, 132–40.

Johnston, M. J. S. & F. D. Stacey 1969. Volcano-magnetic effect observed on Mount Ruapehu, New Zealand. *Journal of Geophysical Research* **74**, 6541–65.

Kienle, J., K. G. Dean, H. Garbeil, W. I. Rose 1990. Satellite surveillance of volcanic ash plumes, application to aircraft safety. *Eos* **71**, 266.

Kling, G. W., M. A. Clark, H. R. Compton, J. D. Devine, W. C. Evans, A. M. Humphrey, E. J. Koenigsberg, J. P. Lockwood, M. L. Tuttle, G. N. Wagner 1987. The 1986 Lake Nyos gas disaster in Cameroon, west Africa. *Science* **236**, 169–75

Krueger, A. J. 1983. Sighting of the El Chichón sulphur dioxide clouds with the Nimbus-7 Total Ozone Mapping Spectrometer. *Science* **220**, 1377–9.

Latter, J. H. 1981. Tsunamis of volcanic origin. *Bulletin Volcanologique* **44**, 467–90.

Lénat, J-F., P. Bachèlery, A. Bonneville, P. Tarits, J. L. Cheminée, H. Delorme 1989. The December 4 1983 to February 18 1984 eruption of Piton de la Fournaise (La Réunion Island, Indian Ocean): description and interpretation. *Journal of Volcanology and Geothermal Research* **36**, 87–112.

Lipman, P. W. & D. R. Mullineaux (eds) 1981. *United States Geological Survey Professional Paper 1250, The 1980 eruptions of Mount St Helens, Washington*. Washington DC: United States Goverment Printing Office.

Lipman, P. W., J. G. Moore, D. A. Swanson 1981. Bulging of the north flank before the May 18th eruption – geodetic data. In *United States Geological Survey Professional Paper 1250, The 1980 eruptions of Mount St Helens*, R. W. Decker, T. L. Wright, P. H. Stauffer (eds), 143–55. Washington DC: United States Government Printing Office.

Lowe, D. R., S. N. Williams, H. Leigh, C. B. Connor, J. B. Gemmell, R. E. Stoiber 1986. Lahars initiated by the 13 November 1985 eruption of Nevado del Ruiz, Colombia. Nature 324, 51-3.

Luhr, J. F. & J. C. Varekamp (eds) 1984. El Chichón volcano, Chiapas, Mexico. *Journal of Volcanology and Geothermal Research, Special Issue* 23, 1–191.

Malinconico, L. L. 1987. On the variation of SO$_2$ emission from volcanoes. *Journal of Volcanology and Geothermal Research* 33, 231-7.

Matson, M. 1989. Monitoring volcanic eruptions using meteorological satellite data. *New Mexico Bureau of Mines and Mineral Resources Bulletin* 131, 178.

McGuire, W. J. 1992. Monitoring active volcanoes: procedures and prospects. *Proceedings of the Geologists' Association* 103, 303–20.

McGuire, W. J., A. D. Pullen, S. J. Saunders 1990. Recent dyke-induced large-scale block movement at Mount Etna and potential slope failure. *Nature* 343, 357-9.

McGuire, W. J., J. B. Murray, A. D. Pullen, S. J. Saunders 1991. Ground deformation monitoring at Mt. Etna, evidence for dyke emplacement and slope instability. *Journal of the Geological Society of London* 148, 577-83.

McKee, C. O., R. W. Johnson, P. L. Lowenstein, S. J. Riley, R. J. Blong, P. De St Ours, B. Talai 1985. Rabaul caldera, Papua New Guinea: volcanic hazards, surveillance, and eruption contingency plans. *Journal of Volcanology and Geothermal Research* 23 195–237.

Miller, C. D., D. R. Mullineaux, D. R. Crandell 1981. Hazards assessments at Mount St Helens. In *United States Geological Survey Professional Paper 1250, 1980 eruptions of Mount St Helens, Washington*, P. W. Lipman & D. R. Mullineaux (eds), 789–802. Washington DC: United States Government Printing Office.

Miller, T. P. & R. L. Smith 1977. Spectacular mobility of ash flows around Aniakchak and Fisher calderas, Alaska. *Geology* 5, 173-6.

Mouginis-Mark, P. J., D. C. Pieri, P. W. Francis, L. Wilson, S. Self, W. I. Rose, C. A. Wood 1989. Remote sensing of volcanoes and volcanic terrains. *Eos* 70, 1567, 1571, 1575.

Murray, J. B. & A. D. Pullen 1984. Three-dimensional model of the feeder conduit of the 1983 eruption of Mount Etna volcano from ground deformation measurements. *Bulletin Volcanologique* 47, 1145–63.

Murray, T. L. & E. T. Endo 1992. A real-time seismic-amplitude measurement system (RSAM). In *United States Geological Survey Bulletin 1966, Monitoring volcanoes: techniques and strategies used by the staff of the Cascades Volcano Observatory 1980–90*, J. W. Ewert & D. A. Swanson (eds), 5–10. Washington DC: United States Government Printing Office.

Nunnari, G. & G. Puglisi 1991. Data related to eruptive activity, Etna: GPS survey. *Acta Vulcanologica* 1, 268–71.

Okamura, A. T., J. J. Dvorak, R. Y. Koyanagi, W. R. Tanigawa 1988. Surface deformation associated with the 1983 Kilauea eruption. *United States Geological Survey Professional Paper* 1463, 165–81.

Omori, F. 1914. The Sakura-jima eruptions and earthquakes. *Bulletin of the Imperial Earthquake Investigation Committee* 8, Nos 1–6, 525.

Oppenheimer, C. M. M. & D. A. Rothery 1991. Infrared monitoring of volcanoes by satellite. *Journal of the Geological Society of London* 148, 563-9.

Oppenheimer, C. M. M. & D. Stevenson 1989. Liquid sulphur lakes at Poás volcano. *Nature* 342, 790–93.

Pierson, T. C. 1985. Initiation and flow behaviour of the 1980 Pine Creek and Muddy River lahars, Mount St Helens, Washington. *Geological Society of America Bulletin* 96, 1056–69.

Pierson, T. C. 1992. Rainfall-triggered lahars at Mt. Pinatubo, Philippines, following the June 1991 eruption. *Landslide News* 6, 6–9.

Punongbayan, R. S. & R. I. Tilling 1989. Recent case histories. In *Short courses in geology, vol. 1: Volcanic hazards*, R. I. Tilling (ed.), 81–101. Washington DC: American Geophysical Union.

PVOT (Pinatubo Volcano Observatory Team). 1991. Lessons from a major eruption: Mount Pinatubo, Philippines. *Eos* 72, 545, 553, 555.

Rothery, D. A. 1992. Monitoring and warning of volcanic eruptions by remote sensing. In *Geohazards, natural and man-made*, G. J. H. McCall, D. J. C. Lamming, S. C. Scott (eds), 25–32. London: Chapman & Hall.

Rothery, D. A. & D. C. Pieri 1993. Remote sensing of active lava. In *Active lavas*, C. R. J. Kilburn & G. Luongo (eds). 203–229. London: UCL Press.

Rymer, H. & G. C, Brown 1989. Gravity changes as a precursor to volcanic eruption at Poás volcano, Costa Rica. *Nature* **342**, 902–5.

Rymer, H., J. B. Murray, G. C. Brown, F. Ferrucci, W. J. McGuire 1993. Mechanisms of magma eruption and emplacement at Mount Etna between 1989 and 1992. *Nature* **361**, 439–41.

Sawada, Y. 1989. The detection capability of explosive eruptions using GMS imagery, and the behaviour of dispersing eruption clouds. In *IAVCEI Proceedings in Volcanology vol. 1: Volcanic hazards - assessment and monitoring*, J. H. Latter (ed.), 233–45. Berlin: Springer.

Scott, W. E. 1989a. Volcanic and related hazards. In *Short courses in geology, vol. 1: Volcanic hazards*, R. I. Tilling (ed.), 9–23. Washington DC: American Geophysical Union.

Scott, W. E. 1989b. Volcanic-hazards zonation and long-term forecasts. In *Short courses in geology, vol. 1: Volcanic hazards*, R. I. Tilling (ed.), 25–49. Washington DC: American Geophysical Union.

Shepherd, J. B. & H. Sigurdsson 1978. The Soufrière crater lake as a calorimeter. *Nature* **271**, 344–5.

Shimada, S., Y. Fujinawa, S. Sekiguchi, S. Ohmi, T. Eguchi, Y. Okada 1990. Detection of a volcanic fracture opening in Japan using Global Positioning System measurements. *Nature* **343**, 631–3.

Siebert, L. 1984. Large volcanic debris avalanches: characteristics of source areas, deposits, and associated eruptions. *Journal of Volcanology and Geothermal Research* **22**, 163–197.

Sigurdsson, H., J. D. Devine, F. M. Tchoua, T. S. Presser, M. K. W. Pringle, W. C. Evans 1987. Origin of the lethal gas burst from Lake Monoun, Cameroun. *Journal of Volcanology and Geothermal Research* **31**, 1–16.

Sigvaldason, G. E. 1989. International conference on Lake Nyos disaster, Yaounde, Cameroon, 16–20 March 1987: conclusions and recommendations. *Journal of Volcanology and Geothermal Research* **39**, 97–107.

Smithsonian Institution 1992. Galeras (Colombia): explosion kills 9 people on the active cone, including 6 volcanologists. *Bulletin of the Global Volcanism Network* **17**, 2–4

Stothers, R. B., M. R. Rampino, S. Self, J. A. Wolff 1989. Volcanic winter? climatic effects of the largest volcanis eruptions. In *IAVCEI Proceedings in Volcanology vol. 1: Volcanic hazards - assessment and monitoring*, J. H. Latter (ed.), 3–9. Berlin: Springer.

Swanson, D. A., W. A. Duffield, R. S. Fiske 1976. *United States Geological Survey Professional Paper 963, Displacement of the south flank of Kilauea volcano: the result of forceful intrusion of magma into the rift zones*. Washington DC: United States Government Printing Office.

Swanson, D. A., T. J. Casadevall, D. Dzurisin, S. D. Malone, C. G. Newhall, and C. S. Weaver 1983. Predicting eruptions at Mount St Helens, June 1980 through December 1982. *Science* **221**, 1369–76.

Swanson, D. A., T. J. Casadevall, D. Dzurisin, R. T. Holcomb, C. G. Newhall, S. D. Malone, C. S. Weaver 1985. Forecasts and predictions of eruptive activity at Mount St Helens, USA: 1975–1984. *Journal of Geodynamics* **3**, 397–423.

Tedesco, D., R. Pece, J-C. Sabroux 1988. No evidences of a new magmatic gas contribution to the Solfatara volcanic gas, during the bradyseismic crisis at Campi Flegrei caldera (Italy). *Geophysical Research Letters* **15**, 1441–4.

Thompson, S. A. 1982. *Natural hazards research working paper 45, Trends and developments in global natural disasters 1947 to 1981*. Colorado: University of Colorado Institute of Behavioural Science.

Tilling, R. I. (ed.) 1989. *Short courses in geology, vol. 1: Volcanic hazards*. Washington DC: American Geophysical Union.

Tilling, R. I. & D. W. Peterson 1993. Field observation of active lava in Hawaii: some practical considerations. In *Active lavas*, C. R. J. Kilburn & G. Luongo (eds), 141–68. London: UCL Press.

Tilling, R. I., C. Heliker, T. L. Wright 1987. *United States Geological Survey General Interest Publication Series, Eruptions of Hawaiian volcanoes: past, present, and future, 1–54*. Washington DC: United States Government Printing Office.

Toutain, J. P., P. Bachèlery, P. A. Blum, J. L. Cheminée, H. Delorme, L. Fontaine, P. Kowalski, P. Taochy 1992a. Real-time monitoring of vertical ground deformations during eruptions at Piton de la Fournaise. *Geophysical Research Letters* **19**, 553–6.

Toutain, J. P., J. C. Baubron, J. Le Bronec, P. Allard, P. Briole, B. Marty, G. Miele, D. Tedesco, G. Luongo

30

1992b. Continuous monitoring of distal gas emanations at Vulcano, southern Italy. *Bulletin of Volcanology* **54**, 147–55.

Ui, T. 1983. Volcanic dry avalanche deposits – identification and comparison with non-volcanic stream deposits. *Journal of Volcanology and Geothermal Research* **18**, 135–150.

UNESCO 1993. Disaster reduction. *Environment and Development Briefs 5*. London: Banson.

Voight, B. 1988. Countdown to catastrophe. *Earth and Mineral Sciences* **57**, 17–30.

Voight, B. 1990. The 1985 Nevado del Ruiz volcano catastrophe – anatomy and retrospection. *Journal of Volcanology and Geothermal Research* **42**, 151–88.

Williams, R. S. & J. G. Moore 1983. *United States Geological Survey General Interest Publications Series, Man against volcano: the eruption on Heimaey, Vestmannaeyjar, Iceland*. Washington DC: United States Government Printing Office.

Yakutake, Y., T. Yoshino, H. Utada, H. Watanabe, Y. Hamano, Y. Sasai, T. Shimomura 1987. Changes in the electrical resistivity of the central cone, Miharayama, of Izu-Oshima volcano, associated with its eruption in November 1986. *Proceedings of the Japanese Academy, Series B* **63**, 55–8.

Yamashita, K. M. 1981. *United States Geological Survey Open File Report 81–523, Dry tilt: a ground deformation monitor as applied to the active volcanoes of Hawaii*. Washington DC: United States Government Printing Office.

Yamashita, K. M. 1992. Single-setup levelling used to monitor vertical displacement (tilt) on Cascades volcanoes. In *United States Geological Survey Bulletin 1966, Monitoring volcanoes: techniques and strategies used by the staff of the Cascades Volcano Obsevatory 1980–90*, J. W. Ewert & D. A. Swanson (eds), 143–50. Washington DC: United States Government Printing Office.

Zablocki, C. J. 1978. Application of the VLF induction method for studying some volcanic processes of Kilauea volcano, Hawaii. *Journal of Volcanology and Geothermal Research* **3**, 155–95.

Zlotnicki, J. 1986. Magnetic measurements on La Soufriére volcano, Guadeloupe (Lesser Antilles) 1976-1984: a re-examination of the volcanomagnetic effects observed during the volcanic crisis of 1976–1977. *Journal of Volcanology and Geothermal Research* **30**, 83–116.

Zlotnicki, J. & J. C. Le Mouel 1988. Volcanomagnetic effects observed on Piton de la Fournaise (Réunion Island): 1985–1987. *Journal of Geophysical Research* **93**, 9157–71.

Zlotnicki, J. & J. C. Le Mouel 1990. Possible electrokinetic origin of large magnetic variations at La Fournaise volcano. *Nature* **343**, 633–6.

31

# 2　Data acquisition and telemetry

Ph. Lesage, J. Vandemeulebrouck and M. Halbwachs

## 2.1 Introduction

Volcano monitoring may be defined as encompassing all the methods used to provide timely information on the evolution of activity at a particular volcanic site. This monitoring is inseparable from the instruments necessary for its performance, few of which are widely available. Seldom can equipment already employed in other fields of science be integrated into a volcanological surveillance system without modification and, more often, the instruments have to be designed, built and tested on site, where, depending on the prelimininary results obtained, they are modified accordingly.

A wide variety of intruments is used to monitor a volcano, including microphones, seismometers, tiltmeters, and gravimeters. Although data processing for the different instruments varies greatly, data analysis has a common purpose, i.e. to detect changes in the character of the transmitted signal. It is largely the notion of temporal variation that distinguishes volcanology from the other earth sciences: *rather than the stable condition of a volcano, it is the record of its evolution over time that is the subject of the monitoring methods.* In other words, it is not the absolute level of a measured parameter, but its derivative, which is significant.

Inseparable from the temporal concept is that of frequency. What is the time-scale of the evolution of the parameter under observation and, consequently, what is the required sampling period? This period can vary between $10^{-4}$ seconds for electromagnetic and acoustic parameters, to over a year for ground deformation, gravimetry and temperatures. Equipment for processing signal variations of several kilohertz will differ from equipment created to sample variations on the scale of an hour or so, not only in terms of material design but also in its software architecture.

When defining a system to provide information from sensors placed on a volcano, a choice has to be made between several basic options. The most important criterion is whether the system will be digital or analogue, if digital, it is important to decide at which point the analogue-to-digital conversion will be made, thus determining which parts of the systems will be analogue and which digital. So it is important to resolve:

(a)  whether the data are to be recorded *in situ* or transmitted directly

(b)  the type (analogue or digital) of recorded data and the means of storage

(c)  the means of transmission (radio, cable, telephone, satellite) of data
(d)  the level of automatic processing of the signal.
The options available are summarized in Figure 2.1.

Selection of appropriate monitoring equipment has to take into account the activity of the volcano, the potential risk to the local population, and the general level of funding available. Dangerous and regularly erupting volcanoes require surveillance by a permanent network which is able to monitor a range of different parameters. Such surveillance normally requires the existence of the local observatory supported by a technical team for maintenance of equipment and interpretation of the data. The definition of a volcano "observatory" is a wide one; at the simplest level, as for example on many Indonesian volcanoes, the "observatory" may comprise a single drum-type seismometer directly connected to the sensor, and maintained by an observer who has received basic training in volcanic seismology. No telemetry is necessary, and the surveillance of one seismometer allows the observer to deliver a rough and somewhat empirical diagnosis of the mean activity of the volcano. This minimum equipment is the cheapest possible for an alert station which is able to detect abnormal changes during the dormant stage of a volcano. In terms of communications, all that is required is a reliable radio or telephone link with the central volcanological institute in order to seek further assistance or advice in case of anomalous activity. A slightly more sophisticated seismic monitoring system would incorporate an analogue radio transmission network to collect the data from stations located around the volcano. In this case, the seismic signal would remain analogue (analogue transmission and analogue recording), and a simple time analysis would result in a rough localization of the earthquake hypocenter.

Increasingly, in modern volcanology, there is a need for intensive digital processing, and more often the observer is assisted or replaced by the computer for signal analysis and temporal-activity diagnosis. Generally these procedures can only be implemented at the expensive monitoring observatories which are maintained by the developed countries (e.g. at Hawaii, Piton de la Fournaise, Mount St Helens, and Mount Etna). The capital cost of the sophisticated equipment required is often high, as are the running costs, because of the skilled technical expertise required for maintenance and repair. The ideal observatory should be able to measure, continuously, and at the appropriate frequency, the greatest feasible number of geophysical and geochemical parameters at different points on the volcano. Such simultaneous multiparameter measurement allows informative correlations to be made which help to explain the general nature of the eruptive phenomenon. In addition to their monitoring function, modern observatories also serve to act as centres for fundamental experimental research into all aspects of volcano behaviour, and can thus be considered as true field laboratories.

Modern observatories make considerable use of both computers and peripherals which are adapted to the reception and processing of real-time, high-rate information, allowing immediate analysis of changes in volcanic behaviour. Because digital processing usually removes some of the information in the signal (especially seismic signals), the data must be recorded in as unprocessed a form as possible so that researchers retain the capability of going back over the data to check if anything was missed.

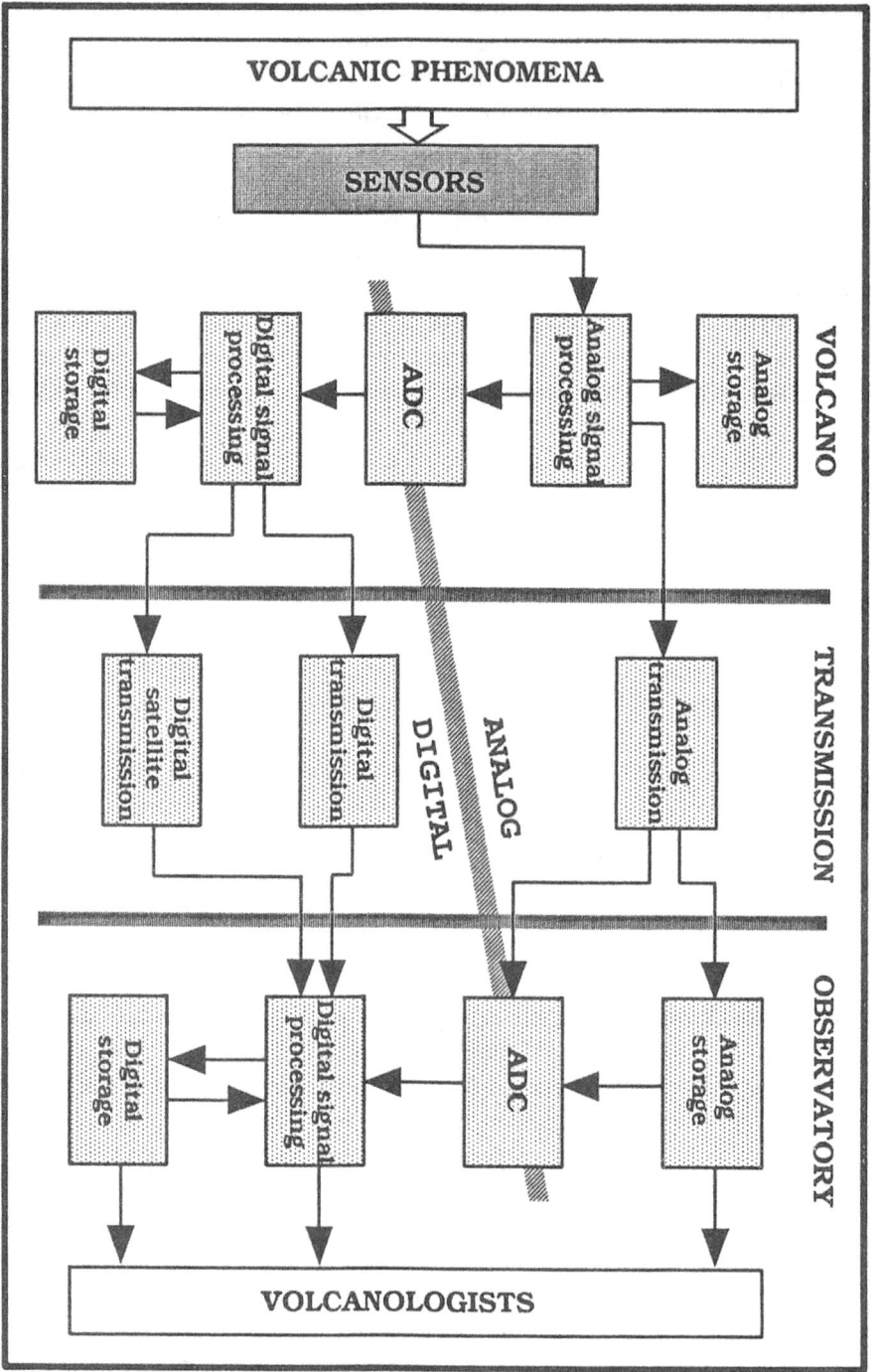

**Figure 2.1** Block diagram of acquisition and transmission systems. Each possible path between the sensors and the volcanologists represents a type of data acquisition and transmission system. The figure does not take into account the design differences linked to the flow of data to be processed. ADC, analogue-to-digital converter.

## 2.1.1 Main features of volcano monitoring instrumentation

The instruments used in volcano monitoring must satisfy a number of requirements which distinguish them from electronic equipment used in the laboratory. On the technical side, equipment must be able to withstand the rigours of the field environment. It must, therefore, be robust; protective cases and connections must be waterproof and, where required, airtight, resistant to temperature extremes, corrosion and volcanic gases, and protected against lightning and electrostatic interference. Where lives may be at risk, reliability is of course of paramount importance in the operation of monitoring systems, and it should be emphasized that there are considerable difficulties involved in maintaining the reliable operation of stand-alone stations for long periods without intervention. Reliability is also affected by financial constraints. In particular, maintaining and repairing remote stations can be costly, and so, to avoid excessive travelling expenses, it is advantageous to install instrumentation that can be repaired by local, often unqualified, personnel. To allow this, instrumentation systems must be modular and, in particular, the electronic systems must be made up of interchangeable and easily switchable boards. Modular systems also facilitate the identification of the causes of breakdown, and make it cost-effective to train a technician to maintain a specific type of equipment. In practice, the current tendency is more towards replacing, rather than repairing, defective units which can then be returned to the factory for refurbishment. The concept of modularity, and also that of standardization, becomes more important as monitoring systems become increasingly complex. The establishment of standards for connections and data-transfer buses facilitates the interconnection of different systems, while the design of user-friendly software and hardware permits the operation of advanced instrumentation by volcanologists not trained in electronics.

The main difficulty encountered in the design of autonomous systems for volcano monitoring lies in energy consumption, and the use of solar panels, and occasionally wind generators, is often recommended to provide extra power. Solar energy is generally satisfactory in tropical areas, where a solar panel of 50 W peak power allows a permanent supply of 3–5 W. In areas with a marked rainy season, the battery storage capacity needs to be increased. Power constraints become increasingly marked at higher latitudes and in areas of greater cloud cover.

The results of vandalism are a major and apparently unavoidable problem in the installation and operation of a volcano monitoring system, the seriousness of the damage varying from one country to another. In some cases, it is advisable to hide material left in the field, although of course this is virtually impossible for solar panels and some radio antennae. In other cases a shelter may have to be constructed to provide protection, or local authorities may be requested to keep watch.

The cost of instruments relative to the available budget is another important constraint, and it is necessary to assess the various costs of a programme before its initiation, including purchase of equipment, its operation, maintenance and personnel. The various needs and budgets of different developing countries will impose compromises on the available choice of volcanological surveillance equipment. The fact that the instruments are not always long-term capital investments must be emphasized. For example, the entire seismic network at Pinatubo was destroyed during the 1991 eruption. If it had comprised state-of-the-art telemetered digital instrumentation, this could have generated a loss of US$150000. As it was, the selection of cheaper, but still effective, equipment limited the loss to only US$35000. Finally, it should be emphasized that any surveillance system should ideally be designed to permit a quick preliminary analysis in real- or near-time. Volcano monitoring does not, unfortunately, allow scientists the luxury of spending months on interpretation before releasing their data.

In this introduction, we have tried to show to what extent the choice of instruments depends on the selected monitoring strategy, we have discussed the principal features of volcanological equipment and have outlined the difficulties likely to be encountered by personnel responsible for volcano surveillance. Later in this chapter, we will explain in some detail the basic principles underpinning the operation of a monitoring system, in particular with regard to analogue-to-digital signal conversion and transmission. In the second part of the chapter, we will describe a number of surveillance systems currently in operation, the selection of which has been made to illustrate the range of techniques used in the acquisition, transmission and storage of data. This certainly does not constitute an exhaustive list of current systems, but rather provides a number of representative case histories designed to aid in the choice of a system suited to the different requirements and circumstances of the potential user. In the conclusion, the advantages and disadvantages of the different systems are discussed, as are possible future trends in data acquisition and telemetry.

## 2.2 Principles of data acquisition and transmission

### 2.2.1 Basic principles

In the monitoring of volcanoes, most of the observations are based on information transmitted by sensors in the form of an electric current or voltage which is proportional to a physical phenomenon. This quantity is called the *signal* and is generally represented as a function of one variable, i.e. time. Both the function and the variable have either continuous values, when the signal is *analogue*, or discrete values, when the signal is *digital*. A data acquisition and transmission chain, i.e. each possible path between the volcanic phenomena and the volcanologist (Fig. 2.1), can be considered as a set of signal-processing systems. The systems are digital or analogue according to the signal that they process. Each link in the chain can be described as a signal processor which modifies the signal in some manner, for example, filtering (i.e. changing the frequency content), introducing a phase delay, or adding noise. The signal processors are characterized by their frequential response which indicates how they modify the spectral content and the phase of the signal. The whole chain is characterized by the product of the frequential responses of its constituent parts. The signal as seen by the volcanologist is then the original phenomenon modified by the response of the whole chain.

A system is also characterized by its *dynamic range*. In analogue systems, it is defined by the ratio, $r$, between the smallest and largest signal amplitude that it can process, the former often being the noise level of the system itself. The dynamic range is generally expressed in decibels (dB level $= 20 \log r$). In digital systems, the dynamic range depends on the number of bits used and on the type of coding. It can also be expressed in decibels. Because each succeeding bit doubles the dynamic range, adding 6 dB to it, a signal coded on 8 bits will have a maximum dynamic range of 48 dB, or 72 dB, if coded on 12 bits.

An analogue-to-digital converter (ADC) transforms a continuously variable input into a succession of digitally coded values. The conversion can be visualized in terms of two processes, the *quantification* of the amplitude, and the time *sampling* of the signal. The process of quantification transforms an analogue value, the amplitude of which may vary continuously, into a discretely variable digitally coded value. The analogue value is rounded up or down to the nearest digital code value (Fig. 2.2). The conversion introduces a noise corresponding to the difference between the analogue values and their coded representation. The distribution of the conversion noise depends on the method of quantification (either by rounding up or by truncation) and its amplitude depends on the number of bits used for coding, the type of code, and the amplitude in the range of variation of the signal. The quantification of analogue values is carried out at finite and regular time intervals, $\Delta t$, called the *sampling period* (Fig. 2.3). The digitization of an analogue signal is thus a combination of the processes of quantification and sampling (Fig. 2.3), while the fidelity of digital representation of an analogue signal depends on the sampling period and the accuracy of quantification.

In choosing the sampling period of an ADC, it is crucial to respect the constraint imposed by the *sampling theorem:*

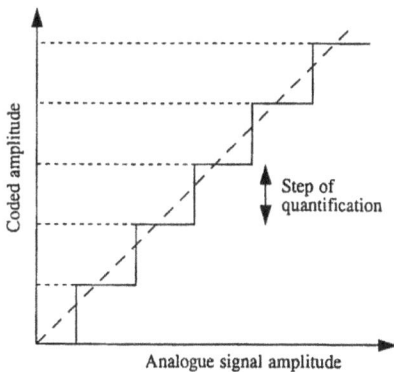

**Figure 2.2**  Quantification by rounding.

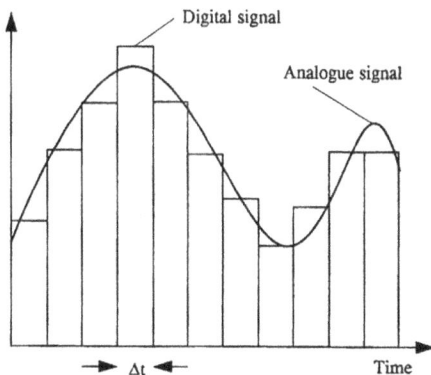

**Figure 2.3**  Digitization of an analogue signal. $\Delta t$, sampling period.

An analogue signal of finite bandwidth and of maximum frequency value $f_b$ can be exactly reconstituted from its samples only if the sampling frequency $f_s = 1/\Delta t$ is equal or superior to $2 f_b$.

The bandwidth of a signal is the range of positive frequencies containing non-zero values. $f_b$ is thus the highest frequency of the range of the spectrum. The condition imposed by the sampling theorem is thus $f_b < f_N$, where $f_N = 1/2\Delta t$ is called *Nyquist's frequency*. If this condition is not met, that is, if the analogue signal contains some frequencies superior to $f_N$, the phenomenon of *aliasing* occurs, causing shifting of information from one band of frequency to another and errors in the reconstitution of the analogue signal. In this case, the measures will be biased, whatever the type of signal and sampling device. Aliasing can even occur with low-frequency data or when the measures are taken manually at regular intervals. It could, for example, happen when an operator measures fumarole temperature at the same time every day, here Nyquist's frequency is $f_N = 0.5$ cycles/day, and if temperature variations with a frequency superior to $f_N$ occur, e.g. diurnal, the corresponding spectral components will contaminate the low-frequency components and distort the measurements. It is important, therefore, to take the sampling theorem into account whenever regular readings are taken, especially if the bandwidth of the phenomenon measured is not known in advance. Several examples of aliasing are shown in Figure 2.4.

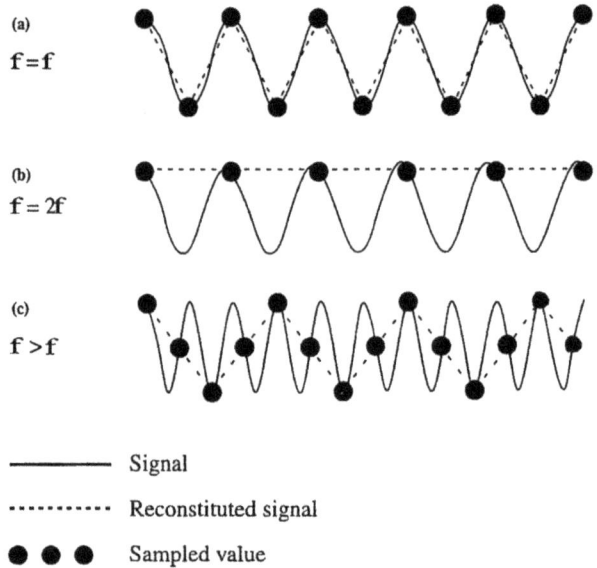

(a) $f = f$

(b) $f = 2f$

(c) $f > f$

——— Signal

·········· Reconstituted signal

●●● Sampled value

**Figure 2.4** Examples of aliasing. (a) A signal with Nyquist's frequency $f = f_N$ can be reconstituted exactly from its samples. (b) A component of frequency $f = 2f_N$ could induce an offset of the digitized signal. (c) A component of the signal, of frequency $f > f_N$, produces low frequency fluctuations of the measures.

To sample an analogue signal correctly, it is necessary either to choose a sufficiently high sampling frequency or to eliminate from the analogue signal all frequencies superior to $f_N$, using a low-pass analogue filter called an *anti-aliasing filter*. The signal can also be sampled at a frequency $n$ times greater than the desired sampling frequency, and the averages (weighted or not) of $n$ successive values kept. This procedure, called *oversampling*, is in effect a digital low-pass filter. In calculating an anti-aliasing filter, four independent parameters must be taken into account:

(a) the expected dynamic range of the signal (this determines the maximum acceptable level of signal interference caused by aliasing in the band of interest)
(b) the slope of the filter's response curve
(c) the cut-off frequency, $f_c$, of the filter, i.e. the frequency for which attenuation is equal to $-3\,\text{dB}$
(d) the sampling frequency.

Let us take the example of a 6th order filter with a $36\,\text{dB/octave}$ slope and a signal for which the desired dynamic range is $72\,\text{dB}$, that is 12 significant bits, in its frequency band. Figure 2.5 shows that it is necessary to use a sampling frequency of at least 5 times the filter cut-off frequency, so that the level of signal aliasing is attenuated by at least $72\,\text{dB}$ inside the band $(f < f_c)$. Generally, the anti-aliasing filter and the sampling frequency must be chosen so that the attenuation of the signal between the cut-off frequency and the sampling frequency is equal to the required dynamic range.

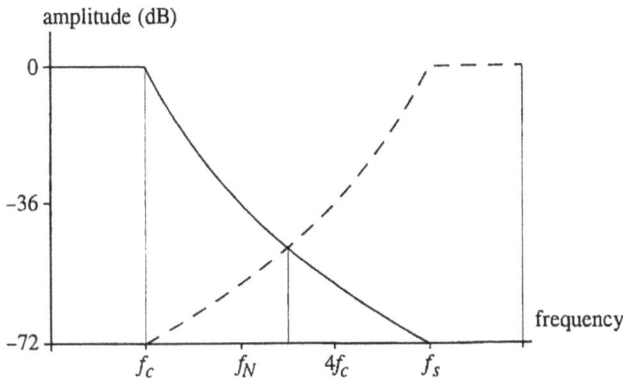

Figure 2.5  Anti-aliasing filter. The dashed curve represents the effect of spectral folding of the filter response. Both curves are symmetrical about the Nyquist's frequency, $f_N$. The sampling frequency $f_s$ must be greater than 5 times the cut-off frequency, $f_c$, of the filter in order to avoid aliasing.

## 2.2.2 Analogue signal processing

We now turn to the part of the acquisition chain between the sensor and the sampling device (ADC). It is a particularly delicate task to design and set up this part of the chain for the measurement of natural parameters, because:

(a) the signal variations may have a very wide amplitude range and the fluctuations are often sudden and apparently random
(b) the signal transmitted by the sensors is sometimes very weak, and superimposed on other diverse signals (noise and interference) with much greater amplitude
(c) this stage of the processing requires a knowledge of analogue electronics, an area now being neglected in favour of digital techniques.

When constructing an acquisition chain, one obvious fact must be borne in mind, i.e. the front end of the chain carries the maximum level of information. Any degradation of the signal by inadequate processing anywhere in the chain cannot subsequently be rectified. Thus the ana-logue processing at the start of the chain is of the utmost importance. The initial link in the chain is the sensor itself and its ability to accurately translate volcanic phenomena into meas-urable electrical phenomena. The output of the sensor goes through an analogue signal process-ing circuit which contains:

(a) a pre-amplification stage to avoid alteration of the signal by electronic components fur-ther down the chain

(b) a selection and amplification stage for the signal frequency bandwidth (this stage may involve the use of diverse filters and amplifiers)

(c) an anti-aliasing filtering before digitization.

It is important to know the particular frequency spectrum of each signal type so as to avoid a serious loss of data in the processing chain. Every analogue electronic assembly has a char-acteristic transfer function and bandwidth. It is essential to design into an analogue chain a transfer function that is suitable for the frequency spectrum of the signal to be processed. This is particularly the case for quantities in which the data are contained by their frequency spectrum, e.g. in the case of seismic, acoustic and electromagnetic signals. The analogue signal processor can be quite complicated, for example, the processing required by an Audio Mag-neto-Telluric, as described by Lebsir (1991), consists of four amplifier stages, a low-pass filter, a high-pass filter, a notch filter and, finally, an anti-aliasing filter. The total amplifica-tion reaches $140\,dB$ with a noise level of less than $4\,nV\,Hz^{-\frac{1}{2}}$.

### 2.2.3 Analogue to digital conversion

The tremendous development of digital techniques over the last two decades has emphasized the fundamental importance of the component at the interface between the analogue and dig-ital "worlds", i.e. the ADC. The use of the ADC is essential to applications involving signal analysis, including convolution, digital filtering, correlation function, and Fourier transform. In this section we provide the geophysicist–volcanologist with the basic information neces-sary to help him or her in the choice of a suitable system and, in particular, the definitions which will allow him to understand the often off-putting technical specifications in the data sheets for electronic components. A brief summary of several conversion techniques should give an idea of the pros and cons of each component currently on the market.

An ADC transforms an analogue quantity (generally a voltage) into a digital quantity com-patible with the entry ports (serial or parallel) of a microprocessor. The analogue voltage to be converted must be suitably conditioned, in particular filtered, to avoid problems of signal aliasing in the relevant frequency bandwidth. A converter is represented in schematically in Figure 2.6. The main characteristics of a converter are its available sampling frequency and its dynamic range. A number of ADCs are in general use for volcano monitoring, and their characteristics are briefly outlined below.

(a) *Successive approximation ADC.* This is the most widely used type of converter. A logic system generates a digital code applied to the input of an ADC which delivers a voltage proportional to the code. This voltage is compared with the voltage to be converted, and

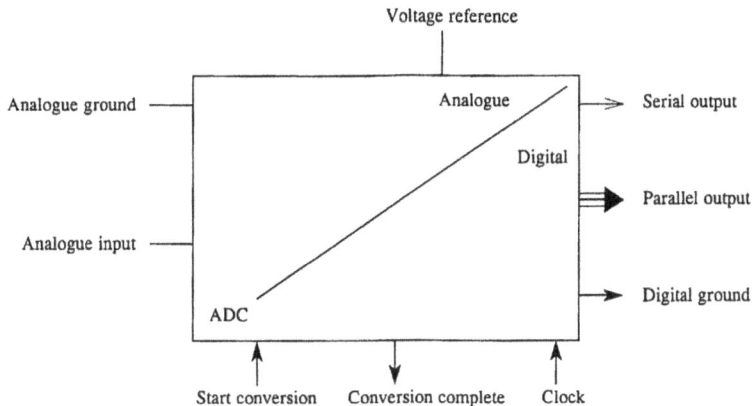

**Figure 2.6** Diagram of an analogue-to-digital converter (ADC). The analogue input signal can be positive, negative or bipolar relative to ground. The voltage reference is usually integrated into the converter and must be both precise and very stable. The input of the clock signal synchronizes the commands of the converter and microprocessor. The digital output signal can either be transmitted via a parallel bus (connected, for example, to the entry port of a microprocessor) or via a serial line.

the logic system modifies the code as a function of the result, proceeding from the most significant bit to the least significant bit. The specific applications of these converters concern data acquisition with rapid frequency multiplexing.

(b) *Dual slope integrating ADC.* Here, the voltage to be measured serves to charge a capacitor, generating an analogue "ramp" the height of which is proportional to the reference voltage. A clock is used for the duration of this ramp, to increment a counter. This ADC periodically samples and averages, it is slow but can be highly accurate.

(c) *Voltage controlled oscillator (VCO).* The VCO transforms the voltage to be converted into a signal with a frequency proportional to its amplitude. The number of periods of this signal are then counted over a fixed duration. By adjusting the counting duration, the conversion time can be shortened (shorter duration) or the resolution increased (longer duration). In a measuring system, the main advantage of a VCO is that it allows the conversion of the signal close to the sensor and the transmission of the digital signal in a form largely unaffected by noise, on a simple two-wire line to the place where it will be analyzed.

In its application to volcanological data logging, the choice of a converter is based on the following general criteria:

(a) sampling frequency should be easily determined in relation to the frequency of the signal (sampling theorem)

(b) the resolution of the converter (number of bits) should be a compromise between the desired sensitivity and the extent of possible fluctuation of the parameter being measured.

The use of switchable gain amplifiers makes a high level of sensitivity compatible with a wide

41

range of measurements. The choice of resolution level should take into account the quality of the analogue processing. A 12-bit converter would be superfluous if all the precautions to avoid noise and interference are not taken (e.g. screening, well-designed ground circuit, and uncoupling of power supply). As for all electronic components used for measurement in natural environments, the choice of converters is subject to certain particular constraints:

(a) *Power consumption.* The use of solar panels or batteries as the power source may impose electrical consumption limitations. Generally speaking, power consumption rises with the sampling rate, and with the level of resolution chosen. Dual-slope converters, however, have low power consumption, with excellent accuracy for parameters of low-rate variation (frequency less than approximately 10 Hz).

(b) *Performance relative to temperature.* Electronic components are generally available in different ranges according to the ambient temperature during use (standard range 0 to 70°C; industrial range –25 to 85°C; military range –55 to 125°C). Within each temperature range, it is also necessary to check the performance of the converter relative to temperature. In the data sheets, the variations of coefficients such as gain, accuracy and offset are generally given in relation to temperature.

(c) *Sensitivity to power supply voltage.* The stability of the voltage reference is very important. A 16-bit ADC combined with a reference voltage the fluctuation of which is 0.05% produces an ADC with only 12-bit accuracy. The data sheets indicate the component's sensitivity to the supply voltage expressed as a percentage of the input voltage for a variation of 1% of the supply voltage.

(d) *Temporal variation.* For ADCs with integrated voltage reference, temporal variation is mainly caused by variation in the reference. This is rarely indicated quantitatively (long term stability in per cent per year).

## 2.2.4 Transmission systems

Analogue transmissions have been and continue to be widely used to transmit data by radio to observatories, because they constitute a proven and reliable method of communication. This transmission, largely used for seismic data, is generally effected by frequency modulation, a well-established means which is insensitive to non-linear distortions and has an acceptable signal/noise ratio. Up to eight seismic signals can be carried along one channel by using frequency division multiplexing (FDM). The transmission channel is divided into separate audio bands (typically 8 bands from 555 to 3185 Hz, each with a band 250 Hz wide). A voltage-controlled oscillator converts the analogue seismic signal into a frequency-modulated signal in one of the bands. Signals from multiple seismometers are carried on a single line by summing them together. Adding a channel into the line is easy and involves a single op-amplifier to sum the new signal with the others. At the receiver, bandpass filters extract the individual signals. A phase lock-loop tracking the frequency of the incoming signal produces a voltage proportional to the original analogue signal. Although the dynamic range is limited (about 40–60 dB), the frequency content is accurately transmitted by this means.

The development of computer technology, the falling cost of printed circuits, and the integration of multiple functions into a single chip have all favoured the construction of digital systems which are more efficient (in terms of, for example, speed and power consumption), increasingly reliable (better coding) and easy to integrate. The digital equipment deployed in the field encourages the use of digital transmission which has several advantages over analogue communication, namely, relative immunity to interference, a greater dynamic range (about 100 dB), and transmission security if error detection and correction procedures are used. The disadvantage, compared with analogue, is that the digital transmission requires a channel with a larger bandwidth and a greater signal/noise ratio. For transmitting seismic signals, this lack of bandwidth has hampered the growth of radio-telemetered digital data. The improvements in coding schemes are being countered by a continuing shrinkage of available bandwidth due to pressures from all users of the radio spectrum. We will thus only discuss digital transmission which will doubtless become much more widespread in the years to come.

### 2.2.4.1. Principle of digital transmission

Most existing transmission lines (e.g. telephone lines) are designed for analogue transmission. When used for data transmission, there must be devices at each end of the line to convert the data to an analogue format. A digital transmission liaison can be represented thus:

SOURCE → ENCODER → TRANSMITTER → TRANSMITTING MEDIUM → RECEIVER → DECODER → USER

The encoder transforms the digital signal of the source into a low frequency signal adapted to the characteristics of transmission (e.g. the transmitting medium or noise). A high-frequency auxiliary signal, the carrier wave, is used to carry the information. The transmitter is composed of a modulator, a carrier-wave oscillator and an amplifier. The modulator combines the modulating signal and the carrier, to obtain the modulated signal. The modulated signal is both transmitted into and received from the medium by an aerial or terminal. Three phenomena affect propagation in the medium: attenuation, distortion and noise. The receiver extracts the signal from the medium and reconstitutes the modulating signal by demodulation; it also amplifies the signal which has been attenuated by the medium and filters the noise. The decoder receives a noisy and distorted signal and elaborates the signal for the user. A line is characterized by two interdependent parameters: the binary rate, which is the maximum number of bits per second that the circuit can transmit, and the bandwidth, the range of transmittable frequencies.

The modems (modulator/demodulator) perform the conversions which are necessary to transmit digital bits on an analogue line. They encode, decode, filter, modulate and demodulate the data. The different modulation techniques used in digital transmission are amplitude modulation, phase modulation, frequency modulation, or both phase and frequency modulation. Amplitude modulation of a carrier $p(t)$ by a signal $m(t)$ consists of multiplying $p(t)$ by a factor in relation to $m(t)$: if $p(t) = \cos(wt+f)$, then the modulated signal will be $s(t) = (k+m(t)).\cos(wt+f)$. Transmission of a digital signal by frequency shift keying (FSK) modulation consists of associating each symbol (two for binary) with a particular frequency of the carrier (Fig. 2.7). The choice of modulation method depends on the digital throughput, bandwidth, noise and cost.

For analogue signals, the parameter of transmission quality is the signal/noise ratio, defined

Phase continuity

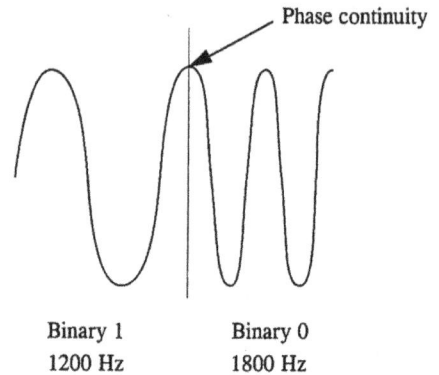

**Figure 2.7** An example of frequency modulation used in digital transmission: the frequency shift keying (FSK). With a bit rate of 1200 bit s$^{-1}$, the frequency of the modulated signal is usually 1800 Hz for binary "0" and 1200 Hz for binary "1". Each binary "0" and "1" waveform starts with either phase 0° or 180°, to ensure phase continuity.

Binary 1
1200 Hz

Binary 0
1800 Hz

as the ratio between the mean powers of signal and noise and expressed in decibels. For digital signals, the parameter of quality is the error rate, i.e. the number of false bits relative to the number of bits received, which generally varies between $10^{-3}$ and $10^{-10}$. The effect of noise on the line is to lose some of the information transmitted in the messages. In digital transmission, it is possible to detect and correct errors in the message. A simple method of preventing errors is to repeat the data in the message, but this greatly reduces the throughput. Usually, an error detecting code is used, with re-transmission of the incorrect part of message. The technique used for detection is that of redundant encoding, if the throughput is less than the capacity of the channel, a certain number of control bits can be added to the message. When the signal is received, the control bits are checked; any errors are identified and can then be corrected. A parity check allows the rejection of erroneous words, but rarely their reconstitution, the receiving terminal would request partial or total re-transmission of the message. Cyclical codes based on polynomes are simple to use and well adapted to the detection of rapid bursts of errors. In practice, such codes allow a reduction in the non-detected error rate on a line from $10^{-6}$ to $10^{-10}$ (Nussbaumer 1987).

The problems involved in the use of relay stations for returning information from monitoring stations spread out in the field can be resolved by comparison with local data networks having an arborescent or net-like structure. Certain recently developed complex radio liaison systems around a volcano use remote data processing modes of operation because of their standardization and because of the existence of network management software. Such systems use different standard protocols to transmit data in the field: RS-232C covers the material aspect of communication; high level data link control (HDLC) covers point-to-point liaison and governs the format of messages; and X25 covers the terminal connection to a network. The use of elaborated communication procedures has many benefits, among which are: reliable and bi-directional transfer of information (data and controls) between the field stations and the control site; modification of information routes in the event of a relay breakdown; and synchronization between stations.

In networks with many stations, where communication is costly, it is helpful to increase communication possibilities with multiplexing. In frequency division multiplexing, the bandwidth is divided into adjacent channels (Fig. 2.8) which is more appropriate for analogue signals. In time division multiplexing (for digital data only), the time is divided into slots,

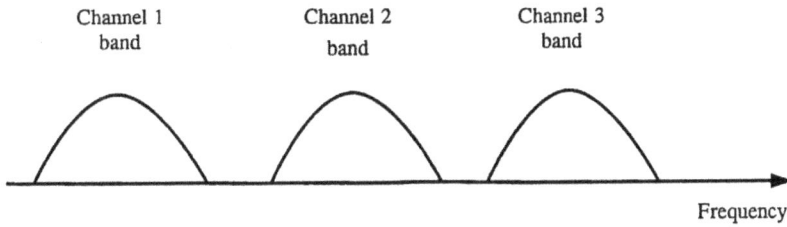

**Figure 2.8**   In frequency division multiplexing, the channels are distributed in separate frequency bands over the spectrum, and superimposed in time.

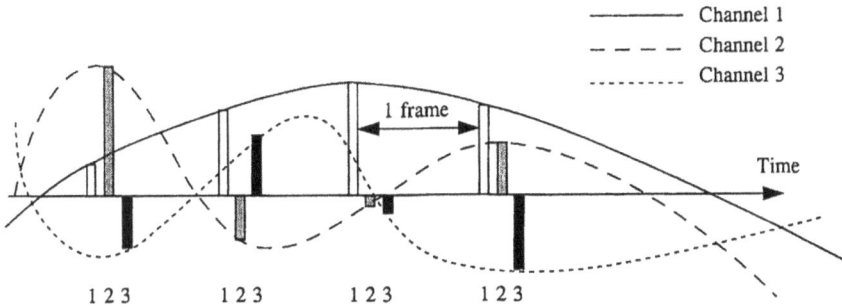

**Figure 2.9**   Time division multiplexing consists of time sharing the channels using a pulse modulation. The channels are juxtaposed in time, and superimposed in frequency (adapted from Fontolliet 1990).

and each channel is allocated the entire transmission capacity of the line for a time slot (Fig. 2.9). Line multiplexing can be static or dynamic, depending on the network structure, the nature of the line, and the throughput. With a radio network, there must be an access procedure, so that stations do not transmit simultaneously. Line-sharing is effected either by polling, where a master station controls access, or by contention, where collisions are permitted, but with an algorithm of conflict resolution to prevent repetition of the collisions. MARS-88, a system developed by Lennartz Electronic for seismic networks, is an example of non-continuous digital transmission controlled by a central station, in which network traffic is reduced to a minimum. The remote stations store the whole seismic signal in memory, and detect the events. One after another, they send trigger-status information for the elapsed period. The central station compares the information it receives from the remote stations and requests information only for periods in which several stations triggered.

*2.2.4.2 Data-transmission material*
In the electromagnetic spectrum, three areas are favoured for data transmission: VHF (30–300 MHz) is for distances of up to 100 km, UHF (300 MHz–3 GHz) is used for line of sight transmission and little affected by atmospheric interference and industrial noise; and SHF (3–30 GHz) is used for satellite communication, owing to the low level of cosmic interference in this band and the large bandwidth necessary for satellite telemetry. In a VHF or UHF data link, the error rate is of the order of $10^{-3}$ to $10^{-5}$. The bandwidth depends on the type of

45

modulation used and on the throughput. For example, a frequency shift keying (FSK) modulation, and for a given bit rate, the required radio frequency bandwidth in hertz is about twice the bit rate in bits per second. The use of radio frequencies is, however, subject to national and international regulations which put considerable constraints on transmissions. In general, the authorized rate is low (1200 Bd), the bandwidth is limited to about 10 kHz, and some frequency bands are reserved. In isolated areas, the use of radio telemetry for monitoring can sometimes benefit from preferential treatment or from non-existent or seldom applied regulations. Care should be taken, however, in the choice of frequencies; in some countries such as Indonesia, the proliferation of cheap portable transmitters and the absence of regulations has led to an anarchic use of the radio waves.

The simplest and most widespread mode of radio operation is simplex. There is a single transmission channel, so only one station can transmit at a time. For full duplex, simultaneous transmission and reception are possible, so two frequencies are needed, and they should be sufficiently separated to avoid interference; more expensive equipment is thus required. Certain equipment developed for amateur radio use is well suited to the constraints of transmission in the field, particularly because of their low power requirements and small size. The units generally comprise a modem, a low power (several watts) transceiver and have communication management software in their microprocessor.

When acquisition stations can be installed near a telephone system, data transmission can be effected by telephone, on public network or leased lines. Public network lines, with a bandwidth of 300–3200 Hz, allows a transmission rate of up to $2400\,\mathrm{bits^{-1}}$, and even 4800 or 9600 with elaborate modems. Leased lines are generally more expensive, but have the advantage of lower interference, and in some cases they can be used for high-rate transmission. With a fixed bandwidth, the only way to increase the bit rate is to use more elaborate modulations: FSK modulation for $300–1200\,\mathrm{bits^{-1}}$, phase modulation for $1200–2400\,\mathrm{bits^{-1}}$, and phase and amplitude modulation for $2400–9600\,\mathrm{bits^{-1}}$. The transmission error rate depends greatly on the line interference, but can be as low as $10^{-5}$ in favourable circumstances.

Digital cellular telephone networks are being established throughout the world, as for example the GSM system in Europe, or, in the future, the TETRA project which is better adapted to telemetric applications. Using this access to the telephone system, cellular phones allow data transfer at $9600\,\mathrm{bits^{-1}}$. Such systems, restricted to within cities and along main roads, seem to be currently of little use around volcanoes, which are in general in isolated situations. Of much more potential for the future could be the worldwide telephone systems, like the IRIDIUM which, using 77 satellites, would be much more suitable for volcanological applications.

Certain short-distance (several kilometres) links can be effected by electric cables, which can be either twisted pairs or coaxial cables, or by optical fibres. In the case of electric cables, transmission is made on baseband, i.e. without frequency transposition of the signal. The maximum rate, $R$ (in bits per second), on a twisted pair depends on the range, $r$ (in kilometres) with $Rr = 10^8$ to $10^9$. Coaxial cables are slightly better with $Rr = 5 \times 10^8$ to $5 \times 10^9$. Data transmission by optical fibre involves the propagation of light in a glass fibre. Optical fibres offer dramatically improved performance over electrical wires. At the moment, however, their installation cost and fragility limit their use to short runs to protect equipment against lightning, or to use at well-established observatories

### 2.2.4.3 Satellite telecommunications

Radio communication between a satellite and the Earth has some advantages over propagation between two terrestrial points, there is no ground interference and no need for a line-of-sight for communication. This eliminates the phenomena of reflection and diffraction and, as the angle with the Earth is large, the passage through the atmosphere is greatly reduced. The primary disadvantage lies in the amount of energy needed for transmission over large distances, e.g. 36000 km for geostationary satellites. A satellite transmission system consists of a data collection platform (DCP), a platform terminal transmitter (PTT), a satellite as relay, and a receiving station. There are several systems of satellite data collection, including geostationary satellites, satellites in other orbits, and satellite telephone systems.

The geostationary meteorological satellites of the GOES, GMS and METEOSAT series collect environmental data. These platforms usually time slice the availability so that each field station has a 1-minute time window for transmission every 3–4 h. For GOES, the transmission rate is 100 Bd. METEOSAT allows the transmission, in quasi-real time of 5192 bits every 3 h. As geostationary satellites are always "visible" from stations on the ground, these systems permit the transmission of alert messages in real time. The systems have been used for both low-frequency and high-frequency data (Endo et al. 1975, Clark & Medina 1976, Webster et al. 1981, Poupinet et al. 1989b, Silverman et al. 1989).

In order to use the INTELSAT satellites, transmitting and receiving stations both have to be equipped with parabolic antennas, which allows a link with possible throughputs in excess of 1200 Bd. The INMARSAT A and C and EUTELTRACS systems, developed for communications with mobile units, permit two-way data transfer. INMARSAT C, although better adapted to the power constraints imposed on remote stations, still needs a minimum transmission power of 25 W to transmit at $500 \, \text{bits}^{-1}$ messages whose length may be about 10 kbit.

The ARGOS system localizes and collects environmental data by satellite. The equipment is installed on satellites of the United States National Oceanic and Atmospheric Administration (NOAA). Two satellites are currently in service in circular, polar orbits at an altitude of 800 km, and with their orbital planes oriented perpendicular to each other. Each satellite simultaneously receives the transmissions from all transmitting platforms within a terrestrial radius of 2500 km. The number of times a satellite passes in sight of a point on Earth varies from an average of seven per day at the equator to 28 per day at the poles. A volcanological monitoring station transmits data to the PTT in the form of a message of 256 bits maximum, via a RS 232 link. The data are transmitted every 100–200 s by the PTT. The data received by the system on the satellite are immediately re-transmitted to the ground to a local receiver, and stored to be send later to the three control and reception stations (in France and the USA). The data received by these stations are made available to the user within 8 h of reception. The advantages of ARGOS are the ease of setting up a system, the global coverage, the possibility of synchronization with 12 ms precision, and the low power consumption of the transmitters. The main disadvantages are the reduced quantity of data and the limited transmission frequency. With no local receiving station, data are not available in real time and have to be sent by computer networks or telephone modems. The cost of a transmitter is low (less than US$1000), but the annual user fee is prohibitive for a system with a large number of stations.

## 2.2.5 Data storage

The choice of data storage depends on the data type (analogue or digital) and the storage duration required. Other important criteria include the volume of information to be treated, the cost of the systems and the storage media, the time necessary for access to the information, and, for autonomous stations on site, their power consumption. Recording equipment using paper (ink, smoke, or thermally sensitive), photographic film (microfiche), or magnetic tape, all store analogue data. Although these methods are used less nowadays, they are still important for certain applications, such as continous recording of seismic activity. The life expectancy of the medium varies, magnetic tape lasts for 10–20 years, film for a century or so, and paper for several centuries. The main disadvantages of these analogue systems are the low dynamic range (linked, for paper and film, to the relationship between the size of the medium and resolution), an unimpressive storage capacity, and an energy consumption which is generally incompatible with an isolated situation.

The main systems of digital data storage are semiconductor memories (e.g. dynamic random access memory (DRAM) and static random access memory (SRAM)), magnetic tape (tape, cassette or cartridge), floppy disks, hard disks, optical and optical-magnetic disks. These systems all have the advantages of high (and increasing) storage capacity, low (and decreasing) costs, great adaptability in their use, and a practically unlimited dynamic range. Several types of semiconductor memory can be used for data storage. The DRAM, used in computers, has a temporary storage capacity of several megabytes. SRAM cards, when used with a lithium safeguard battery, have a permanent storage capacity of 5–10 years, and a current (1992) capacity of 8 MB. "Flash" cards are electrically erasable by blocks and the manufacturers claim that their life will exceed 100 years. These cards already have capacities of up to 20 MB, are robust, relatively shockproof and can be used in temperatures up to 70°C. With the standardization of their connections, removable memory cards could become as widely used with microcomputers as floppy disks, their speed and robustness compensating for their higher cost.

Magnetic tape, streamer and digital audio tape (DAT) are still the most commonly used means of storage for backing up information. Their capacity is generally several hundred megabytes for tape and cartridge, but some storage systems do better than 1 GB per medium. Like all "magnetic" storage devices, their life expectancy is limited to 10–20 years, depending on the type of tape, the density of the stored information, and the storage conditions. Certain cassette or cartridge recorders can be integrated into autonomous acquisition systems for data storage *in situ*.

Floppy disks have capacities in the order of 1 MB, but the new floptic drive can stock more than 20 MB on a 3.5 in. Their advantage lies in great flexibility for low cost, but their life expectancy is less than 10 years. Floppy drives can be also be used for storage in autonomous stations. Hard disks can store tens, hundreds, or even thousands of megabytes. As for life expectancy, the limiting factor is the life (a few years) of the mechanical and electronic components. Small, low consumption, removable hard disks have been in use for a few years. When incorporated into an autonomous station, these can provide an impressive storage capacity, and could be particularly suited to portable seismic systems.

The several types of optical disk available are all read by the reflection of a laser beam on their surface. Although these systems are still undergoing rapid evolution, and certain types are not yet standardized, one can distinguish the compact disk read-only memory (CD-ROM), the write once, read many (WORM) disks, and the optical-magnetic disks as having application to volcano monitoring systems. The technology of the CD-ROM is similar to that of audio CD, 520 MB can be stored per disk, and the life expectancy is measured in tens of years. The method of manufacture, and of information recording in the factory, makes this storage format appropriate for data banks, such as the world seismic data networks. WORM disks allow a once-only recording of 300–600 MB of data. The technology is too recent to be sure of the life expectancy, but it would seem to be about 10–15 years. Optical magnetic disks are re-recordable and can store 128 MB. Slightly slower than low-performance hard disks, but with a removable and resistant medium, and much more rapid than floppy disk drives, the optical magnetic disks are suited to file archiving, although they currently have a life of only 10 years. The drives themselves are still relatively expensive, but their high capacity and the low cost of disks make the cost of storage per megabyte attractive.

The rapid improvement in digital data storage (higher flexibility and capacity accompanied by lower costs) gives it a considerable advantage over analogue storage. Recording on paper will remain important for visualizing information and because paper lasts. Digital storage units are, however, becoming increasingly easy to integrate into autonomous acquisition stations, thanks to continuing miniaturization, portability, and a reduction in their power requirements. Semiconductor memories are the preferred method to store data on-site at field stations because they require no moving parts and because other forms of mass storage are not rated to run at temperatures much below 0°C.

## 2.3 Examples of acquisition and transmission systems

Certain phenomena associated with volcanic activity, such as ground deformation and surface temperature vary slowly. Readings can thus be digitized at low sampling frequencies of about 1 point/min maximum. Several multichannel acquisition systems fulfill the specific technical constraints imposed by this type of measurement, which are, low power consumption, the ability to be put on stand-by, low sampling rate, and a resolution of 12 bits. After acquisition and conversion, the quantity of digital data to be transmitted or stored is reduced. In the case of storage on site, a capacity of 1 MB allows the safeguarding for 1 year of the coded values on 12 bits of $n$ parameters with a maximum sampling period of ($47n$ seconds). For data transmission, the reduced quantity of information permits a lower telemetry rate. Satellite transmission is possible and the platform can transmit the values of several parameters at an hourly rate.

Several volcanoes have been equipped with ARGOS-transmitted, low-frequency measuring systems: Etna in Italy (Briole et al. 1992) and Colima in Mexico (Lesage & De la Cruz-Reyna 1992) for recording deformation using tiltmeters; and Momotombo in Nicaragua, Lake Nyos in Cameroon (Sabroux et al. 1990), and Matthew and Hunter volcanoes (M. Lardy, personal communication) for temperature monitoring. As an example, since 1989, in collaboration with the Instituto Nicaraguense de Energia, we have been observing variations in fumarolic activity at Momotombo, mainly because of a spectacular temperature increase since the 1970s. Every 3 hours, the acquisition station records the temperature of five fumaroles (ranging from 100 to 700°C) together with the oxygen fugacity of one fumarole (Fig. 2.10). The 256-bit message is transmitted via ARGOS, and includes the readings of the latest monitoring, some technical parameter values (battery and reference voltages) and certain readings repeated from the previous sampling.

**Figure 2.10** The ARGOS-transmitted station installed in the Momotombo (Nicaragua) fumarolic field measures geochemical parameters (temperatures and $O_2$ fugacity) every 3 hours.

Numerous radio or telephone links, and digital acquisition and transmission systems are currently being used to monitor active volcanoes. For example, the Nordic Volcanological Institute uses a standardized data acquisition system in the active volcanic zones of Iceland (the Krafla, Askja, Hekla and Grimsvötn volcanoes). This is an eight input channel digital system compatible with various sensors (including temperature, ground tilt, and crater-lake level), in which the data are either stored *in situ* or transmitted (via radio or telephone) to a central computer system for analysis (Halldórsson 1991). Another system of low frequency data transmission was developed by T. L. Murray at the Cascades Volcano Observatory (Murray 1988, 1992) and is compatible with the constraints of power consumption, cost and robustness imposed by monitoring of Mount St Helens. The system's main distinguishing feature is the ability to time-multiplex one of the channels of the analogue seismic system.

This enables low-frequency data from multiple transmitters to be sent on the same line with analogue seismic data.

The transmission security and the low minimum power consumption of a monitoring system have been studied in detail on a system used at the Taal volcano (Philippines). In co-operation with the Philippines Institute of Volcanology and Seismology (PHIVOLCS), we recently installed a station using digital radio transmission to measure geochemical parameters. Every 2 hours the station acquires data on temperature, conductivity and the pH of the water in the crater lake. Transmission is effected using a small network of three radio beacons: one at the monitoring station; one at the observatory of Buco, about 10 km distant; and one on a mountain summit acting as a relay. In order to reduce power consumption, the whole communication system operates for only a few minutes each day. The transmitting beacon linked to the station is alerted by the latter via a serial port and remains on until the end of the transmission. The relay beacon comes on alert automatically at a fixed time and resets its internal clock to the transmission time, while the terminal beacon, linked to a personal computer PC, operates continuously. When the radio channel is free, i.e. when the carrier wave is not detected by a transmitter, the transmitting station sends the data message. During the transmission, if an error is detected by the use of detection codes, the receiving station demands repetition of the message until it conforms (Fig. 2.11). The receiving station then confirms its receipt, the transmitting station reverts to stand-by mode, and the message continues in the network. It is planned in future to enable commands to be sent from the observatory to the monitoring station after transmission of data, so that the acquisition frequency can be altered from the PC.

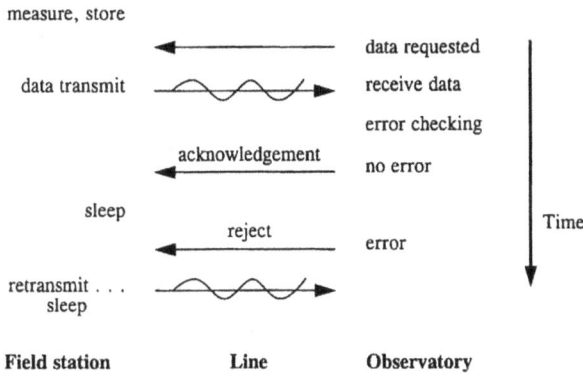

**Figure 2.11** Acquisition and transmission of geochemical data on Taal lake, Philippines. Simplified diagram showing the sequence of the main tasks of the field station and the observatory.

## 2.3.1 Seismic systems

In this section we describe the principal types of seismic monitoring systems used on volcanoes. These all reduce, by some means or other, the amount of data produced by the sensors, because of the physical limits of the storage, processing or transmission systems. So, when defining a seismic system, it is important to decide which is the required information and how unnecessary data can be stripped away.

Seismic networks involving the analogue transmission of data to an observatory are most commonly used for seismic and volcanic monitoring. Until the early 1980s, data were generally stored on paper, using drum-type recorders. In certain cases, computerized digital acquisition systems have been developed, but these have been relatively costly (of the order of tens of thousands of US dollars) and have necessitated a considerable investment of time in programme development. Thanks to microcomputers, to the IBM/PC compatibility standard and especially to the increasing power and falling prices of these micro-computers and of the compatibility of analogue-to-digital (A/D) conversion boards, we are now witnessing a profound change in systems of seismic data acquisition and storage. Several research groups have developed PC acquisition systems which are not only relatively simple, cheap, reliable and efficient, but which also can be easily transferred onto any IBM/PC compatible microcomputer (Green et al. 1987, Horiuchi et al. 1987, Lee 1989, Hurst et al. 1989, Snissaert 1991). Because of these advantages, such systems are becoming widespread and in a few years it is likely that most volcanological observatories will be equipped with them.

The best-known seismic data-acquisition system is distributed at a very low price by the International Association of Seismology and Physics of the Earth's Interior (IASPEI) (Lee 1989). The minimum equipment necessary for this system is a PC/AT of at least 8 MHz speed with 1.5 MB random access memory (RAM), an Intel 80287 maths coprocessor, a floppy drive, a Hercules screen and a Data Translation DT2821 or DT2824 A/D conversion board. The software provided by IASPEI controls the A/D board, gives a continuous digital read-out on the display, detects events by a short-term average/long-term average (STA/LTA) type algorithm and stores them on disk (Tottingham et al. 1989). Locating the events in real time is possible with a program option. The software can then be used off-line to analyze the recordings, pick the arrivals of seismic waves and calculate the spectra or estimate the quality factor (Valdés 1989, Valdés & Novelo-Casanova 1989). This system is capable of digitizing 16 channels at up to 500 samples per second. The 12-bit conversion used is adapted to conventional systems with analogue transmission, and the total cost of hardware and software can be less than US$5000. Accurate synchronization is effected by entering an external time-code signal on one of the input channels. On other systems, for example in the SISMALP network, the PC clock is permanently synchronized by an external time signal (J. Fréchet & J-P. Glot, personal communication).

The limitations of MS-DOS mean that the acquisition PC has to be dedicated, and so the acquisition process has to be interrupted to allow the processing of the stored data, or to transfer it to another PC for processing. The problem can be solved by using a PC network to move data without disturbing acquisition. This limitation does not occur with multitasking systems, for example those operating in UNIX workstations (Endo & Smith 1992).

To overcome the limitations met in analogue transmission of seismic data (e.g. dynamic range, noise, distortion, and intermodulation) several digital transmission systems have been developed. For example, SISMONET 90 (Holl & Speisser 1992) by the Institut de Physique du Globe (IPG), Strasbourg, France, which can transmit by radio up to 16 channels, each at $1200 \text{ bits}^{-1}$, using frequency bands conforming to European radio transmission standards. Another possibility is to transmit the digital signals by telephone either using expensive, leased lines or using the public network. The solution is only feasible where the telephone network is sufficiently developed. This is the case for SISMALP in the western Alps, where a network of 30 autonomous stations transmit collected data, via modem, to a central station (Poupinet et al. 1989a). This is also the solution adopted by the Sakurajima Volcanological Observatory in Japan to receive data collected at the volcanoes of the Satsunan and Kyushu Islands, which extend over several hundred kilometres (Iguchi 1991). Each remote station in the system is composed of: a PC (80206, 8 MHz) with 1 MB of RAM; an A/D conversion board (16 channels, 12 bit); an extra clock board with time-base stability of $6 \times 10^{-7}$; and a modem allowing a transmission rate of 9600 bit s$^{-1}$. The central station consists of a more powerful PC and some extra peripheral equipment. The remote stations record seismic and infrasonic signals and also ground tilt, strain and temperature data. The software on the remote PC is responsible for the following tasks:

(a) digitizing analogue data
(b) detecting seismic and infrasonic events
(c) counting the number of events detected
(d) calculating the mean amplitude of seismic and infrasonic signal
(e) averaging over 1 min the data of tilt, strain and temperature
(f) storing the results of the tasks described above on the hard disk
(g) transmitting the data to the central station
(h) synchronizing the remote stations' clocks with the central station's clock.

Program parameters and the programming of remote stations can also be modified by telephone from the central station.

The development of microcomputers, along with improved storage capacities and component reliability have allowed autonomous digital seismic acquisition stations to become increasingly efficient. Microprocessors allow high levels of processing of digital data, and the software can be modified to the needs of the user. Most often the systems are modular and can thus be connected to various peripheral devices, including transmitters. As an example let us take a station developed under the French programme LITHOSCOPE and constructed by CEIS-ESPACE (Poupinet 1987, Poupinet et al. 1989a & b). The analogue board is used with a Mark Products L-4C 1 Hz vertical seismometer, and includes a 1, 10 or 100 gain switch selectable preamplifier, a sixth order Butterworth anti-aliasing filter with cut-off frequency of 25 Hz, a variable-gain amplifier (2 to 256) controlled by software, and a current generator to calibrate the seismometer. The digital board is built around an Intel 80C31 microprocessor. It includes a 10-bit A/D converter (sampling rate 12.5, 25, 50 or 100 Hz), 32 kB of EPROM, 64 kB of RAM, a clock which can be synchronized to external timing signals (DCF or OMEGA) and an RS-232 port for communication with an external microcomputer. The system consumes about 14 mA, so it can function for more than a month on a 12 V car battery. Events are detected using an STA/LTA algorithm, then 1024–32784 samples are stored temporarily in the RAM, including the 256 samples preceding the trigger. These events are classed in order of a pseudo-magnitude equal to the sum of the absolute values of the samples. When the 64 kB stack memory is full, the smallest event is replaced if the pseudomagnitude of the latest event is greater than it. At regular intervals (0–48 h), the data in the 64 kB memory are transferred to a 1 MB memory board in order to keep the strongest events in each time interval. The station can be programmed to start up at a set time or be activated by radio. These possibilities are useful for seismic refraction and for recording volcanic tremor or background noise simultaneously at several stations. During a visit to the station, the data can be transferred in seconds to a portable memory board from which they can be copied onto a lap-top computer. The station can, if required, be interfaced to a METEOSAT or equivalent transmitter (GOES or GMS). In this configuration, 630 byte of the strongest event in the stack of 64 kbyte are sent to the geostationary satellite every hour (Poupinet et al. 1989b). This system allows seismo-logical studies to be undertaken in places where access is difficult.

A network of stations similar in concept to those of LITHOSCOPE, and transmitting via the ARGOS satellite system operated for several years on the Mount Etna volcano, complementing the network of the University of Catania (Glot et al. 1984). The stations only transmitted arrival time, the polarity of the first movement, the quality of detection and the duration of the first arch for seven events per message; 20–25 messages per day are sent at the Etna latitude. Another seismic alert station was developed in 1992 by the Laboratoire d'Instrumentation Géophysique of the Université de Savoie, France. Constructed around an HC MOS 68HC11 microcontroller, it includes a preamplifier and an amplifier, both with programmable gain control and a 12 bit plus sign A/D converter with a sampling frequency of 100 Hz. The station is connected to an ARGOS transmitter. The peculiarity of the system is the software, which extracts from a seis-mic event the information necessary to signal the re-activation of a volcano. Using an STA/LTA algorithm, the station detects seismic events and classifies them according to their dura-tion, their mean frequency, and their energy. Two histograms, one of which is bi-dimensional, represent the number of classified events, and are transmitted with the ARGOS message. The program can automatically modify certain parameters, defining the classification in order to adapt to the level of recorded activity, and to avoid saturation of the histograms. The mean level of the seismic signal between two transmissions is also calculated and coded onto the message. The analysis of received data gives the possibility of discriminating between the various types of seismic events observed at volcanoes (e.g. tectonic, volcanic, and tremor) and thus to trace the evolution of a number of features of seismic activity at a volcano.

Most seismic acquisition stations are not well adapted to the analysis of signals such as volcanic tremor, seismic swarms or noise, so it has been necessary to conceive specific sys-tems for recording either the mean amplitude of the seismic signal (Endo & Murray 1991, Iguchi 1991, Murray & Endo 1992) or spectral parameters of the signal (Hurst 1985, Cosentino et al. 1989, Schick 1991, Marso & Murray 1991). For example, a real-time seismic ampli-tude measurement system (RSAM), developed at the Cascades Volcano Observatory, USA, digitizes and averages the seismic signal over 1–2 min. This system is installed in the observa-tories of Cascades, Alaska and Hawai and gave invaluable information during the Pinatubo crisis in 1991 (Murray et al. 1991, Murray & Endo 1992).

At the Chateau Observatory in New Zealand, a specialized volcanic tremor monitoring system has been established. Here, the analogue seismic signal, sent from stations near the Ruapehu and Ngauruhoe volcanoes, is digitized at 50 samples per second and transmitted to a personal computer. The PC performs a permanent fast Fourier transform on 256 points with 50% overlap, then stores the average of the spectra every 3 min. A pattern-recognition test is then applied to the averaged spectra to distinguish the harmonic tremors from earthquakes and noise (A. W. Hurst 1985, personal communication).

54

## 2.3.2 Discussion

The continuous and real-time monitoring of volcanic activity requires a data-transmission system, although these remain limited. Despite a low dynamic range, analogue transmission will still be used for a long time for seismic data because of its simplicity and because its characteristics are good enough for routine volcanic monitoring. In any case, the topography of volcanic regions make radio transmission difficult and it is sometimes necessary to use relay stations. Cable transmission can be a good solution for larger distances or where there are topographical problems, but it requires a well-developed telephone network which is not common in most volcanic areas. Furthermore, the transmission of large volumes of data requires expensive, leased telephone lines. Transmission via the public telephone network is much cheaper, but the rate of data transfer is lower.

The continuous digitization of seismic signals produces a large amount of data, for example, digitizing three seismic components at 100 Hz with 16 bit, gives 52 MB of data per station per day, i.e. 1.6 GB per month. It is necessary, therefore, to reduce the amount of seismic data in order to avoid saturation of the storage system. Nevertheless, the improvement of data storage capacity will probably soon make the storage of the whole of the seismic signal feasible, even in stand-alone stations. In any case, the improvement of algorithms is essential in order to extract useful information on volcanic activity automatically. At present, the earthquake detection algorithms, which are based generally on the STA/LTA ratio, have some limitations when used on volcanoes. In particular, volcanic tremors can saturate the LTA and impede the detection of discrete events. A sequence of many earthquakes during a short time period, for example a seismic swarm or an aftershock sequence, can also saturate real-time detection and seismic signal-processing systems. In addition, automatic P-waves pick-up and localization algorithms do not work very well for volcanic earthquakes for which the P-wave first arrivals are generally emergent.

Because of the limited capacities of available algorithms and of the features of volcanic seismic activity, it is difficult to find a balance between the risk of saturation of the detection and storage systems, and the risk of loss of useful information. It is possible to reduce these risks by using simultaneously several parallel processing and storage systems; for instance:

(a)  an analogue paper recorder *and* an acquisition system on PC
(b)  a detection algorithm of discrete events *and* a RSAM
(c)  the monitoring of spectral features of earthquakes *and* of volcanic tremors
(d)  an automatic processing of seismic activity *and* a more elaborate off-line analysis.

In the many cases of volcanoes that are difficult to reach or located in regions which lack even basic equipment, the use of monitoring stations with satellite transmission can be an interesting solution from a technical and economic point of view. Nevertheless, due to the low data flow of this kind of transmission, it is necessary to extract *in situ* some characteristic parameters of volcanic activity from the raw data. The interest in satellite stations is increased by the development of international networks of data transmission and electronic mailing (e.g. Internet and Bitnet), which facilitates and speeds up data transfer from the receiving sites to the national institutions in charge of volcanic monitoring. The development of satellite telephone global networks could also provide interesting possibilities for the monitoring of remote volcanoes.

## 2.4 Conclusions

In this chapter, we have reviewed the existing techniques in data-acquisition and telemetry applied to volcanic monitoring and briefly described a limited number of representative systems. We have examined the principal trends in monitoring-system characteristics and especially the current technological changes involving the development of digital and computerized systems.

The volcano monitoring profession does not constitute a large potential market for industry, so there is little commercial development of surveillance devices. Nevertheless, the new low-power and miniaturized chips intended for general use, in the form of hand-held phones or lap-top computers, will enable volcanologists to use smart and easy-to-use systems that support high-level programming languages. The cost, processor speed, and power consumption of such equipment will undoubtedly continue to improve. As systems become cheaper, it will be possible to install more stations on a volcano to improve monitoring. In this case, the telemetry systems used to collect data from the stations will become more complex and will require network-management software adapted to the particular conditions of volcano monitoring. Certain available telemetry equipment permits bi-directional links between the measuring stations and the control site. The software and the hardware of the systems will therefore need to be designed in order to allow:

(a) remote maintenance of the stations
(b) remote management of data-acquisition, processing, and transmission parameters, and even
(c) remote software modification.

For low-data-transmission rates, numerous solutions, using current technology, exist for acquisition, transmission and in situ storage. The digital systems currently in use have several advantages:

(a) higher dynamic range all along the acquisition and transmission chain
(b) transmission security
(c) possibility of easy to use bi-directional links
(d) lower power consumption by using sleeping mode or temporal mutiplexing for radio transmission.

Volcano monitoring based on high-data-rate methods, such as seismic or magnetotelluric, means that the quantity of data coming from the sensors is too great. At any stage of the chain, it is necessary to extract the required information in order to evaluate and characterize the volcanic activity. This extraction can be performed either at the field station or the observatory, and consists of either selecting parts of the time series, or processing the signal. A preprocessing stage at the source makes the transmission and storage easier. Nevertheless, the rejected part of the signal cannot be visualized or analyzed later owing to a loss of information. Some low-cost and low-power commercialy available systems can perform signal processing *in situ*, provided that the acquisition frequency is not too high and the processing not too sophisticated. So, several characteristics have to be improved, including processing speed, the quantity of data to be processed, and the power consumption of systems.

The availability of PC compatible mini-modules and complementary metal oxide semiconductor (CMOS) chips with full PC capacities, impel the developer to elaborate the new monitoring station hardware based on PC architecture. In the near future, we visualize volcanologists without any knowledge of microelectronics developing, testing, and modifying acquisition and data processing software, written using a high-level language on their own PCs. The probable increasing use of PC systems standards in the field will provide confirmation that the software tasks, previously totally confined to the observatory, will be progressively executed in the field stations.

From a technical point of view, we can consider the two following options in designing a monitoring system.

(a) Systems with the simplest possible electronics, containing cheap and widely available components, so as to be easily tested, repaired or changed in all observatories, even those in developing countries. Such systems involve a well-trained technical team, but do not take full advantage of up-to-date technology.

(b) Systems composed of more complex modules, which can be linked together to form a complete chain. Full standardization is necessary to allow the modules to communicate easily with one another, and repairs are effected by simply changing the defective module. Such user-friendly systems are more expensive, but do not require an electronic specialist for installation. They do, however, require some knowledge of microcomputer software. For such reasons, institutions in charge of monitoring may have difficulty in implementing new technology. This is an especially crucial point for developing countries where technology-transfer problems, personnel training and tight budgets have to be taken into account.

There is no doubt that technological advances are providing numerous ways of modernizing volcanic monitoring. Worthy of special mention is the automation of some data-processing tasks and the detection of alarm thresholds, the remote control of field stations, and the development of alert stations using satellite transmission. It is imperative that such new instrumental strategies be used to continually upgrade existing volcano surveillance networks. It is already apparent that the improvements in instrumental techniques, by their positive influence on the quality and quantity of measurements carried out on volcanoes, are contributing to raising the standards of volcano monitoring and to our knowledge of volcanic phenomena, and later chapters in this book will include many examples of this progress.

# Acknowledgements

We would like to thank the persons who kindly sent us very useful information about data acquisition and transmission systems designed in their institutions: J. Ewert, J-M. Holl, A. Hurst, K. Kamo, G. Luongo, G. Poupinet, G. Suarez, G. Sigvaldason and E. Wolfe. We are also grateful to an anonymous reviewer who carefully improved the manuscript. D. Hunter greatly helped us in writing the English version of this paper.

# References

Briole, P., R. Gaulon, G. Nunnari, G. Puglisi, J.C. Ruegg 1992. Measurements of ground movement on Mount Etna, Sicily: a systematic plan to record different temporal and spatial components of ground movement associated with active volcanism. In *Volcanic Seismology*, P. Gasparini, R. Scarpa, K. Aki (eds), 120–9. Berlin: Springer.

Clark, H. E. & E. S. Medina 1976. Tsunami seismic system. *USGS Albuquerque Seismology Laboratory Report*, 77, 777. USGS: Albuquerque.

Cosentino, M., G. Lombardo, E. Privitera 1989. A model for internal dynamical processes on Mt Etna. *Geophysical Journal* 97, 367–79.

Endo, E. T. & T. L. Murray 1991. Real-time Seismic Amplitude Measurement (RSAM): a volcano monitoring and prediction tool. *Bulletin of Volcanology* 53, 533–45.

Endo, E. T. & G. Smith 1992. Seismic data-acquisition systems at the Cascades volcano observatory. In *US Geological Survey Bulletin 1966, Monitoring volcanoes: techniques and strategies used by the staff of the Cascades volcano observatory 1980–1990*, J. W. Ewert & D. Swanson (eds), 45–52. Washington DC: United States Government Printing Office.

Endo, E. T., P. L. Ward, D. H. Harlow, R. V. Allen, J. P. Eaton 1975. A prototype global volcano surveillance system monitoring seismic activity and tilt. *Bulletin of Volcanology* 38, 316–44.

Fontolliet, P. G. 1990. *Systèmes de communications*. Lausanne: Presses Polytechniques Romandes.

Glot, J. P., S. Gresta, G. Patanè, G. Poupinet 1984. Earthquake activity during the 1983 Etna eruption. *Bulletin of Volcanology* 47, 953–63.

Green, R. W., J. Borchers, W. Theron 1987. A PC/XT based seismic data acquisition system. *IUGG, XIX General Assembly Abstracts* 1, 380.

Halldórsson, H. 1991. Monitoring systems on Icelandic volcanoes. *Cahier du Centre Européen de Géodynamique et de Séismologie* 4, 25–34.

Holl, J. M. & E. Speisser 1992. SISMONET 90. *Cahier du Centre Européen de Géodynamique et de Séismologie* 5, 11–13.

Horiuchi, S., T. Matsuzawa, A. Hasegawa 1987. A personal computer system for automatic detection, location of seismic events and for recording of multichannel seismic signals. *IUGG, XIX General Assembly Abstracts* 1, 380.

Hurst, A. W. 1985. A volcanic tremor monitoring system. *Journal of Volcanology and Geothermal Research* 26, 181–7.

Hurst, A. W., S. Sherburn, V. M. Stagpoole 1989. The Taupo seismic system. In *IAVCEI Proceedings in Volcanology vol. 1: Volcanic hazards – assessment and monitoring*, J. H. Latter (ed.), 513–9. Berlin: Springer.

Iguchi, M. 1991. Geophysical data collection using an interactive personal computer system (part 1). *Bulletin of the Volcanological Society of Japan* 36, 335–43.

Lebsir N. 1991. Réalisation d'un appareillage de sondage audio-magnétotellurique à traitement numérique intégré. Thesis, University of Savoie, Chambéry, France.

Lee, W. H. K. 1989. Getting started. In *Toolbook for seimic data acquisition, processing, and analysis*, W. H. K. Lee (ed.), 1–46. International Association of Seismology and Physics of the Earth's Interior. El Cerrito, California.

Lesage, P. & S. De la Cruz-Reyna 1992. La primera estacion inclinometrica del volcan Fuego de Colima. *International Reunion of Volcanology Abstracts*, 7. Colima, Mexico: Universidad de Colima.

Marso, J. N. & T. L. Murray 1991. Real-time display of seismic spectral amplitude measurements: Examples from the 1991 eruption of Pinatubo volcano, Central Luzon, Philippines. *Eos* 72, 67.

Murray, T. L. 1988. *United States Geological Survey Open-File Report 88-0201, A system for telemetering low-frequency data from active volcanoes*. Washington DC: United States Government Printing Office.

Murray, T. L. 1992. A low-data-rate digital telemetry system. In *United States Geological Survey Bulletin 1966, Monitoring volcanoes: techniques and strategies used by the staff of the Cascades volcano observatory 1980–1990*, J. W. Ewert & D. Swanson (eds), 11–23. Washington DC: United States Government Printing Office.

Murray, T. L. & E. T. Endo 1992. A real-time seismic amplitude measurement system (RSAM). In *United States Geological Survey Bulletin 1966, Monitoring volcanoes: techniques and strategies used by the staff of the Cascades volcano observatory 1980–1990*, J. W. Ewert & D. Swanson (eds), 5–10. Washington DC: United States Government Printing Office.

Murray, T. L., J. A. Power, G. D. March, A. B. Lockhart, J. N. Marso, A. Miklius 1991. Application of a real-time data-acquisition and analysis system in response to activity at Mount Pinatubo, Philippines 1991. *Eos* **72**, 67.

Nussbaumer H. 1987. *Informatique industrielle*. Lausanne: Presses Polytechniques Romandes.

Poupinet, G. 1987. Seismic data collection platforms for satellite transmission. In *Seismic Tomography*, G. Nolet (ed.), 239–250. Dordrecht: Reidel.

Poupinet, G., J. Fréchet, F. Thouvenot 1989a. Portable short period vertical seismic stations transmitting via telephone or satellite. In *Digital Seismology and fine modeling of the lithosphere*, R. Cassinis, G. Nolet, G. F. Panza (eds), 9–26. New York: Plenum.

Poupinet, G., M. Pasquier, M. Vadell, L. Martel 1989b. A seismological platform transmitting via METEOSAT. *Bulletin of the Seismological Society of America* **79**, 1651–61.

Sabroux, J-C., A. Villevieille, E.Dubois, C. Doyotte, M. Halbwachs, J. Vandemeulebrouck 1990. Satellite monitoring of the vertical temperature profile of Lake Nyos, Cameroon. *Journal of Volcanology and Geothermal Research* **42**, 381–4.

Schick, R. 1991. The calculation of "volcanic activity parameters". *Cahier du Centre Européen de Géodynamique et de Séismologie* **4**, 249–54.

Silverman, S., C. Mortensen, M. Johnston 1989. A satellite based digital data system for low-frequency geophysical data. *Bulletin of the Seismological Society of America* **79**, 189–98.

Snissaert, M. 1991. PC-system acquisitions and the Belgium seismic network. *Cahier du Centre Européen de Géodynamique et de Séismologie* **4**, 277.

Tottingham, D. M., W. H. K. Lee, J. A. Rogers 1989. User manual for MDETECT. In *Toolbook for seimic data acquisition, processing, and analysis*, W. H. K. Lee (ed.), 49–88. International Association of Seismology and Physics of the Earth's Interior. El Cerrito, California.

Valdés, C. M. 1989. User manual for PCEQ. In *Toolbook for seimic data acquisition, processing, and analysis*, W. H. K. Lee (ed.), 175–201. International Association of Seismology and Physics of the Earth's Interior. El Cerrito, California.

Valdés, C. M. & D. A. Novelo-Casanova 1989. User manual for QCODA. In *Toolbook for seimic data acquisition, processing, and analysis*, W. H. K. Lee (ed.), 237–56. International Association of Seismology and Physics of the Earth's Interior. El Cerrito, California.

Webster, W. J., W. H. Miller, R. Whitley, R. J. Allenby, R.T. Dennison 1981. A seismic signal system processor suitable for use with NOAA/GOES satellite data collection system. *IEEE Transactions on Geoscience and Remote Sensing*, **GE19-2**, 91–94.

59

# 3 Seismic monitoring at active volcanoes

F. Ferrucci

## 3.1 Introduction

On 20 March 1980, an earthquake of magnitude $M = 4.1$ occurred at a depth of about 5 km beneath Mount St Helens (Washington State, USA). Five days later as many as twenty-four $M = 4$ earthquakes were recorded in only 8 h, together with several tens of lower energy events, preceding by 2 days a major phreatic explosion. Seismic activity thus proved to be an efficient indicator of what Mount St Helens might do two months later on 18 May 1980, when a slightly shallower, $M = 5.1$ earthquake triggered the climactic explosive eruption that released 2 km$^3$ of ejecta up to an altitude of 30 km (Rice & Watson 1981), covered half of Washington State with visible ash fall, and caused some US$900 million of damage in the surrounding region (Bates et al. 1982). Later seismic monitoring at Mount St Helens was also successful in predicting some of the smaller dome-related eruptions which followed the 18 May event (Swanson et al. 1982).

Although the spectacular eruptions of Mount St Helens constituted a recent and important opportunity to undertake seismic surveillance of a reactivated volcano, the earliest application of seismological techniques to volcanic phenomena dates from the second half of the 19th century, when an experimental electromagnetic seismometer was installed at Vesuvius (Bay of Naples, Italy) (Fig. 3.1). Quantitative routine observations started several years later with, for example, Bosch-Omori seismographs being installed at Mont Pelée (Martinique) 1 year after the 1902 *nuée ardente* catastrophe, and at the Usu (Japan) and Kilauea (Hawaii) volcanoes in 1910. Because data could be collected using sensors located at sites remote from the source, seismic surveillance of volcanoes developed more quickly than other geophysical techniques. In particular, it underwent substantial changes during the 1950s and 1960s when rapid developments in electronics and computerization made the acquisition, transmission and storage of data increasingly simple, reliable and cheap.

Over a century of seismological observation has borne witness to the fact that seismicity, in the form of earthquakes, volcanic tremor, or both, nearly always precedes, accompanies or follows unrest at all type of volcanoes (andesitic, dacitic, and basaltic; central volcanoes, and restless calderas). Seismic activity is consequently considered to be the most dependable indicator, and often a reliable short- to mid-term (days to weeks) predictor, of the type, level, and evolution of volcanic activity. In contrast to purely tectonic dynamics, where all seismic

**Figure 3.1**  Vesuvius was the first volcano to be monitored using seismological equipment. Giuseppe Palmieri's electromagnetic seismograph, built in 1862, wrote data on a telegraph tape. (Courtesy of Osservatorio Vesuviano, Naples, Italy.)

failure can be ascribed to (double-couple) systems of shear forces acting on two orthogonal planes, the seismic source at active volcanoes may be highly non-linear, since it often involves interactions between gas and liquid or liquid and solid. The role of gas and melt may be either active, giving rise to pressurized intrusions of magma into pre-existing or newly formed zones of weakness, or to sustained vibration of melt and included material, or passive, with brittle failure and the consequent stress readjustments modifying the distribution of melt within the crust. Furthermore, since volcanic media are characterized by dense systems of pores, fractures and faults at all scales, sudden modification of the local stress field may induce seismic failure independent of whether or not melt propagates through them.

For these reasons, volcano seismologists are often required to solve difficult and ambiguous problems which make accurate eruption forecasting, based on seismic data, a testing vocation. As eruption forecasting is fundamentally based on the repeatability of phenomena, and the reliability of quantitative models developed to describe them, it is vital that seismic arrays be adapted in order to gather the required data both on the type and the style of expected activity. Not only is this essential to the utility of any volcano surveillance programme, but it also provides the potential for more effective advance warning of impending eruption. At large basaltic volcanoes, for example, where the principal risks are largely confined to lava flows, arrays are best tailored to forecasting the sites where effusive boccas will open, i.e. to locating and characterizing the shallow seismic swarms that typically herald and/or accompanying the opening of flank fractures. In contrast, at explosive volcanoes surveillance is more usefully focused on the timing of an expected eruption. When considering seismic network design, it is also worth considering the nature of the volcanic structure. At restless calderas, for example, all seismic foci will be located beneath the array, while at high central volcanoes earthquakes may equally well occur beneath or above the array.

There is little doubt that hardware development has played a vital role in the seismic surveillance of volcanoes, since most recent advances in volcano seismology have relied heavily on technological advances, both on the instrumental side, and in the design and organization of networks. As demonstrated by Mount St Helens, data provided by a minimal number of modern seismic sensors may often be sufficient, at least for determining whether or not a volcano is in a quiescent state, or for recognizing the basic features of its activity (e.g. frequency of seismic events, rates of energy release, and spectral characteristics of volcanic tremor). Quantitative modelling of the seismic phenomena observed in volcanic environments, however, requires much more extensive data which, in terms of both type and quantity, needs to match the scale and character of the observed or expected seismicity, and which can be provided only by suitable seismic arrays. Such topics are addressed here with the aid of some recent case studies. Although not exhaustive, this account is designed to give the reader some idea of the importance and usefulness of volcano seismology and some of the problems encountered in the seismic monitoring of active volcanoes. In particular, some stress is placed upon experience gained during the period of major activity at Mount Etna between 1989 and 1993. This, it is hoped, will help in providing the reader with an up-to-date picture of the state-of-the-art, and the trade-off between the interests of pure seismological science and the civil defence needs of improved volcano surveillance.

## 3.2 Locating seismic events: some constraints

Seismic networks are designed for a primary task, i.e. the location of earthquakes. Hypocentral location is a strongly non-linear problem, but even though non-linear searches have been attempted (e.g. Rabinowitz 1988), linearization is used in most cases. This is done by assuming that the Earth model and the identification of each raypath are known and fixed, which allows reducing the model parameters to only four (origin time, the depth, and two horizontal coordinates). With such an approach, however, significant trade-off exists between the solution of the location problem and the chosen Earth model, since the origin time depends on the latter and computed travel times depend on the origin time. When the medium displays significant departure from horizontal homogeneity, location attempts become subordinate to the determination of a much larger number of parameters, and the operation of 3-dimensional ray tracing when undertaking forward modelling of travel times (e.g. Virieux et al. 1988).

Practically, however, these limitations do not greatly affect the average reliability of determined locations, and the need for authentically non-linear and three-dimensional algorithms is subordinated to the need for robustness. Indeed, widely available location programs of the HYPO generation (eg. HYPO 71, Lee & Lahr 1975; HYPOINVERSE, Klein 1978) are robust, which compensates some structural limitations. The latter include the requirement for a stack of constant-velocity horizontal layers in the a priori model, the need for velocity increasing with depth, the fixed-ratio dependence of the S-wave velocity on the P-wave velocity, and difficulties with taking into account rugged topographies or near-surface structural heterogeneities which do not immediately underlie a seismic station.

The reliability of computed earthquake locations depends on, (and can therefore be biased by): (a) the areal density of stations; (b) the contingent positions of sources and stations; and (c) the model approximation of the (mostly hidden) real heterogeneity of the medium. While heterogeneity is constant at the time-scale of seismic observations, the number and positions of stations may either grow with time (Table 3.1) or be subjected to sudden modification because of technical failures, alteration of the topographic surface due to catastrophic events (e.g. explosions, slope failures, or the growth of large lava flow fields), or a change in the seismically active zones with time. It may be, therefore, that seismicity occurring in an assigned area will be equally poorly constrained and well constrained as a function of time and in dependence of (a) and (b), while its absolute position will be biased by (c). Such problems may significantly affect the rapid interpretation of major seismic sequences involving magma, which may presage eruptions only a few days or even hours later. Since earthquake location, like all inverse problems, can be very ambiguous when the available data are inadequate, effective solutions rely strongly on the choice of a suitable recording array. This is crucial for constraining coherent temporal changes of focal positions, either in terms of depth or epicentre, a feature characteristically associated with dyke propagation and magma intrusion in general.

Table 3.1 The approximate areal densities of sensors for some permanent seismic arrays in volcanic areas illustrating the increased levels of monitoring at many volcanoes (temporary (i.e. portable) stations are not included).

| | Year | No. of Stations (N) | Network area (km$^2$) | Density (N km$^{-2}$) | Source |
|---|---|---|---|---|---|
| Campi Flegrei caldera (Italy) | 1982 | 8 | 80 | 0.1 | Barberi et al. (1984) |
| Hawaii | 1970 | 22 | 11 000 | 0.001 | Klein et al. (1987) |
| Hawaii | 1985 | 50 | 11 000 | 0.005 | Klein et al. (1987) |
| Kilauea rift zones | 1970 | 10 | 2400 | 0.004 | Klein et al. (1987) |
| Kilauea rift zones | 1985 | 30 | 2400 | 0.01 | Klein et al. (1987) |
| Long Valley, California | 1984 | 30 | 375 | 0.08 | Hill (1984) |
| Etna, Italy | 1978 | 6 | 750 | 0.008 | Cosentino et al. (1982) |
| Etna, Italy | 1992 | 13 | 750 | 0.02 | Ferrucci & Patanè (1993) |
| Pavlov, Alaska | 1985 | 8 | 400 | 0.02 | McNutt (1986) |
| Mount St Helens, Washington | 1980 | 6 | 250 | 0.02 | Shemeta & Weaver (1986) |
| Tokachi, Japan | 1987 | 6 | 40 | 0.15 | Okada et al. (1990) |
| Usu, Japan | 1978 | 12 | 180 | 0.07 | Okada et al. (1981) |
| Usu, Japan | 1988 | 12 | 180 | 0.07 | Okada et al. (1981) |
| Piton de la Fournaise, Réunion | 1986 | 14 | 450 | 0.03 | Hirn et al. (1991a) |
| Rabaul, Papua New Guinea | 1984 | 8 | 120 | 0.07 | McKee et al. (1984) |
| Vulcano, Italy | 1988 | 5 | 32 | 0.15 | Falsaperla et al. (1989) |

It is typically the case that the shallower the seismicity, the larger the probability that an eruption may soon follow. At Etna, for example, seismicity preceding the last two main eruptive episodes (1989 and 1991–1993) was generally confined to depths greater than around 5 km below sea level during the months preceding both eruptions, and became shallow to very-shallow only a few weeks before the eruption outbreaks (Fig. 3.2). Comparable evidence for the upward migration of seismic foci, accumulated elsewhere (e.g. Klein et al. 1987), confirm this depth relationship, and supports the idea that seismicity at active volcanoes is often magma related even though melts do not always contribute directly to seismic failure. Constraining the depths of seismicity during the recent activity at Etna was largely accomplished by means of networks provided with many three-component stations. The importance and need for such effective and versatile hardware has already been demonstrated, however, in the seismological literature (e.g. Aster & Meyer 1988, Hirn et al. 1991a & b, Castellano et al. 1993), and will not be discussed in detail here. The advantages of using three-component, as opposed to one-component stations, are, however, graphically illustrated with reference to Figure 3.3. Besides stressing the need for adequate hardware, it is also worth emphasizing the requirement for network geometries which allow the collection of seismic data as near as possible to likely epicentres. Experimental results show that this requirement is at least as important as the availability of unambiguous S-wave times in locating earthquakes.

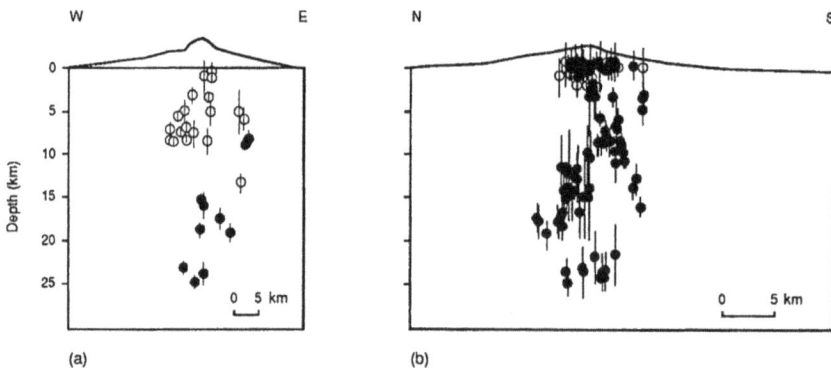

**Figure 3.2**   Hypocentres of the best-located events (error bars shown) recorded recently at Mount Etna, indicate that seismic activity is transferred to the upper volcanic basement and to the volcanic pile immediately prior to eruptions. This supports the idea that seismicity at active volcanoes is magma related, even when melt is not directly involved in the seismic rupture process. (a) Seismicity preceding the 1989 eruption by less than 2 months (O) superimposed on seismicity (●) recorded in 1987–9 (from Castellano et al. 1993). (b) Seismicity immediately preceding and accompanying the 1991–3 eruption (O) superimposed on seismicity preceding the eruption by up to 1 year (●).

**Figure 3.3**   Relocation of the 40 strongest earthquakes belonging to a seismic swarm recorded at Mount Etna during the 1989 eruption (around 300 events with magnitudes of 1–3) demonstrates that adequate constraints on focal positions cannot be obtained without reliable S-wave timepickings and one station at least above the seismically active volume. (a) The array was composed of 13 stations (areal density 0.1 stations/km²), 11 of which were three-component stations. Also shown is the two-arm fracture system, the southern branch of which developed almost aseismically during the eruption. The northeast arm fed effusive flank activity, while the southern arm did not provide any direct evidence for the presence of magma at shallow depths beneath it.

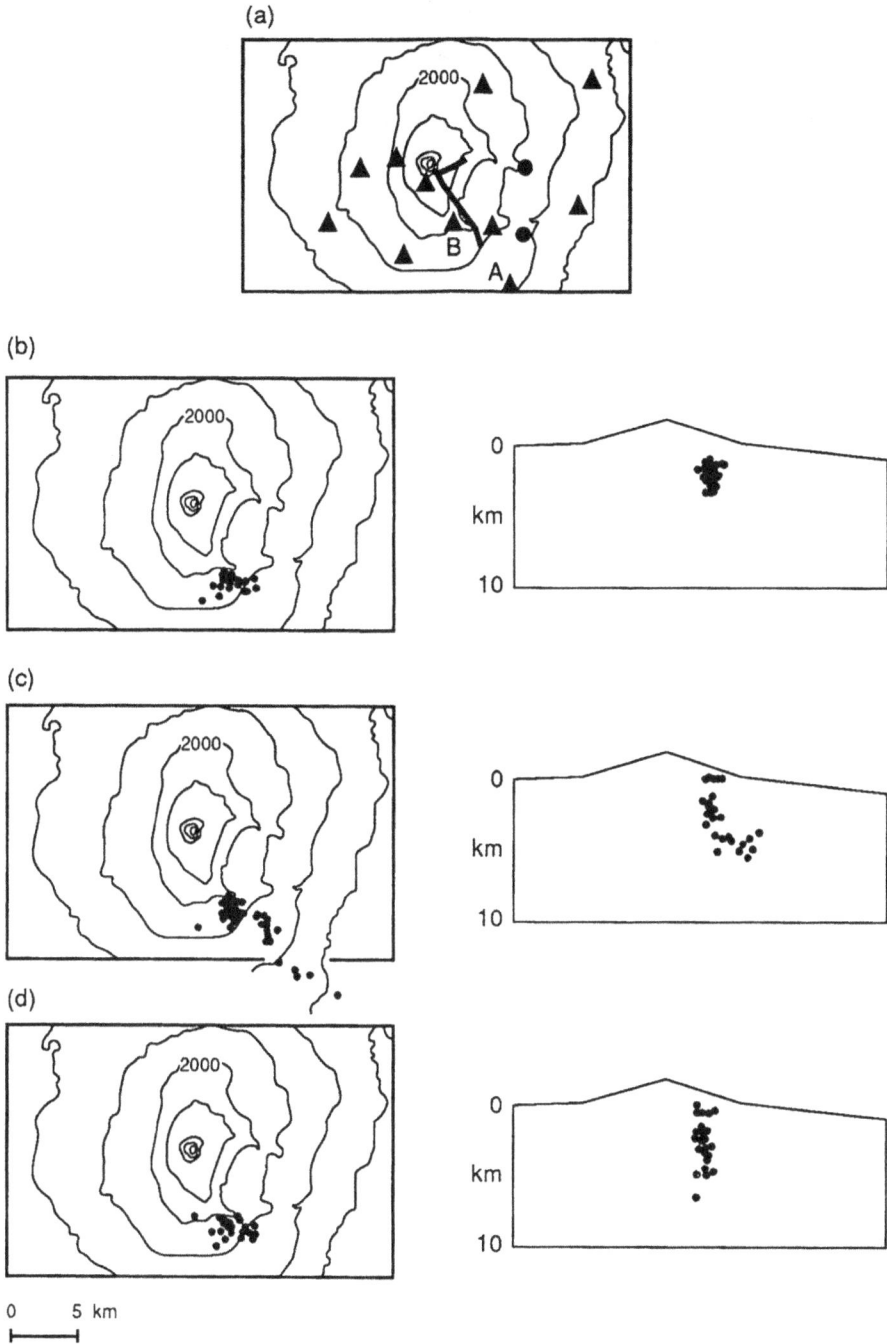

(a)

(b)

(c)

(d)

0    5 km

Data from three-component stations A and B are used in Figure 3.4 for providing direct proof of the migration of epicentres towards the southeast within the small focal zone shown in (b). (b) Epicentres (left) and hypocentres (right, west–east cross-section), computed using all the P- and S-wave data, constrain the seismicity to a focal volume of less than 12 km$^3$. (c) As in (b), but excluding P- and S-wave data recorded at the pair of three-component stations (A & B) nearest to the epicentral area. Focal positions are loosely constrained, even though three stations nearest to the epicentral zone are less than 5 km distant. (d) Location of epicentres and hypocentres using P-waves (all data) only. The epicentral distribution is more reliable than in (c), while hypocentral depths are predictably scattered over a range double that obtained from the complete dataset (b).

To illustrate this, the 40 best-constrained earthquakes recorded during a shallow and intense swarm at Etna in 1989 are relocated in the absence of some essential data (Fig. 3.3). This example simulates the level of constraint that would be imposed on this type of seismic sequence by a slightly looser array or by an array composed of one-component stations. The results are compared with the complete dataset, which was gathered at the digital network shown in Figure 3.3a. This covered around $130\,km^2$ and comprised 13 stations, 11 of which were of the three-component type. The areal density of sensors ($0.1$ stations/$km^2$) is equal to that for the densest networks in Table 3.1, and, even after removal of stations above the focal volume (Fig. 3.3c), three stations nearest to the epicentral zone are still only $5\,km$ distant. Location of recorded events using all the available data (Fig. 3.3b) constrains seismicity to within a focal volume of less than $12\,km^3$, displaying an overall northwest–southeast trend, and confined between sea level and $2\,km$ depth.

The importance of three-component data in fixing focal depths can be seen with reference to Figure 3.3c & d. In Figure 3.3c, the seismic cluster is shown as computed after reduction of the dataset except that a pair of three-component stations nearest to the epicentral area have been excluded. It can be clearly seen that the lack of stations above the seismically active volume causes a dramatic spreading of the epicentres towards the southeast, despite the use of the same reliable P- and S-wave data as in Figure 3.3a. In Figure 3.3d, the three-component stations are included; but only P-wave data have been used; the result being only a weak effect on the epicentral distribution. In both cases, however, focal depths are poorly constrained, as emphasized by the scattering of hypocentral depths over a range double that obtained from the complete dataset. It is arguable that, in the case of deeper seismic sequences, such combined effects of array and raypath geometries (i.e. the velocity increase with depth) would result in even more poorly constrained focal depths, in particular when reliable S-waves (Fig. 3.3c) are not available. In the light of the above demonstration, it is worth remembering that, at least until the mid- to late-1980s, most seismic networks operating in volcanic areas comprised only a few one-component stations, and there is some evidence that this defficiency may often have been responsible for the "clouds" of earthquakes which have been reported spreading at all depths beneath a number of volcanoes. In the context of eruption forecasting, it is worth noting that acceptance of the worst-case seismic interpretation model for the Etna data would lead either to an overestimation, by two orders of magnitude, of the crustal volume affected by potentially upwelling magmas, or to the depiction of conduit-like features from poorly constrained hypocentral cross sections.

As well as accurately constraining the locations of hypocentres, it is also important to have the capability of monitoring the migration of the seismic sources with time, as this may provide a guide as to what the volcano might do next. Migration of the foci during a seismic swarm may provide evidence for the involvement of magma in the rupture process, and also help in determining where effusive vents are likely to open. Such tasks, however, require the acquisition of very-high-resolution constraints on the location of the seismic foci, something that cannot depend on computed hypocentral locations. Depending on the extreme stress concentration and/or heterogeneity of the medium accounting for the seismic swarm events (e.g. Scholz 1968), the swarm-foci volume may be small, and migration of individual foci may take place over path lengths which are shorter than the average spacing of stations, and comparable to or less than the average location error. It should be stressed that the seismic swarm-type release of elastic strain energy is typical of areas characterized by exceptional heterogeneity and, since intruded regions typically comprise bundles of frozen or magma-filled dykes, is typical of volcanic areas without exception. It should also be noted that the location problem may become strongly underdetermined in the event of active or passive involvement of melt, which may modify the Poisson ratio of the medium, and thus lead to a change of P- and (mainly) S-wave travel times, and a consequent unconstrainable bias in computed focal positions.

Such difficulties support the view that attempts to constrain the migration of the foci require direct evidence from the raw data. The seismic swarm recorded during the 1989 eruption of Etna, for example, consisted of around 300 events, only 40 of which (Fig. 3.3 & Table 3.2) satisfied the requirement for high-quality location data. Preliminary evidence for foci migration was already contained in the computed focal positions, since the events accompanying the end of the swarm were located to the southeast of the earliest shocks in the sequence (Table 3.2). Unequivocal evidence for migration could, however, only be provided by utilizing all available data, most of which referred to unlocatable or poorly located earthquakes recorded at the pair of three-component stations sited near and above the epicentral area (A & B in Fig. 3.3a). As shown in Figure 3.4a, the average decrease of the S–P travel time difference observed at station A, and the contemporary increase at station B, are consistent with the foci migrating south towards station A. Sharp changes with time, however, of the P-wave azimuths (Fig. 3.4b) at station B indicate that the epicentres must also have progressively moved eastwards, thereby resulting in an overall migration of foci from northeast to southwest, and the observed travel-time variation independent of any change in the elastic properties of the medium. In total, the calculated, cumulative horizontal movement of foci did not exceed around 2 km in 3 days. Such a detailed insight into the dynamics of a near-stationary swarm agrees quite well with the behaviour inferred by analysis of the 40 strongest events only, supporting the idea that the information content of even a few well-located events is much greater than that which would be available from stacking all recorded events.

Constraining evidence for earthquake migration accompanying intrusions is usually reported for Hawaii, the length of dyke paths from Kilauea caldera towards the adjaceny rift zones typically reaching as much as 20 km (Fig. 3.5a). Seismic evidence for dyke emplacement is confirmed by accompanying cyclic ground deformation and subsequent downrift effusive activity (Klein et al. 1987). Comparable evidence for down-rift migration of dykes and earthquakes was obtained during the 1977 eruption of Krafla (Iceland) (Brandsdottir & Einarsson 1979). Here, only imprecise time-picking accuracy was available due to the type of instruments used (essentially smoked-paper recorders), and the depths of the seismic sources were poorly constrained due to a lack of horizontal-component information. Nevertheless, the spacing of stations, which was comparable to the average focal depth, and was also much less than the epicentral migration path length, provided sound evidence for the displacement of foci from the caldera to the adjacent fault zone (Fig. 3.5b). Such straightforward seismological behaviour in volcanic rift zones has allowed the establishment of robust models for the dynamics of shallow feeding systems and the mechanics of dyke intrusion. Such simple models rarely apply, however, to volcanoes characterized by more complex structures and eruptive

**Table 3.2** Hypocentral locations of seismic events recorded at Etna during the 1989 eruption (see figure 3.3b and text for explanation).

| Date 1989 m.d | Origin time (hour/min/s) | Latitude (north) | Longitude (east) | Depth (km) | Gap (°) | rms (s) | EH (km) | EZ (km) |
|---|---|---|---|---|---|---|---|---|
| 09.30 | 211455.09 | 37°42.46 | 15°01.47 | 2.83 | 162 | 0.06 | .2 | .7 |
| 09.30 | 230100.82 | 37°42.16 | 15°02.14 | 2.70 | 119 | 0.12 | .3 | .9 |
| 09.30 | 230137.31 | 37°42.34 | 15°01.34 | 2.63 | 123 | 0.10 | .4 | .8 |
| 10.01 | 011405.49 | 37°42.20 | 15°01.72 | 3.14 | 120 | 0.06 | .2 | .4 |
| 10.01 | 125354.04 | 37°42.18 | 15°01.89 | 4.13 | 119 | 0.10 | .3 | .3 |
| 10.01 | 143241.14 | 37°42.15 | 15°01.28 | 3.57 | 119 | 0.17 | .5 | .9 |
| 10.01 | 165548.87 | 37°41.15 | 15°01.99 | 3.69 | 106 | 0.14 | .3 | .7 |
| 10.01 | 171635.18 | 37°42.11 | 15°01.43 | 3.17 | 119 | 0.07 | .2 | .4 |
| 10.01 | 193201.23 | 37°42.17 | 15°01.78 | 2.22 | 119 | 0.05 | .2 | .3 |
| 10.01 | 202554.36 | 37°42.11 | 15.01.56 | 2.13 | 126 | 0.04 | .1 | .1 |
| 10.01 | 205433.27 | 37°42.31 | 15°01.93 | 2.78 | 172 | 0.05 | .3 | .4 |
| 10.01 | 213742.47 | 37°42.10 | 15°01.69 | 2.04 | 157 | 0.09 | .4 | .6 |
| 10.01 | 214246.92 | 37°42.31 | 15°01.49 | 2.19 | 122 | 0.05 | .3 | .3 |
| 10.01 | 214833.30 | 37°41.81 | 15°01.21 | 2.52 | 173 | 0.06 | .5 | .4 |
| 10.01 | 220629.49 | 37°42.22 | 15°01.53 | 2.61 | 175 | 0.03 | .2 | .2 |
| 10.01 | 222537.26 | 37°42.05 | 15°01.61 | 1.85 | 117 | 0.06 | .1 | .2 |
| 10.01 | 223315.89 | 37°42.03 | 15.01.62 | 2.33 | 147 | 0.04 | .4 | .4 |
| 10.01 | 230323.42 | 37°41.99 | 15°01.35 | 2.41 | 121 | 0.10 | .3 | .4 |
| 10.01 | 233400.35 | 37°42.23 | 15°01.55 | 2.05 | 176 | 0.02 | .2 | .2 |
| 10.02 | 003013.20 | 37°42.16 | 15°01.78 | 2.15 | 135 | 0.05 | .2 | .2 |
| 10.02 | 005844.20 | 37°42.06 | 15°01.65 | 1.76 | 118 | 0.06 | .1 | .2 |
| 10.02 | 011926.42 | 37°42.17 | 15°01.65 | 2.44 | 167 | 0.04 | .3 | .4 |
| 10.02 | 034552.25 | 37°41.99 | 15°01.58 | 2.20 | 148 | 0.04 | .1 | .3 |
| 10.02 | 083929.04 | 37°42.04 | 15°01.64 | 2.08 | 148 | 0.03 | .1 | .2 |
| 10.02 | 100456.43 | 37°42.06 | 15°01.69 | 2.05 | 117 | 0.04 | .1 | .3 |
| 10.02 | 113139.76 | 37°41.81 | 15°02.00 | 2.79 | 087 | 0.05 | .1 | .3 |
| 10.02 | 124920.90 | 37°42.12 | 15°02.30 | 4.38 | 165 | 0.07 | .3 | .5 |
| 10.02 | 131419.57 | 37°42.01 | 15°01.63 | 2.37 | 113 | 0.07 | .1 | .2 |
| 10.02 | 140850.82 | 37°41.65 | 15°01.51 | 2.82 | 159 | 0.12 | .4 | .5 |
| 10.02 | 171500.64 | 37°41.57 | 15°02.46 | 3.45 | 138 | 0.08 | .3 | .5 |
| 10.02 | 171758.89 | 37°41.76 | 15°02.07 | 4.14 | 143 | 0.02 | .2 | .2 |
| 10.02 | 183844.22 | 37°41.47 | 15°02.36 | 3.77 | 124 | 0.04 | .3 | .5 |
| 10.02 | 203155.65 | 37°41.96 | 15.01.82 | 2.67 | 138 | 0.04 | .2 | .2 |
| 10.02 | 235651.65 | 37°41.84 | 15°01.86 | 4.10 | 147 | 0.07 | .4 | .5 |
| 10.03 | 031122.10 | 37°41.93 | 15°01.89 | 3.23 | 145 | 0.07 | .3 | .3 |
| 10.04 | 051711.65 | 37°41.80 | 15°01.21 | 2.88 | 173 | 0.06 | .6 | .4 |
| 10.03 | 054217.77 | 37°41.74 | 15°01.98 | 3.76 | 131 | 0.06 | .2 | .2 |
| 10.04 | 194321.73 | 37°41.86 | 15°01.78 | 2.39 | 149 | 0.05 | .2 | .2 |
| 10.07 | 115110.59 | 37°41.98 | 15°02.04 | 2.86 | 118 | 0.08 | .3 | .4 |
| 10.10 | 133320.77 | 37°41.85 | 15°01.80 | 2.66 | 140 | 0.09 | .3 | .4 |

Hypocentral locations of the events illustrated in Figure 3.3b, with focal co-ordinates referred to the free surface. The high quality of locations is testified by squared travel-time residuals (rms) of less that 0.2 s. Note that epicentral latitudes decrease slightly, and longitudes increase slightly throughout the swarm, consistent with migration of epicentres towards the southeast. These events are fully representative of the space trend of the seismicity (the swarm was composed of around 300, mostly unlocatable, events with magnitudes of 1–3). Since, however, calculated displacements of the foci are much shorter than station spacing, and slightly larger than the average error in epicentre location (EH being radius of the uncertainty in the horizontal plane and EZ the error in depth), direct proof for migration is needed from raw data, and is provided in Figure 3.4.

(a)

(b)

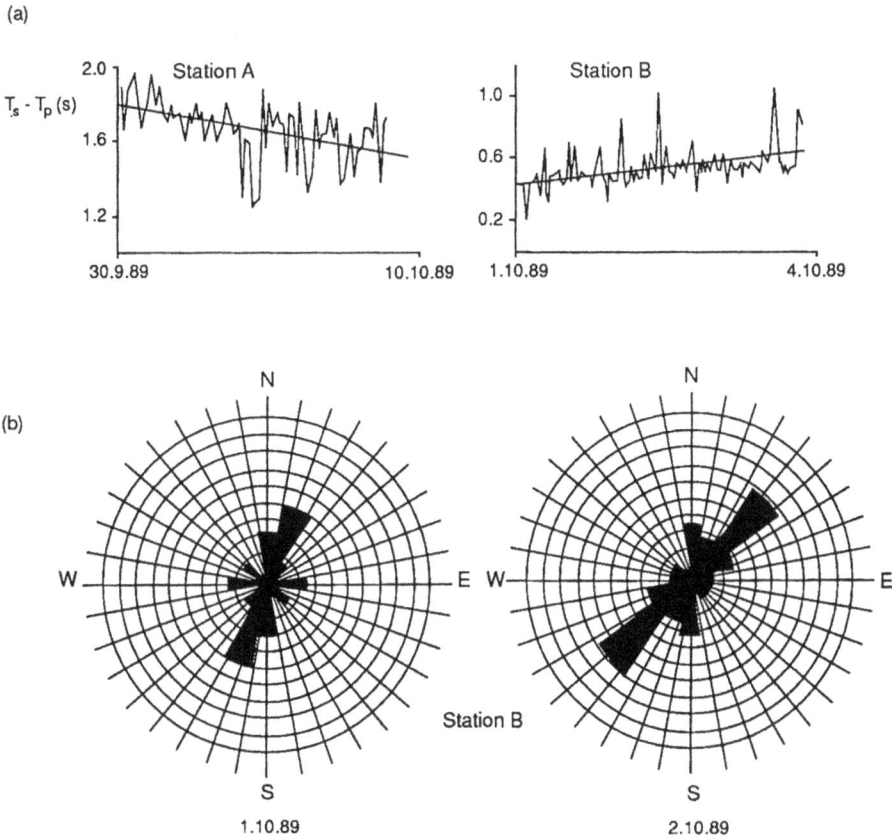

**Figure 3.4**  Direct proof that northwest–southeast migration of seismicity occurred during the swarm shown in Figure 3.3 is obtained by using whole sets of crude data recorded at the three-component stations A and B. The data mostly rely on poorly constrained or non-locatable events that are not included in the location maps in Figure 3.3 nor in Table 3.2 (from Ferrucci et al. 1993). (a) Overall decrease with time of S–P travel time differences at station A (left) and contemporary increase at B (right), are consistent with foci moving southwards towards station A. (b) Overall displacement of the foci towards the east is revealed by an increase with time of P-wave azimuths at station B. (Left) Rose diagram constructed from the first 40 events of the swarm. (Right) Rose diagram for 40 events accompanying the end of the sequence.

mechanisms. In such cases, the presence of major fault structures along which magma rises, or the presence of significant morphological barriers to shallow dyke intrusion, may limit dyke propagation to short distances, or result in dyking being accompanied by discontinuous seismic behaviour with the potential to mask different intrusive mechanisms.

Seismological evidence for the onset and the mode of emplacement of the intrusion which fed the 1989–93 eruptive period at Etna, for example, was certainly not completely consistent with models established for rift zones on the basis of the Hawaiian and Icelandic observations. In 1989, the non-effusive fracture system illustrated in Figure 3.3a developed aseismically

(a)

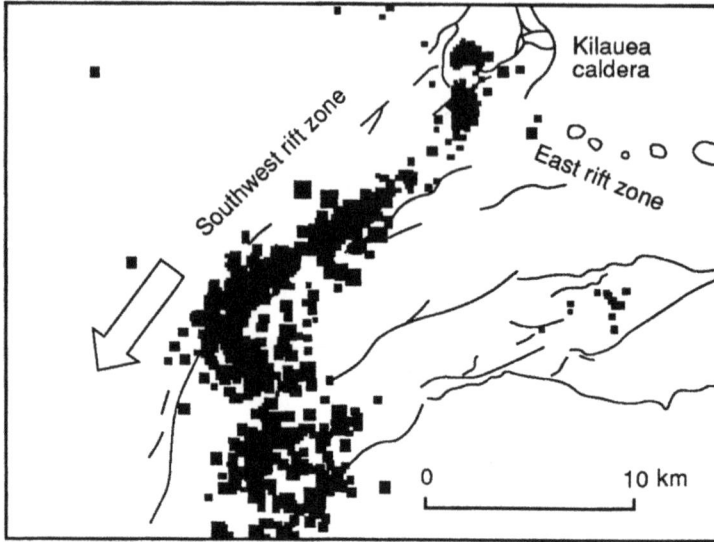

(b)

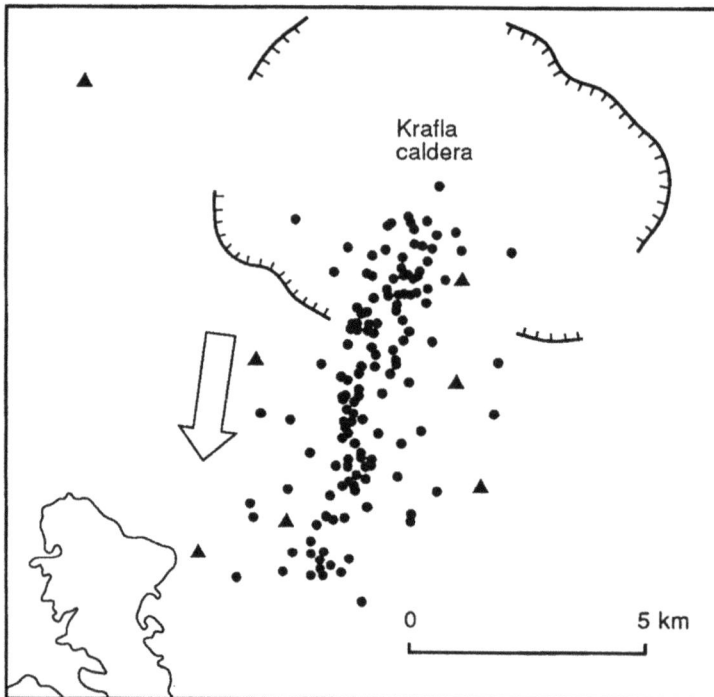

**Figure 3.5** Seismicity relating to characteristic mechanisms of dyke emplacement in rift zones. (a) At Hawaii (1983 episode), intrusions are usually accompanied by seismicity marking the whole dyke path, inflation relating to cyclic withdrawal of magma from the Kilauea caldera, and effusive activity down rift (from Klein et al. 1987). (b) In Iceland, earthquakes marked the migration of magma from the Krafla caldera towards the adjacent rift zone during the 1977 eruption (from Brandsdottir & Einarsson 1979). (▲) portable seismic stations.

over a distance of some 7 km across the upper southern flank of the volcano, with seismicity confined beneath the active end of the fracture (Figs 3.3 & 3.4, Table 3.2) only after the surface failure event was nearly complete. Such behaviour, which is not compatible with the expected brittle response of a volcanic edifice to a radial dyke originating at the summit feeder system and continuously propagating towards the south, suggested that a genetic relationship might exist between the seismic swarm and the emplacement of a near-vertical dyke offset by a few kilometres further south of the summit region (Ferrucci et al. 1993). In late 1991, eruptive vents opened along the 1989 fracture system, although some 2 km north of the 1989 swarm area (Fig. 3.6), and fed the major flank eruption of 1991–93. With reference to Figure 3.6 it can be seen that the space-time distribution of the seismicity affecting the volcanic pile between 1989 and 1991, reveals a gap encompassing the inferred top of the intrusion. This suggests that the swarm accompanying the end of the 1989 eruption might be considered a posteriori to be a forerunner of the major eruptive event which followed 2 years later.

The sequence of seismic crises illustrated above indicates that, in order to properly locate earthquakes and follow their evolution in space and time, the type of seismological equipment and the geometry of the network have to be determined in advance to suit, as much as is feasible, the locations and styles of expected seismic and volcanic activity. Such a strategy will always be a compromise, particularly in regard of the logistical problems encountered in establishing optimal network geometries on large and complex volcanoes which may be active at a number of different locations within a short time-frame. Perfect networks are likely

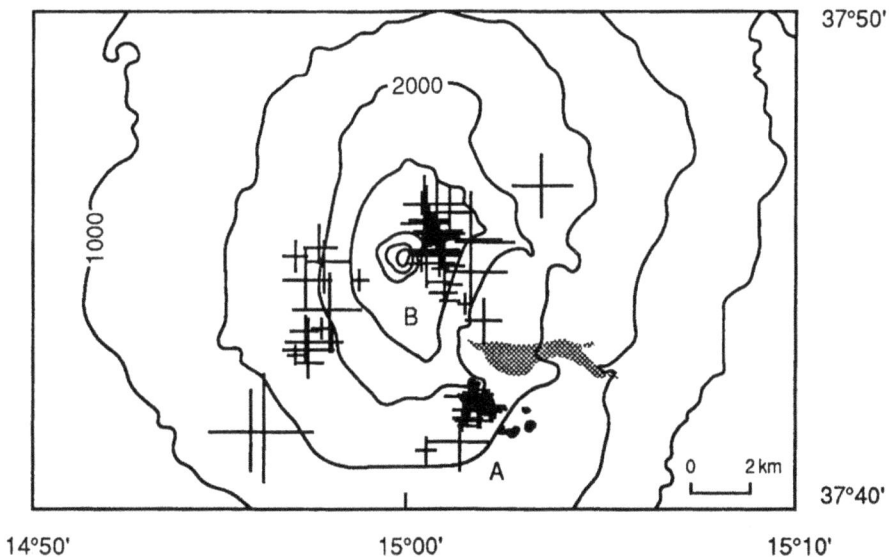

**Figure 3.6** Consistent with the location of the eruptive vents (1991–92 lava flow), the areal gap in epicentres of strongest (magnitude > 1.8) shallow seismicity (< 1 km below sea level) recorded at Mount Etna between the end of the 1989 (A) and the outbreak of the 1991–93 (B) flank eruption, marks the inferred position of the top of the intrusion.

71

to remain elusive, with the quality of accumulated data varying greatly from one network to another, or even at the same network from one period to another, depending on the number and type of sensors used, the geometry of the arrays, and technical and logistical problems. The examples reported here may, therefore, be useful in assessing the implications of comparing the variable quality data accumulated using sparse, and generally outdated networks, with those acquired by means of dense, contemporary, seismic arrays, and evaluating these differences in terms of updated hazard models for the relevant volcanoes.

## 3.3 The seismic source

Although it can be reasonably assumed that seismic failure at active volcanoes originates primarily from magma-induced stress fields, evidence for the presence of magma at the source is infrequent. This has led to most of these earthquakes being interpreted simply in terms of shear failure caused by a double-couple system of forces as a result of the stress distributions in the source area. It should also be considered, however, that: (a) a single seismic-source area might undergo both shear failure and tensile failure as a function of the involvement of variable quantities of liquid, at variable pressures, in the rupture process (Fig. 3.7); and (b) degassing which provides magma bodies with the thrust necessary to reach the free surface, typically also gives rise to the sustained oscillation of the conduit and the magma body (volcanic tremor). Despite some remarkable attempts at constructing a unified seismic model to explain the various transport mechanisms of magma (e.g. Aki 1984), these are still largely discussed separately in the literature, and are treated as such here.

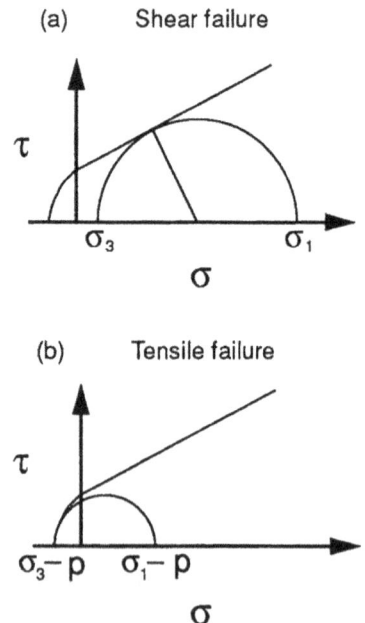

(a)    Shear failure

(b)    Tensile failure

**Figure 3.7** The Mohr diagram with assigned Griffith–Coulomb failure envelope can be used to explain why tensile failures, seldom observed in the Earth, are likely to occur in volcanic or geothermal areas (from Julian & Sipkin 1985). (a) In the absence of fluid, or in the presence of fluid at negligible pressure with respect to the overburden pressure, only shear failure can occur. Conversely, tensile cracks can open if fluid under high pressure flows through them and holds them open. In such a case (b) the fluid pressure, $p$, lowers the principal stresses $\sigma_1$ and $\sigma_3$, and pure tensile failure can occur at low stress differences (hydraulic fracturing). $\tau$ and $\sigma$ are the shear and the normal stresses, respectively, acting on a point source lying on the fault plane.

### 3.3.1 Shear failure and double-couple patterns

Two major uncertainties affect the significance of focal mechanisms in volcanic areas, and both involve the applicability of the double-couple system of forces. The first relates to the highly incoherent level of computed double-couple solutions for events clustered in space but separated in time, or belonging to the same temporal sequences but differing slightly in location. The second relates to the fact that it is frequently impossible to fit observed first-motion polarity patterns using simple double-couple models. Such difficulties suggest that the use of classical analysis, based on polarity and the duration of first motions, should not always be trusted when applied to the determination of focal mechanisms at active volcanoes. In cases of shear failure, seismic ruptures displaying near-horizontal, intermediate ($\sigma_2$) and least ($\sigma_3$) principal stresses can be explained in terms of dyke-like intrusions, with the dykes oriented normal to the average direction of $\sigma_3$.

Such mechanisms are typical, for instance, of the tensile regimes that characterize rift zones, where seismic ruptures are strongly conditioned by the orientations of both the principal axes of the regional stress tensor, and by the value of the deviatoric stress field (Fig. 3.8). In contrast, if the near-horizontal components of the stress tensor are the largest ($\sigma_1$) and intermediate ($\sigma_2$) principal stresses, significant stress increases in the horizontal plane are indicated, capable of overcoming the lithostatic pressure, and compatible with reverse-faulting mechanisms and with fracturing by sill-like magma bodies. In practice, the $\sigma_1$, $\sigma_2$, and $\sigma_3$ components of the stress tensor are often assumed to coincide with the average orientation of the $P$

(a)                                              (b)

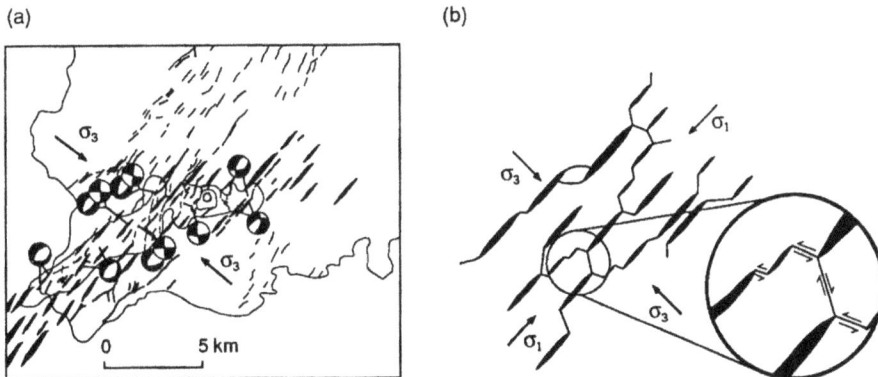

**Figure 3.8**  In volcanic areas that contain clusters of solidified or magma-filled dykes and "nests" of fractures mostly oriented according to the stress field acting on the region, earthquake swarms can be modelled in terms of en echelon failure sequences occurring along a system of conjugate fault planes joining at oblique angles the tips of adjacent, near-parallel dykes (from Hill 1977). (a) The orientation of the least principal stress ($\sigma_3$) inferred by microearthquakes in the Reykjanes rift zone in Iceland is consistent both with the volcanic and the tectonic lineations of the area. Largest principal stresses ($\sigma_1$) are vertical in the normal-fault zones limiting the rift, and horizontal in the strike-slip fault area at its centre. (b) Idealized representation of the interaction between adjacent dyke tips in a homogeneous medium subjected to horizontal $\sigma_2$ and $\sigma_3$.

(pressure), $T$ (tension), and $N$ (neutral) axes respectively (Raleigh et al. 1972, Gephardt & Forsythe 1984). This is questionable (McKenzie 1969), and holds true only in cases of newly formed fractures in a homogeneous isotropic medium, or where rupture mechanisms differ from those normally allowed by the deviatoric stress acting on the source region. Most of the ambiguities, however, originate from poor constraints on the focal position or unsatisfactory coverage of the focal sphere, both resulting from the unsuitability of the arrays used, and insufficient knowledge of the transmitting medium. Accurate reconstruction of the incidence angle of the ray at the source (takeoff angle) is essential to the correct plotting of rays on the focal sphere. This demands accurate determination of the incidence angle at the free surface, which depends strongly both on the velocity distribution in the Earth model used, and on the extent of lateral heterogeneity in the real Earth. Constraining earthquake mechanisms can be significantly improved by combining P-wave onset polarities and the information carried by S-wave polarizations. The different radiation pattern of S- with respect to P-waves allows for significant reduction of the ambiguities in the orientation of the P and T axes, provided that the S-wave polarization is not exceedingly perturbed by the heterogeneity or the anisotropy of the medium. This has been shown to hold true within a few kilometres of the source (Iannaccone & Deschamps 1989), and is therefore suitable for data gathered at most seismic networks operating in volcanic areas. An elegant solution for such a class of near-source data (Zollo & Bernard 1991) involves using the P-wave polarity information as the a priori probability density function, and the S-wave polarization set as the conditional probability. Assuming then, that P- and S-wave data are independent, and expected errors are normally distributed, maximum-likelihood solutions are located by an extensive search for the a posteriori probability of model parameters (strike, dip-, and slip-fault angles). The use of even a few additional S-wave polarization directions can then significantly increase the constraints acting on the set of fault parameters, as emphasized by sharp clustering on the focal sphere of the solutions for the P and T axes (see Fig. 3.9).

The earthquake illustrated in Figure 3.9 is representative of seismicity recorded during the 1982–84 episode of unrest at the Campi Flegrei caldera (Bay of Naples, Italy), where some 15000 shallow earthquakes (Fig. 3.10) accompanied major ground uplift (Barberi et al. 1984, De Natale et al. 1991). The 2-year seismic sequence at Campi Flegrei displayed a strong non-random character, both in space and time, and the distribution of earthquakes remained constant throughout the 2-year period of unrest. Substantial use of three-component seismographs allowed 80% of the foci to be constrained, with mostly negligible errors, within a volume of less than $15\,km^3$ (Aster & Meyer 1988) overlying the inferred top of the magma chamber (Dvorak & Berrino 1991, Ferrucci et al. 1992), although a few stronger events ($M_{max} = 4.0$) were randomly distributed within temporal clusters. Quiescent stages were not systematically followed by large magnitude shocks, and the majority of the energy in the most intense swarms was typically released in their central zones (De Natale & Zollo 1986). As generally accepted for seismic swarms (Scholz 1968, Hill 1977, Okada 1983), such evidence suggested failures relating to stress concentrations in a strongly heterogeneous medium (Okada et al. 1981). As regards the nomenclature of seismic events in volcanic areas (e.g. Minakami 1974, Latter 1979) none of the earthquakes recorded at Campi Flegrei was of the "B type" (low-frequency) and did not display any "volcanic" character. However, established classifi-

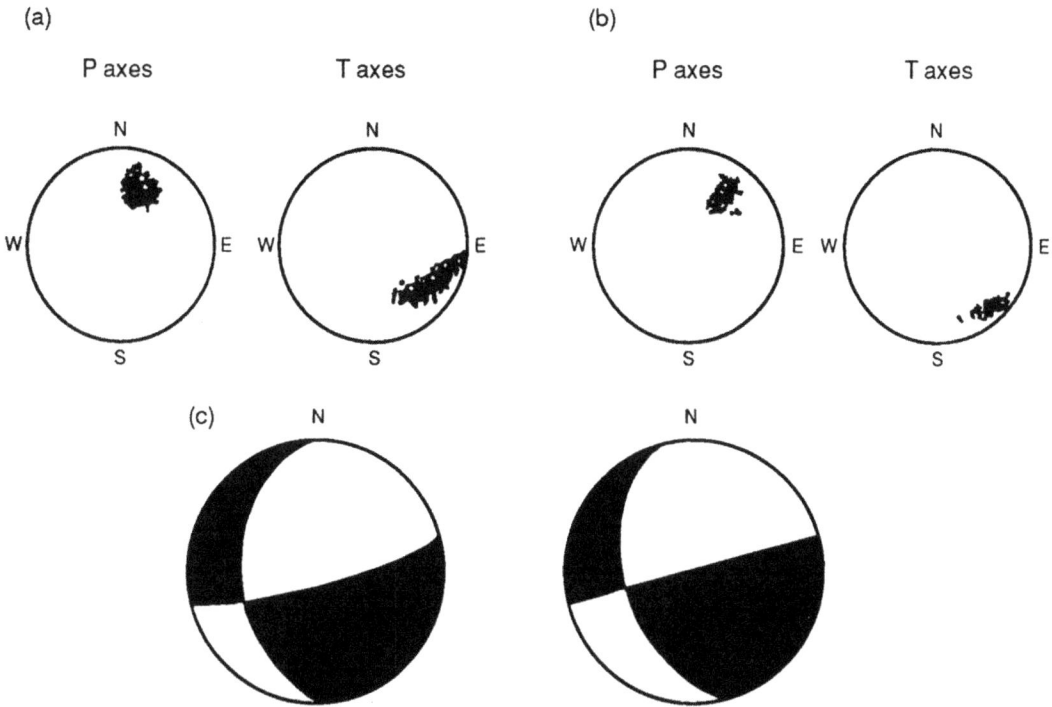

**Figure 3.9** The use of even a few S-wave polarizations in addition to P-wave polarities may significantly improve the constraints on the model probability of the orientations of P and T axes (from Zollo & Bernard 1991). The earthquake data shown are representative of a swarm composed of around 500 events, recorded at an array of three-component stations on 1 April 1984 at Campi Flegrei caldera (Italy). (a) Lower hemisphere projection of solutions for the P and T axes model probability, obtained using only the P-wave polarities. (b) As (a), but using both P-wave polarities and S-wave polarizations. (c) Faulting mechanisms obtained using only P (left) and P and S data (right). Even though mechanisms do not differ substantially, reliability of the fault-plane solution is ensured by sharp clustering of computed solutions for P and T axes shown in (b).

cation schemes for volcanic seismicity are by no means all-encompassing, and have little quantitative significance. Nevertheless, the forms of the Campi Flegrei seismograms showed an absence of the usually accepted evidence for the participation of fluids in the rupture process. The large numbers of seismic events recorded at Campi Flegrei might, therefore, be largely attributed to shear failures, and to the strongly heterogeneous nature of the host medium at all scales, a combination of these factors accounting for the relatively poor spatial coherence of computed focal solutions (Gaudiosi & Iannaccone 1984) (Fig. 3.11).

### 3.3.2 Tensile failure and non-double-couple mechanisms

Quite different evidence was gathered during the the Long Valley caldera crisis (Hill, 1984), which occurred at around the same time as those of the Campi Flegrei and Rabaul (Papua New Guinea) calderas. Such restless behaviour captured the attention of both scientists and the media because of the high level of risk posed by the vicinity of the calderas to densely

(a)

(b)

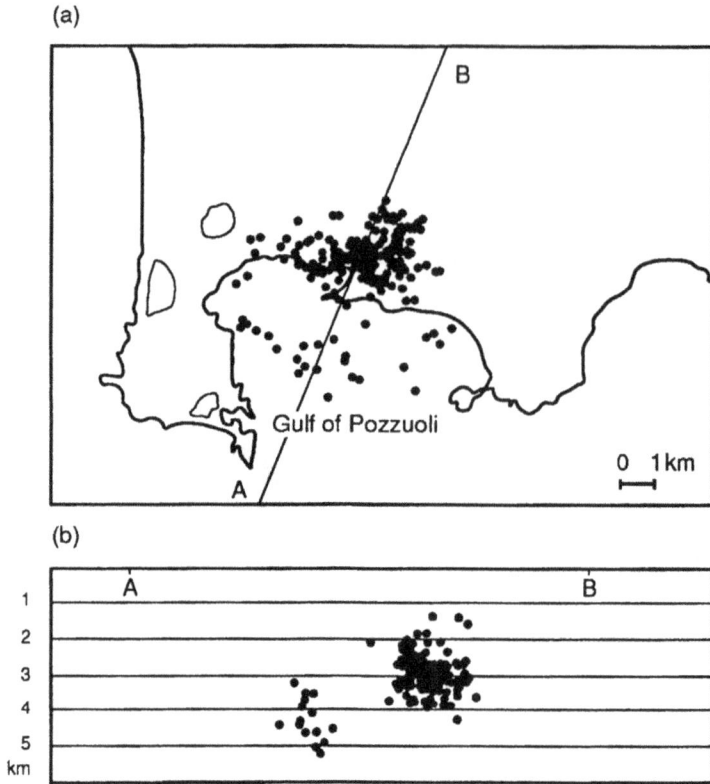

**Figure 3.10** Epicentres (a) and hypocentres (b) along the A–B cross section of the best located earthquakes recorded at Campi Flegrei between 1983 and 1984 (from Aster & Meyer 1988). During caldera unrest the distribution of earthquakes remained invariable and no low-frequency events were detected.

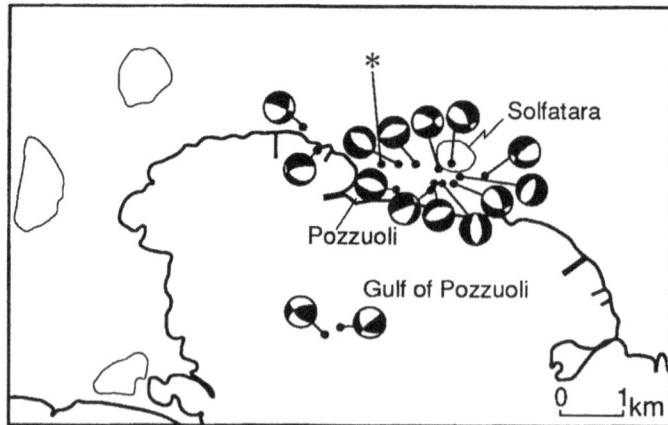

**Figure 3.11** Focal mechanisms of selected events recorded at Campi Flegrei in 1983 (determined using only P-wave polarities) display poor spatial organization. This can be ascribed to the strong heterogeneity of the medium, but also to weaker constraints than for Figure 3.9 in computing focal solutions. Compressional quadrants are marked in black (from Gaudiosi & Iannaccone 1984). The epicentre of the event illustrated in Figure 3.9 is marked by an asterisk.

Long Valley caldera

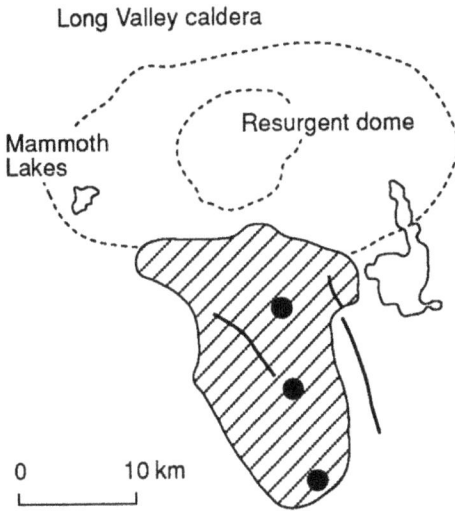

Figure 3.12  Locations of three $M = 6$ earthquakes which occurred in May 1980 at Long Valley Caldera, California. The hatched zone marks the seismically active area. The early 1980s saw near-simultaneous unrest at the Campi Flegrei, Long Valley and Rabaul (Papua New Guinea) calderas. In spite of considerable ground uplift and seismic energy release, largely in the form of swarms, no tremor has ever been observed and no eruption has occurred, at the time of writing (October 1993), at any of these large silicic calderas.

populated areas. No eruptions have, as yet occurred at any of the sites, despite the major release of seismic energy and accompanying large-scale ground deformation. The seismic unrest at Long Valley (Fig. 3.12) began in 1978 with an earthquake of magnitude 5.8. Between 1979 and 1984 a resurgent dome was uplifted by around 50 cm, accompanied by frequent earthquake swarms involving thousands of events, and several large-magnitude earthquakes in the Mammoth Lakes area, four $M = 6$ events in mid-1980, two $M = 5$ earthquakes in 1983, and one $M = 5.7$ event during late 1984. The strong earthquakes of May 1980 fed rapidly growing arguments amongst seismologists about the type and complexity of the source, comparative suitabilities of short-period versus broadband data, misidentification of the main motion as the initial motion, and the inconsistency of accumulated data with the theoretical radiation pattern of either typical double-couple or non-double-couple compensated linear vector dipole (CLVD) mechanisms (after Knopoff & Randall 1970) (Fig. 3.13).

Problems of interpretation arose early, once the first seismological studies of the May 1980 sequence pointed to strike-slip motions occurring on near-vertical faults (Ryall & Ryall 1981), a mechanism not supported by surface ruptures. In fact, long-period data recorded worldwide (Given et al. 1982) provided evidence for a moment tensor formed by the melding of three orthogonal force dipoles, a conclusion which was inconsistent with previous short-period data and consistent with large non-double-couple components in the total moment (Julian & Sipkin 1985). This could not be explained by recursive, accidental occurrence of simultaneous shear faulting on multiple faults, but might instead agree with tensile failure under fluid pressure (Julian 1983) (Fig. 3.7).

Breakdown of the deviatoric part of the moment tensor into double-couple and CLVD components (Fig. 3.13) led to an alternative view of the rupture mechanism consistent with the near-vertical intrusion of a dyke. Modelling of the radiation pattern which would be observed in the case of sudden propagation of a fluid-filled crack predicted, however, compressional first motions radiated all over the focal sphere (Chouet & Julian 1985). This was inconsistent with dilatations in the focal sphere of short-period data (see Fig. 3.14), and was also not in

(a)   Quadrantal          (b)   Conical

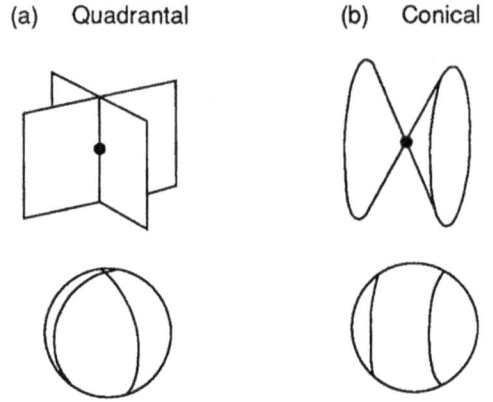

**Figure 3.13**  Nodal surfaces for P-wave radiation patterns associated with (a) double-couple and (b) non-double-couple (CLVD) rupture mechanisms (from Julian & Sipkin 1985).

N

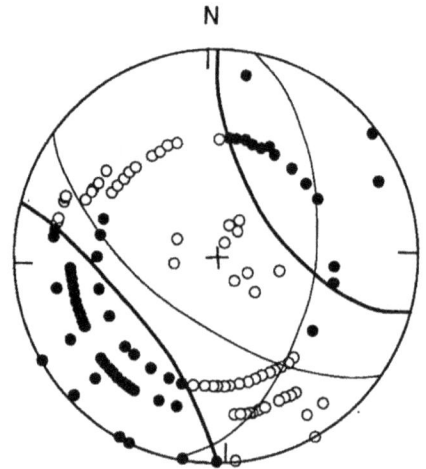

**Figure 3.14**  Nodal surfaces for the Long Valley earthquakes shown in Figure 3.12, fitted by the two types of radiation patterns illustrated in Figure 3.13. Thin lines indicate the double-couple solution (Given et al. 1982), and thick lines the CLVD solution (Julian & Sipkin 1985). (●) indicate compressions and (O) indicate dilations, as recorded worldwide at stations of the Global Digital Seismograph Network (Julian & Sipkin 1985).

agreement with long-period data which at several stations displayed unequivocal dilatations, whereas short-period data reported compressions, the latter systematically occurring earlier in time than the former. Aki (1984) suggested that such scattered behaviour of short-period versus long-period data might be explained by separating the first motions from the main motions, that is admitting that short-period and long-period data each contain dominant information only about a part of the displacement given by a very non-impulsive source function. In Aki's view, fluid pressure increasing at one end of a crack-like system (Fig. 3.15) might have resulted in: (a) the sudden opening of a narrow channel connecting the fluid-filled cracks; (b) the transfer of pressure to the other end of the crack system; (c) a consequent fall in the magma pressure; and (d) the resulting closure of the open tip. In such a seismic-source mechanism, stages (a) and (b) involve a change of volume, the presence of a large CLVD component in the general moment tensor, and radiation of high-frequency compressional motions in all directions. In contrast, relaxation accompanying the end of viscous fluid flow accounts for dilatations generated in stages (c) and (d), the delays of which with respect to

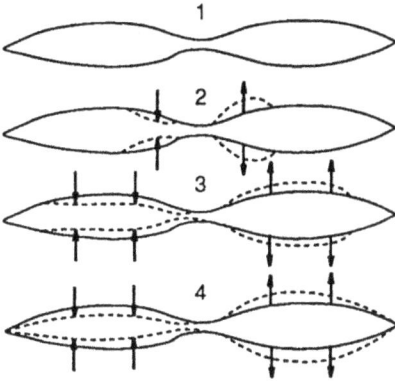

**Figure 3.15** Idealized, four-stage view of magma transport between a pair of cracks connected by a narrow channel joining their tips, a mechanism which may represent the typical propagation of melt in the brittle upper crust (from Chouet et al. 1987). Movement is from left to right. Such a model source has been proposed to explain most of the low-frequency signals observed in volcanic environments, as well as the mechanisms of the major, May 1980 earthquake sequence of Long Valley caldera (Aki 1984).

short-period first motions may increase as a direct function of the source length.

Although the debate about CLVD mechanisms has died down somewhat, small earthquake sources that are poorly fitted by orthogonal planes have been reported from, for example, Iceland (Foulger & Long 1984), and Piton de la Fournaise (Réunion) (Hirn et al. 1991a), but are often disregarded because of the poor quality of the first motions and/or the spatial or technical inadequacy of the recording arrays. Ambiguities originate from the fact that significant departures from the quadrantal geometry of the nodal planes (Fig. 3.13) might also be obtained by a combination of two or more double-couple force systems. For example, the mechanism illustrated in Figure 3.14 is roughly equivalent to that which would be obtained by stacking, in time and space, a strike-slip and a normal-fault rupture, with the horizontal-tension axes both tending roughly northeast (Barker & Langston 1983). This was unlikely for the Long Valley sequence, because of the repeated occurrence of similar time–space seismicity patterns, and because of the exceptional seismic energy involved. At lower magnitudes, however, near-simultaneous ruptures on adjacent segments of *en echelon* fault systems characterizing zones of intense dyking, are not unlikely when sudden stress changes are driven by magma emplacement. If, for example, a number of events constituting Hill's (1977) model for earthquake swarm formation in volcanic areas (Fig. 3.8) occurred simultaneously, they would provide non-double-couple patterns similar to those observed at Long Valley but involving much lower energies.

The Long Valley case demonstrates the complex problems encountered in understanding seismic sources in the volcanic environment, and demonstrates that increased knowledge about highly non-linear seismic failures remains crucial if a better understanding of volcano dynamics is to be achieved. Because of the ambiguities associated with seismic sources at active volcanoes, however, computed focal solutions determined during routine surveillance should not be used in isolation in interpreting the dynamics of magma intrusion.

### 3.3.3 Low-frequency earthquakes and volcanic tremor

Aki's (1984) model for Long Valley was designed to synthesize the more peculiar features of seismicity in volcanic environments, namely tensile failure, the radiation of low-frequency wavetrains and sustained oscillation of the buried melt the surrounding rock. However, since neither low-frequency events nor volcanic tremor were observed in the Mammoth Lakes region of the caldera, it was argued that either magma viscosity or the radiation loss of the chamber was insufficiently low to allow the excitation of harmonic oscillations of melt following large stress falls such as those characterizing the main shocks of the Mammoth Lakes seismic sequence.

The association of volcanic (harmonic) tremor and low-frequency events with magmatism is quite clear, given that such signals are recorded at nearly all volcanoes during periods of significant activity. Furthermore, at most volcanoes outstandingly similar spectral patterns are encountered (Figs 3.16 & 3.17), suggesting that a similar source mechanism, or a similar combination of concurrent factors including the type and modes of vibration of the source and similar involvement of the medium, might be responsible for both phenomena. On the basis of quantitative evidence collected worldwide, most seismologists share the idea that volcanic tremor may simply be a swarm of low-frequency events. This model makes assumptions, however, about the roles played by source and medium in determining the harmonic characteristics of tremor and low-frequency signals. Minakami (1974), for example, suggested that low-frequency characteristics resulted principally from the shallow depth of the source, and attributed the harmonics to the propagation of wavefields in very absorbent shallow layers typical of volcanic edifices. Tremor, he proposed to be the result of a stack of short-last-

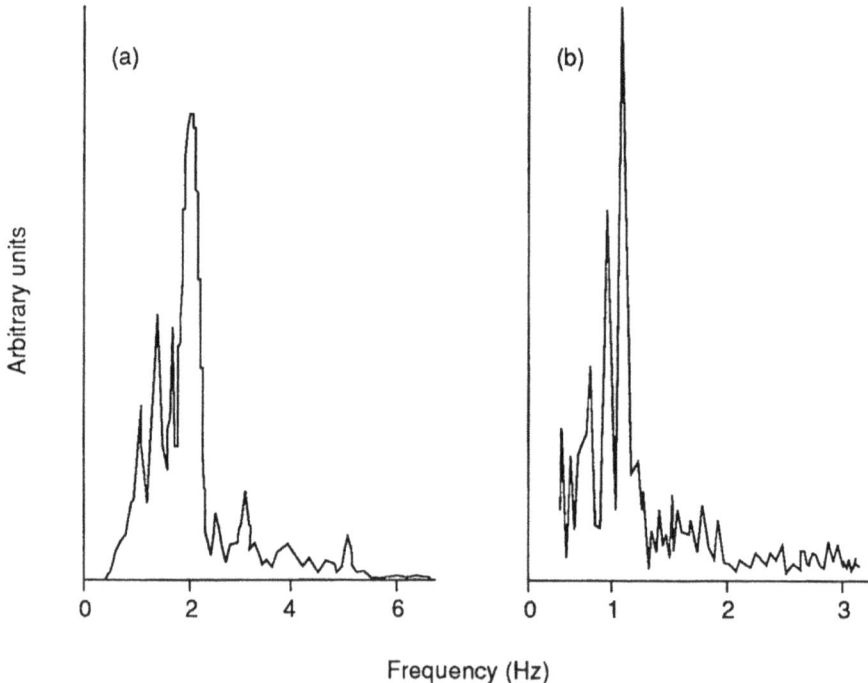

Figure 3.16  The striking similarity between frequency spectral patterns of harmonic tremor recorded at (a) Kilauea (from Chouet et al. 1987) and (b) Etna (courtesy of G.P. Ricciardi) suggests similar transport mechanisms and magma rheologies for both volcanoes.

(a)

B type

5 sec

(b)

Explosion

(c)

Tremor

0    1    2    3    4    5

Frequency (Hz)

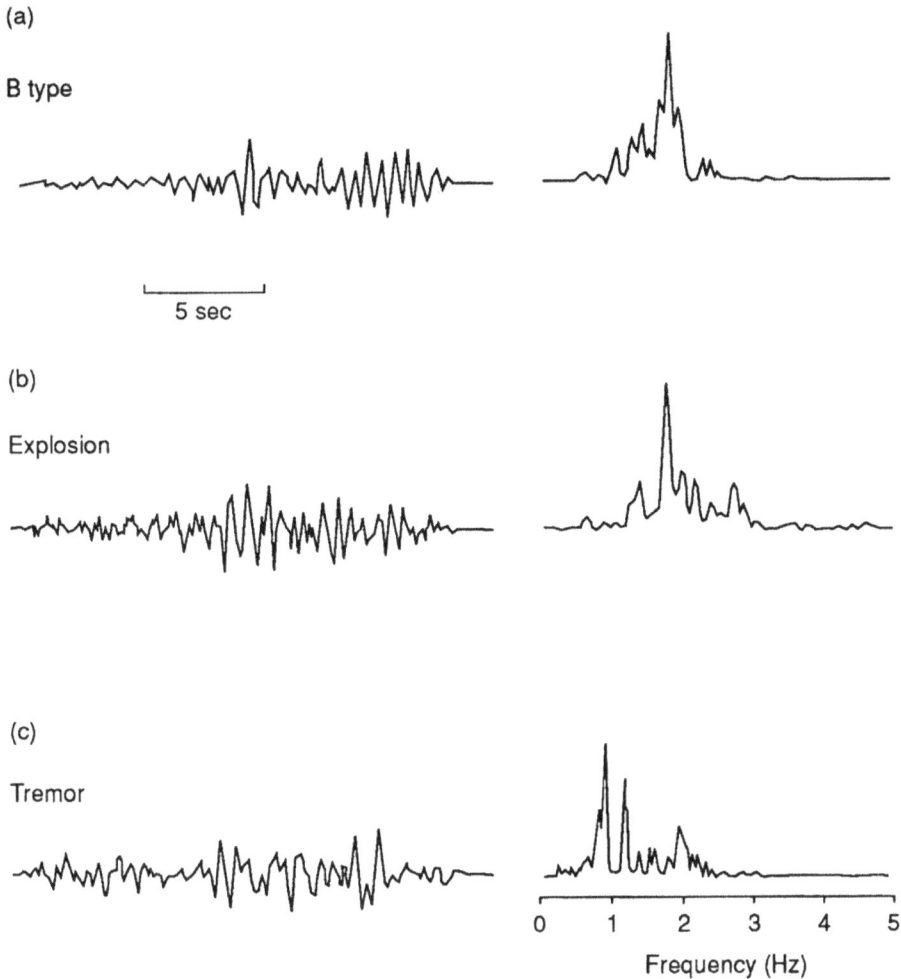

**Figure 3.17**  Seismograms (left) and frequency spectra (right) of Minakami's B-type events (a), explosion quakes (b), and volcanic tremor (c) recorded at Pavlof volcano (Alaska), emphasize the strong similarity in character which leads many seismologists to infer a similar, shallow origin for both low-frequency earthquakes and tremor (from McNutt 1986).

ing low-frequency events (B type). In this model, the harmonic content of both tremor and low-frequency earthquakes would depend only marginally on the type and dynamics of the source. If this model is accepted, then increasing numbers of low-frequency events would be interpreted as being consistent with increased activity at shallow depths in a volcanic edifice, a feature which, although ubiquitous, is contradicted by at least one example (Kilauea) of both low-frequency events and tremors sharing a probable deep origin (Aki & Koyanagi 1981, Shaw & Chouet 1991).

In opposition to the Minakami model, several authors suggest that the dynamics of the source is important in determining the harmonic characteristics of low-frequency events, and a number of subtly different models have been proposed. St Lawrence & Qamar (1979), Seidl et al.

(1981), and Ferrick et al. (1982) amongst others, propose that such characteristics can be entirely accounted for by the resonance of volcanic conduits (Schick et al. 1982a), and that the type of radiation is conditioned by conduit length and the rheological properties of the fluid and the host medium (Chouet 1985). Seidl et al. (1981) propose that continuous fluid flow through volcanic pipes is responsible for causing them to resonate at characteristic eigenfrequencies, a view which has led to the modelling (e.g. Schick et al. 1982b, Gresta et al. 1991) of the shallow feeding system of Etna in terms of a plexus of conduits of different length, the vibrations of each accounting for one spectral line of the frequency spectrum shown in Figure 3.16. A further refinement of the Seidl et al. (1982a) model involves the proposition that changes with time in the spectral shape of volcanic tremor (in its amplitude–frequency content) might be consistent with magma ascent and degassing involving different conduits (e.g. Cosentino et al. 1989). On the basis of such working hypotheses, and holding to the assumption that the tremor is stationary or quasi-stationary, tremor signals are usually routinely sampled and analyzed in terms of frequency for surveillance purpose.

In contrast to the above, tremor models based on the excitation to resonance of cracks filled with viscous fluid (Aki et al. 1977, Aki 1984, Chouet et al. 1987, Ferrazzini & Aki 1987) best account for melt rising or propagating through the brittle upper crust through "nests" of fractures. Chouet (1985) proposed that the frequencies of the principal spectral peaks are influenced by crack stiffness (a dimensionless parameter which is inversely related to the rigidity of the medium and directly related to its bulk modulus and to the aspect ratio of the crack) and their energies reduced with increasing viscosity of the transmitted fluids. Fluid-filled crack models provide a quantitative explanation for a number of the dynamic features associated with low-frequency events and tremor, but they cannot easily be used to directly model tremor for surveillance purposes since: (a) they involve numerous unknown parameters; and (b) cracks or "nests" of fractures typically undergo major evolution with time during the intrusive process.

All attempts at explaining low-frequency sources share the assumption that although a number of phenomena may be responsible, there is one mechanism which clearly overcomes the others (Aki 1992). Such thinking demonstrates a limited understanding of low-frequency sources at active volcanoes, which makes it difficult to draw unequivocally on the behaviour of such phenomena for purposes of volcano forecasting. This can again be demonstrated with reference to the 1989–93 eruptive period at Etna. The 1989 eruption lasted slightly less than a month and erupted around $40 \times 10^6 \, m^3$ of lava (Barberi et al. 1990). Activity was characterized by 16 short episodes of strong tremor activity (Figs 3.18b & 3.13a) accompanying violent strombolian explosions which fed both lava fountains and overflows. Although activity switched from the summit to subterminal and then to flank effusion during the course of the eruption, significant changes in the frequency spectrum of the tremor were not detected in routinely analyzed signals recorded at stations located 5–7 km from the summit craters (Ferrucci et al. 1990).

Such steady behaviour of the frequency spectrum over time clearly argues against the idea that certain harmonics might relate to particular feeding conduits (e.g. Schick et al. 1982b) and suggests, rather, that either (a) the shallow plumbing system of Etna has a single resonance mode or (b) that tremor activity is due to forced fluid flow through a "nest" of cracks

(a)

10 s

(b)

1 min

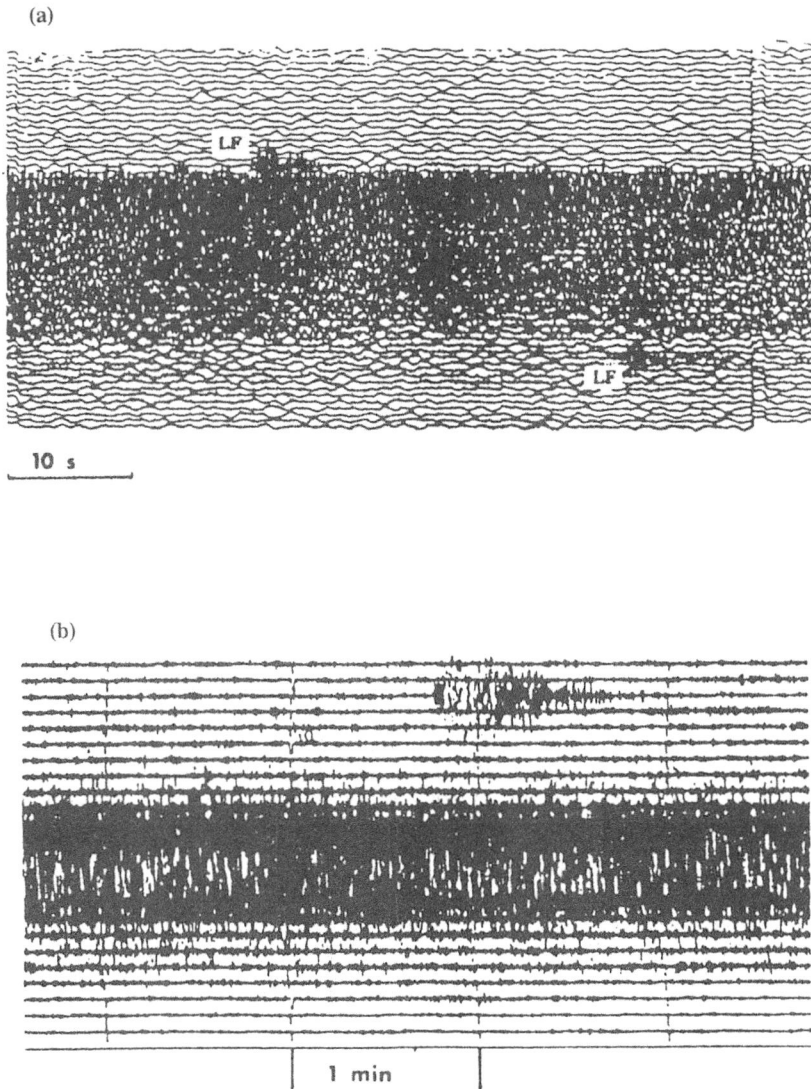

**Figure 3.18** Similar tremor bursts observed during (a) the 1986, non-eruptive unrest at Mount Tokachi (from Okada et al. 1990), and (b) during a lava fountaining episode in the 1989 Etna eruption (Barberi et al. 1990), indicate that violent degassing through volcanic conduits is of primary importance in engendering tremor activity. LF, low-frequency.

of invariable configuration. In contrast to 1989, the 1991–93 flank eruption, the largest of this century in terms of erupted volume, was neither preceded nor accompanied by significant tremor activity. Minor tremor episodes (Fig. 3.19b) were observed only in association with the weak and short-lived strombolian activity, which accompanied the outbreak of the eruption, and during infrequent episodes of minor degassing through the summit crater conduits. The absence of tremor in the presence of huge magma masses available for eruption

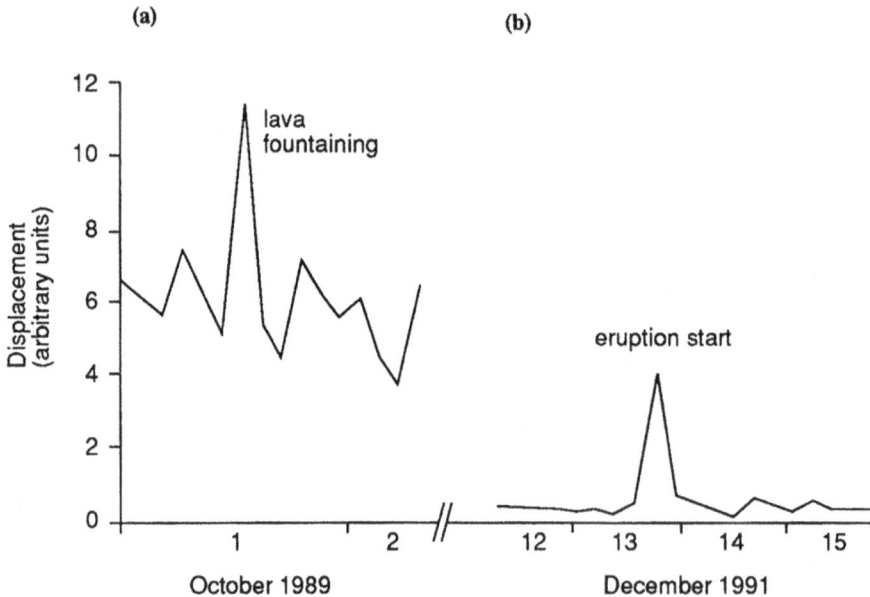

**Figure 3.19** Evidence provided in Figure 3.18 is supported by data collected during the 1991–93 eruption of Etna. In 1989 (a) average tremor amplitude was 10 times that of the 1991–93 eruption (b) in inverse relationship to erupted volumes. While the 1989 eruption lasted 25 days and erupted around $40 \times 10^6$ m$^3$ of lava, the 1991–93 eruption, the largest on Etna this century, has produced some $300 \times 10^6$ m$^3$ of lava. In both cases, however, sharp increases in tremor amplitude were observed, essentially during episodes of degassing through the summit craters.

cannot be accounted for either by source models involving the propagation of fluids through cracks or by tremor sources which are anything other than extremely shallow. Indeed, sporadic resumption of tremor in conjunction with short-lived episodes of degassing through the summit crater conduits suggests that sustained vibration of the conduits and magma contained therein was primarily linked to gaseous flow. A speculative model might then envisage gas velocity and/or the turbulence of a two-phase (gas/magma) fluid (e.g. Montalto et al. 1992) having been responsible for the tremor episodes observed in 1989, but seldom observed during the 1991–93 activity, with the conduits acting, during the earlier eruption, as a booster to the magma-exsolved gases and melt rising through it. Such contrasting evidence, collected at the same volcano during eruptive events which were closely associated in time and likely to be related to the same intrusive episode (see Fig. 3.6), indicates that we are still some way from developing a convincing, all-encompassing model to explain volcanic tremor. Nevertheless, on the basis of data gathered at volcanoes around the world, it can be clearly demonstrated that tremor and low-frequency earthquakes precede and accompany eruptive episodes of all types (Fig. 3.18). Generally speaking, an increase in the rate of low-frequency events constitutes an increased probability that tremor episodes may follow, while rising levels of tremor energy are consistent with an increased probability that an eruption may occur in a few weeks, days, or even hours.

As in earthquake monitoring, however, the adequacy of recording arrays is of fundamental importance in constraining the properties of the tremor signals. Monitoring tremor at one station only, as is often the case at active volcanoes, will not allow for separating the source from the host-medium effects, whenever the tremor source is constant in space and time. Changes in the frequency content of the tremor signals may result from a change in the source-to-station distance or may represent variations in the geometry of the source, the rheological properties of the host medium, or both. In turn, shallowness of the source may account for the greatest information content being contained in the horizontal plane (Gordeev et al. 1990, Ferrucci et al. 1990) which, in agreement with current trends of research (e.g. Ferrazzini et al. 1991, Chouet & Shaw 1991), highlights the need to undertake measurements using dense recording arrays comprising many three-component stations.

As must be apparent to the reader by now that useful modern volcano surveillance using seismic techniques relies strongly on both new and more sophisticated hardware and on ideas about network organization which were not considered as recently as a decade ago. This overview of seismic monitoring will, therefore, be concluded by focusing on current trends in both hardware development and network design and operation.

## 3.4 The role of instrumentation: past, present and future

Prior to the early 1970s, slow-turning drums equipped with mechanical or electromechanical tracers, constituted the core of all seismological acquisition systems. Data were frequency modulated and transmitted by wire or radio telemetry, or locally acquired on smoked-paper recorders, after magnification of the seismometer output signal. On such records (e.g. Fig. 3.18), the length equivalent of 1 min of seismic signal is typically of the order of 6–12 cm, allowing for best P-wave time-picking accuracy of the order of 0.3 s (i.e. every 0.5 mm). Objective precision in the determination of P-wave time and polarity was relatively poor, and relied on the thickness of the pen, the drum-revolution speed, and the frequency response of the galvanometers (typically limited to less than 5 Hz). Precision might be even further reduced for non-minimum-phase signals, or in cases of low signal-to-noise ratios. Such drawbacks were, however, reasonably consistent with both the low spatial resolution of the sparse seismic networks of the 1960s and 1970s (see Table 3.1), and the poor precision allowed by location procedures mainly based on simple geometrical techniques.

Early attempts at improving time-picking resolution were undertaken at the Hawaiian Volcano Observatory (Klein et al. 1987), and involved the use of a camera recorder (the Develocorder), which increased picking resolution to better than 0.1 s but did not allow for further processing of recorded signals. Such photographic devices rapidly became obsolete, and were replaced first by a magnetic-tape analogue recorder, and later by digitally based recorders. Nearly all contemporary data are gathered, or at least converted to, digital form, allowing for full signal processing in both the time and frequency domains. Time-picking accuracy has improved more than 10-fold with respect to drum-recorded signals, the resolution now being limited only by sampling rates (typically 100 or 125 Hz) and the inherent anti-aliasing low-pass filtering (typically half the Nyquist frequency). Analogue-to-digital conversion may be accomplished either at the field stations, or at a network central recording unit after transmission of analogue signals. Since, however, digital devices can reach the large amplitude ratios between the largest signals prior to saturation and the smallest signals distinguishable from electronic noise (typically more than 120 dB in recent equipment running in gain-range mode), direct digital acquisition is more suitable for the high-quality recording of seismic activity which is likely to spread over a wide magnitude range.

The principal inconvenience resulting from the use of digital systems arises from the large physical space required for data storage. The information content grows as a direct function of the analogue-to-digital converter resolution, the sampling frequency, and the number of channels involved. Dataflows of the order of $50–100\,\text{kbit}\,\text{s}^{-1}$ may be easily achieved at standard central arrays. Therefore, in order to avoid rapid saturation of the storage units, almost all systems operate in a trigger mode which is based on the detection of seismic events. Triggering may be accomplished in a number of different ways, such as by means of an amplitude threshold which must be crossed, or by continuous comparison of the signal with the average background noise to look for discrete events. All methods require the use of a segment of pre-event memory, while a mechanism for comparing average signal with average noise is generally used to avoid false alarms caused by triggering on spurious signals. This averaging is based on the continuous computation of the ratio between the energy enveloped by a few cycles of the current signal (short-term average (STA)) and that associated with some cycles tens of seconds immediately preceding it (long-term average (LTA)). The ratio is near unity when no event is present in the pre-event memory segment (i.e. when noise is compared with itself). Conversely, when a sudden increase in energy is detected in the STA, a trigger is allowed if the current STA/LTA value equals or exceeds a pre-programmed value. Such criteria apply to each seismograph channel separately, and multichannel systems are usually allowed to record only when triggering occurs simultaneously at an assigned number of channels.

The use of trigger algorithms requires fine tuning of parameters in order to avoid either frequent false alarms at stations operated in local recording mode, or infrequent triggering of central networks during lower magnitude seismicity. In volcanic areas, data loss may arise either in the presence of strong tremor activity because the detection algorithm fails once the background signal has exceedingly increased LTA values or, in cases of poor P-wave onsets, where the required trigger threshold is reached only after the first part of the signal has been removed from the pre-event memory. Since seismic information would be lost in both cases, it is worth combining high dynamic range triggered recording with continuous, low dynamic range, digital or analogue monitoring, at least at a few seismic stations.

Automated analysis, introduced within the last decade at some volcanoes (e.g. Long Valley and Etna), is a logical development of triggered systems, and includes arrival-time-picking, location of the events and evaluation of some source parameters using spectral methods. Algorithms for time-picking are mainly based on the original work of Stewart (1977) and Allen (1982), and make use of the short-term and long-term averages defined above. In general (Fig. 3.20), the seismic signal is non-linearly transformed by continuously computing the amplitude difference, $\Delta(\tau) = S_t - S_{t-1}$, between the current sample and the sample immedi-

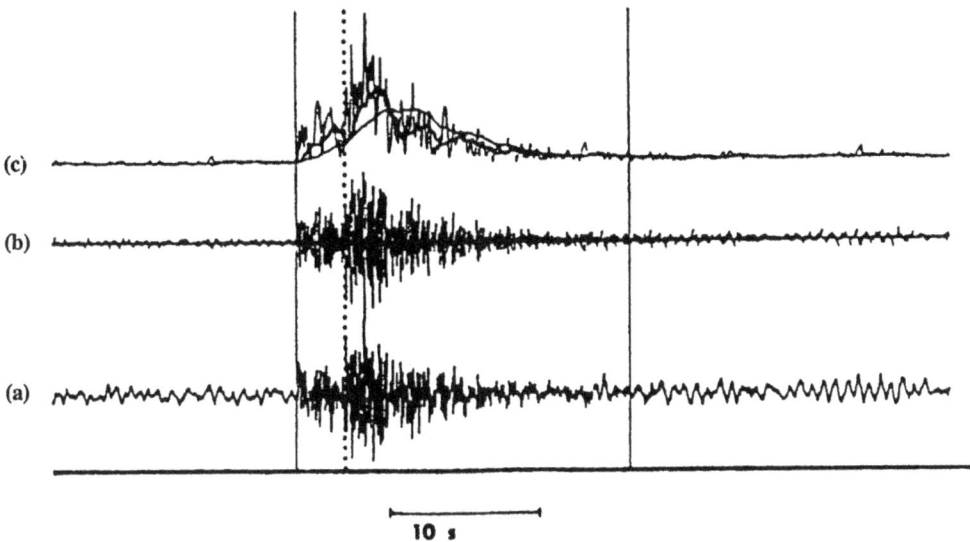

**Figure 3.20**  Example of the processing of a digital seismic trace for automated picking of the P-wave (left-hand vertical line) and S-wave times (dotted vertical line) (from Cattaneo & Augliera 1990). (a) The original seismogram. (b) A non-linear transformation of the signal, carried out according to the criteria described in the text. (c) The short-term (STA) (normal trace), long-term (LTA) (thin line), and mid-term (MTA) (thick line) running averages. The P-wave arrival time is set when the continuously computed STA/ LTA ratio overtakes an assigned threshold value. The same criteria apply for S-waves (STA/MTA). The signal-to-noise ratio being typically much better for P- than for S-waves, automated time-picking is best applied to determination of the onset time and polarity of the P-wave. Such parameters are not affected by eventual clipping of signals.

ately preceding it. Amplitude difference, $\Delta(\tau)$, is compared with its previous value $\Delta_{(\tau-1)}$, and then differences are summed up to $\Sigma\Delta\tau_i$ if signs agree. When $\Sigma\Delta\tau_i$ overtakes the current average value of STA, and the STA/LTA ratio is beyond the detection threshold, picking is performed with a best theoretical resolution of two samples. Once the P-wave arrival is pinpointed, the source parameters may be easily obtained using a fast Fourier transform search for the corner frequency and the displacement spectral amplitude. With minor changes, the same picking procedures may apply to S-waves, even though precision is poorer than for P-waves because of the typically worse signal-to-noise ratios (the noise being, in this case, scattered P-wave energy).

Automated detection and processing (ADP) of seismic events is probably the most significant recent innovation in volcano surveillance-oriented seismology, and it is easy to envisage this procedure playing an increasingly important role in computer-assisted civil protection decisions related to eruptive activity. In terms of their application to volcanic areas, however, ADP systems still need substantial improvements, since: (a) the large data streams expected due to intense seismic swarms may severely affect the efficiency of automated systems, both because of the number and the complexity of the analytical procedures required; and (b) the reliability of real-time operations strongly depends on the effectiveness with which real events can be distinguished from false alarms. At Long Valley caldera, for example, a considerable backlog of data was responsible for a seismic monitoring system falling behind real-time by as much as 11h during the 1986 Chalfant Valley earthquake sequence (McNutt 1991). Since this kind of problem originates from the high occurrence rate of earthquakes, remedies rely mainly on the power and architecture of the central processing unit. Future use of massive-parallel-processing machines is likely to significantly improve the performances of ADP-based systems during major volcanic swarms, an interim solution for conventional computers being to rely on the selective processing of stronger events only. Distinguishing real events from false alarms in volcanic areas remains a problem, since the background noise often displays spectral characteristics which are the same as or similar to low-frequency earth-quakes. Consequently, frequency-based discrimination of the seismic phases will often fail. Enhancement of the discriminating power of picking algorithms can be accomplished by incorporating error-predictive filters (e.g. Granet 1983), although these may also fail in cases of major volcanic swarms when hundreds of events occur that are closely spaced in both space and time, and which display little difference in magnitude, and may result in the magnifica-tion of coherent seismic noise. Analysis a posteriori of the coherence of pickings in space and time might, however, allow for more efficient discrimination of events and the more reliable processing of swarm sequences.

The effectiveness of all such automated procedures depends on the adequacy of the den-sity and geometry of the seismic arrays in relation to the extent and depth of the seismically active zones. At a large basaltic volcano such as Etna, for example, where flank eruptions and the associated shallow seismicity are likely to occur in more than one area at almost the same time, conventionally sized networks (see Table 3.1) provided with evenly spaced sen-sors would not allow the routine, high-resolution constraining of seismic sources in space wherever they occur. Furthermore, since lack of continuity of shallow layers and strong lat-eral variations of the quality factors of the medium may cause dramatic losses in the charac-ter and the frequency content of seismic signals travelling across the volcanic body, automated processing of signals would provide poor results unless a considerable number of sensors was placed above and near the source. A crude estimate of the requirements, determined from typical depths of relevant seismic sequences recorded between 1989 and 1991, indicates that areal densities as high as one sensor per square kilometre are needed in order to best con-strain seismicity linked to shallow magmatic intrusions at Etna.

Instead of increasing the overall numbers of sensors, which is expensive, areal distribu-tions of recording stations could be changed in order to fit the circumstances. Densities 10 times higher than those reported in Table 3.1 could be achieved in areas where shallow intru-sions and shallow seismicity are most likely to occur, with the remainder of an array being devoted to the geometrical strengthening of the network and the location of deeper events. Because the large amount of information contained in three-component data is not easily reducible by automated routines, and since eruption forecasting requires rapid information on epicentre locations and the seismic energy released, it may be acceptable for the avail-ability of accurate focal data to be delayed by a few hours. Consequently, one-component stations would be suitable for supplying the bulk of the data to real-time systems, with the more refined seismological analysis being undertaken off-line by specialized personnel using additional three-component data. Given such a scenario, the choice between acquisition of digital data or analogue-to-digital conversion of data following analogue transmission (i.e.

the difference of some tens of decibels in the dynamic range) becomes of secondary importance. The choice of the dynamic range depends primarily on the largest magnitude expected in that area, and increasing it might be more expensive and require changing the type of sensors. It is also worth pointing out that, while energy-frequency analyses or time-picking of later phases are obviously made impossible by the clipping of signals, automated P-wave picking and polarity determination of P-wave onsets will paradoxically benefit from it. All the points discussed above were considered in planning improvements of the seismic surveillance network at Etna (scheduled for completion in 1994), and should be of general interest to those involved in establishing or upgrading seismic networks in other volcanic areas.

In conclusion, little mention has been made of the use of portable arrays and their integration within volcano surveillance networks. Modern portable arrays, largely based on high dynamic range, digital, three-component stations operated in either telemetry or local-recording mode, may be useful in better constraining seismic sources and improving our understanding of volcano dynamics (e.g. Fehler & Chouet 1982, Aster & Meyer 1988, Hirn et al. 1991a, Ferrucci & Patanè 1993). In terms of their application to volcano surveillance tasks, however, the theoretical efficiency of such portable arrays may be a problem. This has been demonstrated to be affected by: (a), typically long delays between acquisition and processing of data; (b) the need for adequate logistics to support the deployment and recovery of several instruments in often rugged terrains; and (c), significant periods of non-operation during the movement of the array from one area to another. Because of points (a) and (b) in particular, portable arrays are not generally successful in recording the, often most important onsets of seismic swarm sequences. Therefore, although portable arrays can be considered to play a useful supporting role, they do not at present offer an alternative to volcano seismic surveillance undertaken by means of permanent arrays.

## Acknowledgements

The author of this chapter has benefited from the constructive criticism of Stefano Gresta, Steve McNutt, Domenico Patanè and Riccardo Rasà and, in particular, from past and recent exchanges of views with Alexandre Nercessian.

## References

Aki, K., M. Fehler, S. Das 1977. Source mechanism of volcanic tremor: fluid-driven crack models and their application to the 1963 Kilauea eruption. *Journal of Volcanology and Geothermal Research* **2**, 259–87.

Aki, K. & R. Y. Koyanagi 1981. Deep volcanic tremor and magma ascent mechanism under Kilauea, Hawaii. *Journal of Geophysical Research* **86**, 7095–109.

Aki, K. 1984. Evidence for magma intrusion during the Mammoth Lakes earthquakes of May 1980 and implications of the absence of volcanic (harmonic) tremor. *Journal of Geophysical Research* **89**, 7689–96.

Aki, K. 1992. State-of-the-art in volcanic seismology. In *Volcanic Seismology*, P. Gasparini, R. Scarpa, K. Aki (eds), 3–12. New York: Springer.

Allen, R. 1982. Automatic phase pickers: their present use and future prospect. *Bulletin of the Seismological Society of America* **72**, 225–42.

Aster, R.C. & R. P. Meyer 1988. Three-dimensional velocity structure and hypocenter distribution in the Campi Flegrei caldera, Italy. *Tectonophysics* **149**, 195–218.

Barberi, F., G. Corrado, F. Innocenti, G. Luongo 1984. Phlegrean Fields 1982-1984: brief chronicle of a volcano emergency in a densely populated area. *Bulletin Volcanologique* **47**, 175–86.

Barberi, F., A. Bertagnini, P. Landi (eds) 1990. *Mt Etna: the 1989 eruption*. Pisa: Giardini,

Barker, J.S. & C. A. Langston 1983. A teleseismic body-wave analysis of the May 1980 Mammoth-Lakes, California, earthquakes. *Bulletin of the Seismological Society of America* **73**, 419–34.

Bates, C. C., T. F. Gaskell, R. B. Rice 1982. *Geophysics in the Affairs of Man*. Oxford: Pergamon Press.

Brandsdottir, B. & P. Einarsson 1979. Seismic activity associated with the September 1977 deflation of the Krafla central volcano in northeastern Iceland. *Journal of Volcanology and Geothermal Research* **6** 197–212.

Castellano, M., F. Ferrucci, C. Godano, S. Imposa, G. Milano 1993. Upwards migration of seismic focii: a forerunner of the 1989 eruption of Mt Etna. *Bulletin of Volcanology* **55**, 347–51.

Cattaneo, M. & P. Augliera 1990. The automatic phase-picking and event location system of the IGG network, NW Italy. In *Cahiers du Centre Européen de Géodynamique et de Sismologie, vol. 1: Seismic networks and rapid digital data transmission and exchange*, 65–74. Luxembourg: Centre Européan de Géodynamique et de Sismologie.

Chouet, B. 1985. Excitation of a buried magmatic pipe: a seismic source model for volcanic tremor. *Journal of Geophysical Research* **90**, 1881–93.

Chouet, B. & B. R. Julian 1985. Dynamics of an expanding fluid-filled crack. *Journal of Geophysical Research* **90**, 11187–98.

Chouet, B. & H. R. Shaw 1991. Fractal properties of tremor and gas-piston events observed at Kilauea volcano, Hawaii. *Journal of Geophysical Research* **96**, 10177–89.

Chouet, B., R. Y. Koyanagi, K. Aki 1987. Origin of volcanic tremor in Hawaii (Part II): Theory and discussion. In *United States Geological Survey Professional Paper 1350, Volcanism in Hawaii*, R. W. Decker, T. L. Wright, P. H. Stauffer (eds), 1259–80. Washington DC: United States Government Printing Office.

Cosentino, M., G. Lombardo, G. Patanè, R. Schick, A. D. L. Sharp 1982. Seismological researches on Mount Etna: state-of-the-art and recent trends. *Memorie della Società Geologica Italiana* **23**, 159–202.

Cosentino, M., G. Lombardo, E. Privitera 1989. A model for internal dynamical processes on Mt. Etna. *Geophysical Journal International* **97**, 367–79.

De Natale, G. & A. Zollo 1986. Statistical analysis and clustering features of the Phlegrean fields earthquake sequence. *Bulletin of the Seismological Society of America* **76**, 801–14.

De Natale, G., F. Pingue, P. Allard, A. Zollo 1991. Geophysical and geochemical modelling of the 1982-1984 unrest phenomena at Campi Flegrei caldera (southern Italy). *Journal of Volcanology and Geothermal Research* **48**, 199–222.

Dvorak, J.J. & G. Berrino 1991. Recent ground movement and seismic activity in Campi Flegrei, Southern Italy: episodic growth of a resurgent dome. *Journal of Geophysical Research* **96**, 2309–23.

Falsaperla, S., G. Frazzetta, G. Neri, G. Nunnari, R. Velardita, L. Villari 1989. Volcano monitoring in the Aeolian Islands (Southern Tyrrhenian Sea): the Lipari-Vulcano eruptive complex. In *IAVCEI Proceedings in Volcanology vol. 1: Volcanic hazards – assessment and monitoring*, J. H. Latter (ed.), 339–56. Berlin: Springer.

Fehler, M. 1983. Observations of volcanic tremor at Mt. St Helens volcano. *Journal of Geophysical Research* **88**, 3476–84.

Fehler, M. & B. Chouet 1982. Operation of a digital seismic network on Mt St Helens volcano and observations of long-period seismic events that originate under the volcano. *Geophysical Research Letters* **9**, 1017–20.

Ferrazzini, V. & K. Aki 1987. Slow waves trapped in a fluid-filled infinite crack: implication for volcanic tremor. *Journal of Geophysical Research* **92**, 9215–23.

Ferrazzini, V., K. Aki, B. Chouet 1991. Characteristics of seismic waves composing Hawaiian volcanic tremor and gas-piston events observed by a near-source array. *Journal of Geophysical Research* **96**, 6199–209.

Ferrick, M.G., A. Qamar, W. F. St Lawrence 1982. Source mechanism of volcanic tremor. *Journal of Geophysical Research* **87**, 8675–83.

Ferrucci, F. & D. Patanè 1993. Seismic Activity accompanying the outbreak of the 1991-1993 eruption Mt

Etna (Italy). *Journal of Volcanology and Geothermal Research* **57**, 125–35.

Ferrucci, F., C. Godano, N. A. Pino 1990. Approach to the volcanic tremor by covariance analysis: application to the 1989 eruption of Mt Etna (Sicily). *Geophysical Research Letters* **17**, 2425–8.

Ferrucci, F., A. Hirn, G. De Natale, J. Virieux, L. Mirabile 1992. P-to-SV conversions at a shallow boundary beneath Campi Flegrei Caldera (Italy): evidence for the magma chamber. *Journal of Geophysical Research* **97**, 15351–9.

Ferrucci, F., R. Rasà, G. Gaudiosi, R. Azzaro, S. Imposa 1993. Mt. Etna: a model for the 1989 eruption. *Journal of Volcanology and Geothermal Research* **56**, 35–56.

Foulger, G. & R. E. Long 1984. Anomalous focal mechanisms: tensile crack formation on an accreting plate boundary. *Nature* **310**, 43–45.

Gaudiosi, G. & G. Iannaccone 1984. A preliminary study of stress pattern at Phlegrean Fields as inferred from focal mechanisms. *Bulletin Volcanologique* **47**, 225–31.

Gephart, J. W. & D. W. Forsyth 1984. An improved method for determining the regional stress tensor using earthquake focal mechanism data: application to the San Fernando earthquake sequence. *Journal of Geophysical Research* **89**, 9305–20.

Given, J. W., T. C. Wallace, H. Kanamoori 1982. Teleseismic analysis of the 1980 Mammoth Lakes earthquake sequence. *Bulletin of the Seismological Society of America* **72**, 2381–7.

Gordeev, E. I., V. A. Saltykov, V. I. Sinitsin, V. N. Chebrov 1990. Temporal and spatial characteristics of tremor wavefields. *Journal of Volcanology and Geothermal Research* **407**, 89–101.

Granet, M. 1983. Automatic seismic signal detection based on linear prediction filter theory. *Annales Geophysicae* **1**, 109–14.

Gresta, S., A. Montalto, G. Patanè 1991. Volcanic tremor at Mt Etna (January 1984–March 1985): its relationship to the eruptive activity and modelling of the summit feeder system. *Bulletin of Volcanology* **53**, 309–20.

Hill, D. P. 1977. A model for earthquake swarms. *Journal of Geophysical Research* **82**, 1347–51.

Hill, D. P. 1984. Monitoring unrest in a large silicic caldera, the Long Valley-Inyo craters volcanic complex in east central California. *Bulletin Volcanologique* **47**, 371–96.

Hirn, A., J. C. Lépine, M. Sapin, H. Delorme 1991a. Episodes of pit-crater collapse documented by seismology at Piton de la Fournaise. *Journal of Volcanology and Geothermal Research* **47**, 89–104.

Hirn A., A. Nercessian, M. Sapin, F. Ferrucci, G. Wittlinger 1991b. Seismic heterogeneity of Mt Etna: structure and activity. *Geophysical Journal International* **105**, 139–53.

Iannaccone, G. & A. Deschamps 1989. Evidence of shear-wave anisotropy in the upper crust of Central Italy. *Bulletin of the Seismological Society of America* **79** 1905–12.

Julian, B. R. 1983. Evidence for dyke intrusion earthquake mechanisms near Long Valley caldera. *Nature* **303**, 323–5.

Julian, B. R. & S. A. Sipkin 1985. Earthquake processes in the Long Valley caldera area, California. *Journal of Geophysical Research* **90**, 11155–69.

Klein, F. W. 1978. *United States Geological Survey Open-File Report 78-694, Hypocenter location program HYPOINVERSE*. Washington DC: United States Government Printing Office.

Klein, F. W., R. Y. Koyanagi, J. S. Nakata, W. R. Tanigawa 1987. The seismicity of Kilauea's magma system. In *United States Geological Survey Professional Paper 1350, Volcanism in Hawaii*, R. W. Decker, T. L. Wright, P. H. Stauffer (eds), 1019–186. Washington DC: United States Government Printing Office.

Knopoff, L. & M. J. Randall 1970. The compensated linear-vector dipole: a possible mechanism for deep earthquakes. *Journal of Geophysical Research* **75**, 4957–63.

Latter, J. H. 1979. *New Zealand Department of Science and Industrial Research, Geophysics Division, Report 150, Volcanological observations at Tongariro National Park, types and classification of volcanic earthquakes 1976-1978*. Wellington, New Zealand: New Zealand Department of Science and Industrial Research.

Lee, W. H. K. & J. C. Lahr 1975. *United States Geological Survey Open-File Report 75-311, HYPO 71: a computer program for determining hypocenter, magnitude and first motion pattern of local earthquakes*. Washington DC: United States Government Printing Office.

McKee, C. O., P. L. Lowenstein, P. De St Ours, B. Talai, I. Itikarai, J. J. Mori 1984. Seismic and ground deformation crises at Rabaul Caldera: prelude to an eruption? *Bulletin Volcanologique* **47**, 397–411.

McKenzie, D. P. 1969. The relationship between fault-plane solutions for earthquakes and the directions of the principal stresses. *Bulletin of the Seismological Society of America* **59**, 591–601.

McNutt, S. R. 1986. Observations and analysis of B-type earthquakes, explosions and volcanic tremor at Mt Pavlof (Alaska). *Bulletin of the Seismological Society of America* **76**, 153–75.

McNutt, S. R. 1991. Problems with automated seismic monitoring of volcanic areas, using examples from Long Valley caldera and elsewhere (abstract). *Proceedings of the International Conference on Active Volcanoes and Risk Mitigation*, Naples, Italy.

Minakami, T. 1974. Seismology of volcanoes in Japan. In *Physical volcanology*, L. Civetta, P. Gasparini, G. Luongo, A. Rapolla (eds), 1–27. Amsterdam: Elsevier.

Montalto, A., G. Distefano, G. Patanè 1992. *Journal of Volcanology and Geothermal Research* **51**, 211–20.

Okada, Hm. 1983. Comparative study of earthquake swarms associated with major volcanic activity. In *Arc volcanism: physics and tectonics*, D. Shimozuru & I. Yokoyama (eds), 43–61. Tokyo: TERRAPUB.

Okada, Hm., H. Watanabe, H. Yamashita, I. Yokoyama, I. 1981. Seismological significance of the 1977–1978 eruptions and the magma intrusion process of Usu volcano, Hokkaido. *Journal of Volcanology and Geothermal Research* **9**, 311–34.

Okada, Hm., Y. Nishimura, K. Miyamachi 1990. Geophysical significance of the 1988-1989 explosive eruptions of Mt Tokachi. *Bulletin of the Volcanological Society of Japan* **35**, 175–203.

Rabinowitz, N. 1988. Microearthquake location by means of nonlinear simplex procedure. *Bulletin of the Seismological Society of America* **78**, 380–4.

Raleigh, C. B., J. H. Healy, J. D. Bredehoeft 1972. Faulting and crustal stress at Rangely, Colorado. In *Flow and Fracture of Rocks*. H. C. Heard et al. (eds), 257–84. Washington DC: American Geophysical Union.

Rice, R. B. & D. K. Watson 1981. Satellite Observations of Mt St Helens. *Eos*, **62**, 577.

Ryall, A. & F. Ryall 1981. Spatial temporal variations in seismicity preceding the May 1980, Mammoth Lakes, California, earthquakes. *Bulletin of the Seismological Society of America* **71**, 747–60.

Schick, R., G. Lombardo, G. Patanè 1982a. Volcanic tremors and shocks associated with the eruptions at Etna (Sicily), September 1980. *Journal of Volcanology and Geothermal Research* **14**, 261–79.

Schick, R., M. Cosentino, G. Lombardo, G. Patanè 1982b. Volcanic tremors at Etna: a brief description. *Memorie della Società Geologica Italiana* **23** 191–6.

Scholz, C.H. 1968. The frequency-magnitude relation of microfracturing in rocks and its relation to earthquakes. *Bulletin of the Seismological Society of America* **60**, 399–415.

Seidl, D., R. Schick, M. Riuscetti 1981. Volcanic tremors at Etna: a model for hydraulic origin. *Bulletin Volcanologique* **44**, 43–56.

Shaw, H. R. & B. Chouet 1991. Fractal hierarchies of magma transport in Hawaii and critical self-organization of tremor. *Journal of Geophysical Research* **96**, 10191–207.

Shemeta, J.E. & C. S. Weaver 1986. Seismicity accompanying the May 18 1980 eruption of Mt St Helens, Washington. In *Mount St Helens: five years later*. S. A. C. Keller (ed.), 44–58. Washington DC: Eastern Washington University Press.

Stewart, S. W. 1977. Real-time detection and location of local seismic events in central California. *Bulletin of the Seismological Society of America* **67**, 433–52.

St Lawrence, W. F. & A. Qamar 1979. Hydraulic transients: a seismic source in volcanoes and glaciers. *Science* **203**, 654–6.

Swanson, D. A., T. J. Casadevall, D. Dzurisin, S. D. Malone, C. G. Newhall, C. S. Weaver 1982. Predicting eruptions at Mt St Helens, June 1980 through December 1982. *Science* **221**, 1369–76.

Virieux, J., V. Farra, R. Madariaga 1988. Ray Tracing for earthquake location in laterally heterogeneous media. *Journal of Geophysical Research* **93**, 6585–99.

Zollo, A. & P. Bernard 1991. Fault mechanisms from near-source data: joint inversion of S polarizations and P polarities. *Geophysical Journal International* **104**, 441–51.

# 4 Real-time ground deformation monitoring

J. P. Toutain, P. Bachèlery, P. A. Blum,
H. Delorme and P. Kowalski

## 4.1 Introduction

Periodic deformation of the ground surface is a common feature of active volcanoes, and is usually studied by means of traditional geodetic techniques such as precise levelling and trilateration (Kinoshita et al. 1974). Accumulated data are used for comparison between repeated surveys, providing a tool for modelling the behaviour of the internal plumbing system of the volcano. Levelling, distance measurements and angle measurements permit the relative motions in space of a network of periodically measured benchmarks to be calculated. Such studies provide suitable data for predicting eruptions (Decker et al. 1983), investigating magmatic processes (Jackson et al. 1975, Dvorak et al. 1983, Trygvason 1989), constraining the geometry of feeder-dyke or storage systems (Fiske & Kinoshita 1969, Hoffmann et al. 1990, McGuire et al. 1990), and studying tectonic manifestations in volcanic areas (Tarantola et al. 1980, Lipman et al. 1985, Dzurisin et al. 1990). More recently, such methods as photogrammetry and space geodesy have been used to support studies using traditional geodetic techniques. The former allowing precise measurements of the motion of numerous points (Zlotnicki et al. 1990), and the latter, in the form of the global positioning system (GPS), providing more precise data with respect to classical geodetic measurements, particularly for large distances (Briole 1990).

On basaltic shield volcanoes, vertical ground deformation is mainly related to the accumulation, injection or withdrawal of magma at relatively shallow depths, producing surface displacements and ground-tilt changes. Because rapid inflation of the volcanic pile and upheaval of the summit region are generally recorded as the result of rising magma prior to the onset of eruptions, monitoring of such changes is therefore expected to provide suitable data for eruption prediction. Electronic tilt monitoring is particularly suited to such a goal, because it uses highly sensitive sensors for short-period measurements. Moreover, tiltmeters supply continuous data and therefore allow the monitoring of both rapid ground deformation associated, for example, with dyke emplacement, and the slower deformation caused by magma movement at depth. The method does, however, have its limitations and drawbacks, mostly resulting from the recording of extraneous tilts due, for example, to poor siting, meteorological fluctuations, and long-term instrumental drift (Goulty 1976, Harisson 1976, Henbest et al. 1978, Wyatt & Berger 1980, Briole 1990, Endo et al. 1990), and these are discussed in more detail in the following section. Descriptions and analyses of deformation patterns determined by means of electronic tilt arrays, and related to either dyke emplacement and subsequent subsidence or to lava-dome dynamics, are presented in a number of papers, including, for example, Wadge et al. (1975), Henbest et al. (1978), Zharinov et al. (1983), Trygvason (1986), Dvorak & Okamura (1985), Okamura et al. (1988), Briole (1990), and Endo et al. (1990). Usually, data are stored on site, and periodically processed, although the real-time transmission of data is performed at some volcanoes such as Kilauea (Okamura et al. 1988), Mount St Helens (Dvorak et al. 1981), Sakurajima (Kamo & Ishihara 1989) and Long Valley caldera (Mortensen & Hopkins 1987).

A prime candidate for intensive real-time ground deformation monitoring is the basaltic shield volcano, Piton de la Fournaise (Réunion Island, Indian Ocean), one of the most active volcanoes in the world. During historic times, a dyke-fed eruption occurred, on average, once every 14 months, providing exceptional conditions for establishing and improving ground-tilt monitoring strategies. Consequently, a network of eight telemetered electronic-tilt stations has been progressively installed on the volcano, equipped with highly sensitive fused-silica sensors developed by Pierre A. Blum at IPGP. A number of intrusive episodes have been documented using this network (Lénat et al. 1989a & b, Delorme et al. 1989, Saleh et al. 1991, Toutain et al. 1992). In response to the characteristically rapid nature of dyke intrusions, an acquisition and processing system has also been developed which allows the real-time display of both tilt vectors and inflation centers during eruptions (Toutain et al. 1992). In this chapter, we describe real-time ground deformation monitoring at Piton de la Fournaise using both electronic tiltmeters and automatic electronic distance meters (AEDMs). In addition we relate observed ground deformation patterns to recent eruptions, and show how these data may be used to document the propagation rates of feeder dykes.

## 4.2 Real-time tiltmeter monitoring at active volcanoes

As pointed out by Dzurisin (1992) in his recent excellent assessment of their suitability for volcano monitoring, there has been a rapid proliferation of tiltmeters on many volcanoes. On the plus side, such instruments can be bought off-the-shelf, are relatively cheap, and are easy to establish and maintain. In addition, they have the capability to undertake continuous monitoring, and many research groups now use such instruments to determine the real-time tilt record which is telemetered to the local volcano observatory. Tiltmeters do, however, have their drawbacks, and may, as a consequence of, for example, poor installation or maintenance, or the interpretation of data by inexperienced researchers, generate spurious results which can contradict data gathered using other techniques. Or even worse, in the absence of alternative data, incorrect decisions leading to a major disaster or, conversely, to unnecessary evacuation of the danger region around a volcano, may be based purely on an inaccurate tiltmeter record. Care must, therefore, always be taken in the selection of instruments and particularly in choosing those with the appropriate precision and accuracy for expected movements. The locations for the installation of tiltmeters should always be judiciously selected, and the instruments correctly emplaced. A single instrument is of little use, particularly if there is no other ground deformation technique in operation to provide a control, and three tiltmeters probably represent the very minimum useful number for a volcano under surveillance. Once established, it is imperative that enough baseline data are obtained to filter out any extraneous effects that are not associated with volcanic unrest, but which are likely to be the result of external, often meteorological, factors. Such signals may range from diurnal to annual, are often cyclic, and must be recognized before tiltmeters can usefully be employed to monitor the behaviour of volcanic systems. Both Banks et al. (1989) and Dzurisin (1992) discuss the potential pitfalls that can result from tiltmeter use for volcano monitoring, and Dzurisin (1992) offers sound advice on all aspects of tiltmeter acquisition and operation, based on experience of extensive tiltmeter monitoring at Mount St Helens.

Most tiltmeters used to monitor volcanoes use a fluid to measure tilt changes along a baseline which, in borehole tiltmeters, is only a few centimetres in length, but which in instruments installed at the surface may be several metres long. Traditionally, water was the most commonly used working fluid, and is still successfully used today in water-tube tiltmeters at a number of volcanoes. These commonly consist of U-shaped tubes which are partly filled with water, the level of which changes in the two arms of the tube in response to deformation of the ground surface, and which is usually electronically converted to a tilt value. Mercury is also a commonly used working fluid, and at Kilauea a mercury-pool tiltmeter has been successfully used to record ground tilt since 1967 (Decker & Kinoshita 1972, Banks et al. 1989). Such an instrument records variations in the levels of mercury in two connected containers by monitoring differences in capacitance between the surfaces of the mercury bodies and an overlying capacitance plate. Once again, recorded variations are electronically converted to tilt values, and are usually continuously recorded. Mercury tiltmeters have a number of drawbacks, mainly related to their limited measurement range and servicing problems (Banks et al. 1989), and have been largely superceded by more manageable and less expensive bubble tiltmeters. In these instruments, which can be located in either boreholes or at the surface, tilting of a container holding an electrolytic liquid containing a bubble, causes the bubble to move. This is detected electronically and converted to a tilt value. While both water-tube and mercury tiltmeters are uniaxial, therefore requiring two instruments at each location to give a measure of radial and tangential (relative to the summit region or the expected focus of future tilting) tilt, biaxial bubble tiltmeters are available, thereby significantly reducing installation and maintenance times.

In contrast to the above-mentioned types, the tiltmeters used at Piton de la Fournaise do not utilize a working fluid of any sort, but instead consist of a pendulum located in a vacuum chamber (Saleh 1986, Briole 1990, Saleh et al. 1991). The instrument is damped by a magnet and a photoelectric device inside the vacuum chamber transforms the rotation of the pendulum into a linear voltage variation (Saleh et al. 1991).

## 4.3 Tiltmeter and AEDM monitoring at Piton de la Fournaise

Piton de la Fournaise is a basaltic shield volcano which occupies the eastern part of Réunion island, and which first erupted over 0.5 million years BP (Gillot & Nativel 1989). Structural and radiochronological studies reveal a complex history with the formation of at least 3 successive caldera systems (Bachèlery 1981, Chevalier & Bachèlery 1981), huge landslide episodes, and the eastward displacement of the centre of activity over the last 0.2 MA BP (Gillot & Nativel 1989, Bachèlery & Mairine 1990). This migration has resulted in the formation of a broad shield, in the eastern part of which lies the currently active zone. Most of the recent activity (about 97% of historic eruptions) has occurred within the youngest caldera known as the "Enclos Fouqué", which probably formed less than 5000 years ago (Bachèlery 1981), and its seaward extension, the "Grand Brulè" (Fig. 4.1). This depression is thought to have been caused by a huge landslide that formed in response to repeated dyke injection within two poorly developed rift zones, aligned northeast and southeast of the principal fracture zone in the upper part of the volcano (Vincent & Kieffer 1978, Duffield et al. 1982, Lénat et al. 1989). Two craters occupy the summit of the volcano, the Bory crater, inactive since 1760, and the Dolomieu crater where many collapses and eruptions have taken place during recent years. The few eruptions occuring outside the caldera are located along the two rift zones with an average of two events per century. The most recent such events occurred in the northeast rift zone in April 1977 (Kieffer et al. 1977) and along the southeast rift zone in March 1986 (Delorme et al. 1989).

Both geophysical and geological arguments, as well as instrumental monitoring data obtained by the Piton de la Fournaise Volcano Observatory (OVPF), support the existence of a shallow magma-storage complex operating below the Dolomieu crater (Bachèlery & Lénat 1989). Seismic foci related to the documented eruptions (1981–92) are generally located 1–3 km below the crater. Pre-eruptive inflation episodes are also centred in approximately the same area, with calculated foci located at 1–2 km depth. Geological, dynamical and petrological data together with structural observations of the eroded adjacent massif of Piton des Neiges reveal the general characteristics of the summit storage complex, suggesting that the main reservoir is capped by a complex of active sills and small magma pockets from which vertical dykes and cone-sheet intrusions propagate towards the surface.

Most eruptions at Piton de la Fournaise are of the fissure type, occurring along fractures which are radially distributed around the summit region. "Lateral" fissures (Delorme et al. 1989, Toutain et al. 1992), which occur further away from the summit, are rare but are also fed by dykes propagating from beneath the summit region. Rare phreatic and phreatomagmatic explosions also occurred (about 2 per century) within the Dolomieu crater, which has also experienced a number of collapses related to the withdrawal of magma from the summit reservoir during larger volume lateral eruptions. Morphological studies of the volcano, together with an analysis of the distribution of concentric surface fractures, suggest that the shallow magmatic system has been stable at least over the past 3000 years, although there is some evidence for a recent small eastward displacement of activity.

Ground deformation associated with volcanic events are confined to the summit cone and the associated zone of open fissures. The summit cone, where most of the intrusions and

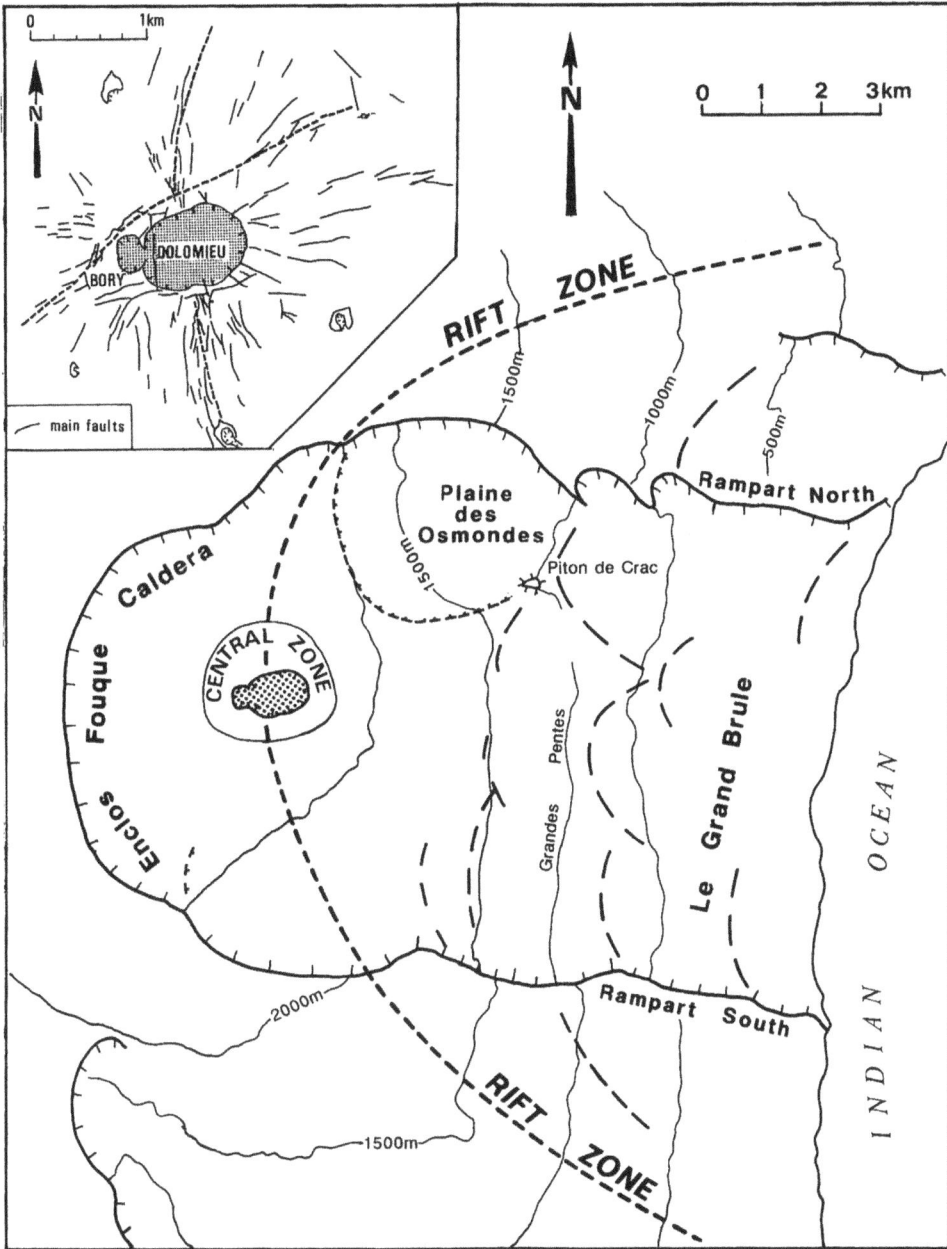

**Figure 4.1** Sketch map of Piton de la Fournaise volcano. The main fracture zones are shown by broad broken lines, and areas possibly subjected to an eastward landslide are delimited by heavy dashed lines. (Modified from Zlotnicki et al. 1990.)

eruptions occur, is essentially constructed from accumulated lava flows, which, around the periphery of the Dolomieu crater, are cut by a dense network of both radial and concentric fractures. Some heterogeneity of the lava pile results from the alternating succession of "pahoehoe" and "aa" lithologies, and the occurrence of occasional pyroclastic units. Of particular importance for facilitating magma ascent are the vertical discontinuities caused by the pre-existing fractures in the summit region. The upper part of the volcano can thus be visualized as a heterogeneous framework made up of blocks bounded by the main fracture zones. The structural setting of the summit region apparently controls the emplacement of dykes in the upper part of the volcano and the development of the eruptive fissures during recent eruptions (Bachèlery et al. 1983), with some intrusions (e.g. the 14 June 1985 intrusion) being trapped by the pre-existing shallow fracture system (Lénat et al. 1989a).

Since the establishment of the OVPF, the study of ground deformation has represented, along with seismic surveillance, the most important method for monitoring volcanic activity. In addition, levelling, dry-tilt, extensiometry and geodetic measurement (Lénat 1989, Lénat et al. 1989a & b, Delorme et al. 1989), and photogrammetric surveys (Zlotnicki et al. 1990) have also been undertaken, allowing deformation patterns to be recorded and interpreted in terms of the combined effect of magma transfer and tectonic activity. Observed deformation has been found to be closely related to eruptive phases and restricted to the summit region and around the zone of open eruptive fissures. Moreover, unlike Krafla and Kilauea, no long-term inflation has been noted since detailed monitoring began in 1980, suggesting discontinuous magma transfer from depth since the last eruption of picrites (oceanites) in 1977 (Lénat & Bachèlery 1988). For some eruptive phases, however, slight deformation patterns (some tens of microradians) have been recorded close to Dolomieu crater during the days preceding the onset of eruptive activity, probably reflecting overpressure in a small, shallow magma pocket which then fed the succeeding eruption (Lénat 1989, Lénat et al. 1989a & b). The distribution of these inflation centres is generally consistent with seismic epicentres and supplies a rough indication of the location of a future outbreak. Minutes to hours before an eruption, the injection of magma towards the surface is characterized by rapid, intense, and irreversible ground deformation, often totalling several hundred microradians.

### 4.3.1 Instrumentation and procedures

At Piton de la Fournaise, both electronic tiltmeters and automatic electronic distance meters are used to monitor ground deformation in real time, and the data are telemetered to the OVPF. The tiltmeters used at Piton de la Fournaise consist of a horizontal Zöllner suspended pendulum made from two silica fibres, carrying a silver sheet as a gravitational mass, and located in a vacuum chamber (Saleh 1986, Briole 1990, Saleh et al. 1991). A magnet installed behind the instrument provides good damping for mass oscillations, and a photoelectric device inside the vacuum chamber transforms, without amplification, the rotation of the pendulum into a linear voltage variation (Saleh et al. 1991). The relation between ground tilt $a$ and the shifting of the spot on the cell is given by the equation:

$$a = d/DKT_o^2 \qquad (4.1)$$

where $d$ is the shifting of the spot, $D$ is the distance of the cell-rotation axis of the pendulum (usually $5.10^{-2}$ m), $K$ is the specific coefficient of the pendulum (usually $5\,s^{-2}$), and $T_o$ is the oscillation time of the pendulum (4–15 s on a volcano).

The resolution is $3.5 \times 10^{-8}$, $10^{-7}$, and $0.9 \times 10^{-6}$, respectively, for $T_0$ values of 15, 8 and 3 seconds. The mechanical amplification (the ratio between the pendulum rotation and the ground tilt) is 1125, 320, and 45, repectively, for $T_0$ values of again 15, 8, and 3 s.

The linear voltage variation supplies tilt values through a calibration curve determined in the laboratory (Saleh et al. 1991). For this operation, the instrument is installed on a tilting platform and a tilt of a known value $a$ is induced. The calibration factor $F$ is given by :

$$F = V/aT_0 \qquad (4.2)$$

where $V$ is the voltage variation and $T_0$ is the oscillation period.

In the field, the value of the ground tilt $a'$ will be:

$$a' = V'/FT_0'^2 \qquad (4.3)$$

where $T_0'$ is the new oscillation period of the installed instrument and $V'$ is the voltage variation induced by ground tilt. If necessary, the voltage variation can be amplified and, if required, offset.

All the mechanical and photoelectric devices are housed in a pyrex sleeve that is plugged at both ends by pyrex disks. This system is sealed to the rock by means of a silica cone (Fig. 4.2). The tiltmeter is protected by a series of four chambers: the first is composed of a silica sleeve which contains the vacuum for the sensitive parts; the second is a protective metal chamber which is enclosed in a stainless steel protective cover with a watertight synthetic seal at the base; the third chamber is composed of polystyrene and provides thermal insulation;

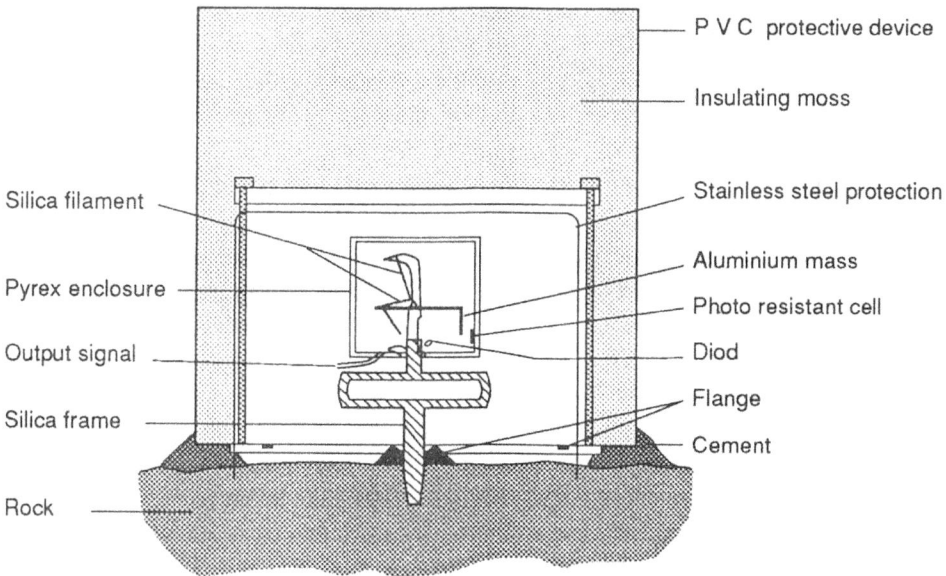

**Figure 4.2**  Cut-away drawing of a tiltmeter installation at Piton de la Fournaise, showing both the internal detail and the successive protective coverings (see text).

and the final chamber is composed of polyvinyl-chloride (PVC) and provides overall protection (Fig. 4.2). Such construction allows great mechanical amplification and thermal inertia, and avoids both mechanical friction and weathering. The long-term annual drift of these silica tiltmeters has been estimated to be only about $0.1\,\mu$rad (Briole 1990), and they have also been found to be insensitive to nearby explosions. Moreover, there is evidence for the absence of any long-period parasitic oscillations within the range of seismic frequencies. Disadvantages do exist, however, and result primarily from the fragility of the instruments during transport, and in the mechanical sensitivity which can be adjusted only at the time of installation. Tiltmeters installed in the Enclos caldera have a period of about 6 s, with output to a 12-digit numerical recorder, allowing high resolution (about $0.3\,\mu$rad per digit) and an operating range of $972\,\mu$rad. The reliability of the tiltmeters at Piton de la Fournaise has been established by comparing both the azimuth and intensity of recorded tilts with ground deformation obtained independently by means of precise levelling (Toutain et al. 1992).

Tiltmeter data at Piton de la Fournaise are supported by information obtained using two AEDM systems which allow continuous measurements of distance changes between benchmarks in a geodetic network. The first AEDM is a GEO 2000 FENNEL distance-meter, installed on the Dolomieu crater rim, which measures a unique east–west distance (approximatively 697 m) with a precision better than $10^{-2}$ m. Due to the high level of electrical consumption of this system, data sampling is only performed every 5 min, and telemetered to the observatory. The second AEDM was established in October 1992, and allows the measurement of five lines extending from the Enclos caldera rim to reflectors installed in a vertical plane across the northern rift zone at distances of 2707, 2314, 2027, 1819 and 1661 m for reflectors 1 to 5, respectively. This second instrument is a PULSAR FENNEL distance-meter which is servo-controlled by a step-by-step motor linked to an infinite screw, and designed to scan a vertical plane. The system is controlled by a CANON X07 microcomputer, which also manages data acquisition and storage. Under ideal meteorological conditions, the maximum length which can be measured is about 6000 m, with a precision of $5 \times 10^{-3}$ m. In the typically poor weather conditions on the volcano, and with the maximum distance being only 2707 m, the precision is considered to be better than $10^{-2}$ m. Due to the velocity of rotation of the step-by-step motor, a complete set of measurements requires at least 20 min. The system remains under test at the time of writing, with data sampled every hour and stored on a memory card. Figure 4.3 shows a 7-week record which, unfortunately, does not coincide with an eruptive episode. Some periods of missing data are obvious, mostly due to poor meteorological conditions, the effects of which increase with increasing distance. Standard deviations ($\sigma$) for the period investigated are low, ranging from $5.7 \times 10^{-3}$ to $7.1 \times 10^{-3}$ m. Such values are highly satisfactory, since changes in the lengths of lines across the rift zone during dyke propagation are expected to be much greater (e.g. up to 2.6 m during Pu'u 'O'o episode of Kilauea volcano) (Okamura et al. 1988). Future improvements of the system are planned, including reducing significantly the duration of measurement sets, and establishing a radio link to the observatory for real-time display of data during dyke propagation episodes.

Data from all the monitoring instrument networks on Piton de la Fournaise (Fig. 4.4) are telemetered to the volcanological observatory installed in 1980 (Bachèlery et al. 1982). These data are supported by periodic ground deformation surveys utilizing a network of spirit-level

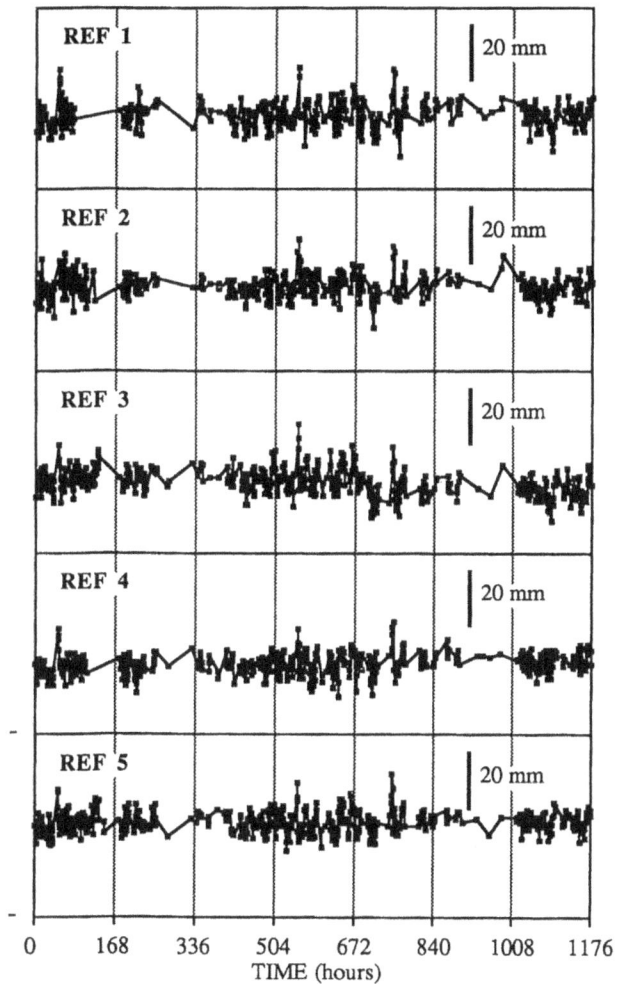

**Figure 4.3** AEDM automatic measurement of five lengths (reflectors 1 to 5) across the northern rift zone. Data are sampled each hour. Absent data are due to "poor weather". Distances to measure are 2707, 2314, 2027, 1819 and 1661 m for reflectors 1 to 5, respectively.

tilt stations, and a trilateration network (Lénat et al. 1989a, Zlotnicki et al. 1990). The electronic tiltmeters are grouped in pairs to form a clinometric station within which one tiltmeter is oriented radially and the other tangentially with respect to the summit crater. Data are sampled at each station every minute by a microcomputer, and consist of the values of the two sensors, of the atmospheric and rock (5 cm depth) temperatures, and of electronic parameters which provide a real-time check on the correct functioning of the station. Blocks of five successive messages are telemetered every five minutes via a relay station, and consist of serial messages of digitized data send by a radio line in very high frequency (VHF) mode to the observatory. In order to increase noise immunity during transmission, each serial message is coded by a frequency-shift keying (FSK) modulation. The clinometric network was first conceived to ensure coverage of the summit cone – the most active part of the volcano. Figure 4.5 displays the theoretical precision of calculated inflation centres as a function of their location with respect to tiltmeters, and demonstrates that, until recently, only summit inflation centers could be determined with sufficient precision. This map highlights the need for a supplemen-

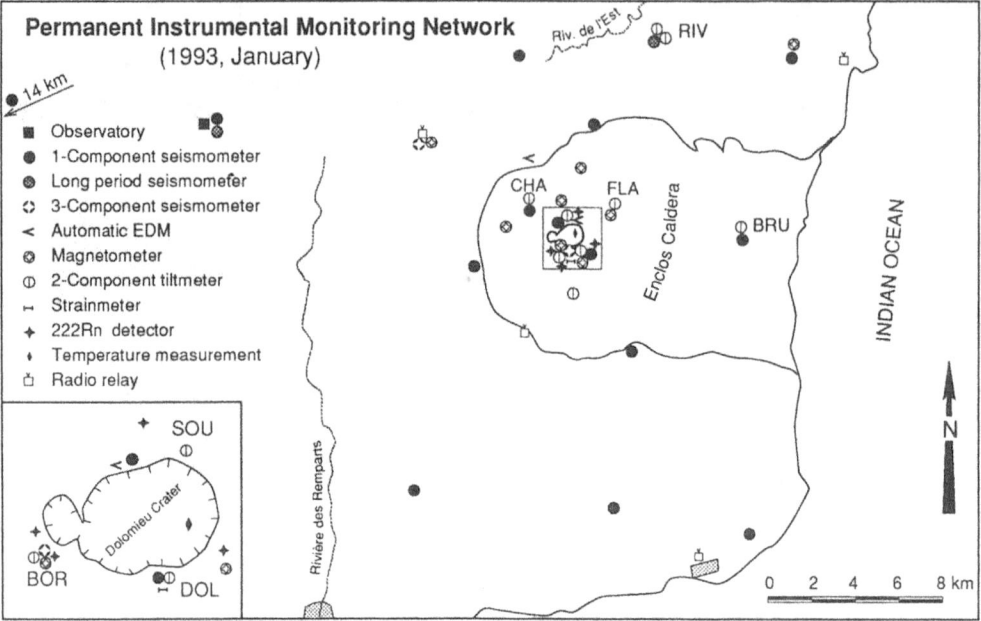

**Permanent Instrumental Monitoring Network**
**(1993, January)**

■ Observatory
● 1-Component seismometer
⊗ Long period seismometer
◌ 3-Component seismometer
< Automatic EDM
⊗ Magnetometer
⊕ 2-Component tiltmeter
⊢ Strainmeter
✦ 222Rn detector
✦ Temperature measurement
⌂ Radio relay

Riv. de l'Est    RIV

CHA    FLA    BRU

Enclos Caldera

INDIAN OCEAN

N

SOU
Dolomieu Crater
BOR    DOL

Rivière des Remparts

0    2    4    6    8 km

**Figure 4.4**    Distribution of the instrumental monitoring network at Piton de la Fournaise.

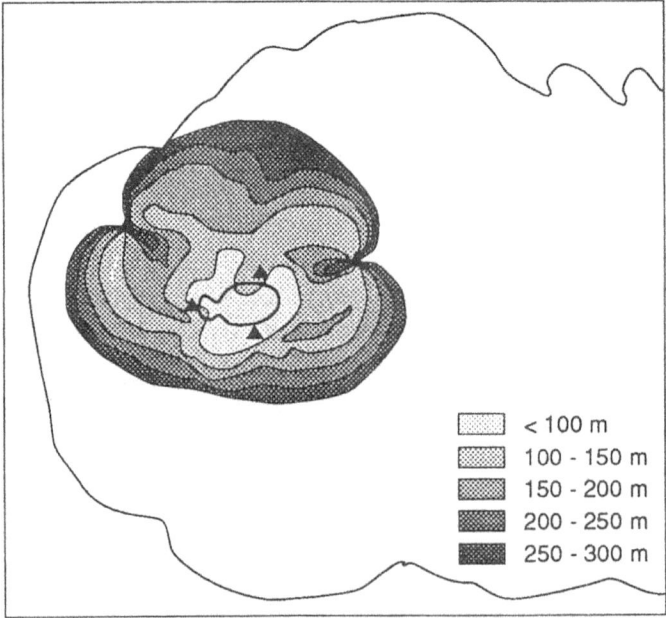

< 100 m
100 - 150 m
150 - 200 m
200 - 250 m
250 - 300 m

**Figure 4.5**    Map of the calculated precision of determinations of inflation centers as a function of their spatial distribution with respect to tilt stations.

tary tilt station in the south of the Enclos caldera, which was installed in June 1992.

At Piton de la Fournaise, the duration of intrusive crisis is generally short (sometimes only 30 min long) (Lénat et al. 1989b, Delorme et al. 1989). This is a result both of the proximity of the magma storage areas to the surface, and the high ascent velocity of the magma. Until 1989, it was rather difficult to control and interpret numerical tilt data during such crises, and processing of seismic data alone was therefore expected to supply constraints on the near-surface location of intruding magma. In late 1989, new software became available that allowed the real-time acquisition and processing of telemetered clinometric data, and the graphical display, in real time, of both tilt vectors and the geometric position of inflation centres. This allowed data from the clinometric stations within the Enclos caldera to be used to locate the changing position of ascending magma beneath the summit region. The resulting inflation centre being computed as the least square adjustment of the intersections of tilt vectors, and successively displayed on a map on the visual display unit (VDU) screen. To our knowledge, only one experiment dealing with real-time processing of clinometric data has been described in the literature, the objective of which was to forecast the precursory ground tilt prior to summit eruptions at Sakurajima volcano (Kamo & Ishihara 1989). It seems, however, that our work is the first attempt to document, in real time with clinometric data, the migration of a magma body during its intrusion in an active volcano.

### 4.3.2 Tilt monitoring during eruptive episodes

Vertical ground deformation patterns associated with some recent intrusive and/or eruptive episodes at Piton de la Fournaise are described here, including events that occurred prior to the establishment of the real-time processing system. Ground deformation data relating to these latter events have been processed using the new software to allow comparison with the more recent activity. Whenever possible, determined inflation centres are displayed graphically.

#### 4.3.2.1 The 18 January 1990 eruption

This event (Fig. 4.6) occurred after 3 months of increasing seismicity. The eruptive crisis began at minute 1410 on January 17 with a 4–5 min swarm of 33 events. Seismicity then decreased until minute 216 on 18 January, when a new swarm of short-period earthquakes was recorded. An $M = 1.2$ shock was detected at minute 408, followed by a seismic swarm. Events from minutes 398 to 408 were found to occur below the Dolomieu's northeast flank, then succeeding events moved below the southeast flank. All were very shallow, the deepest being about 1 km above sea level. Tremor appeared at minute 432 on summit stations, with discrete shocks continuing until minute 440. The opening of an eruptive fissure then occured in the Dolomieu crater at minute 444, associated with both lava fountaining and lava-flow formation. The fissure propagated initially southwards towards the Dolomieu crater rim, and then towards the north-northwest, within the crater itself, to near the northern crater rim. During the 17 h of activity it is estimated (Smithsonian Institution 1990a) that around $1 \times 10^6 \text{m}^3$ of aphyric basalt was erupted.

Figure 4.6a shows the moderate deformations associated with this eruption, and Figure

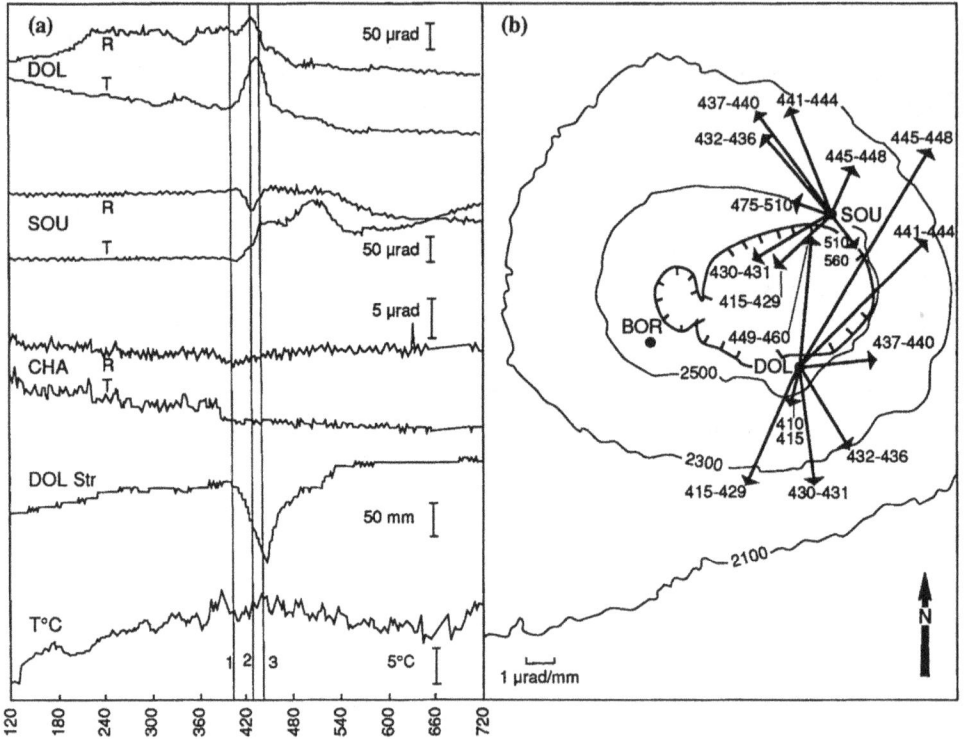

**Figure 4.6** The 18 January 1990 eruption. Tilt and strain curves for a 10-hour period (a) and calculated tilt vectors (b). Vertical lines in (a) correspond to the main phenomena described in text at minutes 408 (1), 432 (2) and 444 (3).

4.6b displays the tilt vectors at the DOL and SOU stations during the course of the eruption (in minutes since midnight on 17 January). Unfortunately, BOR station was out of order and therefore inflation centers were not calculated. The first deformation phase was recorded between minutes 408 and 415, following the $M = 1.2$ shock, and took the form of a very slight inflation phase at DOL ($1.5\,\mu\text{rad}\,\text{min}^{-1}$). Inflation at SOU over the same period took place at a slower rate ($0.7\,\mu\text{rad}\,\text{min}^{-1}$). The second phase was recorded between minutes 415 and 430, marking a clear deflation pattern at SOU ($3\,\mu\text{rad}\,\text{min}^{-1}$), while DOL showed significant tilt towards the south ($5\,\mu\text{rad}\,\text{min}^{-1}$) and a summit strainmeter indicated contraction along a monitored fissure. Tremor appeared at about minute 432, and until minute 444, tilt vectors showed a rotational movement of about 180° and 90° at both DOL and SOU stations, respectively. Significant tilt towards the north-northwest was then demonstrated at both DOL and SOU stations coincident with the emission of lava in the Dolomieu crater close to the DOL station. Dilatation of the fissure was therefore recorded from minute 444 until equilibrium was reached at minute 535. The last movement was recorded by SOU station which showed an eastward tilt from minutes 475 to 510 and a southeast tilt from minutes 510 to 560, over which period no significant changes in ground tilt were noted at the DOL station. This last movement is probably related to the northward migration of the eruptive fissure observed within Dolomieu crater.

### 4.3.2.2 The 18 April–8 May1990 eruption

Eruptive activity in April 1990 commenced after 15 days of increasing seismicity (Smithsonian Institution 1990b). Lavas were extruded in the southeastern area of the Enclos caldera and accompanied by lava fountaining. Effusion rates were between 20 and $30\,m^3\,s^{-1}$, and the total volume erupted is estimated at around $8 \times 10^6\,m^3$ (Smithsonian Institution 1990b). Figure 4.7a shows the considerable vertical deformation recorded from minutes 213 to 230, in two distinct phases (minutes 213 to 224 and 225 to 230). The most significant changes occurred at the stations nearest the summit (up to $400\,\mu rad$), as shown by the tangential DOL tiltmeter, whereas the furthermost station (CHA) displayed very much smaller changes (3 and $5\,\mu rad$ for the tangential and radial tiltmeters, respectively). This pattern suggests short-wavelength deformation due to the shallow, central supply of magma to the summit cone, causing general inflation of the Dolomieu area followed by a strong deflation. A slight subsidence at station CHA, located about 2 km from the summit, is interpreted in terms of either a normal relaxation movement, or a pressure decrease within the magma batch feeding the eruption.

Both tilt vectors and calculated relative positions of the inflation center between minutes 213 and 230 are displayed in Figure 4.7b. Values from CHA have not been used for these calculations, owing to the small scale of the changes observed at this station. The distribution

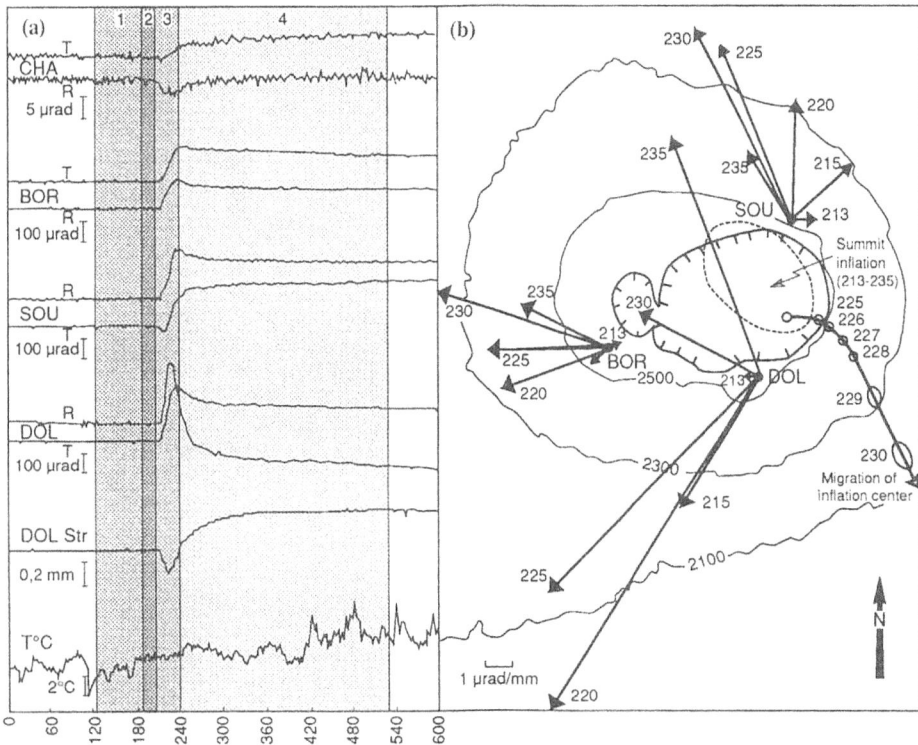

**Figure 4.7**  The 18 April–8 May 1990 eruption. Tilt and strain curves for a 10-hour period (a) and calculated tilt vectors and inflation centres distribution (b). The hatched areas are related to eruption phenomena, as described in text.

of inflation centres shown allows the two main phases of the deformation pattern to be distinguished. Between minutes 213 and 224, a central inflation was located beneath the northeastern part of the Dolomieu crater, and is interpreted as being due to a subvertical dyke intrusion. Between minutes 225 and 230, a migration of inflation centres towards the southeast accompanied a rapid deflation of the summit area (as seen from the decrease of the radial component at the SOU, DOL and BOR stations). Because electronic tilt stations were absent in the southern area of Enclos caldera at this time, it was not possible to document the migration of further inflation centres. It can be inferred, however, both from the well-defined spatial trend of inflation centre between minutes 225 and 230 (Fig. 4.7b), and from the regularly decreasing pattern shown by the summit tilt stations after minute 240, that migration was continuing towards the southeast. This model is supported by the occurrence of the eruption at minute 530 in the southeastern sector of the Enclos caldera.

### 4.3.2.3 The 19–20 July 1991 eruption
Following the small increase in seismicity that began on 24 June, a short eruption occurred between 19 and 20 of July 1991. The event began with a shallow microearthquake swarm comprising nearly 80 low-energy earthquakes ($M < 1.5$) located beneath the southern flank of the summit cone at depths shallower than 1 km (Smithsonian Institution 1991). Lavas were

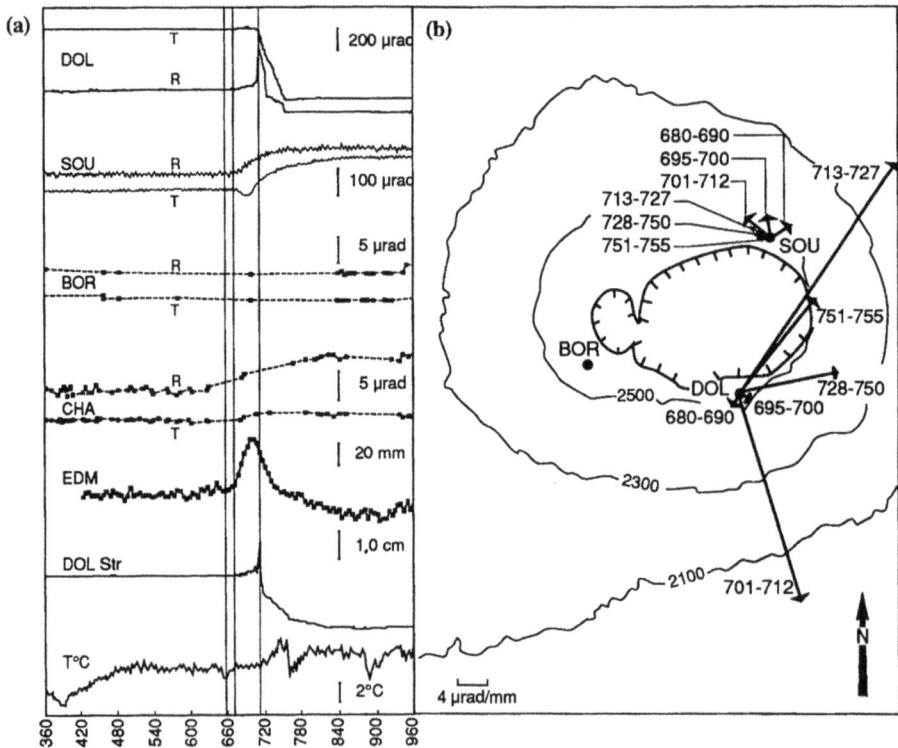

**Figure 4.8** The 18 July 1991 eruption. Tilt, EDM and strain curves for a 10-hour period (a) and calculated tilt vectors (b).

extruded from two vents along an eruptive fissure close to the Dolomieu crater rim, with the total erupted volume estimated at $5 \times 10^6\,\mathrm{m}^3$.

Increasing EDM and radial tilt values shown in Figure 4.8a at station DOL allowed the detection of the first inflation episode at about minute 670, probably related to the early stages of intrusion from the magma reservoir. At minute 703, radial tilt and strain values at DOL began to increase rapidly, with tilt rates of up to $54\,\mu\mathrm{rad}\,\mathrm{min}^{-1}$ being recorded. At the same time, EDM data indicated a rapid decrease in the distance between the two rims of the Dolomieu crater, suggesting both the rapid transport of magma to the surface and a pressure decrease within the magma batch. Figure 4.8b shows the tilt vectors associated this intrusive episode which led to weak northwest and high south-southwest tilting at the SOU and station DOL, respectively, suggesting that the magmatic intrusion was located below the southern area of the Dolomieu crater. A deflation phase was then recorded at minute 712 with north-north-west tilting at DOL station, corresponding with the start of tremor and the emission of lava close to the station.

### 4.3.3 Velocities and mechanisms of dyke emplacement

The precise temporal identification of dyke emplacement during intrusive crises, using ground deformation monitoring, allows dyke migration velocities to be determined directly, both for the intrusive episode as a whole, and for individual phases during emplacement. Velocities are determined for the eruption of April 1990 which was located at a distance of about 3.8 km from the summit, in the southern part of the Enclos caldera, and which was the result of a vertical injection phase followed by a lateral migration phase (Toutain et al. 1992). Figure 4.7 shows that the first phase of deformation was associated with the shallow and subvertical input of a magma batch beneath the Dolomieu crater. Assuming a vertical displacement of 1.5 km for the dyke (this is a reasonable value for the depth of the magma pocket determined from seismic data), and taking into account the 12-min duration necessary for this vertical migration, we can calculate a mean vertical velocity of about $2\,\mathrm{m}\,\mathrm{s}^{-1}$. During the second phase, magma migrates laterally in the summit cone to feed the lateral eruption. Curves from the three summit clinometric stations (Fig. 4.7a) allow the precise determination of the onset of this horizontal migration of magma, and during the first minutes of lateral injection, they provide a good control of the progression of the magma in the summit cone. The calculated lateral velocity of the dyke for this period is about $3\,\mathrm{m}\,\mathrm{s}^{-1}$. If the total horizontal migration of magma from the initial inflation area to the outbreak site is considered, an average migration velocity for the intrusion can be calculated to be about $0.2\,\mathrm{m}\,\mathrm{s}^{-1}$.

Most of the other lateral eruptions during the past 10 years are not as well documented, although in many cases one tiltmeter at least was operating, providing suitable data for similar calculations. Tilt curves associated with earlier eruptions look very similar to those for the April 1990 event, thereby offering the potential to estimate for these both the start time of the vertical migration phase and that of the start of the lateral injection and, knowing the time and location of the eruptions, to determine dyke emplacement velocities. The averaged duration of vertical injections can be estimated for 14 recent eruptions to about 25 min, with extreme values lying between 10 and 50 min. For all these eruptions, it can be inferred that the magma was stored in a shallow reservoir (A. Hirn, personal communication), and therefore a mean vertical migration distance for the eruption feeder dykes of 1.5–2 km can be estimated, resulting in a mean vertical velocity of $1–1.4 \, \mathrm{m \, s^{-1}}$.

Figure 4.9 is a plot of the duration of horizontal magma migration versus distance between the summit and the eruption site for the selected 14 recent eruptions. When these occur at a horizontal distance from the summit of less than 1.5 km, in other words for eruptions opened on the flank of the summit cone, the velocity of migration seems to be around $2 \, \mathrm{m \, s^{-1}}$. This value is very close to that obtained for the first minutes of the April 1990 eruption. It is worth noting that these calculated velocities for dyke injection in the summit zone of Piton de la Fournaise are quite high with respect to the calculations of Okamura et al. (1988) for the velocity of dykes emplaced during Pu'u 'O'o eruption at Kilauea ($0.15 \, \mathrm{m \, s^{-1}}$). However, Okamura et al. (1988) have also documented velocities of up to $0.6 \, \mathrm{m \, s^{-1}}$ as the result of local pulsations caused by contact of the growing dyke with small, shallow magma pockets. In fact, for the later stages of dyke emplacement events at La Fournaise, calculated velocities are significantly lower, averaging $0.2 \, \mathrm{m \, s^{-1}}$, or even $0.1 \, \mathrm{m \, s^{-1}}$ when the propagation distance is over 2.5 km (Fig. 4.9). These values are in good agreement with the estimates of Briole (1990) for the average extrusion rate of dykes on Piton de la Fournaise ($0.21 \, \mathrm{m \, s^{-1}}$), and those obtained by Okamura et al. (1988) for Kilauea.

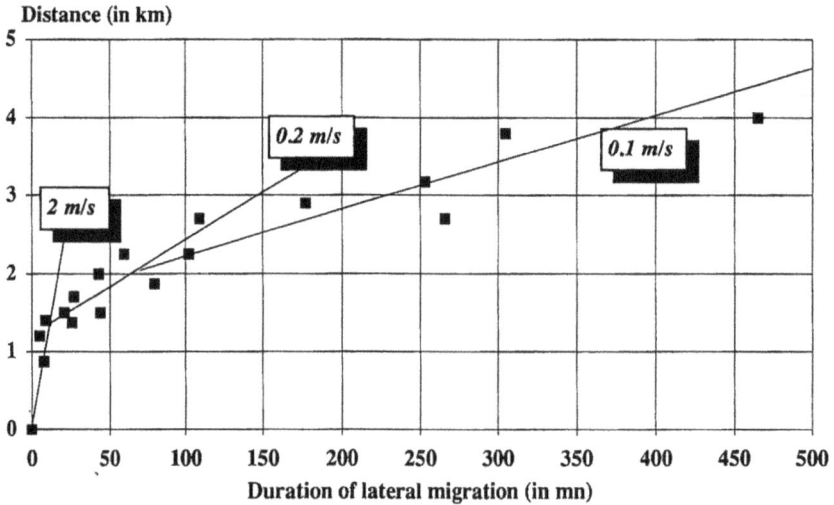

**Figure 4.9** Plot of the duration of lateral intrusions as a function of the distance from the outbreak site to the central zone for 14 recent eruptions.

The decrease in injection velocity with distance from the summit is clearly shown in Figure 4.9. The very high vertical and lateral injection velocities typical of the region immediately around the summit decrease strongly once the intrusion propagates out of this central zone. This general lowering of the injection velocity is probably a function, both of decreasing magma pressures, and of the increasing cross section of the dyke front as it propagates away from the feeding pocket. On the other hand, the very high velocity of injection within the confines of the summit cone can reasonably be related to the near-surface fracture pattern of Piton de la Fournaise. This consists of a very dense network of radial and tangential fissures around the summit craters (Fig. 4.1),which offer rapid routes for intrusions without the need to first fracture the host rock.

In conclusion, both electronic tiltmeters and automatic electronic distance-meters have been very successfully used at Piton de la Fournaise volcano, to record ground deformation associated with both the vertical and lateral emplacement of dykes. Real-time processing of data has allowed changes in the positions of inflation centres to be recognized over very short time-scales, enabling the locations of eruption sites to be predetermined, and dyke injection velocities to be calculated. The techniques employed at La Fournaise have general application to basaltic volcanoes worldwide, and constitute a useful tool for mitigating the effects of effusive eruptions.

## Acknowledgements

We wish to thank J. C. Delmond and Ph. Taochy who ensured the good working of the tilt array since 1985, L. Fontaine who contributed to the real-time software, and A. Legros and Ph. Mairine for producting the figures. We are very grateful to Bill McGuire, who greatly improved the text of this paper.

## References

Bachèlery, P. 1981. Le Piton de la Fournaise (Ile de la Réunion): étude volcanologique, structurale et pétrologique. Thesis, University of Clermont II. Clermont-Ferrand, France.

Bachèlery, P. & J-F. Lénat 1989. Evidence and constraints for the presence of a shallow reservoir at Piton de la Fournaise (Réunion). *Proceedings of the IAVCEI General Assembly. Continental Magmatism*. Santa Fe: New Mexico Bureau of Mines and Mineral Resources.

Bachèlery, P. & P. Mairine 1990. Volcano-structural evolution of Piton de la Fournaise during the last 0.53 M.a. In *Le Volcanisme de la Réunion. Monographie*, J-F. Lénat (ed.). Clermont-Ferrand, France: Centre de Recherches Volcanologiques, Université Blaise Pascal.

Bachèlery, P., P. A. Blum, J. L. Cheminée, L. Chevallier, R. Gaulon, N. Girardin, C. Jaupart, F. Lalanne 1982. Eruption at le Piton de la Fournaise volcano on 3 February 1981. *Nature* 297, 395–99.

Bachèlery, P., L. Chevallier, J. P. Gratier 1983. Caractères structuraux des éruptions historiques du piton de la Fournaise. *Compte Rendu de l'Académie des Sciences de Paris* 296, 1345–50.

Banks, N. G., R. I. Tilling, D. H. Harlow, J. W. Ewert 1989. Volcano monitoring and short-term forecasts. In *Short courses in geology, vol. 1: Volcanic hazards*, R. I. Tilling (ed.), 51–80. Washington DC: American Geophysical Union.

Briole, P. 1990. Mesure de déformations du sol de volcans et zones simiques. Exemples de développement et de mise en oeuvre d'outils de mesure. Interaction mesure-modélisation. Thesis University of Paris 6.

Chevallier, L. & P. Bachèlery 1981. Evolution structurale du volcan actif du Piton de la Fournaise, Ile de la Réunion, Océan indien occidental. *Bulletin Volcanologique*, 44, 723–41.

Decker, R. W., R. Y. Konayagi, J. J. Dvorak, J. P. Lockwood, A. T. Okamura, K. M. Yamashita, W. R. Tanigawa 1983. Seismicity and surface deformation of Mauna Loa, Hawaii. *Eos* 64, 545–47.

Decker, R. W. & W. T. Kinoshita 1972. Geodetic measurements. In *The surveillance and prediction of vol-*

*canic activity*. Paris: UNESCO.

Delorme, H., P. Bachèlery, P. A. Blum, J. L. Cheminée, J. F. Delarue, J. C. Delmond, A. Hirn, J. C. Lepine, P. M. Vincent, J. Zlotnicki 1989. March 1986 eruptive episodes at Piton de la Fournaise volcano (Réunion island). *Journal of Volcanology and Geothermal Research* **36**, 199–209.

Duffield, W. A., L. Stieltjes, J. Varet 1982. Huge lanslide blocks in the growth of Piton de la Fournaise, La Réunion and Kilauea volcano, Hawaii. *Journal of Volcanology and Geothermal Research* **12**, 147–60.

Dvorak, J. & A. Okanura 1985. Variations in tilt rate and harmonic tremor amplitude during the January–August 1985 east rift eruptions of Kilauea volcano – Hawaii. *Journal of Volcanology and Geothermal Research* **25**, 249–58.

Dvorak, J., A. Okamura, S. H. Dietrich 1983. Analysis of surface deformation data, Kilauea volcano, Hawaii. October 1966 to September 1970. *Journal of Geophysical Research* **88**, 9295–304.

Dvorak, J., A. Okamura, C. Mortensen, M. J. Johnston 1981. Summary of electronic tilt studies at Mount St Helens. In *Geological Survey Professionnal Paper, 1250, The 1980 eruptions of Mount St Helens, Washington*, P. W. Lipman & D. R. Mullineaux (eds), 169–74. Washington DC: United States Government Printing Office.

Dzurisin, D. 1992. Electronic tiltmeters for volcano monitoring: lessons from Mount St Helens. In *United States Geological Survey Bulletin 1966, Monitoring volcanoes: techniques used by the staff of the Cascades volcano observatory 1980–90*, J. W. Ewert & D. A. Swanson (eds), 69–83. Washington DC: United States Government Printing Office.

Dzurisin, D., J.C. Savage, R.O. Fournier 1990. Recent crustal subsidence at yellowstone caldera, Wyoming. *Bulletin of Volcanology* **52**, 247–70.

Endo, E. T., D. Dzurisin, D. Swanson 1990. Geophysical and observational constraints for ascent rates of dacitic magma at Mount St Helens. In *Magma transport and storage*, M. P. Ryan (ed.), 317–34. Chichester: John Wiley.

Fiske, R. S., W. T. Kinoshita 1969. Inflation of Kilauea volcano prior to its 1967-1968 eruption. *Science* **165**, 341–9.

Gillot, P.Y. & P. Nativel 1989. Eruptive history of of the Piton de la Fournaise volcano, Réunion island, Indian Ocean. *Journal of Volcanology and Geothermal Research* **36**, 53–65.

Goulty, N. R. 1976. Strainmeters and tiltmeters in geophysics. *Tectonophysics* **34**, 245–56.

Harisson, J. C. 1976. Cavity and topographic effects in tilt and strain measurement. *Journal of Geophysical Research*, **91**, 319–28.

Henbest, S. N., A. A. Mills, P. Ottey 1978. Two tiltmeters and an integrating seismometer for the monitoring of volcanic activity, and the results of some trials on Mount Etna. *Journal of Volcanology and Geothermal Research* **4**, 133–49.

Hirn, A., J. C. Lepine, M. Sapin, H. Delorme 1991. Episodes of pit-crater collapse documented by seismology at Piton de la Fournaise. *Journal of Volcanology and Geothermal Research* **47**, 89–105.

Hoffmann, J. P., G. E. Ulrich, M. O. Garcia 1990. Horizontal ground deformation patterns and magma storage during the Puu Oo eruption of Kilauea volcano, Hawaii. *Bulletin of Volcanology* **52**, 522–31.

Jackson, D. B., D. A. Swanson, R. Y. Konayagi, T. L. Wright 1975. *United States Geological Survey Professional Paper, 890, The August and October 1968 east rift eruptions of Kilauea volcano, Hawaii*, 1-33. Washington DC: United States Government Printing Office.

Kamo, K. & K. Ishihara 1989. A preliminary experiment on automated judgement on the stages of activity using tiltmeters records at Sakurajima, Japan. In *IAVCEI Proceedings in Volcanology vol. 1: Volcanic hazards – assessment and monitoring*, J. H. Latter (ed.), 585–98. Berlin: Springer.

Kieffer, G., B. Tricot, P. M. Vincent 1977. Une éruption inhabituelle (avril 1977) du Piton de la Fournaise (Ile de la Réunion): ses enseignements volcanologiques et structuraux. *Compte Rendu de l'Académie des Sciences de Paris*, **285**, 957–60.

Kinoshita, W. T., D. A. Swanson, D. B. Jackson 1974. The measurement of crustal deformation related to volcanic activity at Kilauea volcano, Hawaii. In *Physical volcanology*, L. Civetta, P. Gasparini, G. Luongo, A. Rapolla (eds), 87–115. Amsterdam: Elsevier.

Lénat, J-F. 1989. Dynamics of magma transfers at Piton de la Fournaise volcano (Réunion Island, Indian ocean). In *IAVCEI Proceedings in Volcanology vol. 1: Volcanic hazards – assessment and monitoring*, J. H. Latter (ed.), 585–98. Berlin: Springer.

Lénat, J-F. & P. Bachèlery 1988. Dynamics of magma transfers at Piton de la Fournaise volcano (Réunion

Island, Indian ocean). In *Modeling volcanic processes*, Chi-Yu & R. Scarpa (eds), 57–72. Braunschweig: Vieweg.

Lénat, J-F., P. Bachèlery, A. Bonneville, A. Hirn 1989a. The beginning of the 1985-1987 eruptive cycle at Piton de la Fournaise (La Réunion), new insights in the magmatic and volcano-tectonic systems. *Journal of Volcanology and Geothermal Research* **36**, 209–32.

Lénat, J-F., P. Bachèlery, A. Bonneville, P. Tarits, J. L. Cheminée, H. Delorme 1989b. The December 4 1983 to February 18 1984 eruption of Piton de la Fournaise (La Réunion, Indian ocean): description and interpretation. *Journal of Volcanology and Geothermal Research* **36**, 87–112.

Lénat, J-F., P. Vincent, P. Bachèlery 1989c. The offshore continuation of an active basaltic volcano: Piton de la Fournaise (Reunion Island, Indian Ocean). *Journal of Volcanology and Geothermal Research* **36**, 1–36.

Lipman, P. W., J. P. Lockwood, R. T. Okamura, D. A. Swanson, K. M. Yamashita 1985. *United States Geological Professional Paper 1276, Ground deformation associated with the 1975 magnitude 7.2 earthquake and resulting changes in activity of Kilauea volcano, Hawaii*, 1-45. Washington DC: United States Government Printing Office.

McGuire, W. J., A. D. Pullen, S. J. Saunders 1990. Recent dike-induced large-scale block movement at Mount Etna and potential slope failure. *Nature* **343**, 357–59.

Mortensen, C. E. & D. G. Hopkins 1987. Tiltmeter measurements in Long Valley caldera, California. *Journal of Geophysical Research* **92**, 13767–76.

Okamura, T. A., J. J. Dvorak, R.Y. Konayagi, W.R. Tanigawa 1988. Surface deformation during dike propagation. In *United States Geological Survey Professional Paper 1463, Pu'u 'O'o eruption of Kilauea volcano, Hawaii: episodes 1 through 20 January 3 1983, through June 8 1984*, E. W. Wolfe (ed.), 165–81. Washington DC: United States Government Printing Office.

Saleh, B. 1986. Dévelopement d'une nouvelle instrumentation pour les mesures de déformations. Application au génie civil. Thesis, University of Paris 6, Paris

Saleh, B., P. A. Blum, H. Delorme 1991. New silica compact tiltmeter for deformations measurement. *Journal of Surveying Engineering* **117**, 27–35.

Smithsonian Institution 1990a. Lava fountains and flows from summit area fissure, with seismicity and deformation. *Bulletin of the Global Volcanism Network* **15**, 5–7.

Smithsonian Institution 1990b. Dyke injection, then eruption from fissure vents near S caldera wall. *Bulletin of the Global Volcanism Network* **15**, 4–7.

Smithsonian Institution 1991. Brief lava production follows seismicity, deformation and magnetic changes. *Bulletin of the Global Volcanism Network* **16**, 17–19.

Tarantola, A., J. C. Ruegg, J. C. Lepine 1980. Geodetic evidence for rifting in Afar, 2. Vertical displacements. *Earth and Planetary Science Letters* **48**, 363–70.

Toutain, J. P., P. Bachèlery, P. A. Blum, J. L. Cheminée, H. Delorme, L. Fontaine, P. Kowalski, P. Taochy 1992. Real-time monitoring of vertical ground deformations during eruptions at Piton de la Fournaise. *Geophysical Research Letters* **19**, 553–56.

Trygvason, E. 1986. Multiple magma reservoirs in a rift zone volcano. Ground deformation and magma transport during the september 1984 eruption of Krafla, Iceland. *Journal of Volcanology and Geothermal Research* **28**, 1–44.

Trygvason, E. 1989. Ground deformation in Askja, Iceland: its source and possible relation to flow of the manle plume. *Journal of Volcanology and Geothermal Research* **39**, 61–71.

Vincent, P. M. & G. Kieffer 1978. Hypothèses sur la structure et l'evolution du Piton de la Fournaise (Ile de la Réunion) après les éruptions de 1977. *Proceedings of 6ème Reun. Ann. Sci. Terre*, 407. Orsay: Societé Geologique de France.

Wadge, G., J. A. Horsfall, J. L. Brander 1975. Tilt and strain monitoring of the 1974 eruption of Mount Etna. *Nature*, **254**, 21–23.

Wyatt, F. & J. Berger 1980. Investigations of tilt measurements using shallow borehole tiltmeters. *Journal of Geophysical Research*, **85**, 4351–62.

Zharinov, N. A., Y. S. Dobrokhotov, M. A. Magus'kin, S. V. Enman 1983. Tilt of the Earth's surface during the formation of Cone II of the Great Tolbachik Fissure eruption in 1975. In *The great Tolbachik fissure Eruption*, S. A. Fedotov & Ye K. Markhinin (eds), 301–06. Cambridge: Cambridge University Press.

Zlotnicki, J. & J. L. Le Mouel 1990. Possible electrokinetic origin of large magnetic variations at la Fournaise volcano. *Nature* **343**, 633–36.

Zlotnicki, J., J. C. Ruegg, P. Bachèlery, P. A. Blum 1990. Eruptive mechanisms on Piton de la Fournaise volcano associated with the December 4 1983 and January 18 1984 eruptions from ground deformation monitoring and photogrammetric surveys. *Journal of Volcanology and Geothermal Research* **40**, 197–218.

# 5 Ground deformation surveying of active volcanoes

J. B. Murray, A. D. Pullen and S. Saunders

## 5.1 Introduction

In the previous chapter, instrumentation that continuously monitors the changes in ground tilt at one or more sites on a volcano was described. In this chapter, we describe the various methods by which the changing positions of large numbers of stations can be monitored by standard surveying methods. These methods include levelling, which gives relative vertical height, distance measurement and theodolite angular measurement, which give horizontal measurements of position, and dry tilt, which gives ground tilt only. Once the network of fixed stations has been measured, the detection of displacements due to ground deformation will depend upon a later re-measurement of the same stations. The deformation is then usually expressed relative to one of the stations in the network and in terms of its azimuth to another (the reference station). The reference station is normally the station furthest from the volcanically active areas, and preferably off the volcano altogether, but for the study of a particular problem it may be necessary to choose a much closer station.

It takes a finite amount of time to fix all the stations on a volcano using these surveying methods – typically from a couple of days to a month, depending upon the methods used, the size of the measuring network, the manpower available and the weather. It is therefore not usually practicable to measure the network more than once a month; financial and logistical considerations often mean that once or twice a year is nearer the norm for many volcanoes.

Surveying methods complement tiltmeters and continuous monitoring techniques very effectively in that they provide a high spatial resolution which, combined with the high time resolution of tiltmeters, can provide an overall picture of events taking place within the volcano. They are particularly suited to studies of internal structure, and also to long-term monitoring of ground deformation. As a predictive tool, they have been effective in predicting position, as well as giving an indication of the likelihood of future eruptions.

## 5.2 Measuring horizontal deformation

The traditional method of surveying a large network of stations is by triangulation, or the measurement of horizontal angles to visible points in the network. This method has been used by all national survey institutions throughout the world to create a country-wide, and later continent-wide network of control points, the positions and altitudes of which are known. The network is then scaled by measuring one or more distances between points in the network. Traditional techniques of measuring this distance were taping or chaining. Taping is only capable of low accuracy ($\pm 1$ mm + 20 ppm at best) and only short distances (around 30 m) can be measured at a time. Longer distances involve repeated set-ups and are quite impracticable for measuring volcanic deformation, as the techniques are slow and laborious, and can only be carried out in calm weather. On a small scale, however, tapes are often used to monitor the width of visible cracks or fissures in the ground surface.

The advent of the various types of electronic distance meter since World War II has meant that it is now quicker and more accurate to measure distances rather than horizontal angles in a network, a technique known as "trilateration". Distance meters have become progressively smaller and cheaper since their first appearance, and in the last few years the total station has appeared, a theodolite that also incorporates a distance meter. These have now become widely used on volcanoes to measure horizontal movements of the ground associated with volcanic activity. Triangulation only really remains a viable option where expense or inavailability of equipment are a problem, for example in Third World countries. Accuracies will depend upon the quality of the theodolite used, but will normally be inferior to those obtained by trilateration.

### 5.2.1 Trilateration

Trilateration survey determines the relative horizontal positions of points on the Earth's surface by measuring the distance between intervisible points. These slope distance measurements must be corrected to a horizontal distance using the otherwise determined altitudes of the points or the measured vertical angle of one point from another. The minimum number of such distance measurements required to fix the positions of all such points is twice the number of points minus three, with each point having at least two measurements to it.

Electro-optical and electromagnetic methods are used to measure the slope distances between points. These techniques use a known wavelength of light or electromagnetic waves through air to determine the distance between two intervisible points. Most modern medium-range ($<3$ km) instruments use infrared light, with lasers being used for longer ranges ($>3$ km). Typically, the accuracy of such an instrument varies from $\pm 3$ mm, + 2 ppm to $\pm 5$ mm + 5 ppm though improved accuracy may be possible with great care. Infrared based systems can be purchased for a few thousand pounds sterling and are also available incorporated into a theodolite (a total station), the principal advantages of such a system being that vertical angles can be measured almost simultaneously with distances. Once the instrument is set up and aimed at the reflector, a single measurement may take a few seconds and require the operator to simply push a button and record the measurement. Some systems will also record the measurement details electronically so that they can be rapidly transferred to a computer (possibly in the field) for analysis, although it is advisable to make a hardcopy back-up as soon as possible.

Electro-optical or electromagnetic distance measuring equipment is often referred to as an "electronic distance meter" (EDM). All EDMs use broadly the same principle of operation. The principle will be briefly described here as utilized in optical systems, but for a more complete description see Burnside (1991). The term "reflector" will be used throughout, though for microwave systems this would be a transponder or receiver/transmitter.

## 5.2.2 Principle of operation

The EDM transmits a modulated light signal to a distant reflector, which returns the signal to the EDM. In theory, this reflector could be a plane mirror, but the accuracy and stability of alignment required makes this impractical except under laboratory conditions for calibration purposes. Instead, a corner-cube or retroreflector is used, consisting of a glass prism with three orthogonal faces. This configuration has the property of reflecting an incoming light signal to its source without the need for accurate alignment.

A constant, known frequency modulation is applied to the amplitude of the transmitted light (a rotating plane of polarization is used instead in some instruments). During one complete cycle of modulation, the light signal covers a distance of one wavelength, $U$, given by:

$$U = C/F$$

where $C$ is the velocity of light, and $F$ is the frequency of modulation.

In travelling from the EDM to the reflector and back to the EDM, the light signal covers a distance, $D$, equal to an integer number, $n$, of wavelengths plus a fraction of a wavelength, $L$:

$$D = L + nU = 2S$$

where $S$ is the actual distance (of interest) between the EDM and the reflector. Therefore, to determine $S$ we must first establish $L$, $n$ and $U$.

The frequency of modulation, $F$, is usually supplied by an electronic oscillator. These can be made extremely stable using modern techniques. The velocity of light through air, $C$, is dependent upon the prevailing meteorological conditions, primarily the air temperature and pressure, but can be derived if these are known. For a typical system with an infrared source, $C$ will vary by about 0.7 ppm per degree celsius and 0.35 ppm per mmHg.

If $L$ is not zero then it is exhibited at the EDM as a phase difference between the transmitted and received signals. A calibrated electronic circuit can be used to add a further phase shift, $e$, to the received signal until $L + e = U$. Alternatively, the frequency of modulation, rather than being held constant, can be varied until $L = 0$. The frequency required to achieve this must then be measured.

The number of wavelengths, $n$, is determined by making gross variations to $F$, usually substantial reductions in $F$ and hence increases in $U$. In this way, several values of $L$ are determined, each associated with a known value of $U$: e.g.

$$D = L_1 + n_1 U_1$$

$$D = L_2 + n_2 U_2$$

$$D = L_3 + n_3 U_3$$

In each case, $D$ is the same, hence a set of consistent and reasonable values of the integers $n_1$, $n_2$, $n_3$, etc., can be determined (where unreasonable values will either be negative or beyond the range of the instrument). This is a potential source of gross error, particularly in older EDMs or any instrument operating at the limits of its capabilities, either because of poor meteorological conditions or because of extreme range.

### 5.2.2.1 Corrections

The distance $S$ thus determined is in fact the distance between some reference point in the EDM and a similar point in the reflector, these points being determined by their construction. What is desired is the distance between the true or useful references of the equipment, which are generally the vertical axes of rotation. The correction required to achieve this can be found in one of three ways:

(a) It can be calculated using knowledge of the construction of the EDM and the reflector.
(b) It can be determined by measuring a (short) line of known length.
(c) It can be found by making measurements of the distances between three points A, B and C such that B is a point on a line joining A and C. Hence, if $E$ is the error which will be removed by this correction, then the distance measured between A and B will be (AB + E), that between B and C will be (BC + E) and A to C will be (AC + E). Since AC = AB + BC then (AB + E) + (BC + E) – (AC + E) = E and the correction is $-E$. In practice, this method is often used since it requires no predetermined distance and can be easily carried out in the field.

Reflectors supplied with EDMs often claim not to require such a correction, but this may be good only to the nearest centimetre. While this may be adequate for construction site purposes, for more accurate work it is a possible source of error which can be removed.

## 5.2.3 Practicalities of implementation

### 5.2.3.1 Basic survey technique

The EDM is set up on a tripod above one survey marker and a reflector is set up above the other. The EDM is aimed at the reflector, either optically with a total station, or by maximizing the reflected signal strength. A series of distance measurements are then made and recorded. Scatter in the recorded measurements is assessed and further measurements made if necessary.

Other peripheral but nevertheless essential measurements are also recorded. These include air temperature and air pressure, preferably at several points along the line, but in practice in the vicinity of both markers. The velocity of light through air is dependent on these parameters and an appropriate meteorological correction can be calculated if they are known. For improved accuracy, the relative humidity can also be recorded. When measuring extremely long lines, and where accuracy is paramount, meteorological measurements can be made along the length of the line by instruments trailed from aircraft.

Once the distance measurement is corrected it represents the distance from the EDM to the reflector. In order to calculate the marker-to-marker slope and horizontal distances, the heights of the EDM and reflector tilt axes above the survey marker must also be accurately measured and allowed for.

Unless the relative altitudes of the two markers are to be determined by some other means such as levelling (see below), vertical angles also need to be measured. With a total station, this is straightforward. Otherwise, the EDM must be replaced by a theodolite and the relative heights of their tilt axes must be accounted for.

### 5.2.3.2 Accuracy

The accuracy of a distance measurement may be affected by the following factors.

(a) *Confidence in the air temperature and pressure measurements* and the degree to which these are representative of those along the entire line. This is often difficult to quantify since the measurements will generally have been made in the vicinity of the two survey markers and, hence, fairly close to the ground surface. The temperature of the air along the sight line is what is required, and this may cross varying topography such that on steep volcanoes it is in places hundreds of metres above the ground. While air pressure is unlikely to be greatly affected (except under extreme weather conditions), air temperature may drop rapidly with height above the ground surface if the temperature of the surface has been raised, either through volcanic activity or by sunlight. This effect is greatest on calm sunny days. The thermometer should therefore be erected as far off the ground as possible, but always totally in the shade, and with the bulb in a position where air can circulate freely round it. The best system is to use a tall cane with a white cylindrical shade open on one side. If this is too unwieldy to be carried long distances, then the thermometer can be suspended from the shaded side of the instrument, but care should be taken that no sunshine falls on any part of it. Most thermometers take a few minutes to equilibrate to the ambient temperature, so for total stations they should always be set up before the vertical angles are read and the latter read before the distances. Pressure is read to the nearest milliibar at both ends of the line. Problems may occur if the survey line passes very close to the ground, as the air temperature may rise above that measured close to the survey markers. Where possible, it is best to avoid sight lines which pass close to the ground when designing a trilateration network. Uncertainties will be greatest if the line crosses an active fissure or lava flow, and this may often be the case since ground deformation across such an area will be of interest. While funding or safety considerations may not permit the routine measurement of air pressure and temperature along the line using aircraft, the occasional use of this technique will allow the effect on accuracy of using measurements at the line ends to be assessed.

(b) *Stability of the frequency standard and associated electronics in the EDM* itself. This is probably best assessed by measuring a stable baseline at frequent intervals. The baseline may need to be sited some distance from the volcano in order to ensure stability.

(c) *Cyclic or delay line errors.* In modern instruments, such errors have been virtually eliminated, but for older instruments, errors of the order of $\pm 10$mm have been noted over the primary modulation wavelength. This can be checked for most instruments by moving the reflector along a graduated scale, preferably under laboratory conditions, i.e. in a stable atmosphere. The scale should be longer than the primary modulation wavelength and should be aligned along the line of sight of the EDM. After subtracting a reference distance (e.g. that measured to the zero reference mark on the scale) the EDM indicated distance can be compared with that indicated by the scale. If a cyclic error is significant it will appear as a sinusoidal type difference between the two distances and an appropriate correction can be calculated.

(d) *Errors in setting up the EDM and reflector* directly above the survey marker may also occur. Both the EDM (preferably a total station) and the reflector will usually be mounted on a tripod, except in cases where survey pillars are used or the reflector is mounted on a pole (the latter for rapid, low accuracy survey or tachymetry). The tripod will have been set up approximately over the marker to within a few centimetres, depending upon tripod design. The head of the tripod can then be aligned more accurately using an optical plummet or a centring pole. A plumb-bob is not practical for precise positioning, as wind is too frequent on most volcanoes. A centring pole is only fitted to certain types of tripod and requires that the benchmark have a small (1 mm diameter) depression in which the point on the lower end of the telescopic pole can rest. A levelling bubble attached to the pole is used to make the pole vertical by moving the tripod head. The height of the tripod head can be read from a scale on the pole (but be cautious of using this for a precise instrument height since the levelling screws are above the tripod head). An optical plummet is incorporated into the design of most total stations, and consists of a right-angled telescope within the tribrach, which is aimed at the survey marker. In older EDMs without optical plummets, it may be necessary to use a separate one. The plummet is levelled until the line of sight is vertical, then the tripod head is moved to precisely position it over the marker. A built-in optical plummet allows the setting up to be checked at intervals when the marker is occupied for a prolonged period. For precise setting up, the optical plummet is to be preferred and can be used on centring pole tripods by unscrewing the pole. Typically, a pole set up will have an accuracy of $\pm 1$ mm (but check for bent poles occasionally by comparing with an optical plummet). With care, the optical plummet can be set up with an accuracy of better than 1 mm.

(e) *Errors in the measurement of the slope angle of the line and/or the determination of the relative altitudes of the two markers.* The former will depend strongly on the weather conditions, and errors will be greatest during windy weather, or sunny weather, when differential heating of the ground will create slight differences in refractive indices of the air, causing heat shimmer and making measurement difficult.

### 5.2.3.3 Constructing a network

In the earliest days of EDM measurement on volcanoes, simple distances between markers (slope distances) were used in critical regions such as across fissures. The danger here is that in subsequent analysis, there is often a tendency to assume that change in line length means that movement has occurred horizontally along the direction of the line. However, as shown in Figure 5.1, an increase in line length across a fissure can be due to a number of distinctly different geological events. For a meaningful volcanological analysis, it is therefore essen-

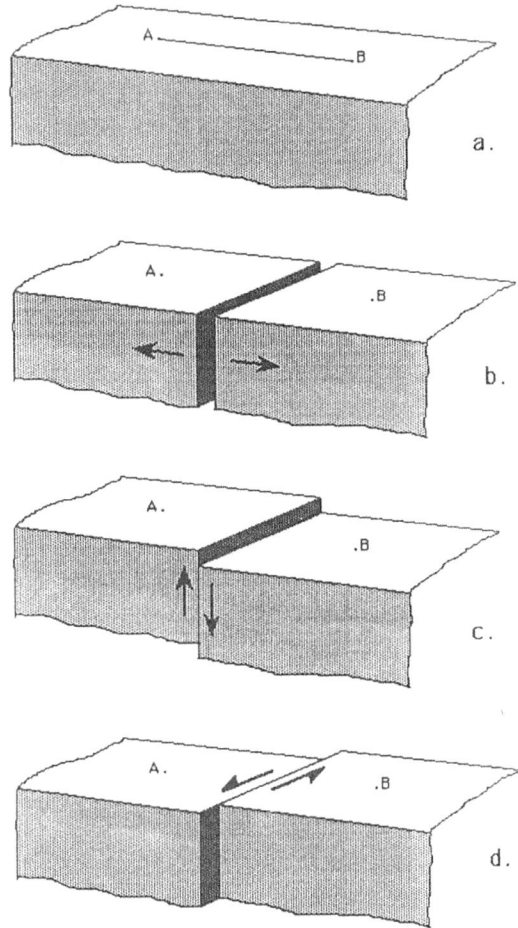

**Figure 5.1** (a) Increases in the measured line AB may be brought about by (b) fissure widening, (c) normal faulting in either direction, (d) strike-slip faulting in either direction, or a combination of these events. Increases may also be brought about by elastic/plastic deformation of the same type which does not actually cause faulting. It is therefore important to interlink a network of line measurements, so that the different types of possible movement shown in (b), (c) or (d) can be distinguished.

tial to derive vectors of movement for each point, and this can only be done by obtaining good height data, and constructing a rigid trilateration network of points.

The most basic trilateration survey network consists of a triangle, the sides of which represent measured distances. The coordinates of the three points at the corners can be determined if, say, the coordinates of one point and the azimuth of one line are defined. This definition may be arbitrary, but should be related to a wider coordinate system if possible. The coordinates of an additional point can be calculated after measuring the distance from it to any two previously determined points. In this way, the network can be extended as required. Thus the number of distance measurements required to determine the relative coordinates of $n$ points is $2(n - 3) + 3 = 2n - 3$.

However, if one distance measurement is in error, the calculated coordinates of many or all of the points in the network may be affected. Consequently, it is better to use the slightly more complex braced quadrilateral as the basic element of a network, rather than a triangle. This has six distance measurements where only five are required, hence the sixth measurement can be predicted by the other five. Such a network is said to have a redundancy of one

(6 – 5 = 1). In general, for a network of $n$ points defined by $m$ line measurements, the redundancy is $m - (2n - 3)$. (A note of caution is required here: it is possible for a network with apparent redundancy to actually contain too few distance measurements to calculate the coordinates of all of the points if the redundancy is not spread evenly over the network.) In practice, volcanoes tend to be conical in shape, so that measuring lines of sufficient length concentric to the summit are few and far between, due to lack of intervisibility. It is therefore often not possible to construct a network rigorously using only braced quadrilateral elements, but a large redundancy can still be built up by measuring all intervisible lines within the range of the instrument.

Redundancy within a network serves as a check for gross errors in the measurements and permits an estimate of measurement accuracy to be made. Since no measurement will be entirely error free, the coordinate set calculated using one non-redundant subset of measurements will differ from that calculated using an alternative subset. The entire set of measurements can be used to calculate a "best-fit" set of coordinates, where the measurement error has been methodically distributed around the network. This process is best carried out using a computer. The technique starts with an approximate set of coordinates, perhaps calculated from a subset of measurements. The distance between two intervisible points is calculated using these coordinates and compared with that actually measured. The coordinate set is adjusted to improve this comparison. This process is repeated using the resultant coordinate set as its input, until no further improvement is possible. The remaining "lack of fit" gives an estimate of the average accuracy of measurement. Provided there is sufficient redundancy, one or a few gross errors will become obvious and can be eliminated from the measurement set.

### 5.2.3.4 Streamlining procedures

In places where large movements are taking place rapidly, it may become more important to make a large number of measurements quickly, but without the need for such high standards of accuracy. During the 1983 Mount Etna eruption, for example, where movements of more than 2 m occurred, it was sufficient to hold the reflector over the station by hand without erecting and centring the tripod. In this way, a much larger number of distances could be measured in a short time to an accuracy of better than 5 cm, without moving the distance metre (Murray & Pullen 1984).

More ingenious methods were adopted during the crisis building up to the May 1980 eruption of Mount St Helens, where extraordinarily rapid rates of deformation of more than $2.5\,\mathrm{m\,day}^{-1}$ were occurring. The problem here was that many of the sites in the upper part of the volcano were only accessible by helicopter, or under threat of snow and rock avalanches or volcanic eruptions. They were also often hidden by bad weather, so that measurements had to be made in short gaps between cloud and hill fog. The most practicable solution was therefore to install permanent reflectors, so that it was unnecessary for anyone to visit the stations during measurement. Reflectors were installed at 19 locations over the upper slopes of the volcano, and were measured from five instrument sites near the foot (Lipman et al. 1981).

Because losses were likely to be high, it was decided to use ordinary traffic reflectors, 8 cm in diameter, which were mounted in arrays of seven on orange boards bolted to steel signposts. This created some reduction in maximum range for the EDM, but on the other hand

it meant a considerable saving in cost, so that the large losses that occurred were not finan-
cially important. Where sandblasting by windblown ash is likely to be a problem, plastic
reflectors will quickly become unusable. This will also eventually happen to standard glass
corner cubes, though at a slower rate.

Both horizontal and vertical angles were read onto the orange targets, and distances read
to the reflectors, so that repeated measurements would give displacement vectors in three
dimensions. Five base stations were used, but angles and distances were not read between
these. It was sufficient regularly to measure some distances to points off the volcano, and
these showed no movement. It was therefore assumed that the base stations were stable, at
least compared with the rapidly moving stations on the north flank.

Meteorological data could only be measured at one end of the line, but distances accurate
to about $\pm 1$ cm were obtained, whereas the theodolite angular measurements gave an uncertainty of about $\pm 10$ cm, too high for normal volcano surveillance, but quite sufficient in this
case in view of the large movements observed. In this way, the very important dataset of
precursory movements prior to the major event of 18 May 1980 (described later) was obtained.

### 5.2.4 Deformation from air photographs

A more complete coverage of deformation at an even lower accuracy was obtained for Mount
St Helens by photogrammetry using repeated stereo aerial photographs. This enabled five
contour maps of the volcano to be produced before the main eruption, from which the large
displacements could be measured with a relative accuracy of about 1.5 m, which is very poor
by normal standards but again quite sufficient for the very high rates of deformation observed.
This technique also had the great advantage of being a very low-risk operation to those gathering the data, so is particularly suited to extremely dangerous situations like that at Mount
St Helens (Moore & Albee 1981).

Aerial photography and photogrammetry has also been used with great success at a much
higher accuracy at Piton de la Fournaise volcano on Réunion Island in the Indian Ocean. In
this case, a detailed ground control net was laid out and surveyed with theodolite and EDM.
The survey markers were surrounded with 20-cm diameter painted circles so that they would
be visible from the air. Low level helicopter flights were used to get very high resolution air
photographs in which the survey markers can be seen.

Using this technique, Zlotnicki et al. (1990) were able to obtain accuracies of a few centimetres from photos taken before and after the December 1983 and January 1984 eruptions.
The advantage of the technique is that a displacement vector is obtainable for any point visible on the air photos. In the case of the 1983 and 1984 eruptions this was particularly useful,
as most of the major deformation took place in an area where there were no stations and access was difficult. The disadvantage is that some of the ground control points have to be re-
surveyed each time air photographs are taken, so that the technique is time consuming and
very costly, and is really only practicable for a large, generously funded organization. Details of aerial photogrammetry technique can be found in standard textbooks such as those by
Wolf (1983) or Burnside (1985).

## 5.3 Measuring vertical deformation

### 5.3.1 Levelling

Levelling is the most accurate method of deriving changes in relative heights on a volcano. It was first used in this way by F. Omori in Japan, to investigate the 1910 eruption of Usu volcano (Omori 1913) and the 1914 eruption of Sakurajima volcano (Omori 1914), but did not come into use as a regular monitoring tool until 1965 at Kilauea volcano, Hawaii (Fiske & Kinoshita 1969). In its simplest form, levelling involves the use of a vertical graduated measuring staff and a level, an instrument consisting of a telescope with a central cross hair set up to read exactly horizontally. Figure 5.2 illustrates the sequence of operations involved in levelling between two fixed markers on a volcano. This diagram deals with ordinary levelling, in which accuracies of 10 mm over 1 km are obtainable.

Levelling accuracy $a$ is normally expressed as $\pm a\sqrt{K}$ mm, where $K$ is the length of the loop levelled in kilometres. Thus an error of 10 mm over 1 km translates into 3 mm over 100 m, or 32 mm over 10 km. On many volcanoes, a higher accuracy is desirable for deformation work, so some or all of the more rigorous methods of precise levelling are adopted.

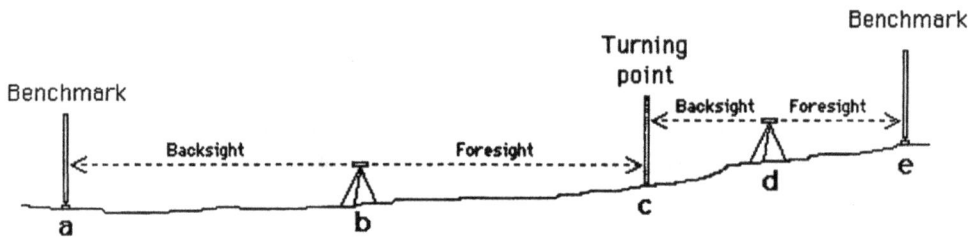

**Figure 5.2** Diagram to show levelling procedure. The staff is set up over the reference benchmark (a), and the level at (b). The value on the staff at the horizontal cross-hair is read and noted; this first value is the "backsight". The staff is then moved forward to a temporary position (c), known as a "turning point", where it is placed on the ground, not on any fixed station, and the staff value read and recorded again. This forward observation is called the "foresight". The instrument is then moved to position (d), and the staff backsight to (c) read again. Finally, the staff is set up over the second benchmark (e), and the foresight read. The difference in height of (e) relative to (a) is the sum of all the foresight values minus the sum of all backsight values. In this example, only one staff turning point is required at position (c), but in practice any number of turning points can be used between the two permanent benchmarks. It is clear from the diagram that speed of levelling depends upon the steepness of the slope as well as the distance between the two points to be measured, as a greater number of temporary staff turning points will be required.

### 5.3.2 Precise levelling

Major sources of error in ordinary levelling may be summarized as follows:
(i) mistakes in reading the staff
(ii) the spirit level bubble going off centre
(iii) the instrument being out of adjustment

(iv) atmospheric refraction in the air layers near the ground;

 (v) the staff not being vertical;

(vi) settling of the staff into the ground; and

(vii) settling of the tripod into the ground.

Error source (i) can only be corrected by using a modern digital level (see below), otherwise mistakes are bound to occur and can be detected by double-running the traverse. Factor (ii) is a problem in manually levelled instruments, and can sometimes be avoided by using a parasol to stop differential heating by the sun. However, for levelling traverses on active volcanoes, a self-levelling instrument is virtually essential because of the necessity for speed, and will correct small changes in horizontality of the instrument. The other error sources will introduce small and variable inaccuracies into the measurements, and precise levelling involves a number of additional procedures designed to eliminate or greatly reduce these sources of errors.

Precise levelling was defined by the International Geodetic Association in 1912 by the following accuracies:

$$\text{Probable accidental error: } < \pm 1\sqrt{K}\,\text{mm}$$

$$\text{Probable systematic error: } < \pm 0.2\sqrt{K}\,\text{mm}$$

These high accuracies are obtained by the following refinements of the procedures illustrated in Figure 5.2.

(a)  A higher precision instrument with a more powerful telescope and a micrometer are used. This considerably cuts down error source (i) above.

(b)  The foresights and backsights at each position of the level are made equal, i.e. in Figure 5.2, distances (ab) and (bc) are made equal, and so are (cd) and (de), but (bc) need not be equal to (cd). This cancels out error source (iii), and reduces (iv).

(c)  The instrument is regularly field-tested during levelling, to check that it reads level to within tolerance limits, to reduce error (iii).

(d)  A metal baseplate, kicked firmly into the ground, is used as a temporary station to support the staff at all turning points. This eliminates much of error (vi).

(e)  A spirit level is attached to the staff so that it can be kept vertical, greatly reducing (v).

(f)  The traverse between each benchmark is run twice, once in each direction, and the mean height difference used. Systematic errors (vi) and (vii) will therefore tend to cancel out. This procedure also gives an idea of precision in the form of a closing error, and facilitates the detection of blunders in reading or writing down staff values.

(g)  A low expansion invar, non-folding rod is used to increase accuracy. It is regularly calibrated to check for gradual change in length.

(h)  Air temperature is regularly recorded so that any expansion of the staff invar strip on which the graduations are marked can be allowed for.

(i)  The staff is not read closer than 1 m to the ground, so that the light path will not pass through the stronger refraction in the air layers near the ground.

(j)  Sighting distances on to the staff are limited to less than 30 m, as visibility becomes steadily poorer on longer sights.

*5.3.2.1 Speed: streamlined levelling*

An active volcano may be measurably deforming all the time, so a levelling traverse intended for deformation measurements should be occupied in the shortest time possible, and every effort made to streamline the procedures described above. The simplest way of doing this is to have more than one levelling team operating; with two teams the United States Geological Society (USGS) is able to occupy 35 km of levelling lines on Kilauea volcano in 2 days (Decker & Kinoshita 1972). The most time-consuming procedure is (f) (see above), since this effectively doubles the time taken and therefore the cost, (i) can increase the number of set-ups by 30% on steep slopes, and (j) may have a similar effect on shallow slopes, and both (b) and (h) will add considerably to the time taken for each set-up.

Considering the fact that a volcano may be deforming all the time, the practice of double running levelling lines is of dubious value, since it can never be certain whether the closing errors are due to levelling errors or to the deformation of the volcano. Double running may in fact confuse the analysis by doubling the time taken, and therefore doubling the likelihood of deformation taking place during measurement. The detection of gross errors in a deformation survey is not dependent upon double running, as they will show up as abrupt changes in derived height difference from the previous occupation. Any odd values which do not fit in with the overall pattern can be checked by repeating the measurements between the two benchmarks concerned.

Many volcanologists therefore prefer not to double-run levelling lines, but if this decision is taken, it is important to make sure that the lines are run in the same direction each time, and that the same procedures are followed with the same equipment, preferably by the same people. In this way, the systematic errors will remain about the same, and so will not show up in a deformation study. The traverse should ideally be closed, or at least contain some long closed loops, so that an estimate of reliability can be obtained.

Equalizing foresight and backsight distances can be time consuming if carried out correctly with a measuring tape. Where their levelling traverses follow paved roads, the USGS in Hawaii have therefore measured and marked all instrument positions with spray paint, and installed nails for every turning point, thus eliminating the need for a base plate. Unfortunately, not many volcanoes have paved roads to the summit. A more common way of saving time on distance measuring is for distances to be paced by the staff holder, which is far quicker than tape measuring, though clearly less accurate. If pacing is used in this way regular field checking of the instrument ((c) above) becomes all important, unless a digital level is used as described later. Taking temperatures to correct for thermal expansion of the staff is only necessary on steep volcanoes, where the traverse includes vertical height differences of several hundred metres. Under these circumstances, errors of a few millimetres may accrue under extreme conditions if this is not done.

Much time can be saved by using additional manpower, a recorder to note down the numbers and an arithmetician with a calculator to check for errors will both speed things up considerably, and the simple act of running rather than walking between set ups can save up to 30 seconds each time. This means saving a day's work on a 1000-set-up traverse such as that on Mount Etna. Great economies in time and expense can also be achieved by using modern levelling instruments. Self-levelling levels are virtually essential for long levelling networks on a volcano, as they eliminate the time-consuming procedure of adjusting the fine level bubble, but the appearance on the market of the digital level has provided an even more fundamental revolution in streamlining.

### 5.3.2.2 The digital level

As yet (1992), only Leica manufacture digital levels: the Wild Na2000 series. They employ an image-recognition system which reads a special barcode staff automatically. Because of the unique pattern of the vertical barcode on the staff, the scanning system within the level is able to detect where the centre of the cross-hair lies against the staff to an accuracy of 0.1 mm. This means that this value is read and displayed on a liquid crystal display (LCD) screen without the surveyor being required to do anything more than set up, point and focus the level and press the measuring button. The advantages of this system are enormous. Blunders due to incorrect reading or noting down numbers are eliminated. The instrument contains a series of internal programs, and a data logger can be attached, so that a levelling traverse can be run without anything being required to be written down, once the reference height of the first benchmark has been entered at the start of levelling.

The instrument is remarkably consistent, even reading through heat shimmer or wind vibration, and with an invar staff it is capable of accuracies of $\pm 0.8\sqrt{K}$ mm, well inside the specifications of precise levelling. If it goes off level, an error message appears and reading is not possible, so that error (ii) above is eliminated. It also measures and displays the distance to the staff, as well as the excess distance of backsight over foresight readings. This means that if there is a systematic error in foresight over backsight distances paced by the staff holder, it can be corrected in the field at each benchmark. The readings are made remarkably quickly, between 2 and 5 seconds on average, depending on distance, wind strength, light intensity and atmospheric turbulence.

The result is a considerable decrease in time required for levelling, as well as manpower. For example, the 25 km levelling network at the summit of Mount Etna volcano, which used to be measured with a Zeiss Ni2 self-levelling level, took 75% of the time and half the manpower (two people instead of four) with the Wild Na2000 digital level. The saving in time is greater when similar manpower is compared: the 10-km traverse on Colima volcano, Mexico, was measured by two people with a Zeiss Ni2 until 1991. In 1992 the same traverse took 40% of the time with the Na2000 digital level. Although the instrument at present costs about 2–3 times more than an equivalent self-levelling level, this difference can often be regained in a single occupation by the savings in time and manpower.

### 5.3.3 Trigonometric levelling

Where accuracy does not need to be of the first order, and cliffs, steep slopes or ground unsuitable for levelling (such as an Aa lava flow) need to be crossed, trigonometric levelling may be the most suitable technique. Trigonometric levelling is simple in principle, and consists of measuring a vertical angle and a distance between two points, and thus deriving the height difference between them by trigonometry. In practice, the phenomenon of atmospheric refraction means that true vertical angles cannot be measured from a single station, and the curvature of the Earth complicates the trigonometry. Atmospheric refraction varies considerably with time according to the changing temperature, pressure and humidity, and thus to correct for it satisfactorily the vertical angle has to be read from both points simultaneously.

Vertical angles need to be read with a 1″ instrument. The appearance on the market of the total station has greatly facilitated trigonometric levelling, as only two instruments are now needed, instead of two theodolites and an EDM. The procedure is to set up a theodolite over one point to be measured, and a total station over the other. Thermometers are set up in the shade by each instrument as described above under trilateration. The heights of the optical axis above the benchmark are then tape measured. A series of vertical angles is then read, usually two to four pairs of face-right and face-left readings, from both ends of the line simultaneously. The centre of the theodolite object glass is used as a target, under some circumstances it may be necessary to clamp on a concentric target round this (Fig. 5.3) to make it visible. The use of radios is highly desirable in order to synchronize the readings, otherwise a system of hand signals can be used.

Figure 5.3   A total station in use on Mount Etna, Sicily, during the 1991–3 eruption. Note that a target concentric to the objective of the instrument has been affixed, so that simultaneous reciprocal vertical angle measurements can be made between this station and a theodolite at the other end of the line being measured.

Once the vertical angles have been read, the theodolite is separated from its tribrach and a reflecting prism mounted. The details of distance measurement are similar to those already described in Section 5.2.3. The total station is pointed at the centre of the prism and the distance read. Normally 6–10 readings are taken, half on face-left and half on face-right. The displayed distances are significantly dependent upon temperature and pressure, so these are measured twice at both the theodolite and the total station position, once during the first distance measurement and once during the last.

The accuracy of the measurements is greatly dependent upon the state of the atmosphere. Conditions are worst on sunny days, as the temperature difference between the hot air near the ground surface and the higher, colder air causes turbulent air currents, and the differences in refractive indices of warmer and cooler air causes the image to move about or blur. This phenomenon is known familiarly as "heat shimmer". Wind, a frequent problem on volcanoes, can cause the instrument to vibrate irregularly, and reduce the accuracy of the measurements. Conditions are best on calm, cloudy days, provided that the clouds are above the volcano, and are not obstructing the sight line.

After measuring angles and distances between two points, one of the instruments, for example the theodolite, stays put whilst the total station moves on to the next benchmark in the traverse. After this next line has been measured, the theodolite then moves ahead of the total station on to the next point. The two teams must progress in this "leapfrog" fashion to avoid accumulating errors due to slight differences in tape-measure length, or in instrument-height constants. Benchmarks for trigonometric levelling are best sited about 500 m apart, 1 km being the maximum, as poor resolution lowers precision to an unacceptable level over longer distances.

Precision normally varies between about $\pm 5\sqrt{K}$ mm and $\pm 10\sqrt{K}$ mm, depending upon conditions, and with practice the speed of levelling can be brought up to between 1 and 2 km h$^{-1}$,

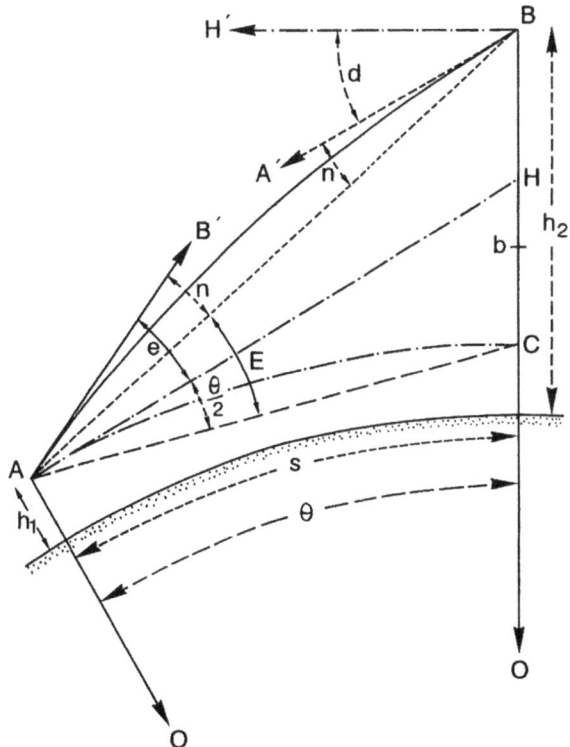

**Figure 5.4** Diagram of the geometry of trigonometric levelling. A and B are two points of height $h_1$ and $h_2$, respectively, above mean sea level, whose difference in height $(h_2 - h_1)$ is required to be known. O is the centre of the Earth, and C is a point of equal distance to A from O. Angle AOC = θ, and the arc AC is parallel to mean sea level. The tangent AH is the horizontal passing through A, and BH′ is the horizontal through B. The arc AB shows the path of light from A to B curved by atmospheric refraction, so that B appears from A to be in a direction B′. The angle of elevation of B observed from A is therefore HAB′ = e. It exceeds the angle HAB by n, the angle of refraction. Correspondingly, BA′ is drawn tangential to the arc BA, thus H′BA′ is the observed vertical angle of A as seen from B, the angle of refraction is again A′BA = n.

depending upon the terrain and the spacing of benchmarks. This means that under optimum conditions, it is possible for one team to level 15 km of traverse in a day with a precision of ±20 mm.

The geometry of trigonometric levelling is quite complex, as can be seen from Figure 5.4. However, reducing the observations can normally be done using a formula which simplifies the situation. In Figure5.4,

$$E = e + \theta/2 - n$$

and the height difference between A and B is

$$h_2 - h_1 = AB \sin E \sec \theta/2$$

These formulae are not quite mathematically rigorous, and neither $n$ nor $\theta$ are directly observable. $\theta$ can be derived approximately from the slope distance measurement, assuming plane geometry, and $n$ can also be similarly derived from differences in the observed reciprocal vertical angles from A and B. Strictly speaking, refraction will cause the distance AB to be measured along an arc, and the vertical detected by the instrument may be slightly off-true due to the gravitational attraction of the volcano, but for deformation purposes, we are not interested in exact values of height difference but in how they change, and for these purposes the above formulae are normally sufficient, as long as distances between stations are within the limits specified above. For a fuller discussion, refer to a standard textbook such as that by Bomford (1980).

## 5.4 Dry tilt

"Dry tilt" is the colloquial name given to a method of measuring changes in tilt of the ground which employs the same optical instruments as levelling. The method was developed in Iceland in 1966 (Tryggvason 1968) and by the USGS in Hawaii in 1968 (Kinoshita et al. 1974), where it came into use as a quicker and less troublesome technique than the use of portable water tube tiltmeters, colloquially known as "wet tilt" from the end state of those making the measurements. The levelling technique came to be known as "dry tilt" by way of contrast, and although some authors have used more formal appelations such as "telescopic spirit-level tilting" (Kinoshita et al. 1974), "spirit-level tilting", "tilt-levelling", "single set-up levelling" or "optical levelling tilt", the cumbersome nature of these descriptions, the lack of semantic rigour and the inability of the volcanological community to come up with a widely accepted alternative make "dry tilt" a preferable term. It has the advantage of being short and universally understood, and is now used in other languages as well.

Dry tilt stations must always be regarded as a poor second to levelling, as they provide no information on altitude changes at widely separated parts of the volcano. Instead they provide a measure of ground tilt only at individual sites. If, for example, ground tilt at such sites around the volcano is always tilting away from the summit, it is assumed that the summit is rising, but such a deduction has to be based on the assumption that the intervening ground between the dry-tilt stations is behaving in the same way, and this may not be the case, particularly in highly faulted terrains. The advantage of dry tilt stations is that they are quick to measure, so that they can be occupied several times a day during periods of crisis.

In its simplest form, a dry tilt station consists of three benchmarks arranged in a roughly equilateral triangle with sides of about 50 m. The level is set up in the centre, and the staff placed on each of the three benchmarks in turn, and their relative altitudes measured to a high precision. Next time the station is measured, any tilt of the ground in the intervening period will show up as a slight difference in relative altitudes of the three benchmarks. Three benchmarks are the minimum, but it is preferable to have additional benchmarks so that there is redundancy in the reductions and the standard error of the tilt can be calculated. At stations where tilts are habitually small, six benchmarks are preferable, as in the analysis this gives two independent measurements of ground tilt.

Because of the limited width of dry tilt stations, measurements need to be as accurate as possible. A higher precision level is therefore normally used for dry tilt work, and several separate readings taken of each benchmark. After each of the benchmarks has been read in this way, a final series of readings is taken on the first benchmark again, so that the tripod settling or instrumental drift can be allowed for. Cloudy, still days are used for measurement if possible, as air turbulence is at a minimum, and accuracy considerably improved. If this is not possible, then readings are made early in the morning. When using the level in sunlight it should always be shaded with a parasol, as heating on one side of the level causes differential expansion which can make the instrument continually drift off level during the observation period. The level is centred over a central marker to within 10 cm, so that sight lengths to the benchmarks are effectively equal, cutting down any error due to maladjustment of the instrument. The staff is supported by stays to keep it vertical and eliminate the small movements of the staff holder.

Such dry tilt measurements are not as sensitive as automatic tiltmeters, but dry tilt stations have a broader base than most tiltmeter designs and therefore show more stability, particularly in the long term. They are also very much cheaper, require no maintenance, and can be occupied in 5–15 min, depending upon the number of benchmarks, so that 20–30 such stations can be set up all over the volcano, and occupied in a couple of days, provided that there are sufficient vehicle tracks. The results of such measurements can give a good indication of the state of the volcano, and the extent of any zones of inflation or deflation.

## 5.4.1 Installing a dry-tilt station

A site for a dry tilt station requires a flat area of ground with a maximum slope of less than 3° and a minimum width of 50 m for optimum design. Three sites for benchmarks are chosen which should be as close as possible to an equilateral triangle, or a square for four benchmarks, a regular pentagon for five benchmarks and so on, but sites with stable ground must be chosen as benchmark stability is all-important. A movement of 1 mm due to an unstable benchmark in a levelling traverse is not important, as it is close to the measuring accuracy, but such a movement in a dry tilt station will show up as a false tilt of 20 μrad. Benchmark sites should therefore be well chosen and well constructed so that local movements do not dominate the tilt signal. Guidelines on benchmark installation are described later.

A permanent marker for the instrument is installed at the exact centre of the triangle of benchmarks. This is a simple matter where only three benchmarks are used, as there is always an exact centre to any triangle, but with more than three benchmarks, problems of station design are encountered, as suitable sites for benchmarks may not exist at the required distances.

Another method of occupying multi-benchmark dry tilt stations that avoids this problem is to level in the ordinary way from benchmark to benchmark in a double-run traverse (Tryggvason 1968) or a closed loop (Wadge 1976), measuring equal foresights and backsights. The advantage of this system is that the station can be made much larger and so more representative of broad tilt, and does not have to be on flat ground. The disadvantage is that it takes longer (30–40 minutes for a 300 m loop) and there may be no corresponding improvement in accuracy. The dry tilt technique was developed in Hawaii and Iceland, where volcano slopes are generally very shallow. It does not always translate so well to steep volcanoes, where flat areas may be few and far between, and not necessarily in the places where they are needed. If this is the case, a levelling loop may be the only possibility.

## 5.5 Benchmarks

Great care should be taken in the installation of benchmarks, to make sure that superficial soil movements are not included. Road and footpath surfaces are subject to varied and unknown loading and should always be avoided for dry tilt benchmarks or other stations where the highest stability is essential. Benchmarks should be installed in old, solid bedrock where possible. If such bedrock is available, the benchmark need not be very elaborate; specially constructed surveying nails can be hammered in where rock hardness will allow, or holes drilled and nails or short metal rods cemented in. However, suitable surface rock outcrops are not numerous on many active volcanoes, but if soils are not too thick a hole can be dug down to the bedrock, and a vertical metal rod secured to it (Fig. 5.5), the soil then being replaced to leave the benchmark at ground level. In the absence of solid bedrock, benchmarks are often installed on lava flows, but caution should be exercised here. Many lava flows are fractured and consist of large slabs, which seem to tilt in response to temperature or other local effects, particularly when the flows are only a few years old. This may render them unsuitable sites for dry tilt work.

In some cases, ash deposits may give a more representative deformation than lava flows, provided that the benchmark is well anchored. In Hawaii, 1 m of metal rod hammered into ash will usually give a stable station, but in some equatorial areas where soils are very thick, and rainfall and vegetation growth are high, a 3 m length of rod is necessary. This normally comprises three 1 m sections which screw into each other, so that each can be hammered in from a convenient height in turn. For extra stability, the near-surface part of the rod can be isolated from the unstable top soil with sand and a vaned sleeve, the inside of which is greased

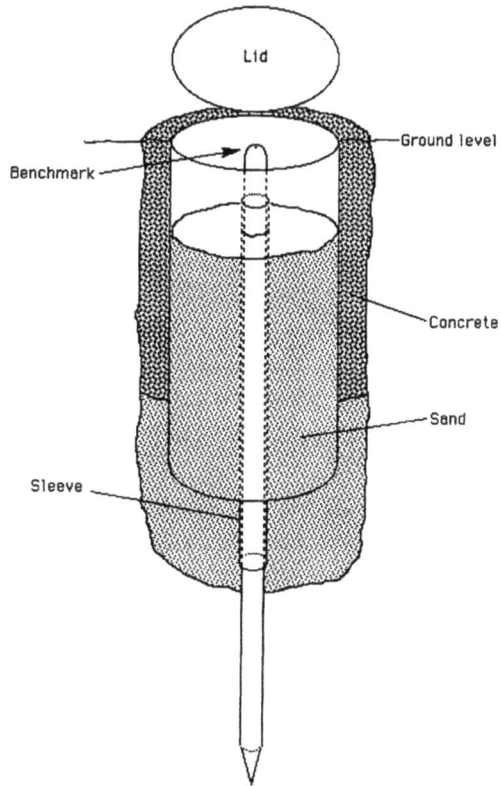

**Figure 5.5** Design of a stable benchmark in soil or volcanic ash. The benchmark itself is of stainless steel on top of a spiked metal rod 1–3 m long, which is driven into the bottom of a pit dug in the ground. The metal rod is then greased and surrounded by a sleeve, and the pit partially filled with sand. An outer casing with a lid can be added for extra protection, and the interior of this filled with sand to just below the top of the sleeve, and the outside back-filled with concrete. Any superficial soil movements can then move the casing and the metal sleeve, but the greased interior will allow the sleeve to slide up and down the metal rod without moving it.

so that it can slide easily up and down the rod with the top soil, leaving the rod itself stable (Fig. 5.5).

The shape of the actual benchmark is critical, as it is important that the same stations should be used for different techniques. Distance measuring stations require a small hole to be drilled at the centre, so that the instrument can be precisely centred over it with sub-millimetre precision. Such benchmarks should not have a flat surface, or grit and dirt may accumulate on top. Ideally, the benchmark itself should be about 2 cm in diameter, with a well domed top and a deep central hole 1–2 mm wide.

For dry tilt or similar accurate work, stainless steel or other low corrosion metals are best for the benchmark top, though any bright and shiny material will tend to attract vandalism (see below). If possible, the benchmark should be recessed and covered with a lid (Fig. 5.5) to keep corrosion to a minimum, though the lid must be wide enough to accommodate the widest staff bases. Corrosion of benchmarks varies greatly from volcano to volcano, according to the composition of gases emitted. On Poás volcano, Costa Rica, steel nails nearly 1 cm thick had virtually disintegrated on the up wind side of the crater rim after only 1 year, whereas on Mount Etna, similar nails have shown a maximum of 1 mm corrosion after 20 years, and some nails of the same type on Colima volcano, Mexico have shown no corrosion in 10 years. Where dry-tilt benchmarks suffer corrosion, this may be unimportant as long as the benchmarks are of the same type, and as long as it takes place at a slow rate. Corrosion is then

131

likely to lower the altitude of each benchmark by a similar amount. Slight variations between benchmarks are not likely to be important if corrosion is only progressing at a rate of 1 mm in 20 years.

Vandalism is a perennial problem, well known to all those who have attempted deformation surveys. Sartorius von Waltershausen (1880) described how, during the first topographic survey of Mount Etna in 1836–42, Sicilian peasants mistook his survey beacons for markers of buried treasure, and not only dug up his benchmarks, but sometimes continued to dig beneath to a depth of several metres. Most volcano deformation surveyors will sympathize with the frustration that Waltershausen must have felt. The only differences are that today the problem is most acute where numbers of tourists are highest, and appears to be motivated by idle curiosity and bedevilment.

The problem is normally dealt with by one of two main approaches. Either the benchmark is made as large and immovable as possible, or else the benchmark is made small and inconspicuous, or designed to look uninteresting or unattractive. As an example of the latter, when the first levelling traverse was set up across the summit of Mount Etna in 1975 (Murray et al. 1977), most of the benchmarks consisted of ordinary steel levelling nails hammered into rock next to the main tourist path to the summit. After only 9 months, 26% of the nails had been removed by tourists. The following year, the nails were beginning to become dull and discoloured, and only a further 9% were taken. After the nails began to go rusty the year after, a further 6% of nails were vandalized in the ensuing 15 years. It is clear that the tourists were only attracted by the nails when they were new and shiny. Since that time nails have, where possible, been installed a few metres away from the path, and hidden beneath stones until the surface has rusted. This has reduced the incidence of vandalism to negligible proportions.

An opposite approach has been adopted by many organizations; the USGS use prominent 7 cm brass disks for many of their geodetic stations, advertising a fine of $300 for removal. The success of such a deterrent depends very much upon the attitude towards authority prevailing in the country concerned, and there is no doubt that a small element within the community would always regard such an advertisement as a challenge. A small notice with a short description of the purpose of the station and its importance to local welfare might be more successful in some cases. Where budget and time restrictions allow, recessed benchmarks covered with a lid (Fig. 5.5) are ideal; the lid can then be made unattractive or hidden beneath stones, etc.

The disadvantage of making inconspicuous or hidden benchmarks is that other workers on the volcano using different methods will not be able to find them easily, and so install their own. The proliferation of markers has reached catastrophic proportions at many volcanoes, and can only lead to confusion and redundancy. It is far better that one network of well-constructed benchmarks be used by all workers, and that any additions to this network are necessary, and carried out with full regard to the benchmarks that are already there.

To avoid problems in finding stations, widely advertised station descriptions should be held by the local organization responsible for the volcano. A station description should always include an accurate grid reference, a description in words referring to local landmarks, a detailed map with sufficient local reference points, a description of the size and type of benchmark, and two photographs, one showing the general area from about 100 m distance, the other closer (from about 5 m). Regarding station designations, it is better not to use numbers or letters, as repetition by different workers from different disciplines working at different times or on different parts of the same volcano is likely. Simple, short names are easier to remember, unlikely to be repeated, and more fun to invent and use.

## 5.6 Designing a deformation network

Every volcano is different in size, shape, eruptive style, chemistry, type of erupted products, frequency of eruptions, volumetric rate of magma emission, depth and size of magma storage space, geometry of eruptive conduits, tectonic situation and many other characteristics. Each of these parameters will affect the deformation regime, and therefore have a bearing on the most suitable design of monitoring system. As well as the purely volcanological questions, logistics may have a critical design input. Accessibility, availabitity of electric power, access roads, vegetation and weather are all vital here. Clearly a volcano that is surrounded by impenetrable rain forest, or is covered for all or part of the year by a thick ice cap, will be severely limited in the choice of monitoring system practicable. Another major constriction is budget. With unlimited funds it would be a simple matter to saturate the volcano with appropriate networks of stations, but in practice the vast majority of active volcanoes have little or no funds set aside for monitoring, and the volcanologist is reduced to trying to guess internal structure or future eruptive activity with the bare minimum of information.

Most deformation networks have in the past been designed on the assumption that there is a shallow magma chamber beneath the volcano, and that the Mogi model (see below) adequately describes the deformation regime. This means that the volcano inflates prior to eruption in response to pressure increase within the magma chamber. Maximum vertical deformation will be above the magma chamber, which is usually beneath the vent region. Figure 5.6 illustrates the situation during an inflation episode, and shows the different responses of different parts of the volcano to the same event. Despite the maximum vertical deformation at the summit, tilt is zero here, and reaches maximum values on the flanks. The same is true of the horizontal displacement when referred to an outside reference, but in practice it is changes in line length on the volcano, i.e. strain, that will be measured. The strain pattern is complex, with maximum line extension occurring at the summit during inflation, but contraction of radial lines taking place lower on the flanks.

The best site for a tilt station or tiltmeter under this kind of deformation regime would therefore be near the point of maximum inclination on the flanks, whereas a network measuring horizontal deformation would be best placed across the summit. A horizontal deformation network on the flanks would be highly ambiguous in interpretation of this kind, for example, a network at point A in Figure 5.6 would show extension during an inflation episode, whereas a network at point B, a short distance away, would show contraction.

This diagram assumes a point source of deformation in an ideal situation. In practice the source of deformation will be of finite dimensions and irregular shape, the topography of the

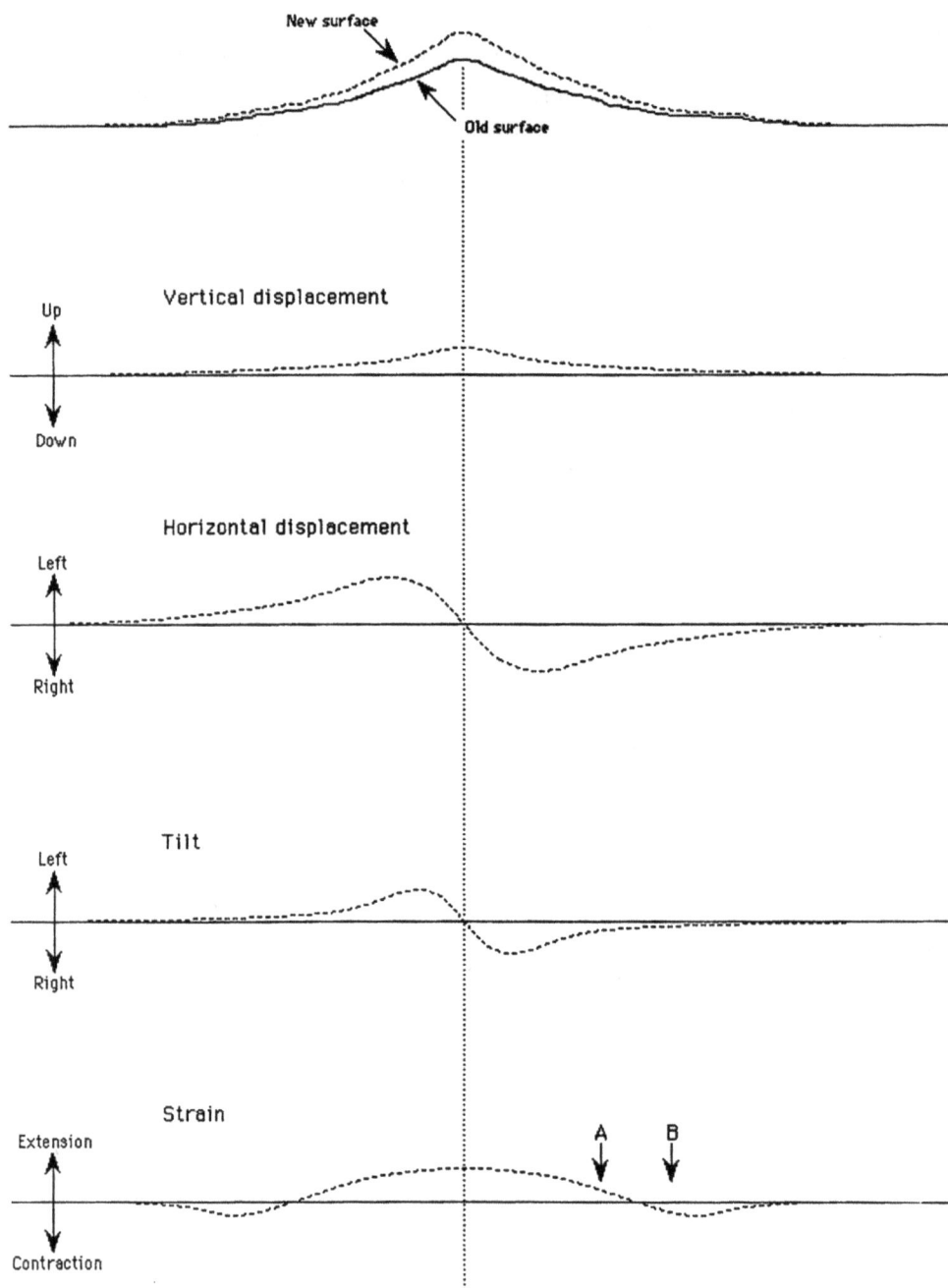

**Figure 5.6** Diagram of the response of various measured parameters to an inflation episode on a volcano (top). Vertical displacement reaches a maximum over the centre of uplift. Horizontal displacement is greatest on the flanks, as is ground tilt, which is zero above the uplift centre. Extension is greatest across the summit, but contraction may occur lower down the flanks.

volcano will contain many irregularities and complexities, and the response of the material is unlikely to be purely elastic. Each of these factors may cause important departures from ideal behaviour.

Leaving these questions aside, the main aim of a deformation network is to provide three dimensional information on the change in position of as many benchmarks as possible in as short a time as possible. The benchmarks should be widely distributed over the volcano, preferably with some benchmarks off the volcano altogether, or at least well outside the zone of significant deformation.

The best way to achieve this information rapidly is with a trilateration network that also includes trigonometric levelling. This means that precise horizontal positions can be derived, and vertical heights at the same time. When two such surveys have been carried out, a vector for each point is then calculated, using the techniques described earlier under trilateration. Such a network will give uncertainties in horizontal position of about 5 mm over 1 km, and rather more for vertical height. To achieve this accuracy in vertical height, it is necessary that the stations in the network be situated about 0.5 km apart, with a maximum of 1 km. At the same time, the horizontal positions will be greatly improved if several lines of maximum length (3–5 km, depending on instrument, conditions and number of reflectors used) are also included, where such points are intervisible.

The accuracy in vertical heights can be greatly increased if some of the stations in the network can be included in a precise levelling traverse. This is usually only practicable along a road or track, but if such a traverse can link opposite ends of the network, heights over the whole network will be improved.

Regarding the measurement of distant base or reference stations, this is often not practicable by the above techniques when the volcano is large and when time, personnel, equipment or funds are limited. In such cases, it is best to have two or more stations in the network which are also monitored as Global Positioning System (GPS) stations. Tying together GPS and conventional surveying is a complex business (see e.g. Seeber 1993), in that the two refer to different reference systems, but once the link is established, the movements of all stations relative to points distant from the volcano can be observed.

If the volcano is sufficiently well covered by trilateration and levelling networks, then dry tilt stations become somewhat redundant in the analysis. However, the fact that they can be occupied much more rapidly (several times a day if necessary) means that it may be advisable to include some in the network to follow the evolution of deformation in more time detail during crises. However, this should only be necessary if there is a lack of continuously recording tiltmeters on the volcano, or if tiltmeters have been recently installed and are therefore settling in.

## 5.7 Analysis

Deformation records of a volcano may reveal a number of different types of behaviour, which in turn may reveal different aspects of its internal structure and future activity. Each of these different types needs to be analyzed in a different way, so diagnosing the type of behaviour observed is critical, and is easiest on volcanoes which have large and comprehensive deformation networks, and therefore most data available. Here we describe four types of ground deformation behaviour which have been encountered on volcanoes: firstly, the rise and fall of the ground associated with the inflation and deflation of a magma storage space; secondly, the permanent deformation of the surface consequent upon the injection of a dyke beneath; thirdly, compaction and subsidence of the ground caused by lava loading and topography; and finally, movements on steep volcanoes caused by slope creep and slope instability.

### 5.7.1 Magma chamber inflation and deflation

The work of Mogi (1958) was a turning point in volcanology, in that for the first time a mathematical model was presented which adequately described the deformation pattern at several volcanoes. It allowed the depth of the source of deformation (presumably the magma chamber) to be quantified. The model assumes that the Earth's crust consists of an infinite elastic half-space, and that the source of deformation is small and spherical, and exerts hydrostatic pressure on the surrounding rocks. Whilst none of these assumptions strictly apply, many volcanoes show deformation patterns close to the Mogi theoretical model, suggesting that, to a first approximation, they behave elastically and have small, roughly spherical magma chambers.

Figure 5.7 shows the geometry of the model. Mogi found that, to a first approximation,

$$\Delta d = \frac{3a^3 Pd}{4\mu\left(f^2 + d^2\right)^{3/2}}, \qquad \Delta h = \frac{3a^3 Pf}{4\mu\left(f^2 + d^2\right)^{3/2}}$$

where $a$ is the radius of the source sphere; $P$ is the change in hydrostatic pressure in the sphere; $f$ is the depth to the centre of the sphere; $\mu$ is Lamé's constant; $d$ ($\equiv$R) is the radial distance at the surface from point A; $\Delta d$ is the radial horizontal displacement of a point at the surface; and $\Delta h$ is the vertical displacement of a point at the surface.

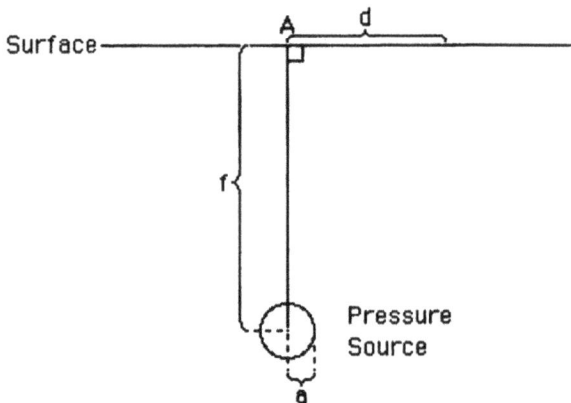

**Figure 5.7**   Diagram of inflation described in Mogi's (1958) model. A spherical pressure source of diameter $a$ and depth $f$ exerts hydrostatic pressure on a semi-infinite elastic body representing the Earth's crust. $A$ defines the point on the surface vertically above the pressure source, and $d$ the distance to a point the displacements of which are described by Mogi's formulae.

Mogi's analysis of the deformation data available for Sakurajima and Kilauea volcanoes (see below) showed a behaviour close to that expected from the model. Later adaptations and extensions of Mogi's model have sought a more rigorous analysis, or have extended it to include shapes other than spherical for the source region (e.g. Yokoyama 1971, Walsh & Decker 1971, Dieterich & Decker 1975, Ryan et al. 1983), but the simplicity of his model means that it is still widely used on volcanoes today, particularly where an absence of data does not justify the use of a more complex model. We will look at examples from three volcanoes (Fig. 5.8) where Mogi's model has been applied.

### 5.7.1.1 Sakurajima volcano, Japan

Following a large eruption in 1914, levelling benchmarks on roads nearby, which had been installed and levelled in 1895, were re-surveyed in June 1914 (F. Omori). The results show local upheaval of several metres on the volcano itself, but further away a broad, roughly circular subsidence more than a metre deep and at least 60 km wide is seen (Fig 5.8a). The subsidence was not centred on the volcano itself, but is roughly concentric to the centre of Kagoshima Bay, an old caldera. These results were analyzed by Mogi (1958), who found that, when plotted against the distance from the centre of subsidence, the data fitted very closely to the deformation expected from his model described above, if $f = 10 \pm 1$ km (Fig 5.8b).

This result was particularly interesting, as it allowed the depth of the magma chamber of an active volcano to be detected for the first time. Omori re-surveyed the area around Sakurajima volcano in 1915 and 1919, and others did so in 1932 and 1946. After slight continuing subsidence in 1915, the trend was reversed in 1919, and the caldera began to inflate until 1946, when there was another large eruption. The inflation also fitted very closely to the Mogi model for a source depth of 10 km. This was the first indication that levelling might be used as a predictive tool, for once the traverse had been occupied for a number of years and the inflation/deflation pattern observed and understood, it was capable of showing when the magma chamber of a volcano is full, and therefore whether a future eruption was likely or not.

Since Mogi, others have re-analyzed the data from the 1914 eruption (Yokoyama 1971, 1986) using more complex models.

### 5.7.1.2 1967–68 eruption of Kilauea volcano, Hawaii

This was the first volcano where levelling was used routinely as a predictive and investigative tool. Although occasional levelling was carried out from 1912 onwards, and Wilson (1935) used the surveys of 1921 and 1927 to show that parts of the summit subsided 4 m during the 1924 eruption, a routine levelling network totalling 35 km in length was established in 1964, and has been regularly occupied and gradually expanded since.

The network was occupied 11 times between January 1966 and October 1967, during the build up to the 1967–68 summit eruption which began in November. The total uplift over this period was just over 0.7 m, and was centred not at Halemaumau crater where the eruption actually occurred, but at a point about 1 km to the southeast. (Fig 5.8c). However, during the 22 months of inflation, it is clear from the 11 levelling surveys that the centre of uplift changed position by up to 3 km as the volcano inflated (Fiske & Kinoshita 1969).

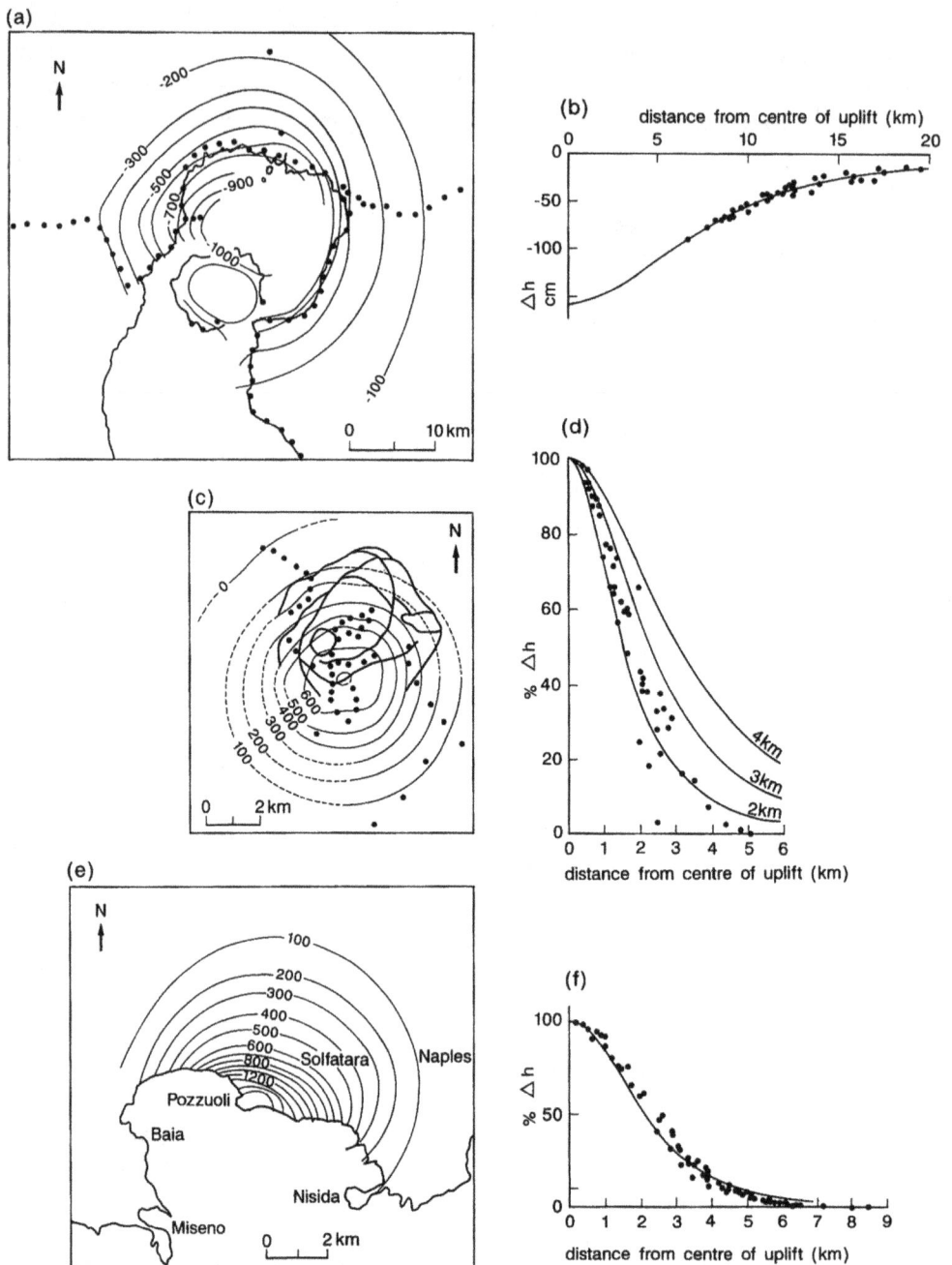

**Figure 5.8** Maps and diagrams of inflation or deflation episodes at three volcanoes where the observed displacements of survey markers correspond reasonably closely to Mogi's (1958) model. In each case the maps (a, c & e) are overlaid with contours of deformation at 100 mm intervals, interpolated between values of vertical displacement measured at benchmarks around the volcano. For each volcano, a benchmark distant from the volcano has been used as a zero reference. In (b, d & f), vertical displacement is plotted against distance from the volcano in kilometres. The black dots show displacement at individual benchmarks, whilst the curves show the theoretical displacement expected from a buried pressure source according to Mogi's (1958) model. (a & b) Sakurajima volcano, Japan: deflation measured between 1895 and June 1914, just after the voluminous eruption of that year. Note the difference in scale between map (a) and maps (c) and (e). The centre of deflation lies not at the volcano itself, but in the centre of Kagoshima Bay, the adjacent caldera. The curve in (b) is for a pressure source at 10-km depth. (c & d): Kilauea volcano, Hawaii: vertical displacement measured

As with Sakurajima volcano, the plot of uplift against distance from centre of uplift gives a curve close to that expected from Mogi's model (Fig 5.8d). In this case, however, the data points lie between the curves expected from a pressure source at a depth of 2 and 3 km.

### 5.7.1.3 Campi Flegrei
Similar patterns of inflation were seen during the crises at Campi Flegrei, near Naples, in 1970–71 and 1982–84 (Berrino et al. 1984). The last eruption in the Campi Flegrei took place in 1538, when Monte Nuovo was formed, but in 1970 uplift began, centred at Pozzuoli, which totalled 60 cm. before ceasing the following year. Then in 1982, a second phase of inflation started, which by the time it stopped in 1984 had added another 160 cm of uplift. Figure 5.8e shows a map of vertical deformation during the 1982–84 episode, illustrating the concentric pattern of deformation contours. A plot of vertical movement against distance from the centre of uplift shows data close to the values predicted by the Mogi model (Fig 5.8f). The depth to the centre of the source region in this case is determined at 2.8 ±0.2 km.

### 5.7.2 Dyke injection

The models of Dieterich & Decker (1975) and Pollard et al. (1983) indicate the type of ground deformation that can be expected on the ground surface above a vertical or near-vertical dyke undergoing dilation, whether during emplacement or during an increase in magma pressure. In general, points either side of the dyke will displace away from the dyke, normal to its long axis, as it inflates. The displacement first increases rapidly with distance from the dyke, reaches a peak value, then decreases slowly toward an asymptote of zero displacement. In the case of vertical dykes, this displacement is symmetrical about the centre of the dyke. For near-vertical dykes, displacements are greater on the side in the direction of dip and the point which experiences zero displacement is above the uppermost tip of the dyke rather than its centre. Horizontal ground strain is the first derivative of this displacement with respect to distance. Its peak tensile value occurs above the tip of the dyke, then decreases rapidly, through zero at the peaks of the displacement, to a relatively low compressive strain which then asymptotes back to zero. The exact shape of the displacement against distance curve depends upon the depth, height and angle of dip of the dyke. In order to simplify the analysis, the models so far derived have been two dimensional, i.e. they apply to a dyke of infinite length and constant cross section, located at a constant distance from the surface. The surface is plane and the medium is semi-infinite, composed of a homogeneous isotropic elastic material.

A real dyke is of finite length, has varying cross section and its path may be curved. Its depth may vary beneath an irregular topography and the material surrounding it will be a mixture of magmatic products. All of these factors will influence the displacement of a point

---

between January 1966 and October 1967. Three curves are shown in (d), showing the expected deformation from sources at 2, 3 and 4 km depth. The data suggest a magma chamber between 2 and 3 km depth. Note that in (d), the vertical displacement is given as a percentage of the maximum displacement at the centre of the swelling. (After Fiske & Kinoshita (1969).) (e & f): Campi Flegrei, Italy: uplift measured between January 1982 and March 1984. The curve in (f) is for a source at a depth of 3 km. No eruption followed this inflation episode. (After Berrino et al. (1984).)

on the surface as the dyke inflates or deflates, and cause it to deviate from the theoretical curve. However, measurements of real dyke-emplacement events (see below) show that the observed displacements can be remarkably close to the simplified models, suggesting that these deviations are not too important in practice.

If the path of the subsurface dyke can be determined, analysis of measured ground deformation (due to an inflation or deflation) perpendicular to a straight section of it may yield the horizontal strain as a function of distance from the dyke. This can be compared with the model to derive the depth, height and dip of the dyke, as well as the dilation which occurs during inflation or deflation. In addition, if the stress–strain relationship of the surrounding material is known, the change in magmatic pressure may be estimated. Such a model has been successfully applied to the 1975–84 activity at Krafla volcano.

### 5.7.2.1 Krafla volcano, Iceland

After 230 years of dormancy, the Krafla fissure swarm in northern Iceland began erupting in December 1975, and there were nine eruptions in the following 9 years. The accompanying deformation patterns showed repeated inflation and deflation sequences, with an uplifted area essentially of the Mogi type, centred on the 9 km wide Krafla caldera (Ewart et al. 1990). It appears that a magma chamber was being supplied with magma over this 9-year period, and that steady inflation of the chamber was punctuated by about 20 rapid deflation events, when vertical fissures crossing the area widened and filled with magma from the reservoir, to initiate dykes laterally which grew rapidly north to south to distances up to 50 km from the caldera centre. Just over half of these injected dykes did not reach the surface, but remained as intrusions.

Analysis of the ground deformation associated with one of these intruded dykes was carried out by Pollard et al. (1983). They computed theoretical displacements and stresses in a homogeneous, two-dimensional elastic half-space caused by pressurized vertical or steeply

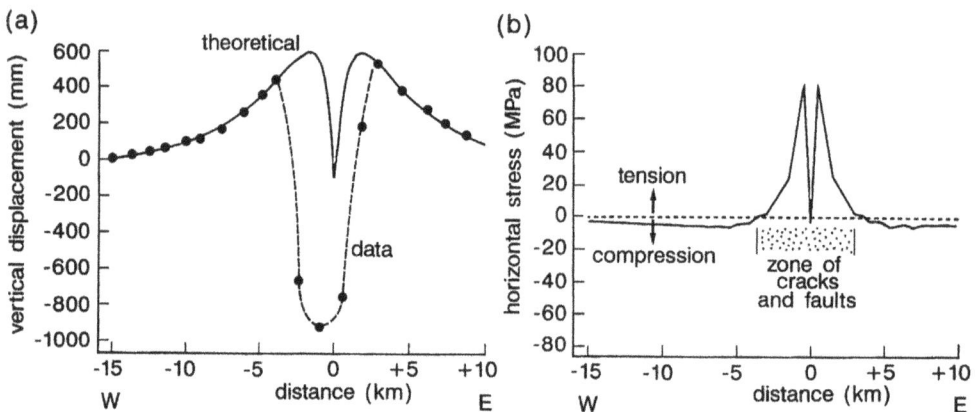

**Figure 5.9** (a) Plot of vertical displacement against distance from a dyke emplaced during a rifting event north of Krafla volcano, Iceland, on 7 January 1978. Data points closely follow the theoretical curve (solid line) for a vertical dyke whose centre lies at 3 km depth, and whose height is 5.5 km. The fit is good apart from close to the dyke itself, where tensional stress is calculated to be very high (see (b)). (b) Calculated horizontal stress plotted against distance from the dyke modelled in (a). (After Pollard et al. (1983).)

dipping cracks within it. Using this general model, they found that the measured vertical displacements accompanying the intrusive event of 7 January 1978 closely fitted the theoretical model for a vertical dyke of 5.5 km height, whose centre lay at a depth of 3 km (Fig 5.9a). The measured values depart from the theoretical curve only near the dyke itself, where the theoretical horizontal stresses are tensile (Fig 5.9b), and many faults and cracks were visible at the surface, indicating that in this zone, the ground was clearly not behaving elastically.

This kind of analysis can be used to build up a record of positions of injected dykes within a volcano, which can aid in studies of subsurface structure and future prediction. Similar analysis of dyke-forming events have been carried out at Kilauea (e.g. Dieterich & Decker 1975, Pollard et al. 1983, Ryan 1988) and Etna volcanoes (Murray & Pullen 1984, Murray 1993a).

An analytical model has also been developed by Ryan et al. (1983) to produce similar model deformation fields for sills. This has been applied to sill-like intrusions or storage compartments beneath Kilauea volcano, Hawaii, revealed by deformation data from the 1924, June to December 1972 and December 1972 to May 1973 eruptions.

### 5.7.3 Lava and edifice loading and subsidence

Subsidence caused by loading, compaction, or gravitational distortion of the volcanic pile may cause large movements at different scales on the same volcano.

*5.7.3.1 Large-scale loading and subsidence: Hawaii*
For groups of very big volcanoes, the subsidence of the entire volcano group, caused by downwarping of the lithosphere under the enormous load placed upon it by the deposition of lavas and other volcanic products, has long been recognized and measured (Moore & Peck 1965). In the case of the island of Hawaii, present subsidence at Hilo, determined by comparing tide guage records with those at nearby Honolulu, amounts to about 2.4 mm per year (Moore 1987).

*5.7.3.2 Lava loading: Mount Etna, Sicily*
At the other end of the scale, the effect of lava compaction causing subsidence in individual lava flows was measured in the case of recent flows on Mount Etna, where downward movements of several tens of centimetres were recorded at benchmarks on 20 m thick flows in the year after eruption (Murray & Guest 1982). After further measurements, it became clear that some of this subsidence was due to downwarping of the previous ground caused by the additional loading of the new lava flows (Murray 1988). Examples of this effect are shown in Fig 5.10. The rate of subsidence varies in direct proportion to the thickness of the flow, and declines asymptotically with time, though it may be revived by tectonic activity. Downward movements of several centimetres per year are quite usual for recent flows.

*5.7.3.3 Whole volcano subsidence: Surtsey, Iceland*
Between these two extremes, compaction and subsidence of a single volcano has been measured for the small volcano of Surtsey, a monogenetic volcano off the coast of Iceland that

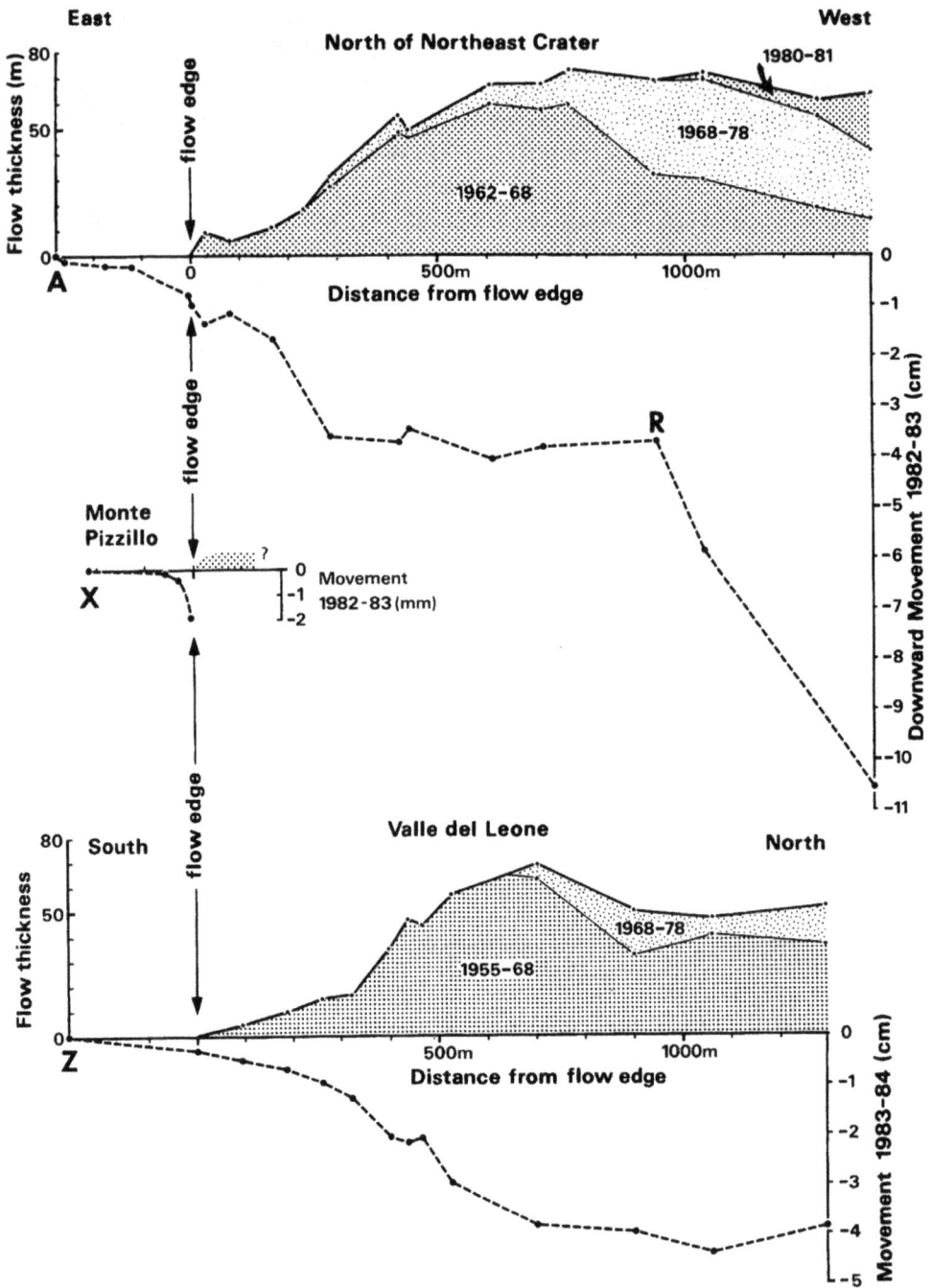

**Figure 5.10** (Top) Plot of thickness of successive flow units against distance from flow edge for lavas north of the Northeast Crater, Mount Etna, Sicily. Below the x-axis, the vertical movement of benchmarks in 1 year (1982–83) is shown for the same locations, referred to benchmark A off the flow. (Middle) Vertical movement of benchmarks against distance from the 1975 flow edge on the north flank of Mount Etna. In this case the flow depth is not known. Benchmark X is used as a reference zero. (Bottom) A similar diagram for lavas flowing into the Valle del Leone, on the upper eastern flank of Etna. Vertical movement is referred to benchmark Z. From Murray (1988).

erupted in 1963–66. It rose to a height of about 250 m above the sea floor, and measured about 3 km × 6 km across the base. Levelling measurements on a 2-km profile across the island since it ceased erupting in 1967 have shown continuing subsidence, which by 1991 had totalled more than 1.6 m near the centre of the island. Like the lava flow compaction measured at Etna, this subsidence has declined asymtotically with time (Fig 5.11).

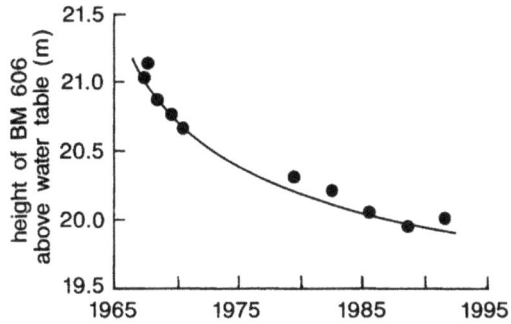

**Figure 5.11** Measured elevation of benchmark 606 between 1967 and 1991, relative to the water table in a pit dug in the north cape of Surtsey volcano, Iceland. The water table corresponds approximately to sea level, indicating that this part of the island has sunk more than 1 m over the period of observation.

This subsidence appears to be partly caused by compaction of the volcanic materials which make up the volcano, and also the sea floor sediments underlying the volcano. Generally speaking, the higher the altitude, the greater the subsidence, though differences in compaction between tephra and sediments mean that other factors dominate in places. It is also possible that some of the movement is due to downwarping of the lithosphere caused by the weight of the new volcano (Moore et al. 1992).

### 5.7.3.4 Whole volcano subsidence: Colima, Mexico

It is not always appreciated that similar subsidence occurs on active volcanoes, and may dominate the deformation pattern completely for steep-sided volcanoes, even during eruption. There is evidence for this from several volcanoes. It has been particularly striking in the case of Colima volcano, Mexico (Murray 1993b), which has been in more or less continuous dome-building eruption, with occasional periods of higher extrusion in the form of lava flows and larger avalanches, since before the first ground deformation network was set up on it in 1982. Yet despite this dome growth, the volcano has shown continuous subsidence since then, roughly centred on the summit, at a rate averaging just under $1 \, cm \, km^{-1} \, year^{-1}$ for stations 1–3 km from the summit (Fig 5.12).

This subsidence continued before, during and after the increase in activity in spring 1991, when increased dome extrusion culminated in the south flank lava flow of March–April and the major dome collapse and avalanche of 16 April 1991.

Evidence for a similar kind of deformation has been observed at Arenal volcano, Costa Rica. It is also a steep-sided edifice, and has been in continuous activity since its first historic eruption in 1968, with spectacularly large strombolian explosions and more or less continuous emission of lava. Dry tilt stations installed on the lower slopes of the volcano near the new lava flows have shown fairly steady deflation (Melson et al. 1979), which Wadge (1983) interpreted as subsidence caused by loading of the recent lava field. However, the fact that trigonometric heighting stations well away from the new lavas have also shown steady down-

Changes
1986 to 1993

Reference

+0.5 to  0.0 cm
 0.0 to -0.5 cm
-0.5 to -1.0 cm
-1.0 to -1.5 cm
-1.5 to -2.0 cm
-2.0 to -2.5 cm
-2.5 to -3.0 cm
-3.0 to -3.5 cm
-3.5 to -4.0 cm
-4.0 to -4.5 cm
-4.5 to -5.0 cm
< -5.0 cm

Summit

0                    1  km

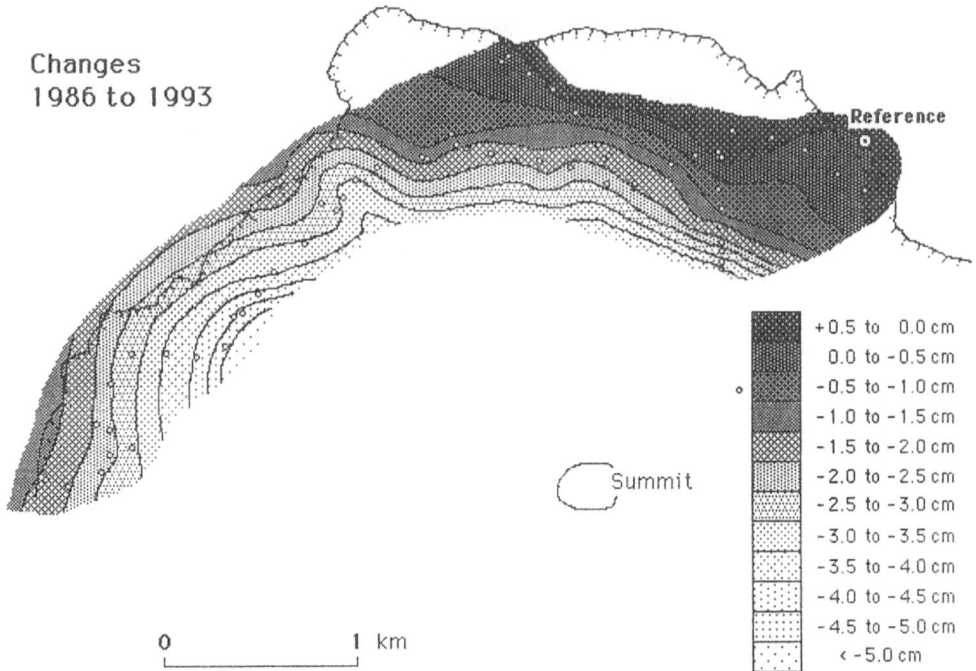

**Figure 5.12**  Subsidence concentric to the summit of Colima volcano, Mexico, measured between 1986 and 1993. The relation to the topography, and the continuous nature of the subsidence over the period of observation, suggests that compaction and down-warping due to loading of the volcano is responsible for this movement.

ward movement since they were installed in 1990 (H. Rymer, personal communication) suggests that the whole edifice may be involved in the subsidence.

### 5.7.4 Slope instability

As a volcano builds up by the steady accumulation of lavas and pyroclastics, its slopes become progressively steeper, and there comes a time when it becomes an unstable structure. If eruptions continue, this instability may ultimately resolve itself in slope failure, where a large section of the volcanic edifice may peel off and slide downslope. The most famous example of this kind of event was the Mount St Helens eruption of 18 May 1980, where nearly 3 km$^3$ of the northern slopes of the volcano collapsed and slid off the volcano, taking the pressure off the magma beneath, and provoking a large pyroclastic eruption. Similar events occurred at Bezymianny volcano, Kamchatka, in 1956 (Gorshkov 1959), and at Bandai San volcano, Japan, in 1888 (Sekiya & Kikuchi 1889).

Before such catastrophic events occur, slope creep and other precursive movements may take place on the unstable slope, which may be capable of indicating both where and when failure will take place. The best monitored of this type of eruption was that of Mount St Helens in 1980.

*5.7.4.1 18 May 1980 eruption of Mount St Helens, Washington, USA*
Slope movements prior to this event were enormous; so great that they were first detected not by sensitive instruments but from air photographs, where the enormous changes were plainly visible. Air photographs had been taken the previous year, on 15 August 1979, and were taken again on 7 and 12 April, and 1 and 12 May.

Between each of these dates, there was subsidence close to the summit, and broad inflation about 1 km downslope on the northern flank. The amount of subsidence had totalled more than 70 m by 12 May, the subsiding area forming an expanding trough or graben on the north side of the summit aligned approximately east–west. This graben was 1.5 km long and grew from 360 m wide on 7 April to about 600 m wide on 12 May (Moore & Albee 1981). The inflation further down eventually raised some areas more than 150 m above the pre-eruption topography (Fig 5.13), forming a prominent bulge which became increasingly obvious down the northern flank. This bulge measured nearly 2 km in diameter, with a zone of maximum uplift that moved northwards as time progressed.

Dry tilt stations were also set up and measured frequently during the crisis, and in some cases showed large short-term fluctuations in tilt, measured at up to $50 \mu rad h^{-1}$ on 10 April.

At the same time, distance measurements to reflectors installed at several points on the upper slopes of the mountain showed that stations on the bulge were moving north at a fairly constant rate of up to $2.5 m day^{-1}$ (Lipman et al 1981). Many reflectors were in fact moving in an almost horizontal direction or even slightly downhill, though the fact that this carried them north into regions well above the pre-eruption slope made them appear as inflated areas on the topographic difference map shown in Figure 5.13. The constant rate of deformation, if extrapolated back in time, leads to the conclusion that the graben and bulge began growing at some time between 20 March, when seismic activity began, and 27 March, when the first phreatic explosions occurred.

Sector collapses and the associated pyroclastic flows that may accompany them are one of the most potentially dangerous and destructive types of volcanic activity, as they occur suddenly and progress very rapidly, and are capable of wiping out towns as far as 100 km distant from the volcano in some cases. The Mount St Helens eruption of 18 May 1980 is the only example of sector collapse on a volcano where monitoring has been carried out both before and after the event, and as such contains important information to bear in mind when facing possible future sector collapses on other volcanoes. As far as deformation goes, it seems that the enormous rates of ground movement began only 8 weeks before the eruption, and continued at a fairly constant rate until the moment of eruption. This indicates a different behaviour from slope creep observed before some types of slope failure, where a progressively increasing rate of deformation is observed, which can be used to predict the time of failure, as described below.

*5.7.4.2 Mount Etna eastern flank, 1982–89*
Apparent signs of slope instability were detected on the eastern flank of Mount Etna, Sicily, between 1982 and 1989 and, although this did not lead to slope failure of any great size, the crisis does illustrate an interesting interplay between eruptive activity and a volcano slope which appears to be close to the limits of stability. Downward movements of a series of benchmarks

145

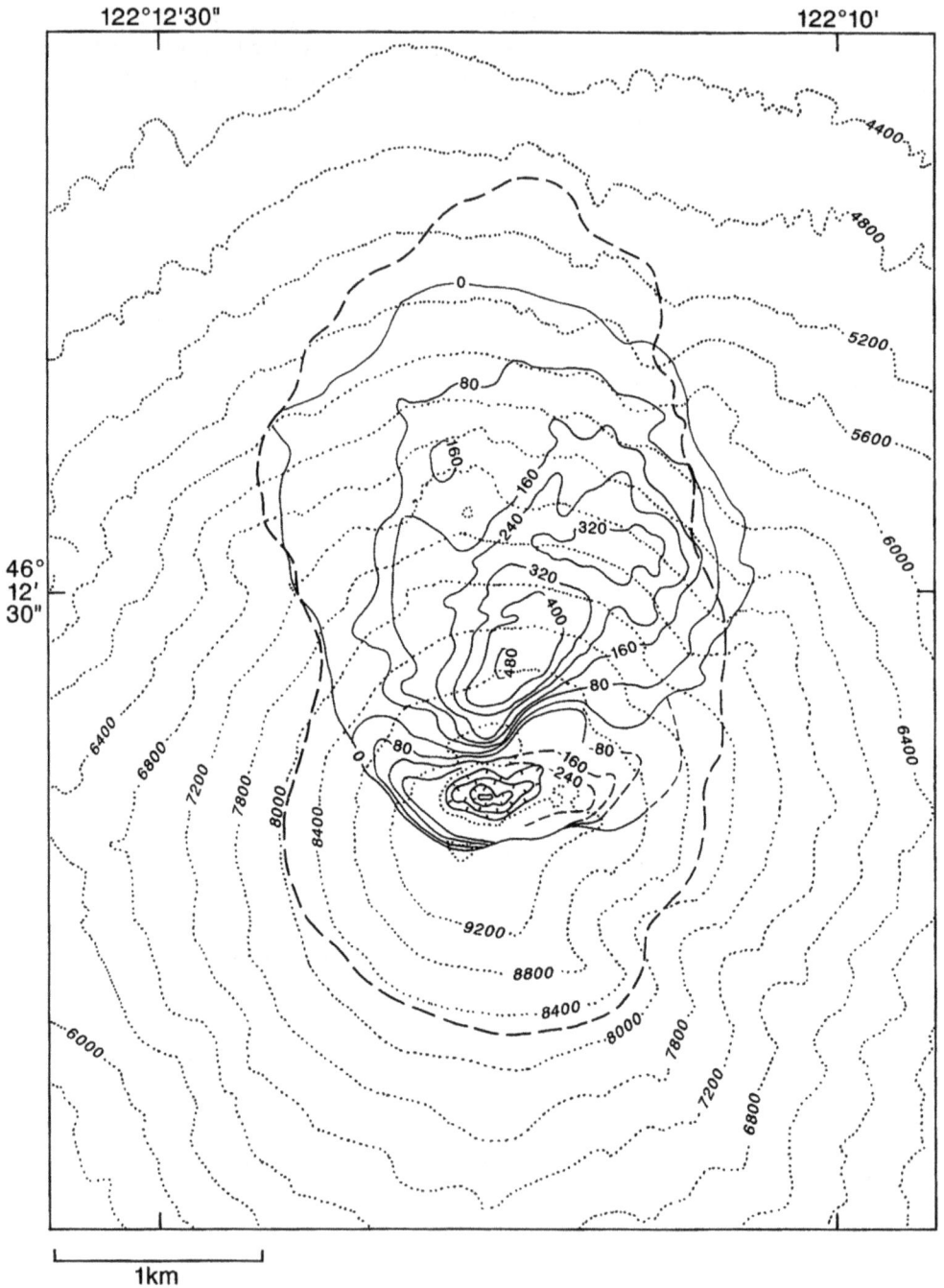

**Figure 5.13** Map showing topography and elevation changes (in feet) between 15 August 1979 and 12 May 1980, at Mount St Helens volcano, USA, before the catastrophic slope failure and pyroclastic eruption of 18 May. The bold dashed line shows the position of the crater rim that formed in the 18 May eruption. (After Moore & Albee (1981).)

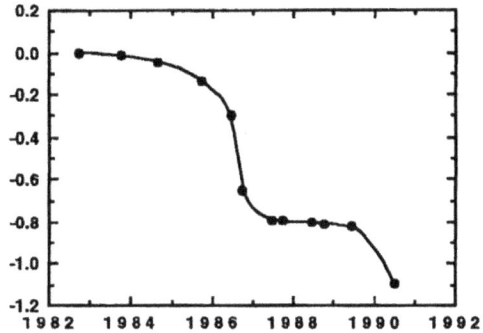

**Figure 5.14** Mean vertical movement since 1982 of four benchmarks across the upper eastern flank of Mount Etna, Sicily. Note the accelerating subsidence over 1982–85, and to a smaller extent over 1987–89.

along the upper eastern flank of Etna caused concern when the annual rate of displacement increased from 2 cm in 1982–83 to 18 cm in 1984–85 (Fig. 5.14). Such accelerating slope creep is a classic precursive sign of slope failure, and since a sector of the volcano about 1 km wide was involved, it seemed as if sector collapse was a real possibility.

The inverse rate method was used to give an estimate of the time of failure. This is based on the simple phenomenological law $\Omega'^{-\alpha}\Omega'' - A = 0$ (where $\Omega$ is an arbitrary quantity such as strain or displacement, and $A$ and $\alpha$ are constants), which adequately describes the terminal stages of failure under constant stress of many different materials. This law has been adapted for the prediction of landslides as well as volcanic eruptions (Voight 1988).

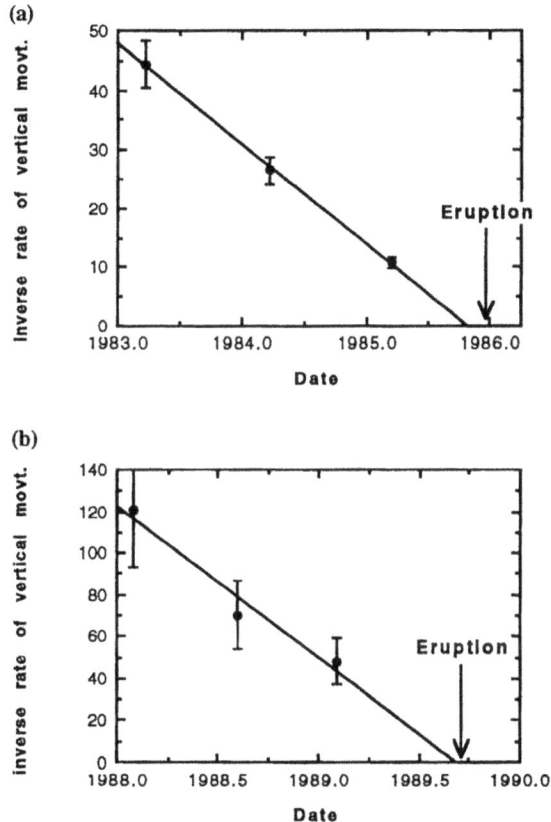

**Figure 5.15** (a) Mean inverse vertical displacement for the benchmarks on Mount Etna's eastern flank (shown in Fig. 5.14). The straight line is a least-squares fit to the data, and cuts the x–axis near the end of 1985. The time of the east flank eruption of 25 December 1985 is indicated. (b) As (a), but for the period 1988–89, prior to the east flank eruption which began on 27 September 1989.

147

The method may be applied by plotting inverse rates of ground movement against time. Failure occurs shortly before the inverse-rate curve intersects the time abscissa, i.e. shortly before the movement becomes infinite. The inverse rate plot for the mean subsidence of the Etna stations during 1982–86 is shown in Figure 5.15a. The number of observed epochs is small (four annual measurements giving only three rates of movement), but they fall close to a straight line which cuts the abscissa towards the end of 1985.

Instead of the expected sector collapse, an eruption occurred on the eastern flank below the line of benchmarks on 25 December 1985, accompanied by only a small slope failure of minor importance. A further eruption occurred down this flank in 1986–87, which seems to have temporarily halted the downslope movement, but this resumed again at a lower rate prior to the eruption of September 1989. Again the inverse rate plot for movements prior to this eruption (Fig. 5.15b) indicates failure in about August 1989, close to the eruption date of 27 September.

It seems that this downslope movement may be occurring when strain is building up on this flank due to a combination of magma pressure and slope instability. Eruptive activity releases the magma pressure and the slope becomes stable again.

## 5.8 Conclusions

Monitoring of ground movements is an important tool which can give information on the internal structure and functioning of volcanoes, and in many cases aid in the prediction of eruptions and other phenomena. Where cost is no limitation, trilateration and levelling networks, supported by GPS, will give the most complete and accurate coverage. However, where funds are low, or if time, accessibility or danger is a problem and displacements are large, then much information may be gained from dry-tilt stations, or from measuring angles and distances from remote stations to unmanned targets, or from aerial photogrammetry.

Ground deformation on volcanoes may be due to one of several causes, including inflation or deflation of a buried magma storage zone, injection of a dyke or sill which may or may not be an eruption conduit, subsidence due to lava loading or gravitational settling or spreading of the whole volcano, and slope movement caused by slope creep prior to failure or to magma pressure variations on steep slopes. Combinations of these causes frequently occur, to produce a complex pattern of deformation.

## References

Berrino, G., G. Corrado, G. Luongo, B. Toro 1984. Ground deformation and gravity changes accompanying the 1982 Pozzuoli uplift. *Bulletin Volcanologique* **47**, 187–200.

Bomford, G. 1980. *Geodesy*. 4th edn. Oxford: Clarendon Press.

Burnside, C. D. 1985. *Mapping from aerial photographs*, 2nd edn. London: Collins.

Burnside, C. D. 1991. *Electromagnetic distance measurement*, 3rd edn. London: Blackwell.

Decker, R. W. & W. T. Kinoshita 1971. Geodetic measurements. In *Earth Sciences, vol. 8: The surveillance and prediction of volcanic activity: a review of methods and techniques*, 47–74. Paris: UNESCO.

Dieterich J. H., & R. W. Decker 1975. Finite element modeling of surface deformation associated with volcanism. *Journal of Geophysical Research* **80**, 4094–4102.

Ewart, J. A., B. Voight, A. Björnsson 1990. Dynamics of Krafla caldera, north Iceland: 1975–1985. In *Magma transport and dtorage*, M. P. Ryan (ed.), 225–76. Chichester: John Wiley.

Fiske, R. S. & W. T. Kinoshita 1969. Inflation of Kilauea volcano prior to its 1967–1968 eruption. *Science* **165**, 341–9.

Gorshkov, G. S. 1959. Gigantic eruption of the volcano Bezymianny. *Bulletin Volcanologique* **20**, 77–112.

Kinoshita, W. T., D. A. Swanson, D. B. Jackson 1974. The measurement of crustal deformation related to volcanic activity at Kilauea volcano, Hawaii. In *Physical Volcanology*, L. Civetta, P. Gasparini, G. Luongo, A. Hapolla (eds), 87–115. Amsterdam: Elsevier.

Lipman, P. W., J. G. Moore, D. A. Swanson 1981. Bulging of the north flank before the May 18th eruption – geodetic data. In *United States Geological Survey Professional Paper 1250, The 1980 eruptions of Mount St Helens*, R. W. Decker, T. L Wright, P. H. Stauffer (eds), 143–56. Washington DC: United States Government Printing Office.

Melson, W. G., J. E. Umana, E. Evans 1979. Arenal volcano: results of dry tilt measurements. *SEAN Bulletin* **4/2**, 13-16.

Mogi, K. 1958. Relations between the eruptions of various volcanoes and the deformations of the ground surfaces around them. *Bulletin of the Earthquake Research Institute (University of Tokyo)* **36**, 99–134.

Moore, J.G. 1987. Subsidence of the Hawaiian ridge. In *United States Geological Survey Professional Paper 1350, Volcanism in Hawaii*, 85–100. Washington DC: United States Government Printing Office.

Moore, J.G. & W. C. Albee 1981. Topographic and structural changes, March-July 1981: photogrammetric data. In *United States Geological Survey Professional Paper 1250, The 1980 eruptions of Mount St Helens, Washington*, 123–34. Washington DC: United States Government Printing Office.

Moore, J. G. & D. L. Peck 1965. Submarine lavas from the East rift zone of Mauna Kea, Hawaii (abstract). In *Geological Society of America Abstracts with Program*, Cordilleran Section, Fresno, California. Washington DC: American Geological Society

Moore, J. G., S. Jakobsson, J. Holmjarn 1992. Subsidence of Surtsey volcano 1967-1991. *Bulletin of Volcanology* **55**, 17–24.

Murray, J. B. 1988. The influence of loading by lavas on the siting of volcanic eruption vents on Mt Etna. *Journal of Volcanology and Geothermal Research* **35**, 121–39.

Murray, J. B. 1993a Elastic model of the actively intruded dyke feeding the 1991–93 eruption of Mt Etna, derived from ground deformation measurements. *Acta Vulcanologia* in press.

Murray, J. B. 1993b Ground deformation at Colima Volcano, Mexico, 1982 to 1991. *Geofísica Internacional* **32**, 659–69.

Murray, J. B. & J. E. Guest 1982. Vertical ground deformation on Mt Etna 1975–1980. *Geological Society of America Bulletin* **93**, 1160–75.

Murray, J. B. & A. D. Pullen 1984. Three-dimensional model of the feeder conduit of the 1983 eruption of Mt Etna, from ground deformation measurements. *Bulletin Volcanologique* **47-4**, 1145–63.

Murray, J. B., J. E. Guest, P. S. Butterworth 1977. Large ground deformation on Mt Etna volcano. *Nature* **266**, 338–40.

Omori, F. 1913. The Usu-San eruption and the elevation phenomena II (comparison of benchmark heights in the base district before and after the eruption). *Bulletin of the Imperial Earthquake Investigation Committee* **5**, 105–7.

Omori, F. 1914. The Sakurka-Jima eruption and earthquakes. *Bulletin of the Imperial Earthquake Investigation Committee* **8**, 106.

Pollard, D. D., P. T. Delaney, W. A. Duffield, E. T. Endo, A. T. Okamura 1983. Surface deformation in volcanic rift zones. *Tectonophysics* **94**, 541–84.

Ryan, M. P. 1988. The mechanics and three-dimensional internal structure of active magmatic systems: Kilauea volcano, Hawaii. *Journal of Geophysical Research* **93**, 4213–48.

Ryan, M. P., J. Y. K. Blevins, A. T. Okamura, R. Y. Koyanagi 1983. Magma reservoir subsidence mechanics: theoretical summary and application to Kilauea volcano, Hawaii. *Journal of Geophysical Research* **88**, 4147–81.

Sartorius von Waltershausen, W. 1880. *Der Aetna*. Liepzig: Engelman.

Seeber, G. 1993. *Satellite geodesy*. Berlin: Walter de Gruyter.

Sekiya, S. & Y. Kikuchi 1889. The eruption of Bandai-san. *Journal of the College of Science, Tokyo Imperial University* **3**, 91–172.

Tryggvason, E. 1968. Measurement of surface deformation in Icelend by precision levelling. *Journal of Geophysical Research* **73**, 7039–50.

Voight, B. 1988. A method for prediction of volcanic eruptions. *Nature* **332**, 125–30.

Wadge, G. 1976. Deformation of Mt Etna 1971–1974. *Journal of Volcanology and Geothermal Research* **1**, 237–63.

Wadge, G. 1983. The magma budget of Volcan Arenal, Costa Rica from 1968 to 1980. *Journal of Volcanology and Geothermal Research* **19**, 281–302.

Walsh, J. B. & R. W. Decker 1971. Surface deformation associated with volcanism. *Journal of Geophysical Research* **76**, 3291–302.

Wilson, R. M. 1935. *Ground surface movement at Kilauea Volcano, Hawaii. University of Hawaii Research Publication No. 10*, 1–56. Honolulu: University of Hawaii.

Wolf, P.R. 1983. *Elements of photogrammetry*. Singapore: McGraw-Hill.

Yokoyama, I. 1971. A model for the crustal deformations around volcanoes. *Journal of the Physics of the Earth* **19**, 199–207.

Yokoyama, I. 1986. Crustal deformation caused by the 1914 eruption of Sakurajima Volcano, Japan and its secular changes. *Journal of Volcanology and Geothermal Research* **30**, 283–302.

Zlotnicki, J., J. C. Ruegg, P. Bachèlery, P. A. Blum 1990. Eruptive mechanism on Piton de la Fournaise volcano associated with the December 4 1983, and January 18 1984 eruptions from ground deformation monitoring and photogrammetric surveys. *Journal of Volcanology and Geothermal Research* **40**, 197–217.

# 6 GPS – Monitoring volcanic deformation from space

G. Nunnari and G. Puglisi

## 6.1 Introduction

The Global Positioning System (GPS) is a space-based network of satellites developed by the United States Department of Defense (since 1976) to facilitate the precise positioning of objects on the Earth's surface. In the context of the monitoring of volcano deformation, GPS constitutes a new technique which does not suffer from some of the drawbacks associated with ground-based methods such as electronic distance measurement (EDM), triangulation, and precise levelling. In particular, surface deformation monitoring using GPS is not constrained, as are ground-based techniques, by the following requirements which together act to hinder the installation and measurement of large geodetic networks on volcanic edifices.

(a) Benchmarks must be installed at the top of hills to allow intervisibility between stations. This often makes the benchmarks less accessible, and their occupation more time consuming.

(b) Distances between benchmarks are usually limited to a few kilometres, requiring a large number of points to cover a given area of interest.

(c) Weather conditions must be suitable to ensure that networks are measured over as short a period as possible.

On large and very active volcanoes these constraints conspire to prevent geodetic networks being reoccupied often enough to reveal short-term temporal changes in the local geodynamic conditions. During the last few years GPS has been widely tested and improved and civilian users have started utilizing the system for navigation and geodetic purposes. In particular, since the end of the 1980s several campaigns have been initiated around the world using the GPS for measuring crustal movements (Jordan & Minster 1988, Smith et al. 1989, Kellogg & Dixon 1990, Hager et al. 1991) in active geodynamic areas. Researchers have been attracted by a number of advantages of GPS, including its all-weather operational capability, portability, the absence of the requirement for benchmarks to be intervisible, and the accuracy of the system in comparison to other terrestrial and space-based measurement systems (Fig. 6.1).

The first GPS survey to be performed in a volcanic area was undertaken in Iceland during 1986 (Foulger et al. 1987). Since then other GPS networks have been established in Hawaii (Sako et al. 1987), Japan (Shimada et al. 1989, Kimata et al. 1991), and Antarctica (Gubellini et al. 1989). Since 1988, the authors have also undertaken GPS monitoring at Mount Etna

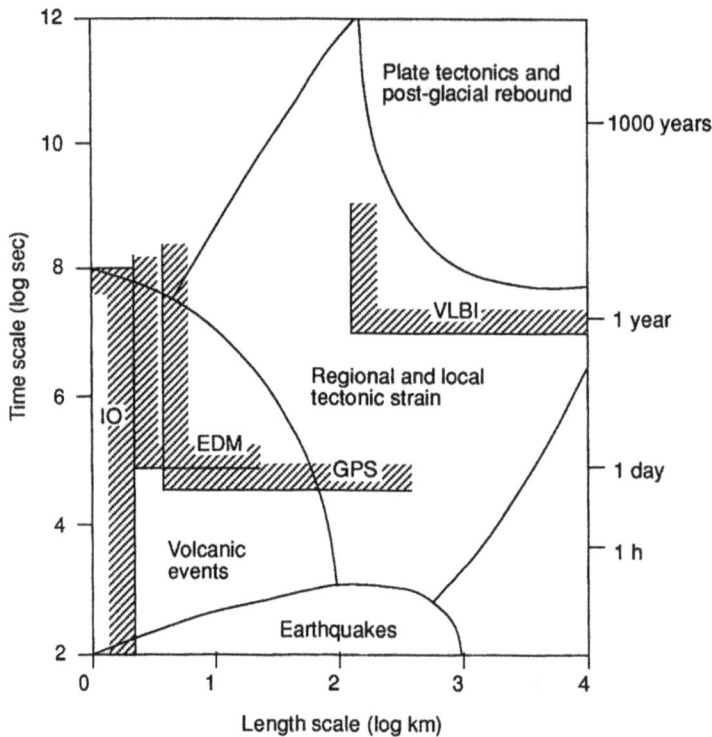

**Figure 6.1** Time and space plots of various geological phenomena versus the degree of accuracy of some geodetic techniques (redrawn from Dixon 1991). Note that EDM and GPS are comparable for distances ranging between 5 and 10 km, while for shorter distances EDM gives the more accurate measurements. Other methods included are VLBI (very long base interferometry), SLR = (satellite laser ranging), and OI (observatory instrumentation, e.g. tiltmeters).

(Briole et al. 1988, 1992), and this work, its implications, and its lessons, are discussed later in this chapter. Following consideration of some basic concepts, and the biases and errors which can affect the accuracy and precision of GPS measurements, potential operational difficulties are discussed, and, with particular reference to Mount Etna, problems associated with the post-processing of GPS data are addressed. The Mount Etna results are also examined in the light of the geodynamic framework of the volcano and its behaviour during the past 6 years. With reference to a prototype currently in use at Mount Etna, attention is also focused on the potential for automated GPS measurement of surface-deformation networks which allows continuous monitoring.

## 6.2 Some basic concepts

The GPS satellite-based positioning system consists essentially of three segments: the space segment, the control segment, and the user segment (Wells 1986). The space segment consists of 21 satellites, arranged in six orbital planes inclined at an angle of about 55° to the equator, plus three more in spare orbits. The orbital period of each satellite is 12 sidereal-hours. At the time of writing (October 1993) the number of operating satellites is 24, which guarantees that for most of the day four satellites are always visible from any point on the Earth's surface. Some of the currently-operative satellites, belonging to the so-called "first block" will be turned off in the near future, and will not form part of the fully functioning system. It is worth noting here that a Russian equivalent of the GPS system, known as GLONASS, is currently under development, and is expected to be completed by 1995. This system will also consist of 24 satellites (including three spares) equally spaced in three orbital planes inclined at 64.8° to the equator. The detailed description of this system is outside the scope of this chapter, but it is briefly mentioned here because it is expected that the two systems will be integrated. Several manufacturers are already producing receivers that have dual GPS–GLONASS capability. For further details concerning comparisons between GPS and GLONASS see, for example, Kleusberg (1990). In GPS, individual satellites are identified by different codes such as the space vehicle number (SVN), or the pseudo-random noise number (PRN). Since all GPS satellites use the same carrier frequencies, they are distinguished by the receivers using the PRN code which is unique to each satellite, and is thus the most common code used (Wells 1986, Hoffman-Wellenoff et al 1992). Each satellite transmits two frequencies for positional purposes ($L_1$ and $L_2$) which are multiples of the fundamental clock frequency ($f_0$ = 10.23 MHz) generated by the satellite's on-board atomic clock. $L_1$ is on 1575.42 MHz and $L_2$ on 1227.6 MHz, which are, respectively, 154 and 120 times the fundamental clock frequency, and correspond to wavelengths of about 19 and 24 cm, respectively. The carrier frequency $L_1$ is modulated by two pseudo-random noise codes referred to as the C/A (Coarse/Acquisition) code and P (Precise or Protected) code, while the $L_2$ frequency is modulated using the P code only. The modulation frequencies of the C/A and P codes are, respectively, $f_0/10$ and $f_0$, corresponding to wavelengths of about 300 m and 30 m, respectively. Both the $L_1$ and $L_2$ carriers are also modulated by a low frequency (60 Hz) stream of data, referred to as the "navigation message", designed to inform the user about the health and positions of all the satellites (broadcast ephemeris) and other useful parameters.

The control segment consists of five monitoring stations and a master control station (located at Colorado Springs, USA) the purpose of which is to monitor the health of the satellites, determine their orbits, observe the behaviour of their atomic clocks, and update the broadcast message to the satellites.

The user segment consists of all military and civilian users who track the satellite signals by using appropriate receivers and antennas, and who perform absolute or relative positioning for various purposes including geodetic control, and local and global deformation monitoring. There are two basic observables using GPS receivers: (a) pseudo-range distance, and (b) carrier beat phase. The first represents the receiver-satellite distance computed from the C/A or P code measurements only. The basic equation of this type of observable is:

$$\rho = R + c \cdot (dt - dT) + d_{trop} + d_{ion} \qquad (6.1)$$

where $\rho$ is the distance between the generic satellite and the antenna expressed as a function of the point ($R$) and the satellite ($r$) coordinates (i.e. $\rho = \lVert R - r \rVert$), $d_{trop}$ and $d_{ion}$, are, respec-

153

tively, the tropospheric and ionospheric delays which act as noise terms, and $c \cdot (dt - dT)$ expresses the clock bias due to the satellite and receiver clock offsets.

As explained in the next section, the satellite clock can be modelled by a polynomial law with coefficients being transmitted in the navigation message. Thus only the receiver clock bias is considered to be unknown.

Considering, for the same epoch, four observables of type (6.1) made by tracking four different satellites, it is possible to compute the receiver-clock effects and the three-point co-ordinates in the reference system, which in the GPS context is the WSG84 (Decker 1986). The precision of pseudo-range absolute positioning is obviously low (of the order of a few tens of metres) but it can be useful where this type of approximation is acceptable, such as for navigation purposes. Pseudo-range absolute positioning is, moreover, often important in the pre-processing phases, and in the differential use of GPS.

The second type of observable is the phase difference between the carrier signal generated by its internal oscillator and the incoming carrier signal from the satellite. In practice this observable, measured at some epoch $t$ can be written as:

$$\phi_{tot} = Fr(\phi) + Int(\phi, t_0, t) + N(t_0) \tag{6.2}$$

where $Fr(\phi)$ is the fractional part of $\phi$ measured by the receiver at epoch $t$ by aligning the receiver clock with the incoming carrier signal, $Int(\phi, t_0, t)$ is the integer phase cycle count from the initial epoch $t_0$ to epoch $t$, and $N(t_0)$ is an unknown integer number of cycles at the initial epoch $t_0$ between satellite and receiver. The sum $Fr(\phi) + Int(\phi, t_0, t)$ is what is measured by a GPS receiver. Equation 6.1 can be written in length units (see Wells 1986) as:

$$\phi = \rho + c \cdot (dt - dT) + d_{trop} - d_{ion} + \lambda \cdot N \tag{6.3}$$

where $\phi$ represents the carrier beat phase in length units, $\rho$ is the distance between the satellite and the receiver, $c$ is the speed of light in a vacuum, $dt$ and $dT$ are, respectively, the satellite and receiver clock offsets, $d_{trop}$ and $d_{ion}$ are, respectively, the tropospheric and ionospheric delays, $\lambda$ is the carrier wavelength, and $N$ is the integer cycle ambiguity. The minus sign in the term $d_{ion}$ indicates, according to the theory of wave reflection, that the carrier phase measurements are advanced, while the C/A code measurements are delayed.

Equations 6.2 and 6.3 both emphasize the presence of the so-called *integer cycle ambiguity*, $N$. As long as the receiver maintains continuous phase lock during an observation session, there is only one ambiguity per satellite/receiver pair. Other ambiguities may occur, however, due to breaks in phase lock known as *cycle slips*. These must be appropriately fixed during the data post-processing in order to obtain reliable baseline solutions. A frame of the file recorded by a receiver in the so-called RINEX (receiver independent exchange) format is shown in Figure 6.2.

Using the basic observables (phase and range in both frequencies), it is possible to form linear combinations in order to solve particular problems (e.g. to remove the ionospheric effects, or to search for integer ambiguities). Let us consider the linear combination based on the phase observables of both the $L_1$ and $L_2$ measurements at the same receivers and time:

$$\phi_c = K_1 \phi_1 + K_2 \phi_2 \tag{6.4}$$

```
2                    OBSERVATION DATA                         RINEX VERSION / TYPE
TRIMVEC              IIV                    1992 SEP 29        PGM / RUN BY / DATE
                                                              COMMENT

NICOSIA                                                       MARKER NAME
G.P.                 IIV-CNR                                  OBSERVER / AGENCY
2156                 4000SST                3.01              REC # / TYPE / VERS

2563                 GEODETIC SURVEY                          ANT # / TYPE

        4887541.1348              1258739.4514      3888063.0658  APPROX. POSITION XYZ
                 1.4525                  .0000             .0000  ANTENNA: DELTA H/E/N
1        2           0                                           WAVELENGTH FACT L1/2
4        L1          C1            L2        P2                   # / TYPES OF OBSERV
15                                                               INTERVAL
1992     9          28            14         0     45.000000     TIME OF FIRST OBS
                                                                 END OF HEADER

92 9  28   14    0 45.0000000 0   4   15  11  14  25 0   0   0   0   0   0   0   .000650667
                 -.60310              20318341.12000             -.45400       20318365.89900
                 -.61010              22348663.85100             -.47200       22348687.88800
                 -.22710              21324813.34100             -.94000       21324838.90600
                 -.17410              24187828.29300             -.65000       24187866.91900
92 9  28   14    1   .0000000 0   4   15  11  14  25 0   0   0   0   0   0   0   .000657855
                 9853.11700           20320218.46700            7677.78000     20320240.90800
                 61248.15300          22360314.25300            47725.71600    22360343.07200
                 -8223.78700          21323248.61500            -6408.87700    21323274.01400
                 -12213.01000         24185518.21500            -9517.09500    24185542.90800
```

**Figure 6.2** Example of the receiver independent exchange (RINEX) format recorded by a GPS receiver during a measuring campaign. The information contained in the upper part of the file (e.g. frequencies used, wavelength and instrument type) is self-explanatory. The lower half of the file lists two samples of data recorded at different epochs. Each sample starts with a row containing the date and time information for each sample, the number of tracked satellites and their identifying numbers, and the receiver clock offsets. The subsequent four rows, each consisting of four columns, represent the $L_1$ phase, the C/A pseudo-range (in metres), the $L_2$ phase and the P pseudo-range, referring, respectively, to the visible satellites (indicated as 15, 11, 14 and 25 in this example).

From this expression several different combinations can be derived depending on the values of $K_1$ and $K_2$. The most useful one is known as the $L_3$ observable (referred to as the "iono-free observable"), where $K_1$ and $K_2$ satisfy the following condition:

$$\frac{K_2}{f_2} + \frac{K_1}{f_1} = 0$$

The values most frequently used for $K_1$ and $K_2$ are:

$$K_1 = \frac{f_1^2}{f_1^2 - f_2^2} \quad \text{and} \quad K_2 = -\frac{f_1 f_2}{f_1^2 - f_2^2}$$

Such values guarantee that the $L_3$ observable is free from ionospheric effects (see e.g. Hoffman-Wellenoff et al. 1982 for the proof). From Equation 6.4, and also taking into account the following relationship between phases and frequencies of the $L_1$ and $L_2$ carriers:

$$\phi_{L_2} = \frac{f_1}{f_2}\phi_{L_1}$$

it is possible to derive for the observable $L_3$ the following expression (for details see e.g. Leick 1990):

$$\Phi_{L_3} = \phi^p_{K,L_1} - \frac{f_{L_1}}{c}\rho^p_k + (K_1 N_1 + K_2 N_2) \tag{6.5}$$

where the term $\phi^p_{K,L_1}$ represents the $L_1$ phase for the satellite p measured at the receiver $K$. In Equation 6.5, the terms for the ionospheric effects have disappeared, while those representing the tropospheric effect have been omitted for the sake of simplicity. The righthand term inside the parentheses in Equation 6.5 represents the "ambiguity term", which in this case is not an integer.

Using GPS data, both absolute and relative positioning are possible. Due, however, to uncertainties in the clock behaviour and satellite positions, and to propagation delays, absolute position can be determined with accuracies limited to only a few tens of metres. Furthermore, for civil applications, the absolute precision is lower than that attainable by the US Department of Defense due to a restrictive policy which denies access to the appropriate signal (see Section 6.3.6). Fortunately, for the relative positioning, which is required for geodetic and geodynamic purposes, the error sources affecting the GPS signals received simultaneously at different stations are correlated, so that their effect can be reduced by differencing the observation equations. Several differencing combinations are possible, the most commonly used being those between receivers, between satellites, and between epochs.

Let us define the following differencing operators: $\Delta$ between receivers, $\nabla$ between satellites, and $\delta$ between epochs. For a pair of stations simultaneously observing the same satellite, the mathematical model for a *between-receiver single-difference carrier phase* can be derived from Equation 6.3 as shown below:

$$\Delta\phi = \delta p - c\cdot\Delta dT + \lambda\cdot\Delta N - \Delta d_{ion} + \Delta d_{trop} \tag{6.6}$$

This kind of difference removes the effects of satellite clock errors. Moreover, if the baseline lengths are short, when compared with the 20000 km altitude of the satellite, then the errors due to satellite orbit and atmospheric delays are greatly reduced.

Let us now consider two receivers and two satellites at the same epoch. It is possible to obtain the *receiver-satellite double-difference equation*, involving both the two receivers and the two satellites. One way to construct this kind of difference is to take two between-receiver, single-difference observables involving the same pair of receivers but different satellites, and difference between the two satellites. From Equation 6.6, by differencing between satellites, the following double-difference observation equation is obtained:

$$\nabla\Delta\phi = \nabla\Delta p + \lambda\cdot\nabla\Delta N - \nabla\Delta d_{ion} + \nabla\Delta d_{trop} \tag{6.7}$$

The double difference removes, or greatly reduces, the effects of misalignment between the two receiver clocks and between the satellite clocks. The *receiver-satellite-time triple-difference observable* is the change in a receiver-satellite double difference from one epoch to the next. The triple difference corresponding to Equation 6.7 is:

$$\delta\nabla\Delta\phi = \delta\nabla\Delta p - \delta\nabla\Delta d_{\text{ion}} - \delta\nabla\Delta d_{\text{trop}} \tag{6.8}$$

It is possible to see that the errors which dropped out in the double-difference observable are also absent from the triple one. In addition, for carrier-beat phase measurements, the initial cycle integer ambiguity terms are cancelled out.

Generally, and regardless of the level (single, double, or triple), the differencing process induces mathematical correlations between the differenced observables which may result in a lower level of accuracy in the solutions. In addition, the triple-difference equation cannot be used to compute the final solution, because integer ambiguities cannot be fixed (they are not present in Equation 6.8). For this reason, triple-difference equations are generally used by most post-processor software packages as an initial step to fixing the cycle slips before baseline processing is begins. Once reasonably good values for stable coordinates have been obtained using triple-difference observables, a search can be carried out automatically on the triple-difference residuals to identify discontinuities in the double differences.

## 6.3 Biases and errors

GPS measurements are affected by several sources of bias and error. Biases are related to characteristics of the satellites, receiving stations, and observations, while errors result mainly from cycle-slips, multipath effects, residual unmodelled biases, and other random observational errors.

### 6.3.1 Biases and errors relating to satellites and receiving stations

Biases originating at the satellites are due to their internal clock offset, as emphasized in Equations 6.1 and 6.3, and by orbital uncertainties. The clock offset $\Delta t$ is generally modelled as a second-order polynomial:

$$\Delta t = af_0 + af_1 \cdot (t - t_0) + af_2 \cdot (t - t_0)^2 \tag{6.9}$$

where $af_0$, $af_1$, $af_2$ and $t_0$ are coefficients contained in the messages that the satellites broadcast. In this way the order of the clock bias is reduced to 30 nanoseconds, corresponding to a range of 10 m in absolute positioning. As previously mentioned, however, this kind of bias can be removed when relative positioning is performed.

Problems arising with uncertainties in the satellite orbit are more difficult to handle than those involving the clock. In some applications, such as kinematic positioning and navigation, these problems are simply ignored by assuming that the broadcast ephemeris is perfect. For differential GPS the effects of this type of bias are limited to about 1 ppm of the baseline length for orbital errors of about 20 m. The normalized baseline error ($db/b$) and the normalized satellite distance ($d\rho/\rho$) are related by the following rule of thumb:

$$\frac{db}{b} = \frac{d\rho}{\rho}$$

Such errors can be further reduced by processing data with the precise ephemeris available from agencies such as the United States Defense Mapping Agency or National Geodetic Survey, that up to now have allowed precision in the satellite positioning of the order of 0.1 ppm. This in turn allows baseline length precisions of around 0.1 ppm (Remondi & Hofman-Wellenof 1989).

Station-dependent biases are due mainly to the receiver clock, and can be reduced in some receivers by use of an external timing source. In the differential use of GPS such biases can, as discussed above (see the comments on Equation 6.7), be reduced. There is, however, some limitation in the resolution of the observation due to the receiver's electronics. As a rule of thumb this resolution is proportional to the wavelength of the signal. Thus, assuming the resolution to be of about 1%, the expected standard deviation is 0.3 m for the C/A code, 3 cm for the P code and 0.2 mm for the carriers. However, recent technical developments suggest that it is now possible to achieve resolutions as good as 0.1%.

### 6.3.2 Observation-dependent biases

Biases dependent upon observations result from the passage of the GPS signal through the atmosphere and to carrier-beat phase ambiguities. Since the different layers of the atmosphere affect the GPS signal in different ways (Fig. 6.3), these effects need to be analyzed separately.

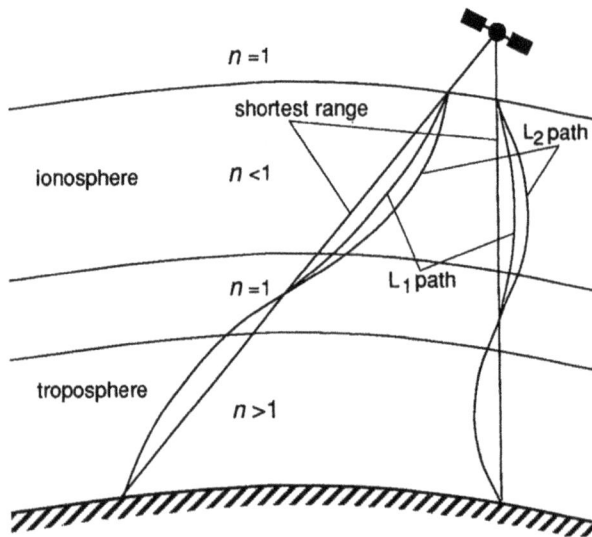

**Figure 6.3** Schematic illustrating the effect of the atmosphere on the GPS signal path. Note the opposing bias effects that the ionosphere and troposphere have on the signals.

It is standard procedure to consider the atmosphere as two layers, the ionosphere and the troposphere. For GPS purposes, the ionosphere consists of the ionized layers of the atmosphere above a boundary located at an altitude of between 50 and 1000 km, while the troposphere is constituted from the layers below 50 km, which are characterized by neutral electrical conditions.

The ionosphere affects the propagation of the GPS signal by introducing a delay which depends on the carrier frequency ($f$) following a non-linear relationship:

$$dI = (K/f^2) \cdot \text{TEC} \cdot \sec(z) \tag{6.10}$$

where $K$ is a constant the value of which ranges between $-40.3$ and $-41$ (measured in electrons$^{-1}$ s$^{-2}$ m$^3$), $dI$ is the delay (in metres), TEC is measured in units of $10^{16}$ electrons m$^{-2}$ (strictly speaking, TEC represents the integral of electronic density along the signal path), and $z$ is the zenith angle of the signal path. The change in satellite-receiver range due to ionospheric effects, plotted against satellite elevation (for values of TEC ranging between $20 \times 10^{16}$ and $50 \times 10^{16}$) is reported in Figure 6.4. According to Hartman & Leitineger (1984), it can be assumed that the lower and higher values for TEC are about $10^{16}$ and $2 \times 10^{18}$, respectively.

The effects ($DI$) on baseline determination follow relationships of the type:

$$DI/l = (K/f^2) \cdot \text{TEC} \cdot g(R,z) \tag{6.11}$$

where $g(R,z)$ is a function depending on $R$ and $z$, with $R$ being the radius of the Earth. Using this model it is possible to estimate a shortening, due to the described effects, which can be estimated using the model of Georgiadou & Kleusberg (1988) (Fig. 6.4). The analysis reported above emphasizes the role of TEC in determining ionospheric effects. The real prob-

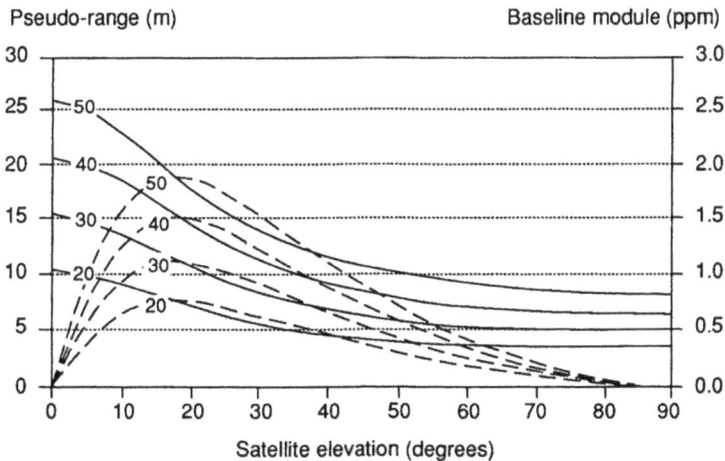

Figure 6.4 Ionospheric effects result in the shortening of pseudo-range and baseline length, although the degree of shortening is significantly affected by satellite elevation. The small numbers in the curves represent the TEC values (in $10^{16}$ electrons m$^{-2}$. The solid curves represent shortening (in metres) of the pseudo-range distance, while the dashed curves represent shortening (in PPM) of the vector baseline module.

lem, however, lies in how to measure, estimate, or, in the most favourable case, to eliminate such effects. A correction for the ionospheric effects can be obtained using the coefficients of a standard model (Klobuchar 1986) contained in the message broadcast by the satellites. However, this model is simply not accurate enough for geodetic applications when single-frequency receivers are used. The model of Klobuchar (1986) allows a reduction of about 50–60% of the rms error (Klobuchar 1991). In most of the available software packages models derived from Equation 6.11, referring to double-frequency observations, are implemented, in order to estimate local ionospheric conditions. They have the following form:

$$d\Phi_i(L_1) = (f_2^2 - f_1^2)/f_2^2 \cdot \{\Phi(L_1) - \Phi(L_2) \cdot f_1/f_2 - [N(L_1) - f_1/f_2 \cdot N(L_2)]\} \quad (6.12)$$

where $N(L_1)$ and $N(L_2)$ are the cycle ambiguities of $L_1$ and $L_2$, respectively, and $f_1$ and $f_2$ are the respective frequencies of the $L_1$ and $L_2$ carriers. From Equation 6.12 it is possible to see that if no cycle slips occur then the first term on the right-hand side reflects the main variation introduced by the ionosphere. The use of local models for the ionosphere are only really important for regional networks, due to the large-scale lateral variations of ionospheric characteristics, while for smaller networks the ionospheric biases assume a secondary role. In the latter case, if all measurements are performed using the same satellite configurations and during the same time interval, then the ionospheric biases can be considered as a scaling factor in baseline measurements. The order of magnitude of this scaling factor can be obtained from models of the type considered for constructing Figure 6.4. Moreover, because values of TEC are independent of both time and geographical area, there are certain areas, and certain times of the day and year that are better suited than others for performing GPS measurements. At middle latitudes such conditions occur, respectively, at night and during July. Variations in the TEC values are also affected by changes in solar activity, including those related to the Sun's 27-day rotational period and to the 11-year sunspot cycle.

As many GPS surveys are often performed using dual-frequency receivers, it is convenient to eliminate the ionospheric effects by using the $L_3$ observable. Although this observable is not affected by any ionospheric refraction, it does suffer from the drawback of a high noise level about three times greater than for the $L_1$ observable. The use of $L_3$ is not, therefore, always the best choice if other dispersive factors (e.g. multipath effects) are significant. This is particularly true for small networks which are characterized by relatively minor differential ionospheric effects compared with the noise level (Bock et al. 1986, Beutler et al. 1989). Another problem related to ionospheric conditions, that involves the "scintillation" effect (Klobuchar 1991), and is due to local irregularities in the ionosphere which can produce short-term signal feeding, and under certain conditions, the formation of cycle slips. According to Klobuchar (1991) the region prone to the larger scintillation effects lies within approximately 30° of the Earth's geomagnetic equator. The problems caused by scintillation can be significantly reduced by avoiding GPS measurement between approximately 19.00h and 24.00h (local time).

Tropospheric effects cannot be eliminated by using double-frequency receivers. Furthermore, unlike effects related to conditions in the ionosphere, they are not dispersive, and thus need to be addressed using different methods. Several models are available to estimate the bias introduced as a result of tropospheric effects (e.g. Hopfield 1969, Saastamoinen 1973).

Such models are generally expressed in the following form:

$$C_t = d_0 \cdot f(z) \tag{6.13}$$

where $d_0$ is the tropospheric zenith path delay, and $f(z)$ is a function of the zenith distance of the station from the satellite, which generally behaves as $1/\cos(z)$. The term $d_0$ can be expressed as the sum of a "dry" and a "wet" delay component. The dry component is essentially a function of the surface pressure, while the wet component depends on the specific atmospheric conditions (particularly humidity and temperature) along the signal path crossing the local troposphere. The minimum magnitudes of the dry and wet delays, corresponding to paths crossing the troposphere along the zenith direction, are 2.2–2.3 m and 20 cm. respectively. Published models give similar results for satellite elevations greater than 20°, which is the range relevant to most applications. For such elevations, the maximum total delay is between 5.3 and 7.3 m, depending on the specific model used. Most available software packages compute atmospheric parameters using standard models of the atmosphere which take into account the altitude of the receiving station. This means that data are normally processed on the basis of environmental conditions which may not be correct, particularly with regard to values of temperature and humidity. Beutler et al. (1989) have shown that an error of 1 °C in the temperature is reflected by an error in $d_0$ of between 3 and 27 mm. Moreover, in the differential use of GPS the errors in estimating and of the tropospheric effects do not cancel out, resulting in a bias in the solution equation (Santerre 1991). The main effects on the vector baseline, for satellite elevations greater that 15°, relate to the vertical component, which has been estimated to range between 3.2 and 4.9 times (for satellite configurations seen from equatorial or middle latitudes, and polar latitudes respectively) the uncertainty in the relative tropospheric zenith delays between two stations. The effects on the horizontal coordinates are, in contrast, negligible (Beutler et al. 1989, Santerre 1991). Hence an error of 1 °C in the temperature will be reflected by an error in the vertical component of the baseline of about 10–130 mm, depending on the satellite elevation and latitude.

For all tropospheric models, it is easier to estimate the dry component rather than the wet one, thereby minimizing its effects. The major problem remains, therefore, in establishing the effects of water vapour in the lower part of the atmosphere. This problem can be solved, at least to some degree, by using meteorological data recorded at the measurement sites or, if feasible, along the signal path (e.g. using a water-vapour radiometer). Some experiences reported by Beutler et al. (1989) and Gurtner et al. (1989) have shown that, in survey areas characterized by large differences in altitude, good results can be obtained if meteorological data are used to compute a model for the local troposphere which replaces the standard one.

### 6.3.3 Integer ambiguities

Integer ambiguity determination, as mentioned in the previous section, is crucial for the successful use of GPS as a geodetic tool. Several strategies have been proposed to solve this problem using both $L_1$ and $L_1/L_2$ observations. The simplest technique involves taking $N$ as an additional unknown variable in a general baseline determination and solving the problem by using Equation 6.7. In this way the values obtained for $N$, which are generally floating-point values, are, in the simplest case just rounded. Other strategies are based on combining both phase and code observables appropriately. It is worth noting that the problem of obtaining an appropriate set of ambiguities is also influenced by the length of the time interval over which the satellites are tracked, and it has been widely and practically demonstrated that operating under normal conditions (e.g. five satellites visible and baselines up to 10 km), about 1 h of observation is sufficient to guarantee the determination of a good set of ambiguities. Recently, a special search-ambiguity strategy has been proposed (Frei & Beutler 1989, 1990), which considers information contained in the variance–covariance matrix of the initial differential position adjustment. Such a technique, which is known in the literature as Fast Ambiguity Resolution Approach (FARA), is implemented in several available software packages allowing the computation of short baselines (less than 15 km) using only 5–20 min of observations.

### 6.3.4 Cycle slips

It has previously been indicated that cycle slips can be "fixed" or "repaired" by making use of triple-difference observations. Other procedures have also been proposed, however, which are able to detect and remove cycle slips using data recorded by a single station. Some of these strategies use appropriate combinations of $L_1$ and $L_2$ data or of code and carrier data (ionospheric-residual and kinematic-mode methods), while others (polynomial approaches or Kalman filtering) handle the observation data as a time series (Lichtenegger & Hofman-Wellenof 1989). In some exceptional cases it is possible to remove cycle slips by simply editing the data file during the pre-processing phase, although such opportunities are rare.

### 6.3.5 Multipath effects

Multipath effects are caused by destructive interference between the direct signal arriving at the user antenna and the signal reflected by other surfaces (e.g. ground or water surfaces) or obstacles (e.g. topographic highs). This results in errors in the determination of the station–satellite distance of between 1.3 and 5 m (Evans & Hermann 1989), and may also cause breaks in satellite tracking. To minimize this problem antennas are carefully designed and manufactured using special absorbent material around their phase centres. Multipath effects can be further limited by judicious site selection.

### 6.3.6 Errors due to signal access constraints

There are two main categories of GPS user, those having access to the Precise Positioning Service (PPS), and those who are relatively disadvantaged by having their access limited to the Standard Positioning Service (SPS). Users of the former category are largely confined to the USA and allied military, the US Government, and a few selected civil users specifically approved by US Government. The second category includes nearly all other civilian users. The PPS provides the user with the most accurate positioning, velocity, and timing information, continuously available on a worldwide basis. In contrast, SPS supplies the user with less accurate data, access to which is felt to constitute less of a threat to United States security. In practice the US Department of Defense use two methods for denying civilian users full use of GPS, selective availability (SA), and antispoofing (AS).

Selective availablity involves the degrading of both the pseudo-range measurement and the satellite position. The first objective is achieved by deterring the satellite clock, using the so-called "δ-procedure", while the satellite position is degraded by introducing errors in the broadcast orbit parameters by means of a method called the "ε-procedure" (Georgiadou & Doucet 1990). Antispoofing involves the eventual replacement of the P code by an encrypted code (the Y code) with similar features. This will be activated when the system is fully operational, and the P code becomes genearally available to all users.

An obvious effect of access denial can be seen with regard to the statistical indicators of the satellite clock and ephemeris prediction accuracy, defined as the User Range Accuracy (URA) or User Equivalent Range Error (UERE), which are transmitted in the navigation message. The worse values range between 2 and 10 m for a satellite without denial of the signal, compared with about 32 m where signal denial is operated.

## 6.4 GPS operating equipment and survey design

The standard equipment required for GPS field operations comprises receivers, antennas, tripods or other means of antenna support, and power supplies (Fig. 6.5). The receivers, after aligning their internal oscillators with the incoming satellite signals, continuously track the satellites, performing measurement both on the carrier frequencies and codes. As GPS measurements refer to the phase-centre of the antenna, this must be accurately and precisely installed over the station or benchmark. Care must be taken to ensure that the orientation of the antenna is maintained between different measurement sessions. The choice of the site is also important, bearing in mind that GPS signals will not penetrate obstacles (including vegetation). Antennas must be located, therefore, in such a way as to guarantee a clear and unobstructed view of the sky. Furthermore, the potential for signal reflections generated from nearby flat surfaces should be minimized as this can can generate multipath interference. Finally, equipment for measuring and recording meteorological data must be used if an accurate benchmark height estimation is required. Whatever the goals of the survey, whether the determination of relative distance changes or of absolute variations in coordinates, it is useful, if at all possible, to fix the GPS network to an existing geodetic network, by using at least three common points.

**Figure 6.5**   GPS equipment in use during the Etna campaign.

### 6.4.1 Planning a GPS survey

A GPS survey must be planned with particular care in order to optimize human and financial resources. To achieve such goals successfully, both technical and logistical aspects must be considered in some detail, and particular attention must be devoted to designing the network configuration. GPS techniques allow measurements to be performed in all weather and visibility conditions, but the geometry of the network may still be constrained by factors such as accessibility and the locations of pre-existing benchmarks which must be used to link the GPS network to other geodetic networks. Even if a GPS network is to be used only for the monitoring of relative distance changes between successive surveys, it is still useful to link up with any pre-existing reference networks in the area of investigation. In this context, several countries are currently developing programmes devoted to updating their existing national reference networks (e.g. the US National Geodetic Reference System, Strange & Zilkoski 1991; or the Italian IGM95 project, IGMI 1992) using GPS to ensure increased precision. Linking GPS and other established networks will also minimize the scale effects which represent a common problem in the operation of GPS networks.

Desirable GPS site characteristics are summarized below:

(a)  A clear view of the sky above a pre-fixed elevation (for example 15°) should be ensured. Station visibility can be determined using "sky-plots" (Fig. 6.6), which are polar or orthogonal plots showing the satellite paths as a function of the elevation angle, at three different latitudes.

(b) Benchmarks should be positioned away from reflecting surfaces.

(c) Vehicle access should be viable to facilitate transportation of the often bulky equipment, while it should be possible to station the vehicle at least 10 m from the benchmark in order to avoid multipath effects.

After selecting sites based on information extracted from a map of the area (a scale of 1:25000 is recommended for this purpose), field reconnaissance is required in order to verify if the selected points meet the above requirements. At this stage station monumentation can be performed. Whenever possible, each station should consist of a solid concrete block mounted on firm bedrock. Another aspect which should be taken into account in planning a GPS survey is the positioning accuracy problem. The effect of satellite geometry on coordinate estimation is usually expressed by a factor referred to as the Dilution of Precision (DOP). Depending on what coordinates or combination of coordinates are considered, several different factors can be defined, e.g. the Vertical DOP (VDOP), and the horizontal DOP (HDOP). Generally speaking, DOP represents the ratio of positioning accuracy to measurement accuracy and can be expressed as:

$$\sqrt{trace(\mathbf{A}'\mathbf{A})} \qquad (6.14)$$

where $\mathbf{A}$ is the design matrix for the instantaneous positioning solution. It is common to view the DOP as the inverse of the volume of the polyhedron the base of which is the instantaneous satellite configuration, and the vertex of which is the point on the Earth's surface. Small DOP values thus correspond to a configuration characterized by satellites with a great angular distance between them. Such a configuration allows positioning with good accuracy. By multiplying the DOP and UERE one can obtain an estimation of the accuracy of absolute positioning.

Another important part of survey planning involves determining the optimum daily observation period and deciding how this will be divided into measuring sessions. For a static GPS survey a session should begin when at least four satellites are above a 15° elevation and should end when the fourth satellite drops below 15°. Of course, sessions when more than five satellites are above 15° are to be preferred, and generally speaking, the more satellites that are visible the better the geometrical considerations and the shorter the length of observation required. This is mainly an effect of a more extensive sampling of the atmosphere (giving a randomization of atmospheric delays) and changing satellite configuration geometry (giving a randomization of orbital errors). A high sampling rate does not necessarily lead to good accuracy, even if a good DOP is used, because this only drives down the random part of the errors but does not reduce biases introduced, for example, by tropospheric or ionospheric effects. In order to reduce such biases, an appropriate mask angle (15–20°) should be maintained during tracking and it is a good precaution to plan sessions (whenever possible) when ionospheric effects are liable to be smallest. Furthermore, it must be remembered that the minimum number of satellites needed to solve the positioning problem is four.

For single-frequency receivers operating in normal ionospheric conditions the session length can be chosen with reference to the Table 6.1 (Hofmann-Wellenhof et al. 1992), and can be greatly reduced if the FARA technique is used. The best method for determining the optimum

165

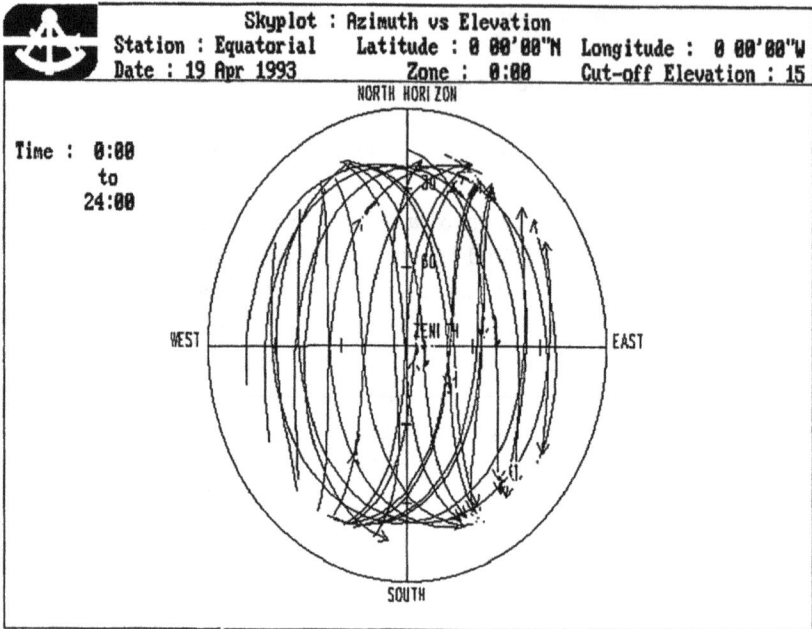

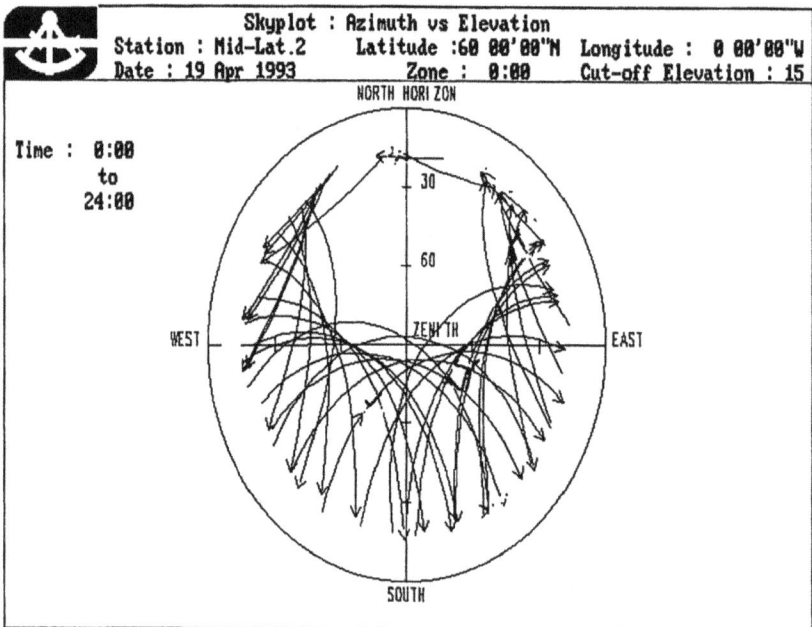

**Figure 6.6** Examples of sky plots for polar, equatorial, and mid-latitudes. Sky plots constitute a representation of satellite visibility, and provide an indication of the available satellite paths at different latitudes.

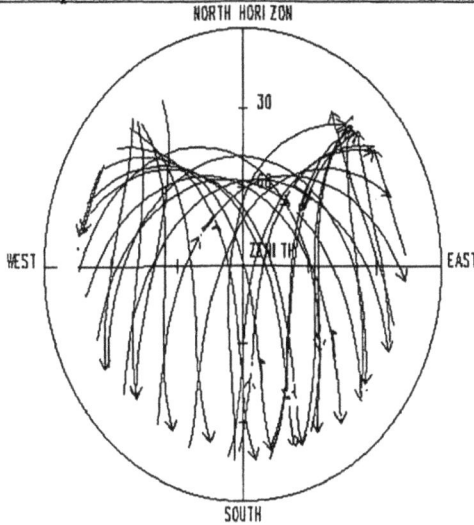

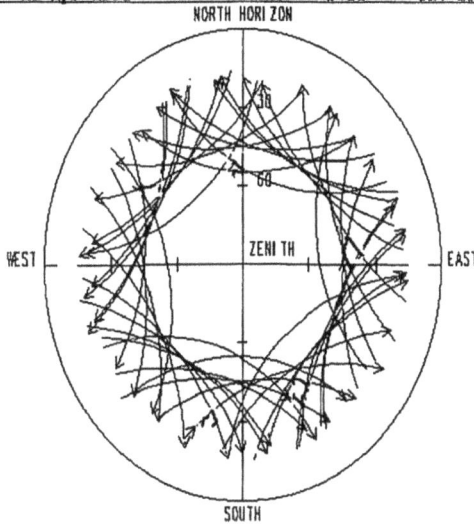

**Figure 6.6** continued.

**Table 6.1**  Session length versus baseline length for single-frequency receiver GPS observations at Mount Etna.

| Baseline (km) | Session (min) |
|---|---|
| 0.1 – 1.0 | 10 – 30 |
| 1.1 – 5.0 | 30 – 60 |
| 5.1 – 10.0 | 60 – 90 |
| 10.1 – 30.0 | 90 – 120 |

observation time, however, involves making longer than normal observations on the first day of the survey and processing data using progressively increasing portions of the recorded dataset, so as to determine when the accuracy of the results can no longer be increased. When a number of sessions have to be programmed during the same day, account must be taken of the time required to guarantee transport of the equipment to another site and to set up the receiver.

The final stage in survey planning ends with the plan of the layout and the definition of the number of sessions. Once the number of sites ($s$) and available receivers ($r$) have been established, the minimum number of sessions can be computed (Hofmann-Wellenhof et al. 1992) as:

$$n = \text{int}\left(\frac{s - o}{r - o}\right) + 1 \qquad (r < 1, o \geq 1) \qquad (6.15)$$

where $o$ represents the number of overlapping sites between the sessions and int(...) denotes the integer part of a number. An example of organizational design is given in Table 6.2.

**Table 6.2**  Example of GPS survey design: this should be organized to illustrate observing times, satellites used, and sites occupied.

| Date | Time (GMT) From | To | Satellites used | Benchmarks occupied |
|---|---|---|---|---|
| 11.6.92 | 10.31 | 12.42 | 03 12 13 20 24 25 | CIS CAP SAP TDF |
| 11.6.92 | 14.20 | 16.30 | 03 17 20 21 23 25 28 | STP VET SAP TDF |
| 12.6.92 | 10.27 | 12.38 | 03 12 13 20 24 25 | VET LAM CIT |
| 13.6.92 | 10.23 | 12.34 | 03 12 13 20 24 25 | CRP CIS STP VET |
| 15.6.92 | 10.15 | 12.26 | 13 12 13 20 24 25 | BEL CRP CAP TDF |
| 15.6.92 | 14.04 | 17.12 | 03 11 17 20 21 23 25 28 | BEL CRP PDG NUN |
| 16.6.92 | 10.11 | 12.22 | 03 12 13 20 24 25 | BEL MGN 832 LAM |
| 16.6.92 | 14.00 | 17.08 | 03 11 17 20 21 23 25 28 | BEL MGN 832 NUN |
| 17.6.92 | 10.07 | 12.18 | 03 12 13 20 24 25 | TDF MIL CIT STP |
| 17.6.92 | 13.56 | 15.18 | 03 17 20 21 23 25 28 | CIS MIL CIT STP |
| 18.6.92 | 10.03 | 12.14 | 03 12 13 20 24 25 | PDG VET CAP 832 |
| 19.6.92 | 09.59 | 12.10 | 03 12 13 20 24 25 | OBS MGN BEL CIT |
| 19.6.92 | 13.48 | 16.56 | 03 11 17 20 21 23 25 28 | OBS MGN LAM MIL |
| 22.6.92 | 09.47 | 11.58 | 03 12 13 20 24 25 | OBS PDG TDF NUN |
| 22.6.92 | 13.35 | 15.38 | 03 17 20 21 23 25 28 | OBS PDG CIS 832 |

### 6.4.2 Field operations

Performing a GPS survey is usually a straightforward task using modern GPS receivers which are easy to transport and are completely automatic. In other words, they are capable of performing all the survey operations (e.g. calibration, satellite tracking, and data recording) without operator intervention. The receiver must, however, be programmed before starting in order to select parameters such as the sampling rate, the minimum number of satellites to track, the particular satellites to be tracked, the cut-off elevation, the start and stop time of the session related to the satellite elevation, and other related parameters. This operation, which can be performed manually using a computer, is usually undertaken in the laboratory in order to avoid mistakes, but it can be carried out in the field if necessary. Provided that the receivers are pre-programmed, then field operations can be undertaken using suitable technical support staff.

Experience has shown that the most critical field operation is the establishment of the antenna, which, if incorrectly executed, may lead to significant errors in the data. In centring the antenna two aspects have to be taken into account. Firstly, due to the fact that the geometric centre of the antenna does not coincide with the electrical centre, all antennas used in a survey must always be aligned in the same direction (usually north). In relative positioning, this uniform orientation will cancel out any systematic offset. Secondly, the antenna support, usually a tripod, must be correctly centred over the station, usually accomplished using an optical plummet, and the antenna height precisely measured. This is particularly important in absolute-coordinate computation. In order to avoid errors which can occur due to mis-measurement of the antenna height, it is recommended that the parameter is measured twice, both at the beginning and the end of the session. It is essential to complete a site-occupation log during the survey, containing such relevant information as the station name, receiver model, observer's name, receiver and antenna serial numbers, antenna height, start and stop time of each session, tracked satellite and the corresponding signal-to-noise ratio at different times, meteorological data, and problems encountered.

GPS receivers store the recorded data from several sessions in their internal memories so that *in situ* data transfer to external storage devices, such as floppy disks, is not necessary. This operation is usually performed in the laboratory at the end of each survey day, when lodging a backup copy of the recorded data files on a different storage media is strongly recommended. Some modern receivers now have the capability of recording to cards making downloading of the data easier to accomplish.

## 6.5 Data processing

Several processing strategies are implemented in existing software packages. Since it is clearly impossible to describe all such packages we will devote our attention to describing the main steps through which the baseline solutions of a GPS survey are obtained. There are three types of processing strategy, choice of which depends on the required solution: single baseline (or vector by vector), multibaseline (or multistation), and multisession. The single-baseline solution is the one most commonly used, at least as a first approach. In contrast to the single-baseline solution, the multibaseline solution takes more than two stations into account during the same session, and a single solution is sought for the baselines involved. The multisession strategy, which is only available on certain software packages (e.g. TOPAS or BERNESE), considers data recorded in different sessions and even on different days. In this case it may not be necessary to include the network adjustment step which is, however, of fundamental importance when the single baseline or multibaseline strategies are considered. Whatever strategy is chosen, a pre-processing phase is always required to determine absolute positioning

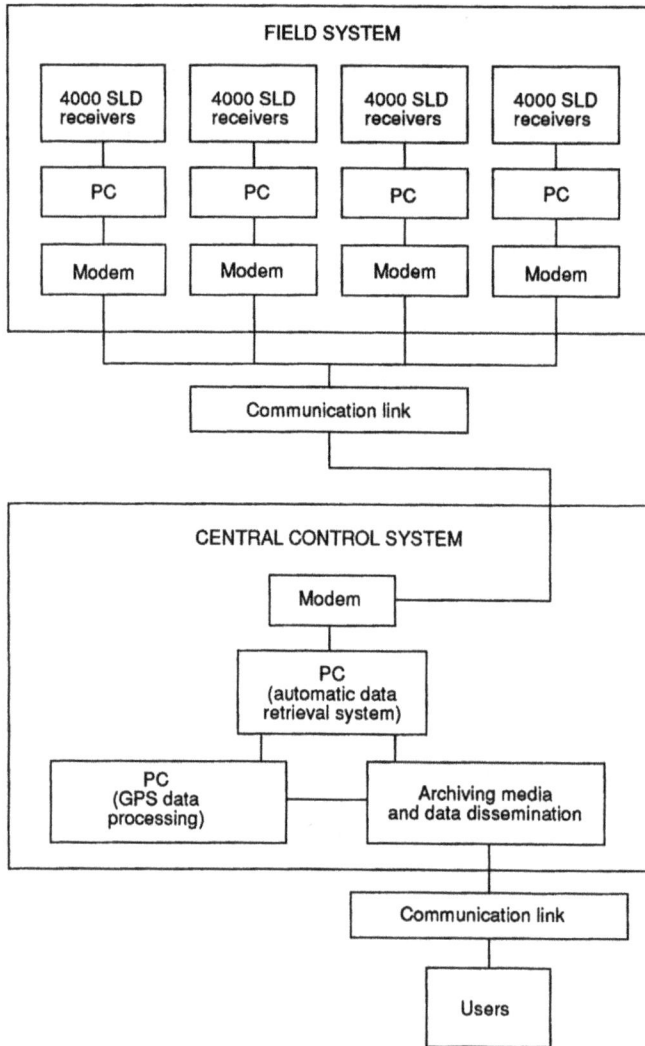

**Figure 6.7** Schematic diagram to illustrate the principal components and interfaces constituting a continuously recording GPS system (see text for explanation).

from the C/A or P code pseudo-range (a set of appropriate satellite ephemera are required in order to accomplish this), and to repair cycle slips.

At the end of the pre-processing phase, a set of "cleaned" data is obtained; these data are then used in subsequent processing steps. At this stage, data are appropriately combined, depending on the processing strategy used, to obtain a set of double-difference equations from which the station coordinates can be obtained, together with additional information concerning, for example, the ambiguities, the clock corrections, and the covariance matrices. The final processing stage, which is usually not needed when multisession strategies are used, involves the adjustment of the absolute coordinates computed in different sessions or reference to vector-by-vector solutions. At this stage station coordinate transformations may also be generated if required; where, for example, the GPS network has ties with reference stations belonging to lower-order networks.

## 6.6 Use of continuous GPS measurements for ground deformation monitoring

The impact that continuous ground deformation measurement can have on mitigating natural hazards, in particular earthquakes, volcanic eruptions, and landslides, is well known. At the same time it is also generally considered that traditional geodetic techniques are not suitable for continuous observations. The advent of GPS has, however, provided the opportunity for continuous geodetic monitoring, using permanent, fixed, receivers which are capable of automatically tracking satellites, recording data, and transmitting the data periodically to a central station where they can be automatically processed, stored for further analysis, and disseminated to interested users. An example of one such continuous GPS monitoring system, similar to that proposed by Bock & Shimada (1989), is illustrated in Figure 6.7, and a prototype system, used to monitor ground deformation at Mount Etna, has been under development since early 1992. Here, two GPS receivers have been maintained in a fixed position on a line (about 22 km long) situated on the south flank of the volcano. The receivers are connected via radio links to a central station located in the nearby city of Catania, and can be fully controlled from the central station where data is collected and processed. The daily distance of the measured line, computed between 14 February 1992 and 11 March 1992 is shown in Figure 6.8. It can be seen that, in spite of the considerable length of the line and the significant height difference between the benchmarks, the distance measurements show a precision of about 1 cm and 2 ppm. It is expected that experience gained through testing of the prototype system will form a sound basis for the construction, within the next few years, of a fixed GPS network for continuous ground deformation monitoring of the volcano (Ferrucci et al. 1993).

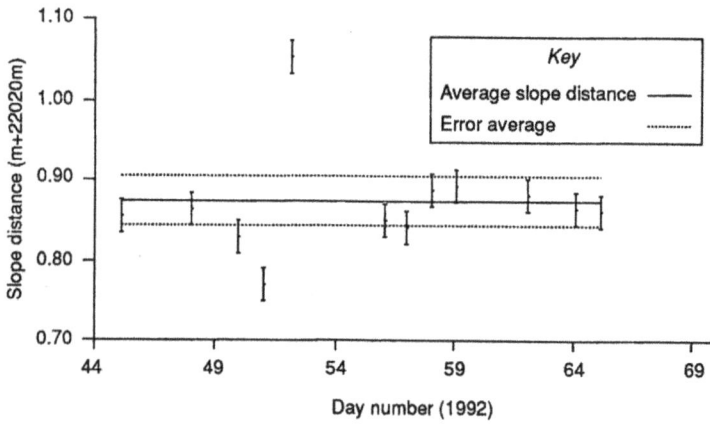

**Figure 6.8** Distance measurements of a baseline (Tetto IIV-Centro Operativo) at Mount Etna measured daily during February and March 1992. Despite being a long line, measurements illustrate a precision of 1 cm and 2 ppm.

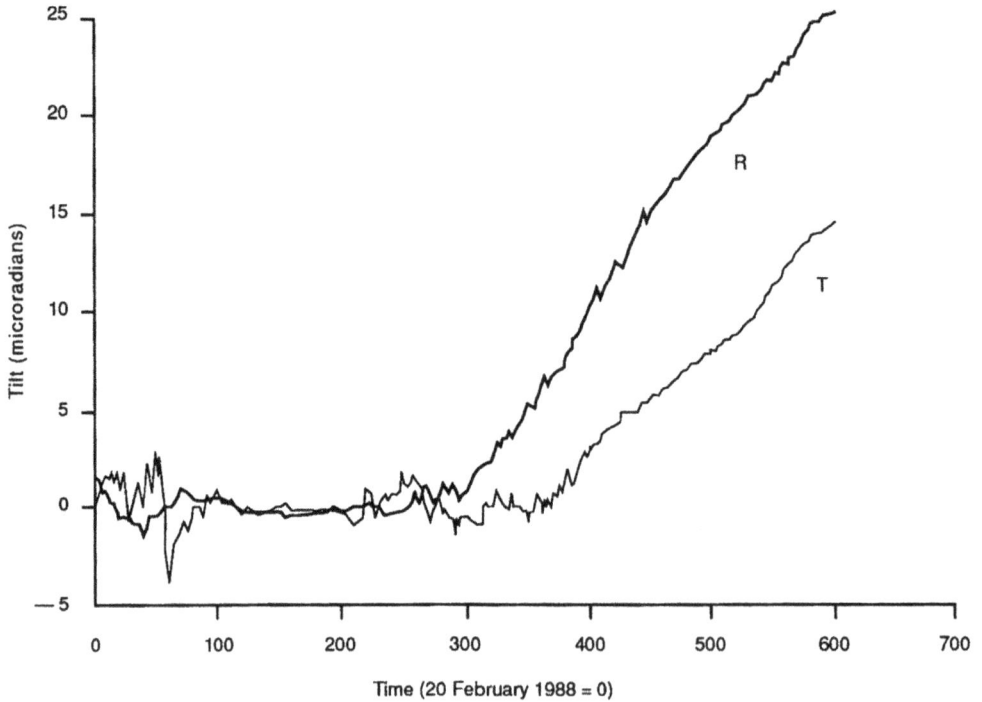

**Figure 6.9** Radial (R) and tangential (T) tilt components recorded on Mount Etna (MAL tiltmeter station) prior to the 1989 eruption. (After Briole et al. 1990.)

## 6.7 Case study: GPS surveys at Mount Etna

In this section the main experimental results obtained using GPS at Mount Etna are described. Over the period of the surveys, volcanic activity was characterized by two main eruptive events. The first commenced on 11 September 1989 after 4 months of strombolian activity at all the summit craters. This eruption started at the southeast crater and was characterized by both lava flows and exceptionally violent lava fountaining. On 27 September, two fracture systems started to open, originating at the Southeast Crater, and trending along bearings 040 and 140. Lavas issued from the northern fracture until 9 October, while the second fracture propagated downslope for about 7 km, reaching an altitude of only 1500 m on 2 October (Bertagnini et al. 1990, Frazzetta & Lanzafame 1990). As shown in Figure 6.9, ground deformation affecting the whole upper part of the volcanic edifice was recorded by automatic tiltmeters 6–7 months before the eruption (Bonaccorso et al. 1990, Briole et al. 1990). The second eruptive event, starting on 14 December 1991 and ending on 1 April 1993, is probably the most important eruptive event on Mount Etna this century, at least in terms of extruded volume. The vents from which the lavas issued are located at an altitude of 2400 m along the same 140-trending fracture system initiated in September 1989

### 6.7.1 Network design and description

The first GPS survey at Mount Etna was performed during 1988 using a geodetic network consisting of nine benchmarks and 18 baselines up to 25 km long. The results obtained showed that the accuracy of the new GPS techniques was comparable with that of the more traditional EDM (Briole et al. 1988). Since this initial survey, the network has been modified and enlarged in order to allow more systematic surveys and integration with other geodetic networks (both trilateration and precise levelling) already established and operated on the volcano. The original number of benchmarks and baselines was subsequently increased to 12 and 36, respectively, allowing a wider region of the volcano to be covered (Fig. 6.10). Two of the new GPS benchmarks (PDG and STP) belong to previously established geodetic networks (Nunnari & Puglisi 1987, Falzone et al. 1988), to allow integration of the different datasets. The data discussed here are based on three GPS surveys performed on mount Etna between 1989 and 1991 (Table 6.3).

Table 6.3  Two types of receiver were used during the Mount Etna GPS campaigns: a Trimble 4000 SL Single Frequency ($L_1$), and a Trimble 4000 SST Double Frequency ($L_1$ and $L_2$).

| Survey date | Receiver no. | Receiver type | No. of baselines measured* | Range of data-file length |
|---|---|---|---|---|
| Aug–Sept 1989 | 2 | Trimble 4000SL | 10 | 2h 26 min to 4 h 04 min |
| July–Aug 1990 | 4 | Trimble 4000SL and 4000SST | 36 | 3 h 45 min |
| June–July 1991 | 3 | Trimble 4000SL and 4000SST | 20 | 1 h 30 m to 1 h 45 min |

* The number of baselines measured was constrained both by the type of receiver used, and by a number of technical problems encountered during the surveys

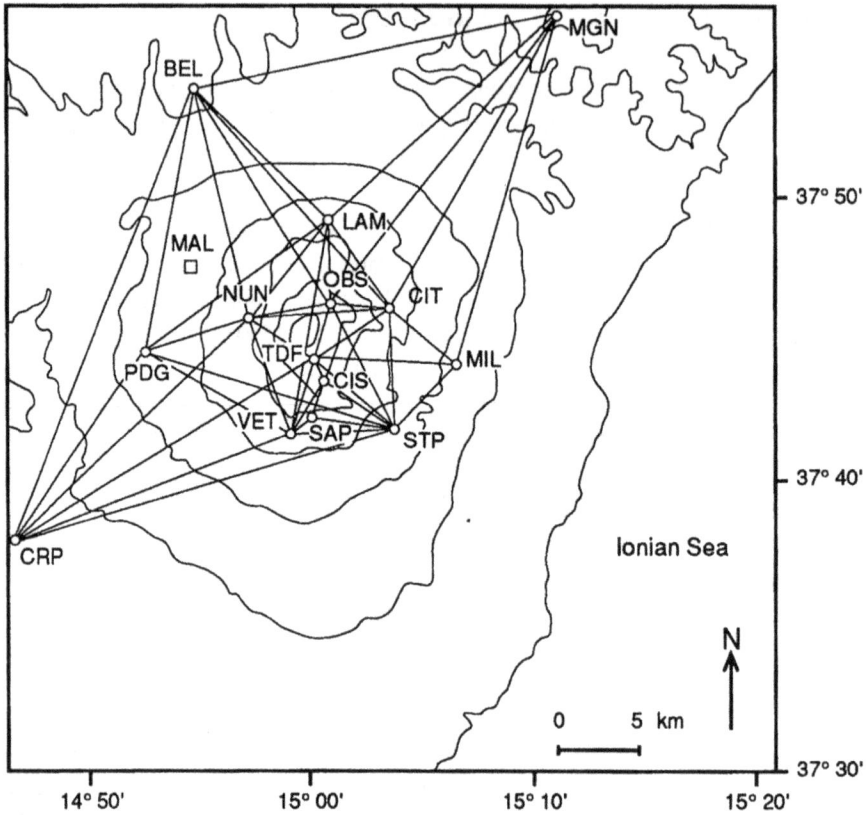

**Figure 6.10**    The Mount Etna GPS Network. This now consists of 12 benchmarks and 36 base-lines.

## 6.7.2 Data processing

Data processing was performed using the TRIMVEC-PLUS (revision D) software package pro-duced by Trimble Navigation Limited (1990). The length and number of measurement ses-sions were different for the three surveys, as shown in the final column of Table 6.1. Recorded data were organized in such a way as to guarantee homogeneous conditions during process-ing, and to make use of all the available information. In particular the sampling frequency was set equal to 0.1 Hz, and the individual measurement sessions obtained each day were merged. For each dataset, a pre-processing phase was performed in order to obtain an ap-proximate set of coordinates for each benchmark and remove errors due to receiver-clock offsets and cycle slips. The first of these goals, which involves the application of a least-squares method to process the pseudo-range observations, is of great importance for relative posi-tioning, and it has in fact been shown (e.g. Santerre 1991) that errors of 10 m in the absolute benchmark coordinates lead to errors of about 1 ppm in the relative distance vector compo-nents. For our purposes a set of benchmark coordinates obtained in a previous adjustment phase (Nunnari & Puglisi 1991) was used to process the datasets considered here, in order to reduce possible scale effects due to data recorded at different epochs. This pre-processing step also made it possible to estimate the receiver clock offset. To eliminate cycle slips, the software package used is based on triple-difference observation (see Eq. 6.8) which characteristically displays a peak corresponding to a cycle slip. In this way cycle slips can be detected and removed from the satellite observation file by using appropriate automatic procedures involving double-difference observations.

After the pre-processing phase, observations were processed using the double-difference model described by Equation 6.7, in order to obtain the baseline vector components. As previously mentioned, the double-difference model involves finding the set of integer cycle ambiguities. The goodness of the found set of ambiguities is evaluated a posteriori using appropriate quality factors related to the rms obtained from the adjustment process, and the set of ambiguities adopted is that which gives the lower rms value. Once a set of baselines was obtained, a compensation phase was performed for each individual survey, which assumed a minimally-constrained condition for the adjusted network. In this hypothesis, no control-point coordinates need to be kept fixed and the software uses the so-called "inner constraints". More precisely, the adjustment considers the centroid of the unadjusted coordinates of the benchmarks as a constraint. This allows both the internal precision and consistency of the GPS solutions obtained to be detected, and the benchmark relative distances over different surveys to be compared, even if vector displacements cannot be computed. For these reasons, the results described here refer only to the ellipsoid distance comparison. However, such a distance usually constitutes the most precise geodetic slope–distance vector component that can be obtained by GPS measurements. The other geodetic components (e.g. azimuth and ellipsoid height), are affected by bias factors, such as atmospheric conditions, which limit their accuracy. Because of such problems, determination of the variation with time of benchmark height will not be considered here, but some further considerations relating to this problem are discussed in the following section.

### 6.7.3 Measurement precision and repeatability

The precision of relative static positioning using GPS depends on a number of factors, including the length of the measurement sessions, the availability of dual-frequency receivers, the particular software used for data processing, and the availability of the precise ephemeris. Here, we report some results, concerning the precision and repeatability of baseline measurement at Mount Etna, using a "minimum-effort" strategy, both in terms of field and laboratory operations. This means, for example, that session lengths were chosen to be as short as possible (although still compatible with baseline lengths and the tracking of at least five satellites at the same time), and standard rather than customized software packages were used. Furthermore, even though both single- and dual-frequency receivers were often used at the same time, only $L_1$ data were processed because this represented the most complete dataset available for analysis. Although resulting in less precision, the minimum-effort strategy is visualized as being realistic for GPS operations on large active volcanoes such as Mount Etna,

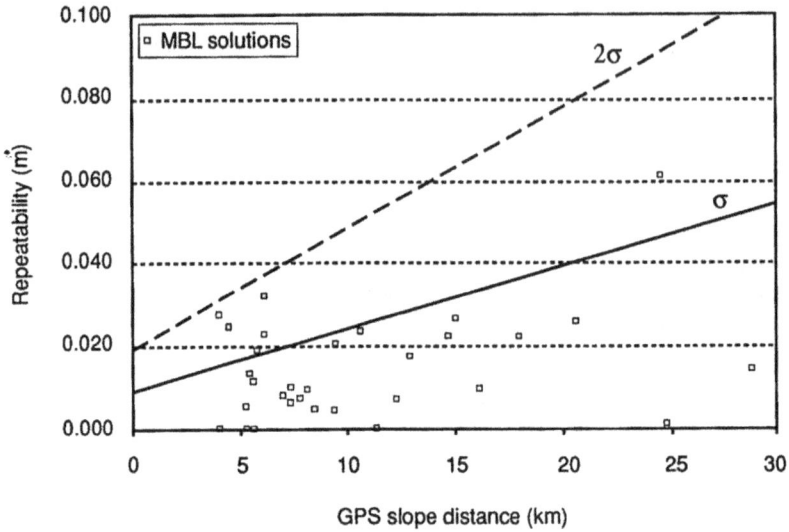

**Figure 6.11** Plot of repeatability versus slope distance obtained using the multibaseline (MBL) procedure. The lines corresponding to $s_1$ = 0.7 cm and 1 ppm times baseline length, and $2s_1$ are shown (see text for explanation).

where there are a number of constraints related, in particular, to logistical and environmental problems, and manpower resources and organization.

In order to evaluate the precision of GPS baseline solutions, data gathered during four surveys performed since 1990 were considered, thus ensuring that the data were recorded over a wide range of operative conditions, including different satellite configurations, session lengths, receiver models, and weather conditions. Two different procedures were used to process the data, which included only those baselines measured at least twice during each survey. The first procedure, referred to as "single baseline" (SBL), uses standard values for all the parameters involved in the computation in order to speed up the whole process. In the second method, multibaseline processing (MBL) some of the parameters, such as the number of iterations and the number of epochs considered, were chosen on an *ad hoc* basis. Care was also taken to ensure that the initial set of benchmark coordinates were sufficiently accurate (i.e. within 10 m with respect to the "true" values). Due to the fact that the results obtained using the SBL and MBL procedures are quite similar, only those generated by the MBL procedure are reported here; and these are illustrated in Figure 6.11. Each point on the figure represents the difference between the minimum and maximum slope distance corresponding to a given baseline measured at least twice during a particular survey. Let us assume the following general expression for the change of precision with baseline length:

$$\sigma = A + Bl$$

where $A$ and $B$ are the two constants defined below. In Figure 6.11, the curves corresponding to $\sigma_1$ and $2\sigma_1$ when $A$ = 7 mm and $B$ = 1 ppm, showing that the repeatability appears generally to be better than $\sigma_1$. Plots above this line correspond to baselines characterized by height differences of between 800 and 1600 m which are strongly influenced by tropospheric effects. This is to be expected when a standard tropospheric model, rather than a local one, is used (Beutler et al. 1989, Tralli et al. 1988). These observations justify why, as described in the following section, the vertical components of the baseline vectors have not been considered in the analysis of ground deformation patterns at Mount Etna using GPS data. It should also be emphasized here that, for the Mount Etna surveys, only the broadcast ephemeris was used; better results would be obtained by utilizing the precise ephemeris.

Assuming that the precision of the measuring process is independent of the choice of baseline, it can be concluded that the considerations reported above for repeatability will also apply to the survey precision. Interestingly, the value obtained for the precision (7 mm and 1 ppm) is almost the same as that obtained by Briole et al. (1992) for the same area using standard EDM techniques. This does not mean, however, that GPS and EDM measurements have the same precision in general. In fact, some authors (e.g. Davis et al. 1989, Beutler et al. 1989, Larson & Agnew 1991), claim higher precision for the GPS method, although such results may only apply to very specific operating conditions and/or processing strategies. Larson & Agnew (1991), for example, report a precision of the order of 0.01 ppm, but this was achieved at night using double-frequency receivers and session lengths of about 7–8 h, and with data processing being done using very sophisticated software.

### 6.7.4 Comparisons between Mount Etna GPS surveys

The Mount Etna GPS surveys have proved useful in revealing ground deformation patterns related to recent volcanic activity. A comparison between the 1989 and 1990 ellipsoidal distances (Table 6.4) shows that no significant variations occurred over this period. Considering that the 1989 survey was performed just before the beginning of the September eruption, this suggests that deformation of the upper part of the edifice had been achieved immediately prior to the eruption, and at the time of the first survey. This is in agreement with the areal dilatation data reported in Bonaccorso et al. (1991) for other geodetic networks on the north-

**Table 6.4** Comparisons between the 1989 and 1990 ellipsoidal distances (in metres) for the Mount Etna GPS campaign shows that little significant variations were recorded over this period (see text for explanation).

|  | 1989 | | 1990 | | Changes | |
|---|---|---|---|---|---|---|
| Line | Distance | Error | Distance | Error | Distance | Error |
| BEL–CIT | 20357.881 | 0.009 | 20357.887 | 0.006 | 0.006 | 0.011 |
| BEL–LAM | 13041.578 | 0.008 | 13041.583 | 0.004 | 0.006 | 0.009 |
| BEL–OBS | 17734.758 | 0.009 | 17734.750 | 0.006 | −0.008 | 0.010 |
| BEL–MGN | 24666.602 | 0.010 | 24666.598 | 0.009 | −0.003 | 0.013 |
| CIT–LAM | 7382.692 | 0.004 | 7382.693 | 0.006 | 0.001 | 0.007 |
| CIT–OBS | 3916.311 | 0.004 | 3916.324 | 0.005 | 0.013 | 0.007* |
| CIT–MGN | 22867.343 | 0.004 | 22867.349 | 0.010 | 0.005 | 0.011 |
| LAM–OBS | 5706.234 | 0.005 | 5706.227 | 0.005 | −0.006 | 0.007 |
| LAM–MGN | 20615.808 | 0.006 | 20615.805 | 0.009 | −0.002 | 0.011 |
| OBS–MGN | 24663.864 | 0.006 | 24663.868 | 0.010 | 0.005 | 0.011 |

* Statistically significant baseline differences

**Table 6.5** Comparisons between the 1990 and 1991 ellipsoid distances (in metres) for the Mount Etna GPS campaign demonstrate an increase in measured ground movement compared to the previous year (see text for explanation).

| Line | 1990 Distance | 1990 Error | 1991 Distance | 1991 Error | Changes Distance | Changes Error |
|---|---|---|---|---|---|---|
| CIT–LAM | 7382.693 | 0.006 | 7382.699 | 0.006 | 0.007 | 0.008 |
| CIT–NUN | 9417.401 | 0.005 | 9417.428 | 0.006 | 0.027 | 0.007* |
| CIT–OBS | 3916.324 | 0.005 | 3916.329 | 0.004 | 0.005 | 0.007 |
| CIT–TDF | 6016.929 | 0.006 | 6016.946 | 0.006 | 0.017 | 0.008* |
| CIT–VET | 10633.918 | 0.006 | 10633.924 | 0.008 | 0.006 | 0.010 |
| CIT–PDG | 16566.685 | 0.006 | 16566.724 | 0.008 | 0.039 | 0.010* |
| CIT–STP | 7781.767 | 0.008 | 7781.791 | 0.023 | 0.023 | 0.024 |
| LAM–NUN | 8171.292 | 0.004 | 8171.291 | 0.006 | –0.002 | 0.007 |
| LAM–OBS | 5706.227 | 0.005 | 5706.210 | 0.004 | –0.017 | 0.007* |
| LAM–TDF | 9200.133 | 0.007 | 9200.131 | 0.006 | –0.002 | 0.009 |
| LAM–VET | 14527.051 | 0.006 | 14527.424 | 0.007 | –0.014 | 0.010* |
| LAM–PDG | 14847.051 | 0.006 | 14847.059 | 0.008 | 0.008 | 0.010* |
| LAM–STP | 14535.191 | 0.008 | 14535.211 | 0.022 | 0.020 | 0.023 |
| NUN–OBS | 5548.101 | 0.004 | 5538.125 | 0.004 | 0.024 | 0.006* |
| NUN–TDF | 5116.834 | 0.006 | 5116.847 | 0.004 | 0.013 | 0.007* |
| NUN–VET | 8452.705 | 0.006 | 8452.703 | 0.007 | –0.003 | 0.009 |
| NUN–PDG | 7310.210 | 0.005 | 7310.220 | 0.007 | 0.011 | 0.009* |
| NUN–STP | 12237.431 | 0.007 | 12237.459 | 0.016 | 0.028 | 0.017* |
| OBS–TDF | 3677.540 | 0.006 | 3677.560 | 0.005 | 0.020 | 0.008* |
| OBS–VET | 9060.838 | 0.006 | 9060.847 | 0.007 | 0.009 | 0.010 |
| OBS–PDG | 12797.680 | 0.006 | 12797.716 | 0.007 | 0.036 | 0.009* |
| OBS–STP | 9128.327 | 0.008 | 9128.352 | 0.007 | 0.029 | 0.021* |
| TDF–VET | 5387.189 | 0.006 | 5387.178 | 0.007 | –0.011 | 0.010* |
| TDF–STP | 7144.149 | 0.008 | 7144.165 | 0.016 | 0.016 | 0.018 |
| VET–PDG | 11286.203 | 0.007 | 11286.224 | 0.006 | 0.020 | 0.009* |
| VET–STP | 6952.416 | 0.007 | 6952.420 | 0.015 | 0.005 | 0.016 |
| PDG–STP | 17449.195 | 0.008 | 17449.233 | 0.014 | 0.038 | 0.016* |
| PDG–TDF | 11241.451 | 0.007 | 11241.476 | 0.006 | 0.025 | 0.009* |

* Statistically significant baseline differences

east and southwest flanks of the volcano. Some deformation was recorded, however, on the southeast flank after the start of the eruption, and related to the development of a 7 km long fracture system in the area. The fact that this was not picked up in the GPS surveys illustrates the need to increase the density of GPS benchmarks in this part of the volcano. Table 6.5 shows comparison between the 1990 and 1991 ellipsoidal distances, while Figure 6.12 illustrates the significant baseline-length variations in order to give a spatial picture of the precursory ground deformation pattern prior to the December 1991 eruption which occurred 5 months after the GPS survey. It is particularly noticeable that there are a large number of deformed baselines with respect to the 1989–90 period. Only three ellipsoidal distances show a shortening (trend north–south), while all the remaining ones show a dilatation (trending roughly east–west). It is unfortunate that logistical problems during the 1991 survey did not allow the ground movements within the area of investigation to be compared with deformation within

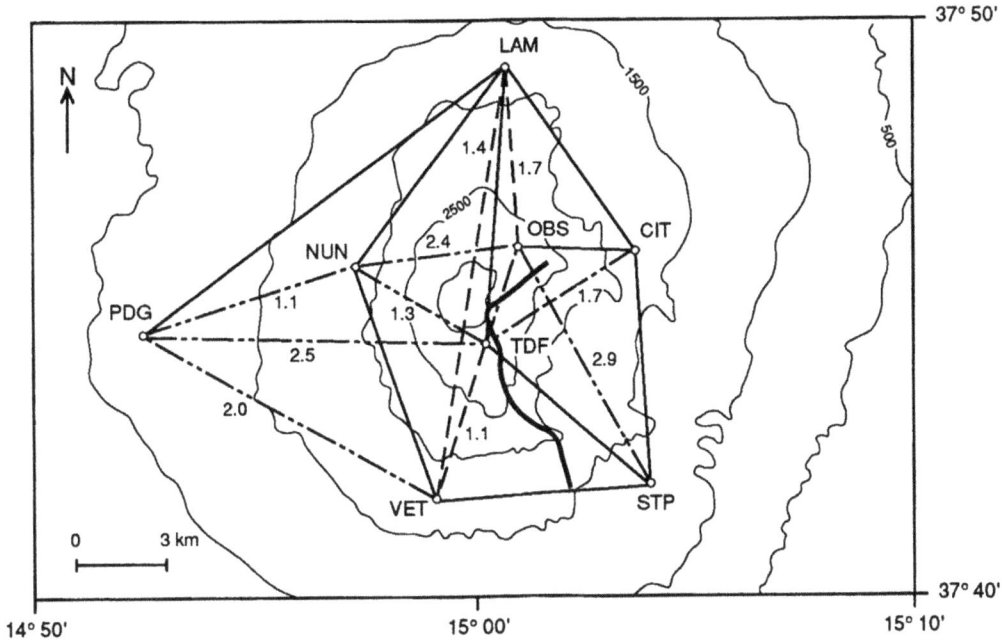

**Figure 6.12** Ellipsoidal distance changes (in cm) between the 1990 and 1991 GPS surveys at Mount Etna. Dotted lines represent extension, while dashed lines represent contraction. The two fracture systems which opened on September 1989 are also indicated. As explained in the text, the data reveal a picture of generally north–south compression associated with extension in a roughly east–west direction.

the more stable surrounding region. It does appear, however, that the ground deformation pattern revealed by the 1990 and 1991 surveys is consistent with a stress field the compressive direction of which is oriented north-northeast–south-southwest, and which is believed to have characterized the Mount Etna region for some time (Lo Giudice & Rasa 1986, 1992, Ferrucci et al. 1992). The ground deformation patterns distinguished by the GPS surveys have been modelled by the authors using a planar-source model of the type proposed by Okada (1985), and the results are reported in Nunnari & Puglisi (1994).

## 6.8 Concluding remarks

In this chapter the main features of the GPS technique as applied to geodetic measurements in volcanic areas have been introduced. GPS is particularly suitable to such terrains because it offers a solution to the many problems associated with the operation of traditional geodetic techniques. On the positive side it provides a precise way of measuring large networks, making the techniques ideal for investigating deep deformation sources, and allowing comparisons to be made between deforming volcanic edifices and the more stable basement and surrounding terrain. In order to optimize the advantages of the GPS method, however, particular care must be taken during all the phases involved in the operation of a GPS survey. Of particular importance are, selecting the appropriate number of satellites, avoiding natural obstacles, ensuring the choice of a good set of coordinates for the reference points, and accurately processing the recorded data. All activities require considerable planning and forethought, and significant computer time. A current drawback, which is likely to limit the use of the GPS method for some time, involves the often prohibitive cost of the satellite receivers. GPS receiver technology is, however, rapidly evolving, and it is likely that these costs will be dramatically reduced in the near future.

The results reported for the Mount Etna surveys are encouraging in terms of both precision and repeatability. The precision of vector baseline determination is generally better than 7 mm and 1 ppm, even using standard processing procedures, while the baseline-determination precision after adjustment processing is even better. The main hindrance to precision improvement results from difficulties in modelling the "wet" part of the troposphere. Although for a typical active volcanic edifice ionospheric effects are minimal, those associated with the troposphere make the determination of vertical positioning particularly difficult. This means that the vertical displacements that are commonly associated with volcanic activity cannot be successfully monitored. One reason for these difficulties lies in the fact that there are usually consistent differences in altitude between benchmarks located on volcanoes, so that the error induced by the different delays of the electromagnetic waves in the troposphere does not allow vertical displacement to be obtained as accurately as baseline lengths, unless appropriate models for the troposphere and accurate local meteorological data are available. Furthermore, the relative differences in height between benchmarks obtained from GPS measurements are computed with reference to an ellipsoidal datum. Volcanic areas are normally characterized by a large undulation of the geoid which should really, therefore, be accurately determined before comparing vertical displacements obtained using different geodetic techniques. Although the problem of successfully modelling tropospheric effects has yet to be solved, in the authors' opinion one can at least partially overcome the problem by measuring a redundant set of data (in different sessions) for those baselines which are particularly sensitive.

A number of other strategies can be adopted to improve the general precision of GPS measurements. A first improvement (of the order of 0.14 ppm) can be achieved by processing GPS data using the precise ephemeris, which is distributed by some specialized centres, such as the Defense Mapping Agency of the US Department of Commerce, instead of the broadcast ephemeris used in the Mount Etna surveys. Secondly, when installing a GPS network for monitoring crustal movements it is advisable to fix it at a minimum of three points considered to be stable, in order to obtain an unbiased estimate of the strain (Van Mierlo 1979, Bibby 1982) which is fundamental for computing vector displacements. The analysis reported in this chapter demonstrates that in this way it is possible to achieve precisions of the order of 1 ppm. Better precisions can be obtained, but at this stage of development to reach precision of the order of, for example, 0.01 ppm requires considerable time and care in both the field and laboratory segments of operations. Our experience suggests that such precision is rarely required when operating on large active volcanoes which are typically characterized often by strains significantly larger than 1 ppm.

In conclusion, it must be emphasized that using GPS for geodetic purposes remains a relatively new but expanding technique. Therefore, it can be expected that, although the precision of this method is even now suitable for geophysical studies in active volcanic terrains, this will undoubtedly be further improved. This, together with the ease of use of the technique, will ensure its successful utilization in volcanology and will encourage an increasing number of users.

# References

Bertagnini, A., S. Calvari, M. Coltelli, P. Landi, M. Pompilio, V. Scribano 1990. The 1989 eruptive sequence. In *Mt. Etna: the 1989 eruption*, F. Barberi, A. Bertagnini, P. Landi (eds), 10–22. Pisa: Giardini.

Beutler, G., W. Gurtner, M. Rothacher, U. Wild, E. Frei 1989. Relative static positioning with the Global Positioning System: basic technical considerations. In *Global Positioning System: an overview*, Y. Bock & N. Leppard (eds), 1–23. New York: Springer.

Bibby, H. M. 1982. Unbiased estimate of strain from triangulation data using the method of simultaneous reduction. *Tectonophysics* **82**, 161–174.

Bock, Y., S. A. Gourevitch, C. C. Counselmann, R. W. King (1986) Interferometric analysis of GPS phase observations. *Manuscripta Geodaetica* **11**, 282–8.

Bock, Y. & S. Shimada 1989. Continuously monitoring GPS networks for deformation measurements. In *Global Positioning System: an overview*, Y. Bock & N. Leppard (eds), 40–56. New York: Springer.

Bonaccorso, A., G. Falzone, B. Puglisi, R. Velardita, L. Villari 1990. Ground deformation. In *Mt Etna: the 1989 eruption*, F. Barberi, A. Bertagnini, P. Landi (eds), 44–7. Pisa: Giardini.

Bonaccorso, A., O. Campisi, G. Falzone, B. Puglisi, R. Velardita, L. Villari 1991. Ground deformation: electro-optical distance measurements and ground tilt. *Acta Vulcanologica* **1**, 268–71.

Briole, P., G. Nunnari, G. Puglisi, J. C. Ruegg 1988. Misure GPS sul Monte Etna: Risultati della Prima Campagna. *Atti del 7° Convegno Nazionale del Gruppo Nazionale di Geofisica della Terra Solida (Rome)* 855–64.

Briole, P. G. Nunnari, G. Puglisi 1990. Ground Deformations: GPS and IPGP Tilt Data. In *Mt Etna: the 1989 eruption*, F. Barberi, A. Bertagnini, P. Landi (eds), 47–53. Pisa: Giardini.

Briole, P., R. Gaulon, G. Nunnari, G. Puglisi, J. C. Ruegg 1992. Measurements of ground movement on Mount Etna, Sicily: a systematic plan to record different temporal and spatial components of ground movement associated with active volcanism. In *Volcanic seismology*, P. Gasparini, R. Scarpa, K. Aki (eds), 120–9. Berlin: Springer,

Davis, J. L., W. H. Prescott, J. L. Svarc, K. J. Wendt 1989. Assessment of Global Positioning System measurement for studies of crustal deformation. *Journal of Geophysical Research* **94**, 13635–50.

Decker, B.L. 1986. World Geodetic System 1984. *Proceedings of the 4th International Geodetic Symposium on Satellite Positioning* **1**, 62–92. Austin, Texas.

Dixon, T. H. 1991. An introduction to the Global Positioning System and some geological applications. *Reviews in Geophysics* **29**, 249–76.

Evans, A.G. & B. R. Hermann 1989. A comparison of several techniques to reduce signal multipath from the Global Positioning System. In *Global positioning system: an overview*, Y. Bock & N. Leppard (eds), 74–81. New York: Springer.

Falzone, G., B. Puglisi, G. Puglisi, R. Velardita, L. Villari 1988. Componente orizzontale delle deformazionei lente del suolo nell'area del vulcano Etna. *Bollettino del Gruppo Nazionale per la Vulcanologia* **4**, 311–48.

Ferrucci, F., R. Rasá, G. Gaudiosi, R. Azzaro, S. Imposa 1992. Mount Etna: a model for the 1989 eruption. *Journal of Volcanology and Geothermal Research* **56**, 35–56.

Ferrucci, F., G. Nunnari, G. Puglisi 1993. Routine measurements of a fixed baseline: towards the continuous mode. *Acta Vulcanologica* in press.

Foulger, G. R., R. Bilham, J. Morgan, P. Einarsson 1987. The Iceland geodetic field campaign 1986. *Eos* **68**, 1801–18.

Frazzetta, G. & G. Lanzafame 1990. The NE and SE fracture system. In *Mt Etna: the 1989 eruption*, F. Barberi,

A. Bertagnini, P. Landi (eds), 23–9. Pisa: Giardini.

Frei, E. & G. Beutler 1989. Some considerations concerning an adaptive, optimized technique to resolve the initial phase ambiguities for static and kinematic GPS surveying techniques. In *Proceedings of the fifth International Geodetic Symposium on Satellite Positioning* 2, 671–86. Las Cruces, New Mexico.

Frei, E, & G. Beutler 1990. Rapid static positioning based on the fast ambiguity resolution approach "FARA": theory and first results. *Manuscripta Geodaetica* 15, 325–56.

Georgiadou, Y. & K. D. Doucet 1990. The Issue of the selective availability. *GPS World* 1, 553–6.

Georgiadou, Y. & A. Kleusberg 1988. On the effect of ionospheric delay on geodetic relative GPS positioning. *Manuscripta Geodaetica* 13, 1–8.

Gubellini, A., M. Marsella, D. Postpischl 1989. Esperienze di misure GPS in Antartide: risultati della campagna 1988–1989. *Atti 8° Convegno Nazionale del Gruppo Nazionale di Geofisica della Terra Solida (Rome)*, 959–70.

Gurtner, W., G. Beutler, S. Botton, M. Rothacher, A. Geiger, H. G. Kahle, D. Schneider, A. Wiget 1989. The use of GPS in mountainous areas. *Manuscripta Geodaetica* 14, 53–60.

Hager, R. H., R. W. King, M. H. Murray 1991. Measurements of crustal deformation using the Global Positioning System. *Annual Review of Earth and Planetary Sciences* 19, 351–82.

Hartmann, G. K. & R. Leitinger 1984. Range errors due to ionospheric and tropospheric effects for signal frequencies above 100MHz. *Bulletin Geodesique* 58, 109–36.

Hofmann-Wellenhoff, B., H. Lichtenegger, J. Collins 1992. *GPS theory and practice*. New York: Springer.

Hopfield, H. S. 1969. Two-quartic tropospheric refractivity profile for correcting satellite data. *Journal of Geophysical Research* 74, 4487–99.

Jordan, T. H. & J. B. Minster 1988. Measuring crustal deformation in the American West. *Scientific American* 48–58.

Kellogg, J. N. & T. H. Dixon 1990. Central and South America GPS geodesy – Casa uno. *Geophysical Research Letters* 17, 195–8.

Kimata, F., M. Satomura, W. Usui, Y. Sasaki 1991. Preliminary result of crustal motion monitoring by GPS in the southern part of central Japan (March 1989–March 1991). *Journal of Physics of the Earth* 39, 649–59.

Kleusberg, A. 1990. Comparing GPS and GLONASS. *GPS World* 1, 52–4.

Klobuchar, J. A. 1986, Design and characteristics of the GPS ionospheric time-delay algorithm for single frequency users. *Proceedings of the IEEE Position Location and Navigation Symposium*. Las Vegas.

Klobuchar, J.A. 1991. Ionospheric effects on GPS. *GPS World* 2, 48–51.

Larson, K. M. & D. C. Agnew 1991. Application of the Global Positioning System to crustal deformation measurement: 1, precision and accuracy. *Journal of Geophysical Research* 96, 16547–65.

Leick, A. 1990. *GPS satellite surveying*. New York: John Wiley.

Lichtenegger, H. & B. Hofmann-Wellenof 1989. GPS data preprocessing for cycle-slip detection. In *Global positioning system: an overview*. Y. Bock & N. Leppard (eds), 57–68. New York: Springer.

Lo Giudice, E. & R. Rasá 1986. The role of the NNW structural trend in the recent geodynamic evolution of North-Eastern Sicily and its volcanic implications in the Etnean area. *Journal of Geodynamics* 25, 309–30.

Lo Giudice, E. & R. Rasá 1992. Very shallow earthquakes and brittle deformation in active volcanic areas: the Etnean region as an example. *Tecytonophysics* 202, 257–68.

Nunnari, G. & G. Puglisi 1987. Elaborazione dei dati geodimetrici sull'Etna: risultati preliminari. *Bollettino del Gruppo Nazionale per la Vulcanologia* 3, 505–20.

Nunnari, G. & G. Puglisi 1991. Data related to eruptive activity at Etna: GPS Survey. *Acta Vulcanologica* 1, 268–71.

Nunnari, G. & G. Puglisi 1994. The Global Positioning System as a useful technique for measuring ground deformations on Mt Etna. *Journal of Volcanology and Geothermal Research* 61, 267–80.

Okada, Y. 1985. Surface deformation due to shear and tensile faults in half-space. *Bulletin of the Seismological Society of America* 75, 1135–54.

Remondi, B.W. & B. Hofman-Wellenhof 1989. GPS broadcast orbits versus precise orbits: a comparison study. In *Global positioning system: an overview*. Y. Bock & N. Leppard (eds). New York: Springer.

Santerre, R. 1991. Impact of GPS satellite sky distribution, *Manuscripta Geodaetica* 16, 28–53.

Saastamoinen, I.I. 1973. Contribution to the theory of atmosperic refraction. *Bulletin Geodesique* 170, 13–34.

Sako, M., P. Delaney, R. Hunatani, A. Okamura 1987. *April monthly report*. USGS Hawaiian volcano observatory.

Shimada, S., S. W. Fujinawa, S. Sekiguchi, S. Ohmi, T. Eguchi, Y. Okada 1989. Detection of a volcanic fracture opening in Japan using Global Positioning System measurement. *Nature* **343**, 631–3.

Smith, R. B., R. E. Railinger, C. M. Meertens, J. R. Hollis, S. R. Holdahl, D. Dzurisin, W. K. Gross, E. E. Klingele 1989. What's moving at Yellowstone? The 1987 crustal deformation survey from GPS, leveling, precision gravity, and trilateration. *Eos* **70**, 113–25.

Strange, W.E. & D. B. Zilkoski 1991. Reference networks (control surveys). United States National Report to the International Union of Geodesy and Geophysics (IUGG) 1987–1990. *Contributions in Geodesy, AGU,* 157–61.

Tralli, D. M., T. H. Dickson, S. A. Stephens 1988. Effect of wet tropospheric path delays on estimation of geodetic baselines in the Gulf of California using the Global Positioning System. *Journal of Geophysical Research* **93**, 6545–57.

Trimble Navigation Limited 1990. *Trimvec-plus GPS survey software user manual and technical reference guide.* Sunnyvale, California.

Van Mierlo, J. 1979. Statistical analysis of geodetic measurement for the investigation of crustal movements. *Tectonophysics* **52**, 457–67.

Wells, D. (ed.) 1986. *Guide to GPS positioning.* Ottawa, Canada: Canadian GPS Associates.

# 7 Infrared thermal monitoring

D. A. Rothery, C. Oppenheimer and A. Bonneville

## 7.1 Introduction

Remote sensing, whether from an aircraft or satellite, offers a way to collect data on any remotely measurable property over a whole volcano virtually instantaneously. Images that record thermal radiation can be used as a basis for measuring such properties as temperature and radiant flux, and to determine how these properties vary spatially, which is not often feasible to achieve by ground-based measurements. Instrumentation has advanced significantly since the classic reviews by Moxham (1971) and Francis (1979). In this chapter we describe remote sensing methods capable of measuring the temperatures of surface phenomena, and consider ground-based techniques somewhat more briefly. We do not discuss detection and monitoring of eruption plumes, which can include thermal studies. For these techniques see Chapters 13 and 16, Sawada (1989), Glaze et al. (1989b), Krueger et al. (1990), and references contained therein. Nor do we discuss thermal remote sensing techniques for mapping lava flows and other deposits, for this see Kahle et al. (1988).

Remote sensing is a discipline beset with acronyms liable to confuse even the experts. Where important to aid comprehension, we have explained acronyms where they first appear in this chapter. The Appendix at the end of the chapter contains a full list of those used.

## 7.2 What can thermal measurements tell us about a volcano?

Volcanic activity is almost always accompanied by heat output, and so at their most fundamental level we might expect thermal measurements to tell us about the amount of power being liberated (e.g. Brown et al. 1991). Thermal data are essential inputs to models of mass transport within magma conduits (Hardee 1993), surface lava flow (Dragoni 1989, Crisp & Baloga 1990, Dragoni et al. 1992), and eruption cloud dynamics (Woods & Self 1992). It has also been proposed that changes in the remotely sensed radiant energy flux of a volcano (Glaze et al. 1989a) or simply the ground surface temperature (Bonneville & Gouze 1992) may, in the right circumstances, be eruption precursor signs. Clearly, therefore, unless we can document the thermal characteristics of a volcano then we cannot expect fully to understand the energetics and dynamics of the events that occur there.

## 7.3 What is remote sensing?

The standard definition of remote sensing is "acquisition of information without being in physical contact". For our purposes, the information we gather is usually in the form of an image of part of the Earth's surface, and the medium through which this is obtained is electromagnetic radiation. Remote sensing is usually done from an aircraft or a satellite. An aircraft has the advantage that you may be able to control exactly where and when it flies, but an airborne campaign may be impracticable to arrange, and in any case usually has to be scheduled months in advance. Purchase of satellite data is usually cheaper than organizing a special aircraft flight, because satellites have normally been put into orbit to perform a wide variety of tasks and each user pays only a fraction towards this cost. However, the opportunities that an individual satellite has to image any particular target on the ground are limited by its orbital cycle, data recording/transmitting capabilities, and cloud cover. A review of the costs, benefits and practicalities of obtaining and using remotely sensed data covering volcanoes is provided by Rothery & Pieri (1993).

A remote-sensing device can record an image of the ground in any part of the electromagnetic spectrum to which the atmosphere is comparatively transparent. Radar can penetrate clouds, but shorter wavelengths require clear skies, and even then transmission is confined to "atmospheric windows", where transmission of radiation is high (Fig 7.1). Usually an imaging system acquires several images of the same area simultaneously in different wavebands (or channels), resulting in what is known as a "multispectral image". When three of these channels are combined for display purposes, using red for one, green for another, and blue for a third (these being the three primary additive colours), the result is a "false colour composite". It is "false" in that the colours do not correspond to true red, green and blue unless the channels fed into the display were recorded in these wavelengths in the first place.

Images are almost always recorded in digital form, and the complete picture in a single waveband is constructed from a series of picture elements, usually known as "pixels", each of which represents (in most cases) an approximately square area of ground. The information recorded for each pixel is an integer, usually in the range 0–255, and conventionally referred to as a "digital number" (DN). This number is proportional to the amount of radiation received from the ground within the pixel in the waveband concerned. This "in-band radiance" can be converted to the more meaningful wavelength-dependent quantity "spectral radiance" by dividing it by the width of the channel. Spectral radiance has the dimensions of power per area of surface per solid angle sensed per wavelength interval (often expressed as $mW\,cm^{-2}\,sr^{-1}\,\mu m^{-1}$). As we explain in Section 7.5, this can be used in the case of thermally emitted radiation to derive information about surface temperatures.

A geologist wishing to use such data usually buys them on magnetic tape (known as computer compatible tape (CCT)) but small extracts can sometimes be obtained on floppy disks. Compact discs (CDs), which are capable of storing large volumes of image data, are becoming an increasingly common medium. These digital data must be displayed on an image processing system. The computing hardware and software for one of these used to cost something like US$100000, but these days little more than US$1000 may suffice (especially if only small volumes of data need to be handled) by using a personal computer (PC) system and taking

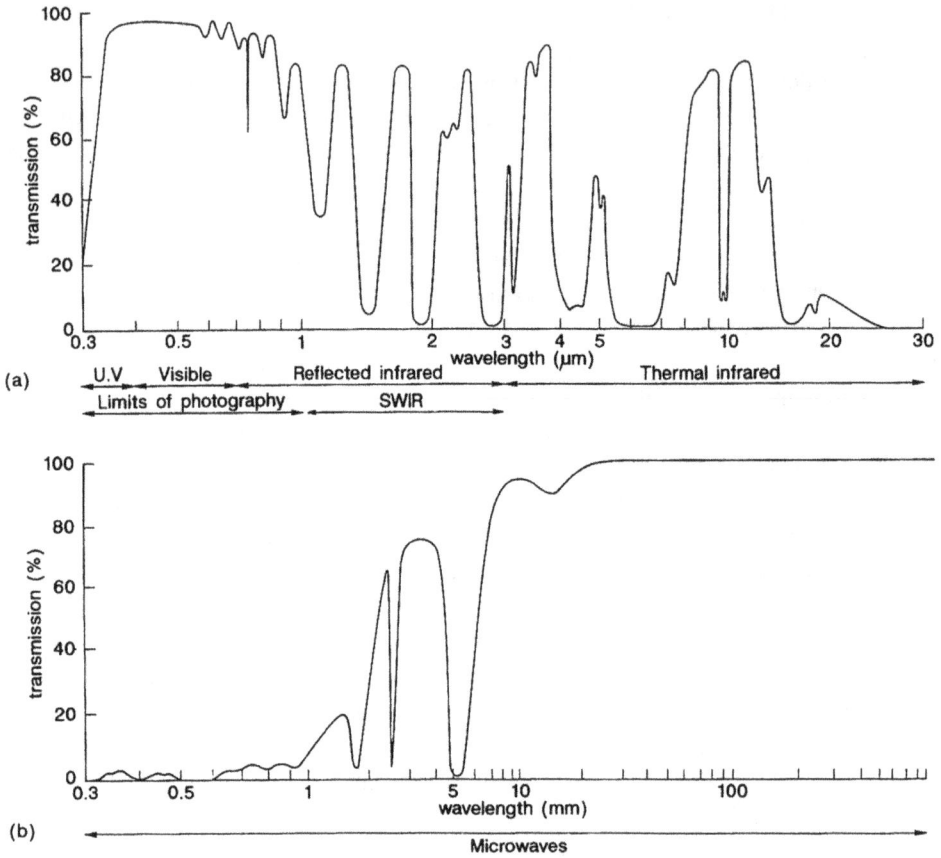

**Figure 7.1** Transmission of electromagnetic radiation through the Earth's atmosphere. Remote sensing from high-altitude aircraft or satellites is usually restricted to spectral regions where atmospheric transmission exceeds about 50%. The reflected infrared is sometimes divided into the very near infrared (wavelength shorter than about 1 μm) and the short wavelength infrared (wavelength greater than about 1 μm).

advantage of cheap or even public-domain software. Images can also be bought already copied into photographic form, these are cheaper and useful for simple mapping, but they cannot be used for quantitative thermal studies. The basics of remote sensing and image processing are described in terms accessible to geologists in several textbooks, most usefully in those by Drury (1993), Sabins (1986) and Rees (1990).

## 7.4 Satellite data

Most of the remote-sensing examples discussed in this chapter rely on data from two instruments: the thematic mapper (TM) carried by satellites of the Landsat series, and the advanced very high resolution radiometer (AVHRR) carried the National Oceanic & Atmospheric

Administration (NOAA) satellites of the TIROS-N series and its successors. However, the techniques we discuss are equally valid for thermal data collected by other spaceborne or airborne instuments. Some essential information about TM and AVHRR data is given below, but advice on obtaining data is deferred until Section 7.9.

## 7.4.1 Landsat

Landsat is probably the most commonly used satellite for geological applications. Landsat-4, launched in 1982, and Landsat-5, launched in 1984, have the TM as their main scanning system. This images simultaneously in seven spectral bands, one of these (band 6) covers the bandwidth 10.4–12.5 µm and is recorded as square pixels 120 m across. The other bands are at shorter wavelengths and are recorded as square pixels 30 m across. As will be seen shortly, it is the two longest wavelength high-resolution bands (band 5, 1.55–1.75 µm; and band 7, 2.08–2.35 µm) that have proved to be of greatest use in remote sensing of hot volcanoes. Landsat-6, lost on launch in 1993, carried an instrument called the enhanced thematic mapper (ETM), recording in the same spectral bands as the TM with the addition of a "panchromatic" channel at 0.50–0.90 µm with 15-m pixels. A replacement is awaited, but meanwhile Landsat 5 is still functioning. The most significant difference as far as thermal measurements are concerned is that the ETM will operate (by request) in a low-gain mode, which should make it possible to record higher spectral radiances from hot targets before overloading the sensors and therefore producing saturated pixels on the image.

Landsat satellites are in a regularly repeating pattern of polar orbits, allowing day-time coverage of any area near the Equator by a single satellite once every 16 days (if there are no clouds in the way), and more frequently at higher latitudes where the degree of overlap between the 180 km wide image swaths increases. If night-time data can be obtained as well, then the repeat frequency is doubled.

## 7.4.2 AVHRR data

The NOAA satellites carrying the AVHRR instrument were designed for global monitoring. By imaging a much wider swath than Landsat, one of these can cover virtually the entire globe twice every 24 h (once by day and once by night). Because there are usually two of them in operation, it is often possible to image a given area once every 6 h. However, concomitantly with the more frequent coverage comes lower spatial resolution, the smallest pixel size for AVHRR data being about 1 km across. The AVHRR does image in the visible, but only channels 3, 4 and 5, which are in the thermal infrared (TIR) region (3.55–3.93 µm, 10.3–11.3 µm and 11.5–12.5 µm), concern us here.

## 7.5 The physics of thermal remote sensing

In this section we first review some general principles of thermal radiation, and then discuss the derivation of radiance and surface temperatures from satellite data, with particular emphasis on the short-wavelength infrared as detected by the Landsat TM. Next we introduce some of the extra physics necessary for a proper understanding of the information that longer wavelength thermal infrared data can reveal about low-temperature thermal anomalies, and finally we consider the effects of the atmosphere.

### 7.5.1 Planck radiation

The spectral radiance, $L(\lambda, T)$ at wavelength $\lambda$, from a surface at temperature $T$ (in kelvin) is given by Planck's formula:

$$L(\lambda, T) = \frac{\varepsilon h c^2 \lambda^{-5}}{\exp(hc/k\lambda T) - 1} \qquad (7.1)$$

where $h$ is Planck's constant ($6.266 \times 10^{-34}$ J s), $c$ is the speed of light ($2.998 \times 10^8$ m s$^{-1}$), $k$ is Boltzmann's constant ($1.380 \times 10^{-23}$ J K$^{-1}$), and $\varepsilon$ is the emissivity of the surface at the appropriate wavelength (an efficiency factor, typically about 0.95 for unweathered basaltic lavas at wavelengths shorter than about 3 μm).

Curves of radiance against wavelength, derived from Planck's formula, are shown in Figure 7.2 for a range of temperatures. These show that the greatest change in spectral radiance between 300 K (i.e. "normal" environmental temperature) and 1400 K (molten basalt) occurs between 1 and 3 μm, a spectral region commonly known as the short wavelength infrared, (SWIR). This therefore is the region of most use in detecting and measuring high-temperature magmatic features by remote sensing. On the other hand, low-temperature anomalies of a few degrees above the ambient ground surface temperature can be detected only at longer wavelengths, the 8–14 μm region of the thermal infared being particularly appropriate.

The first derivative of the Planck formula defines the wavelength, $\lambda_{max}$, where the spectral radiance reaches a peak for a given temperature (the maxima of the curves in Figure 7.2). This is known as "Wien's displacement law", and is expressed:

$$\lambda_{max} = k'/T \qquad (7.2)$$

where $T$ is the surface temperature (expressed in kelvin), $k'$ is a constant (2987 μm K), and $\lambda_{max}$ is in μm.

### 7.5.2 Deriving temperatures from infrared DN

The infrared energy arriving at a detector carried by a satellite or aircraft is composed of varying proportions of radiation emitted and reflected from the target surface (of which only part is transmitted by the atmospheric column), supplemented by upwelling radiation emitted

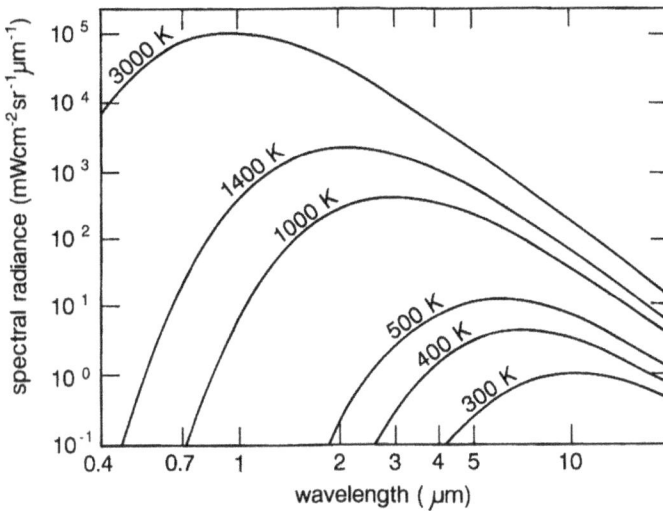

**Figure 7.2** The spectral radiance emitted by surfaces at different temperatures, according to Planck's formula. These curves are drawn assuming the
surface is a perfect radiator (i.e. a "black body"), for real surfaces the radiance has to be multiplied by a factor $\varepsilon$ ($<1$), where $\varepsilon$ is the emissivity of
the surface. Emissivity is likely to vary with wavelength, but for most lavas
it is about 0.95 between 1 and 3 $\mu$m. In practice the 1–3 $\mu$m region has proved
most useful for measuring temperature distributions on lava bodies and hot
fumarole fields. If the whole of a Landsat TM band 7 (2.08–2.35 $\mu$m) pixel
were occupied by material at 1400 K, the radiance received at the satellite
would be more than a 1000 times that necessary to saturate the pixel. However, the band-7 radiance from hot cracks at 1400 K occupying less than 0.1 %
of the area of a crusted lava would be within the measurable range, allowing both the temperature and the total area of the cracks to be measured,
provided that radiance could also be measured in a neighbouring region of
the spectrum (such as TM band 5 at 1.55-1.75 $\mu$m).

or scattered by the constituents of the atmosphere (so-called "path radiance"). The spectral
radiance, $R_\lambda$, reaching the detector at wavelength $\lambda$ may be written as:

$$R_\lambda = \tau_\lambda \varepsilon_\lambda L(\lambda,T) + \tau_\lambda \rho_\lambda R_{\lambda,D} + R_{\lambda,U} \qquad (7.3)$$

where $\tau_\lambda$ is the wavelength-dependent atmospheric transmission coefficient, $\varepsilon_\lambda$ and $\rho_\lambda$ are,
respectively, the spectral emissivity and spectral reflectivity of the surface, $T$ its radiant temperature, and $R_{\lambda,D}$ and $R_{\lambda,U}$ are, respectively, the downwelling and upwelling spectral
radiances of the atmosphere ($R_{\lambda,D}$ is small at TIR wavelengths of 8 $\mu$m and more). Note that
the first term in the right-hand side of this equation describes the thermal radiation from the
surface, the second term is radiance reflected by the surface, and the third term arises directly from the atmosphere.

$R_\lambda$ is obtained from the DN recorded by the detector by radiometric calibration, of the form:

$$R_\lambda = \alpha_\lambda DN_\lambda + \beta_\lambda \qquad (7.4)$$

where $\alpha_\lambda$ and $\beta_\lambda$ are the calibration coefficients of the instrument, and $DN_\lambda$ is the digital number in the waveband centred at wavelength $\lambda$ for a given pixel of image data.

In order to use Equation 7.3 to estimate surface temperatures, we must isolate that part of the satellite response corresponding to the partially transmitted target thermal radiance, $R_{\lambda,thermal} = \tau_\lambda \varepsilon_\lambda L(\lambda,T)$. As a rough approximation, this can be done by subtracting a background non-thermal signal (estimated from neighbouring pixels) from the anomalous signal over a hot spot. Alternatively, by using data recorded at night all except the first term in the right-hand side of Equation 7.3 are effectively zero, in which case there is no need to correct for sunlight at all. Night-time recording is not done routinely by the Landsat TM, and data acquisition must be done by special request. The first example of this is reported in Rothery & Oppenheimer (1991) and Oppenheimer et al. (1993a). "Pixel-integrated temperatures" as measured at a particular wavelength (or within the band-pass of a particular sensor) can then be derived by rearrangement of Equation 7.1:

$$T_\lambda = \frac{c_2}{\lambda \ln\left[1 + c_1 \tau_\lambda \varepsilon_\lambda / \left(\lambda^5 R_{\lambda,thermal}\right)\right]} \tag{7.5}$$

where $c_1 = 1.19 \times 10^{-16}$ W m$^2$ and $c_2 = 1.44 \times 10^{-2}$ m K. The spectral transmissivity of the atmosphere, $\tau_\lambda$, can be estimated from radiation propagation models such as those incorporated in the widely used LOWTRAN computer code (Kneizys et al. 1988). Estimates of spectral emissivity can be based on published data for appropriate surface materials (e.g. Pollack et al. 1973).

If the atmosphere between target and sensor has been well-characterized, this approach can give accurate estimates of the surface temperature of large uniform bodies of water (e.g. Bartolucci et al. 1988). However, on land, surface temperature measurement is far more complex (Wan & Dozier 1989). It may be unrealistic to suppose that thermal distributions, or even terrain types (e.g. ice, snow, rock, vegetation or lava), are uniform across the instantaneous field of view (IFOV) of the sensor. Volcanic thermal features represent an extreme case where surfaces at ambient and magmatic temperatures may occur in very close proximity. Even at high spatial resolutions (for example the nominal 30-m pixel dimensions of TM bands 1–5 and 7), where a pixel lies within such a region, the derived temperature will lie somewhere between those of the hottest and coolest material present, the actual value being weighted according to the spectral radiance integrated across the whole IFOV of the sensor (which is usually similar, but not identical, to the size of the image pixels). In addition, since spectral radiance is a function of wavelength, temperature derivations will vary according to the spectral bandpass of the sensor.

Rothery et al. (1988a) showed how such discrepancies could be exploited in order to measure the temperatures and sizes of the subpixel thermal components of the scene. The starting point is a simplified model of the radiant surface which considers only two thermal components: a hot portion, at a true temperature $T_c$ occupying a fraction $f$ of the pixel; and a cooler part at a true temperature $T_s$ occupying the remaining fraction $(1-f)$. If two different wavebands of a remote sensing device detect a thermal anomaly (and neither is saturated) the values of $L(\lambda,T)$ (from Eq. 7.3) in each bandpass can be related to this idealized thermal model of the surface. By assuming one of the unknown temperatures, the other temperature and $f$ can both be determined. For detailed treatment of the original "dual-band technique" and subsequent developments see Rothery et al. (1988a), Oppenheimer (1991a) and Oppenheimer et al. (1993a).

### 7.5.3 Low-temperature thermal anomalies

The approach outlined above is suitable for measuring temperatures of lava bodies and hot fumaroles (above a few hundred degrees celcius). However, except where magma occurs at, or very near, the surface, or where there is strong groundwater convection between a magmatic intrusion and the surface, the steep thermal gradients that prevail below the soil-air boundary prevent a large surface thermal signature from appearing (Fig. 7.3). In this situation, long-term or transient temperature anomalies may be only a few degrees above ambient temperature. A low-temperature anomaly may represent a magma intrusion at shallow depth, and the evolution of this anomaly over time may provide a good indication of magma movement towards the surface. The study of such anomalies needs a somewhat different approach to the one considered so far, for example if remote sensing is to be used it must be within the conventional thermal infrared, usually taking advantage of the 8–14 μm atmospheric window. In this subsection we summarize some of the additional physics necessary to understand low-temperature anomalies.

First, we have to consider the various terms of the heat budget at the ground surface. The energy exchanges at this boundary between ground and atmosphere are classically put together in an expression for heat budget in which the sum is zero. In this context, it is convenient to deal with radiated flux summed over all wavelengths rather than as spectral radiance at a

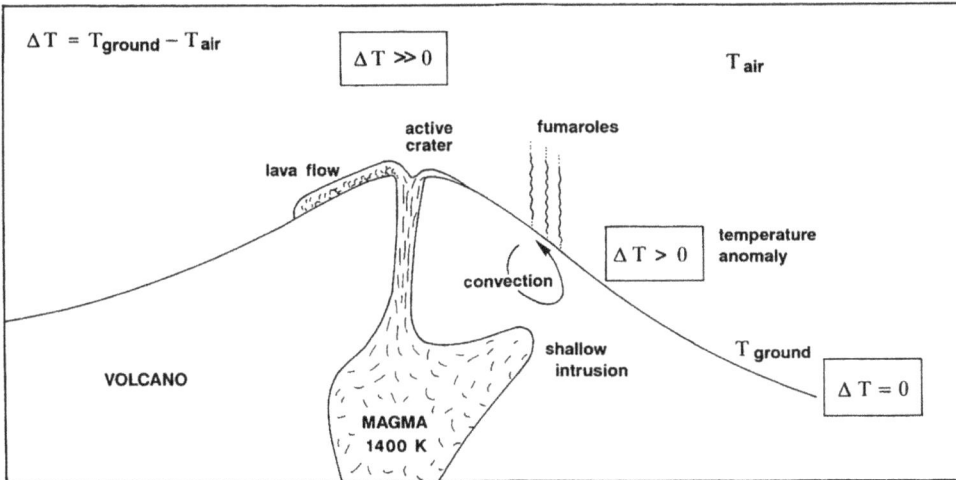

**Figure 7.3**   Sketch of thermal exchange on a volcano. $T_{ground}$ and $T_{air}$ are the temperatures of the ground and air respectively (abbreviated to $T_g$ and $T_a$ in the equations in the text), $\Delta T$ is the difference between ground and air temperature. In normal conditions, to a first approximation, $\Delta T = 0$, but with a large heat supply (magmatic intrusion or convection in a porous or fractured substrate) $\Delta T$ becomes strongly positive. If the anomaly is to be measured by remote sensing, in cases where $\Delta T \gg 0$ it is usually best to use short wavelength infrared data, but where $\Delta T > 0$ the only option is thermal infrared data.

specific wavelength, as previously. If we define as positive the heat received by the ground, this heat budget can be written as (Sekioka & Yuhara 1974):

$$-\Phi_0 + E + X_t + h\,(T_g - T_a) = 0 \tag{7.6}$$

where $\Phi_0$ is the thermal flux from the ground (at depth zero), and is therefore $L(\lambda, T)$ from Equation 7.1 summed over all wavelengths, $E$ is the heat flux due to evaporation and transpiration, $X_t$ is the net radiative input, $h\,(T_g - T_a)$ is the convective exchange flow between ground at temperature $T_g$ and air at temperature $T_a$, and $h$ is an exchange coefficient depending mainly on the wind velocity.

The radiation term $X_t$ can be written as the sum of three terms:

$$X_t = (1 - A)X_i + W_a - W_g \tag{7.7}$$

where $X_i$ is the sum of the direct and and diffuse solar radiative flux, $A$ is the albedo of the surface, $W_a$ is the radiative input from the sky (mainly from clouds), and $W_g$ is the radiative heat output due to ground-emitted energy. $W_g$ is the quantity measured by a ground-based radiometer, or derived from remotely sensed data by means of Equations 7.3 and 7.4 generalized to all wavelengths, and is a function of the ground temperature, thus:

$$W_g = \varepsilon k T_g^4 \tag{7.8}$$

where $k$ is Boltzmann's constant ($5.67 \times 10^{-8}\,\mathrm{W\,m^{-2}\,K^{-4}}$), and $\varepsilon$ is the emissivity of the surface integrated over all wavelengths (effectively equal to $1 - A$). For bare soil or rock, at night and for a cloudless sky, $E$, $X_i$ and $W_a$ can be neglected, in which case Equations 7.6 and 7.7 reduce to:

$$-W_g + h\,(T_g - T_a) = \Phi_0 \tag{7.9}$$

Since the ground is more or less at the air temperature, $T_g$ is usually close to $T_a$, and the expression for $W_g$ can be linearized as:

$$W_g = \varepsilon k T_a^4 + 4\varepsilon k T_a^3(T_s - T_a) \tag{7.10}$$

Thus the heat-budget equation reduces to its simplest possible expression:

$$-\varepsilon k T_a^4 + H(T_g - T_a) = \Phi_0 \tag{7.11}$$

with a new coefficient, $H$, defined as $H = h + 4\varepsilon k T_a^3$.

We are now able to express directly the temperature of the ground surface as a function of both $T_a$ and $\Phi_0$. The latter, being the heat flow in the ground, is the most relevant quantity to deal with. It is composed of one main term of solar origin, $\Phi_s$, and one term of deeper origin, the terrestrial heat flow, $\Phi_g$, which has a global average value of about $80\,\mathrm{m\,W\,m^{-2}}$, over an order of magnitude below the $1500\,\mathrm{m\,W\,m^{-2}}$ of $\Phi_s$. However, in a volcanic domain $\Phi_g$ can become comparable to the solar flux and thus can be detectable at the ground surface.

As an illustration, let us compare two zones on our theoretical volcano (Figure 7.3). The first zone, where $\Delta T > 0$, experiences a thermal equilibrium due to the upward-moving intrusion. Using Equation 7.11, the heat budget can be expressed as:

$$- \varepsilon k T_a^4 + H(T_g - T_a) = \Phi_s + \Phi_g \tag{7.12}$$

The second zone is a reference area where there is no anomaly ($\Delta T = 0$), outside any active part of the volcano but with the same climatic conditions as in the anomalous zone. In this case only the terms $T_g$ and $\Phi_g$ differ, so we can write these as $T'_g$ and $\Phi'_g$ in order to express the heat budget as:

$$- \varepsilon k T_a^4 + H(T'_g - T_a) = \Phi_s + \Phi'_g \tag{7.13}$$

Subtracting Equation 7.13 from 7.12 leads to:

$$H(T_g - T'_g) = \Phi_g + \Phi'_g$$

or
$$H\Delta T_g = \Delta F_g \tag{7.14}$$

where it becomes explicit that the difference in temperature between these two zones, which experience the same external thermal conditions, is due to a difference in heat flow of terrestrial origin.

For example, a positive temperature anomaly of 1°C, with a coefficient $H$ equal to $50\,W\,m^{-2}\,°C^{-1}$ (corresponding to a wind velocity of $10\,m\,s^{-1}$ at an altitude of $2000\,m$) would be associated with an anomaly in heat flow of $50\,W\,m^{-2}$, i.e. over 600 times more than the average terrestrial heat flow. A temperature anomaly of this magnitude is detectable by ground-based or remote sensing methods, as will be seen in Section 7.6. First we have to address the question of how heat is transported in the ground in order to bring about such a temperature anomaly.

If, between a heat source (intrusive dykes or permanent reservoirs) and the ground surface, heat were transported only by conduction, the amplitude of temperature anomalies would be simply too low to be noticeable at the ground surface. Moreover, in the case of an ascending magma body, the peak of the thermal signal would arrive at the surface much later than the magma itself. However, convection is a much more effective mode of heat transport than conduction, especially in volcanic areas where both the temperature gradient and the permeability of the medium can be high. The efficiency of convection in these conditions can be estimated by a dimensionless quantity, the Rayleigh number (Ra), which depends mainly on permeability $K$. For example, if we consider an intrusion at a depth of $100\,m$ and using $K = 10^{-7}\,m^2$, which is an average estimate for a cracked basalt (Bonneville & Kerr 1987), we obtain Ra = 350, well above the threshold value needed to establish stable one-phase convection. The corresponding heat flow at the surface is estimated to be around $160\,W\,m^{-2}$ in good agreement with the order of magnitude of expected heat flow anomalies (Bonneville & Kerr 1987), and the time constants in the case of strong convection could be of the order of days or weeks, compared with tens of years for heat conduction processes. Comparable heat-flow anomalies could be obtained with lower permeability and two-phase convection above a much deeper magmatic intrusion (e.g. Hardee 1982).

### 7.5.4 Removing atmospheric effects

The atmosphere is of considerable importance in modelling thermal phenomena on volcanoes.

As discussed in section 7.5.2, the atmosphere absorbs part of the radiation that leaves the surface before it reaches an airborne or spaceborne detector. In the case of TIR techniques, this provides a strong boundary condition for heat transfer at the ground surface. This effect is due to the absorption of energy at discrete wavelengths by atmospheric constituents such as $H_2O$ and $CO_2$. We have already noted that $\tau_\lambda$ in Equations 7.3 and 7.5 can be estimated from standard atmospheric models such as LOWTRAN. However, the atmosphere over a volcano generally departs from standard models, and it is preferable to use a correction based on the real atmospheric situation at the time of image acquisition. It is especially convenient to do this if the remote sensing data that we intend to use for thermal monitoring themselves contain the basis for such a correction. AVHRR data are probably the best-known example. In this case the two adjacent TIR bands (band 4, 10.3–11.3 μm; and band 5, 11.5–12.5 μm) experience differential (wavelength-dependent) absorption, and a corrected at-ground temperature can be derived from a linear combination of bands 4 and 5 according to the "split window algorithm" (Deschamps & Phulpin 1980, Price 1984), which takes account of water vapour absorptions, though not of specific volcanic emissions such as $SO_2$, $CO_2$ and aerosols. Subject to these limitations, the split window algorithm enables $\tau_\lambda$ to be calculated in each band, by comparing the expected spectral radiance with the actual spectral radiance, and from this we can obtain a map of how $\tau_\lambda$ varies across the image at the AVHRR spatial resolution. We can use this information to correct AVHRR channel-4 data or even TM band-6 data, provided that the TM image was gathered at the same time as AVHRR data (Bonneville & Gouze 1992). However, because the TM records in only a single TIR channel, the split window algorithm cannot be used for TM data alone.

As far as measurements of low-temperature anomalies are concerned, the most important atmospheric effect is due to adiabatic cooling with altitude. This strong inverse correlation between altitude and surface temperature generally obscures the effect of terrestrial heat flux variations, but if the topography of the volcano to be studied is known it is possible to set up a digital terrain model (DTM) that will attribute an altitude to each pixel of the scene. If this is done, the adiabatic cooling effect may be estimated using a statistical approach that consists of determining the correlation between the absolute pixel temperature and its altitude, $z_{pixel}$, as described by Bonneville et al. (1985) and Bonneville & Gouze (1992). We next determine a regional adiabatic cooling gradient thus:

$$\text{Regional adiabatic gradient} = \frac{dT_{ground}}{dz} \tag{7.15}$$

which is generally close to the theoretical adiabatic lapse rate ($0.006\,°C\,m^{-1}$). A value for the corrected surface temperature ($T_{cor}$), which is the best estimate of $\Delta T$ (Fig. 3), can then be calculated thus (Bonneville et al. 1985):

$$T_{cor} = \left(1 - \tau_\lambda\right) \cdot T_\lambda + \frac{dT_{ground}}{dz} \cdot z_{pixel} \tag{7.16}$$

where $T_\lambda$ is calculated from Equation 7.5.

## 7.6 Case studies of remote sensing of high-temperature thermal anomalies

Any distinction between high- and low-temperature thermal anomalies is necessarily rather arbitrary. However, in the discussions that follow, by high-temperature we shall mean anything associated with surface magmatism or fumaroles in excess of about 300°C, whereas the term low-temperature anomaly is reserved for anomalies of only a few degrees or a few tens of degrees above ambient. The 3–5 μm atmospheric window is particularly useful for detecting high-temperature phenomena. AVHRR channel-3 images can provide cheap daily data in this spectral region, which has become known as the "fire channel" due to its ability to dectect fires (e.g. Matson et al. 1987). A team from the Planetary Geosciences Division at the University of Hawaii has used such data to monitor the distribution of active lava in the Kupianaha lava field on a near-daily basis since August 1990 (Mouginis-Mark et al. 1991b), and Harris (1992) has identified 24 cloud-free AVHRR images over a 25-day period on which the development of the September 1984 fissure eruption of Krafla could be monitored. However, the low spatial resolution of AVHRR data often make them unsuitable for studies of hot lava that seek to do more than simply detect its presence (Fig. 7.4), and meaningful temperature measurements can rarely be made (e.g. Weisnet & D'Aguanno 1982, Rothery & Oppenheimer 1994). The same may prove to be true of 1-km pixel data in the fire channel

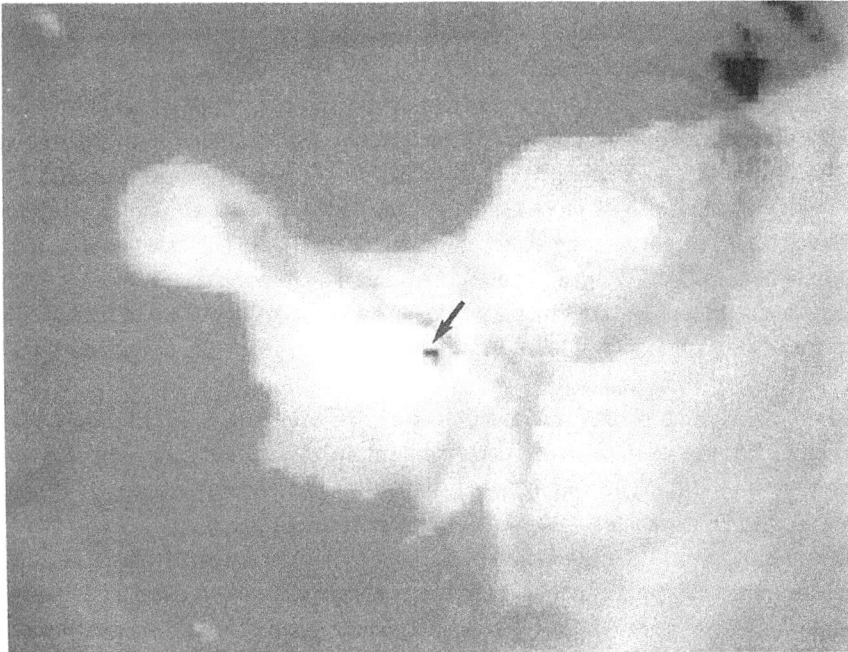

**Figure 7.4** A NOAA polar orbiter AVHRR image, showing detection of the lava lake on Mount Erebus, Antarctica, on 10 October 1980 at 3–5 μm wavelength (1-km pixels). In this image the highest temperatures are in black. The lava lake is revealed by two adjacent hot pixels (indicated by an arrow). Although useful for detecting or confirming the presence of hot lava, such data are inadequate for quantitative thermal studies.

recorded by the ERS-1 along track scanning radiometer (ASTR) that detected a hot-spot associated with the 1992 flank eruption of Nyamuragira (M. Gorman, personal communication).

Attention was first drawn to the greater potential of using shorter wavelength but higher
resolution data, such as the SWIR channels (bands 5 and 7) of the Landsat TM, for detecting
and measuring high-temperature thermal anomalies on volcanoes by Francis & Rothery (1987).
Since then many examples have been studied, most of which are cited below.

### 7.6.1 The problem of interpreting thermal anomalies

In several cases (usually where ground observations of the relevant volcano have been lacking) it has proved difficult to identify with confidence the nature of a particular volcanic phenomenon responsible for a thermal anomaly in TM SWIR data. In particular, discrimination
between lava bodies and high-temperature fumarole fields has been problematic in instances
of intracrater TM SWIR anomalies composed of only a few pixels (see Rothery et al. (1988a)
and Glaze et al. (1989a) on Láscar volcano). Inability to make even such general distinctions
as whether or not lava has reached the surface would obviously limit the value of such investigations, and so we examine in this subsection the criteria that can be used to help identify
the cause of a volcanic thermal anomaly in satellite data. Knowledge of the site in question
clearly aids interpretation (e.g. in the case of a volcano with a known history of lava lake
activity) but there are also clues in the remotely sensed data themselves. All of the following
should be taken into account (Oppenheimer et al. 1993b).

(a) *Spatial attributes.* Size, shape and distribution of thermal anomalies naturally reflect the
    coarse-scale disposition of hot sites on the ground. For example, long, narrow group-
    ings of radiant pixels are suggestive of flows of either lava or pyroclastic material. Smaller
    groupings of pixels can be more difficult to identify, although it has been proposed that
    whereas scattered clusters are likely to be fumarole fields, nucleated groups are more
    likely to be lava lakes (Rothery et al. 1988a). Variation in measured intensity within an
    anomaly should also be taken into account: an anomaly may be more radiant at the mar-
    gins (as in the case of a lava dome reported on a TM image of Láscar by Oppenheimer et
    al. (1993a)), or in its centre.

(b) *Context.* In daytime images, the location of an anomaly with respect to the topography
    and other features of the volcano can be appraised. An anomaly may be sited within a
    crater, or on the flank or foot of a volcano. It may extend downslope, indicative of a
    flow. It may be associated with fumarolic or ash plumes.

(c) *Spectral attributes.* The distribution of thermal radiance between different wavelengths
    depends on the surface temperatures of materials sensed, as described by Equation 7.1,
    and their relative areas within a given pixel. Thus, when remotely sensed observations
    are available in more than one waveband, these offer important clues as to the nature of
    the thermal anomaly. For instance, a broad expanse of ground at 100°C would radiate
    negligibly in the SWIR region, yet may overwhelm the response of sensors in the TIR.
    Conversely, a feature that is very small (compared with the pixel size of a given sensor)
    but is at close-to-magmatic temperature, could saturate SWIR channels while being

undetectable in the TIR. However, temperatures do not provide conclusive evidence as to the nature of a particular feature, for instance it might be impossible to distinguish between a recently deposited pyroclastic flow and a substantially cooled lava flow, using remotely sensed data alone.

(d) *Comparison within a time-series of satellite data.* Temporal changes in spatial and/or spectral attributes of an anomaly can also provide a context for interpretation. For example, radiant temperatures at the surface evolve more gradually for a fumarole field than for a lava-filled strombolian vent (Oppenheimer & Rothery 1991).

To generalize from this discussion, active lava flows are easily recognizable in infrared images, while small groupings of hot pixels are much more difficult to interpret. These matters are discussed by Oppenheimer et al. (1993a & b).

## 7.6.2 Lava flows

The first successful attempts to use Landsat TM data to measure "dual-band" temperatures on active flows were by Pieri et al. (1990) who studied an archived image of a 1984 lava flow near the summit of Mount Etna, and Oppenheimer (1991b) who studied a Landsat TM image of the 1989 lava flow at volcán Lonquimay, Chile, which had been recorded by special request (Fig. 7.5a & b). The prolonged 1992 lava eruption of Mount Etna provided a further opportunity to link remote sensing with field observations (Fig. 7.5c) (Rothery et al. 1992). The aims of these investigations have been to characterize surface temperature distributions across and along active flows with a view to revealing the thermal physics that control lava rheology. However, such data have yet to realise their potential to parameterize and test existing theoretical models of lava flow dynamics (e.g. that for single-lobed basalt flows presented by Crisp & Baloga (1990)). Further discussion of the importance of temperature measurements in understanding and predicting flow behaviour may be found in Chapter 13, Hardee (1993) and Pinkerton (1993).

A frequent problem experienced with Landsat TM measurements is that of sensor saturation. For example, only a small section of the Etna flow studied by Pieri et al. (1990), was amenable to dual-band calculations because most of the upper reaches of the flow were saturated. Saturation problems are also evident in Figure 7.5c, where the bright stripes are lines of data corrupted by the scanner crossing highly radiant sources. This is in marked contrast to the Lonquimay flow (Oppenheimer 1991b) which appears from the TM data to have been substantially cooled at the surface very close to the vent.

An important difference between approaches is that to solve dual-band equations, Pieri et al. (1990) assumed a temperature for the cooler component of a two-component thermal model while Oppenheimer (1991b) assumed a temperature for the hotter fraction.

Other examples of active lava flows recorded by TM in bands 5 and 7 include those reported by Andres & Rose (1994) who describe an anomaly on the northwest flank of Pacaya, Guatemala, Gupta & Badarinath (1993) who were able to detect the May 1991 basaltic eruption of Barren Island using TM bands 5 and 7, and Reddy et al. (1992, 1993) who report radiance from the same eruption in TM bands 5 and 7 by night 2 months later.

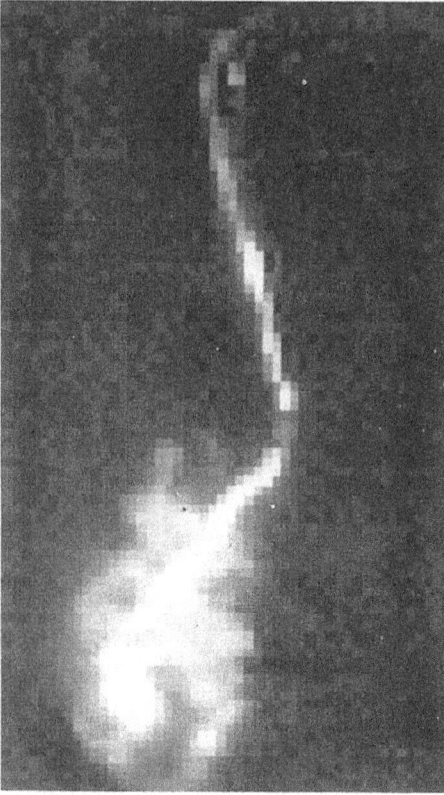 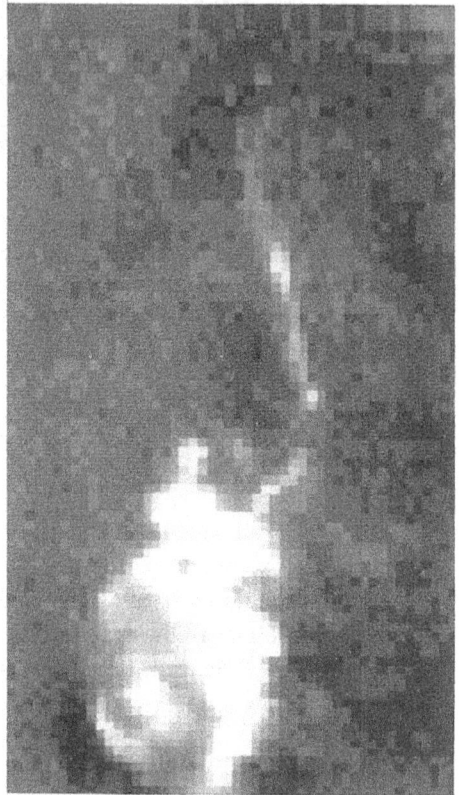

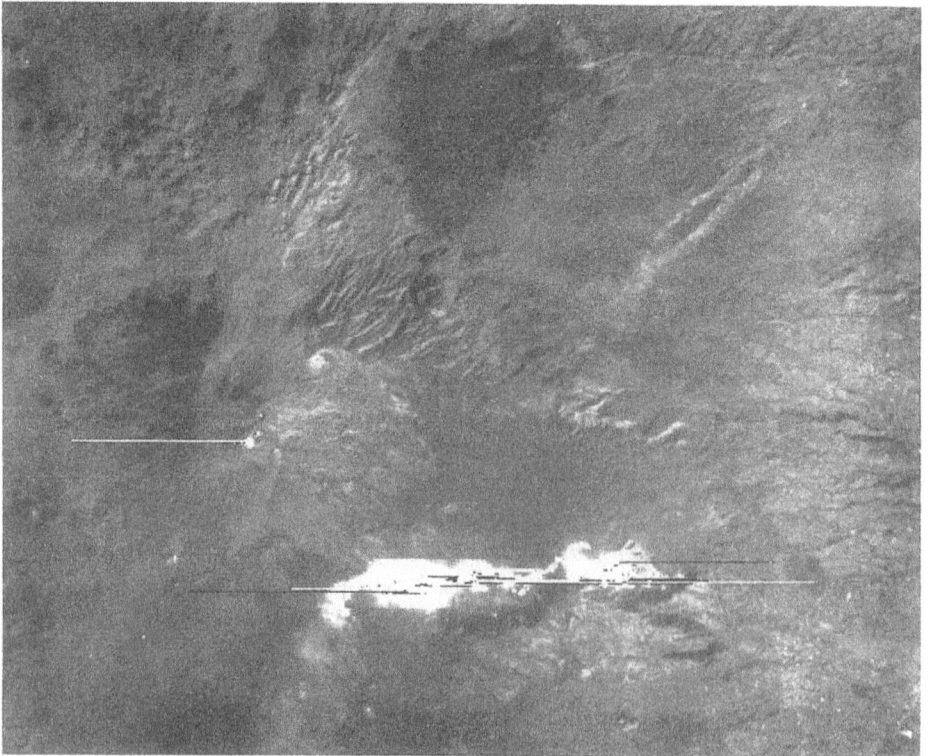

Under the right conditions, there may be measurable thermal radiance in the shorter wavelength TM channels. Images of a 1991 flow from Kupianaha, Hawaii (P. Mouginis-Mark, personal communication), and the 1992 TM image of Mount Etna reported by Rothery et al. (1992) show radiance in band 4 in some of the pixels where band 7 is saturated. Previously, Rothery et al. (1988a) reported an example of a 10 km length of a basaltic flow from Sierra Negra, Galapagos, in 1979 radiating in the Landsat MSS 0.8–1.1 µm channel, much of which was also radiant in the 0.7–0.8 µm channel and 1 km of which was also radiant at 0.6–0.7 µm. Unfortunately, they were unable to derive realistic temperatures from these data by means of the dual-band method. The data did not fit a dual- or triple-band model, perhaps because of errors in correcting for reflected light.

SWIR surveys of active lava flows from aircraft have been carried out only at Hawaii, using NS001 Thematic Mapper Simulator data (Rothery & Pieri 1993). Thermal infrared surveys have been made from helicopters with an Inframetrics I525 video imaging system, and TIMS data have been recorded over active flows on the East Rift of Kilauea. All three data types show the thermal expression of subsurface flows, and TIMS also detects emissivity effects above the active tube flows (Realmuto et al. 1992).

## 7.6.3 Lava lakes/strombolian vents

Active lava lakes have occasionally been observed by satellite. The detection of the Mount Erebus lava lake by weather satellites has already been remarked on (Wiesnet & D'Aguanno 1982, Rothery & Oppenheimer 1994). The first report of the use of Landsat TM data on this target was by Rothery et al. (1988a), but it seems likely that when the image was recorded activity was confined to an incandescent vent rather than a crater-filling lava lake. The same authors also reported the presence of two active lava lakes on Erta 'Ale (Ethiopia) shown in bands 5 and 7 of a TM image recorded on 5 January 1986 (Fig.7.6). It is noteworthy that this observation is the sole basis by which the continued presence of active lava lakes on Erta 'Ale is inferred during 1975–85 in the Smithsonian Institution decade summary of global volcanism (McClelland et al. 1989, pp.16–28, 77). Rothery et al. (1988b) reported the same two lava lakes present on 9 February 1987 based on examination of a photographic version of a TM image (which was much cheaper than digital data). This longest-lived of all lava lake phenomena had been present during all reported field-based observations since at least 1960 until 1973, when field studies ceased, and so far as we are aware the TM data are the only information on the Erta 'Ale lava lakes between that date and visual observations from a low-altitude

---

**Figure 7.5** (a) Extract of Landsat TM band-7 (2.08–2.35 µm) image of the Lonquimay flow on 8 August 1989, showing an area about 2.5 km from top to bottom. (b) The same area shown in TM band-5 (1.55–1.75 µm). The original data for (a) and (b) were obtained in Eosat A-format, but for display purposes, sets of 16 scan lines have been offset sideways (by 40–44 pixels) to remove the distortion caused by the alternating mirror sweeps. (c) Extract of Landsat TM band-7 image of a flow on Mount Etna on 2 January 1992, showing an area about 30 km across. These data were obtained in Eurimage level-1 format, and have received no subsequent geometric processing. Parts of several scan lines have been corrupted due to sensor overload as the TM scanned over the most radiant parts of the flow.

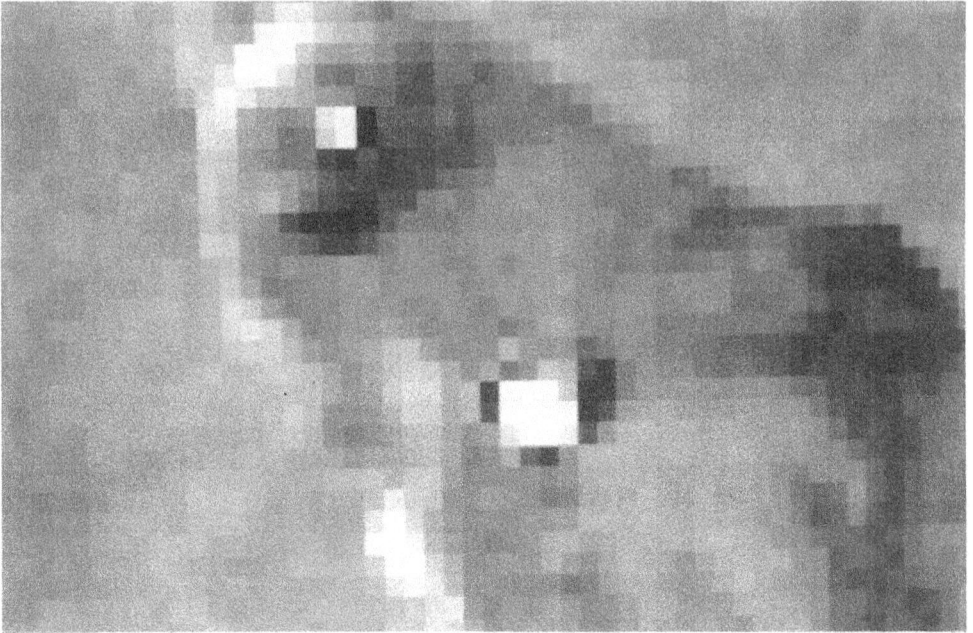

**Figure 7.6**   Extract of a Landsat TM band-5 image showing two lava lakes at Erta 'Ale, Ethiopia, 5 January 1986. Since the last known field observations in the mid-1970s, the northern of the two lakes appears to have shrunk and the southern one to have grown. Pixels are 30 m across.

overflight in September 1992 (Smithsonian Institution 1992a,b) followed up by a descent to the caldera floor in November 1992 (Smithsonian Institution 1992c), at which time only the southern of the two lava lakes was active.

Andres & Rose (1994) report temperature measurements within the active vent of Santiaguito, Guatemala, using TM band-5 and band-7 data. An example of an airborne image spectrometer detecting radiance in the SWIR region from strombolian vents on Stromboli and Etna was reported by Oppenheimer et al. (1992, 1993c).

## 7.6.4 Lava domes

Although there have been occasional attempts at airborne monitoring of lava domes (e.g. Friedman et al. 1981), the longest running series of observations of a hot lava phenomenon covers volcán Láscar, Chile, and is based on Landsat TM data (Fig. 7.7). This extends from December 1984 into 1992, having been begun by Peter Francis and continued by one of us (DAR). Recent work (Oppenheimer et al. 1993a) suggests that the core of the radiant phenomenon since 1989, and probably prior to that, was a lava dome. This is based on concurrent field and Landsat observations, and the similarity of recent and earlier TM thermal anomalies. However, these and previous (Glaze et al. 1989a & b) studies of Láscar emphasise the difficulty in distinguishing the nature of small intracrater anomalies. Strategies for distinguishing between such phenomena are considered in Oppenheimer et al. (1993a & b),

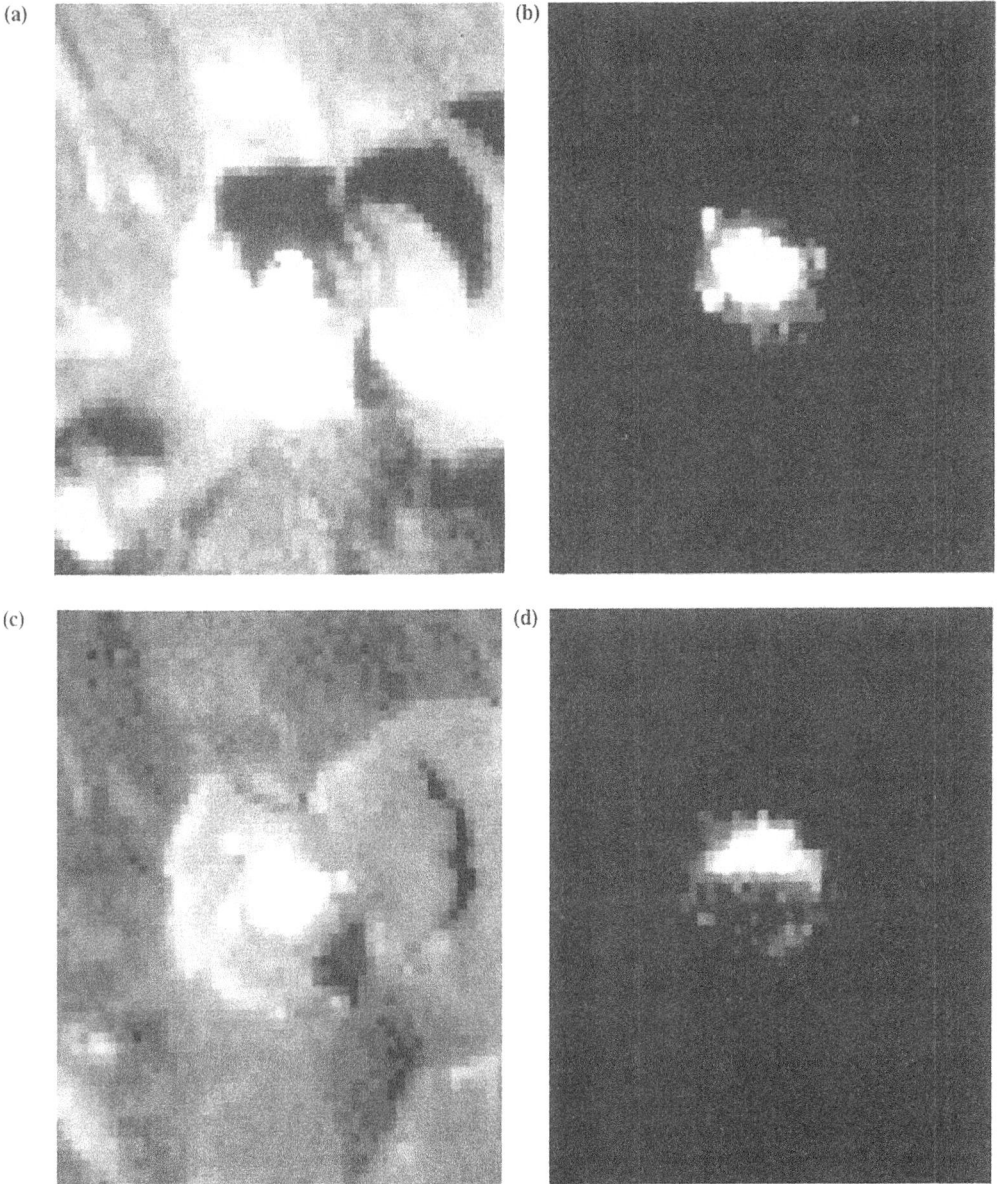

**Figure 7.7** Extracts from Landsat TM band-7 (2.08–2.35 μm) images showing thermal radiance from the lava dome within the active crater of volcán Láscar, Chile: (a) 21 July 1985, (b) 18 November 1989 (night-time), (c) 14 December 1989, (d) 26 March 1990. The lava dome appears to have changed little between 18 November and 14 December 1989, but it can be more clearly seen on the night-time image, and thermal measurements derived from this night-time image are correspondingly more accurate.

though because lavas cool while maintaining sites of hot gas escape, they are by no means foolproof.

### 7.6.5 Fumarole fields

Confirmed fumarole fields have been detected by TM band-5 and band-7 data on several volcanoes: in a 1986 image of Poàs, Costa Rica (Rothery et al. 1988b, Oppenheimer & Rothery 1991), around the walls of the active crater on Lascar 1987–91 (Oppenheimer et al. 1993a) and on the floor of the same crater in the aftermath of an explosive eruption in 1986 (Rothery et al. 1988a), at the summit lava dome of Colima in 1985 and 1986 (Abrams et al. 1991), and on Kinamura cone, Nyamuragira in 1989 (Smithsonian Institution 1989). Because palls of volcanogenic gases and aerosols typically occupy the atmosphere above fumarole vents, the transmission of radiation emitted from hot wallrock may be greatly attenuated, restricting the scope for meaningful radiometry from satellites. Often, fumarole mouths are located on steep crater walls or domes, where they would not be imaged by an overhead sensor. It may prove easier to detect active fumaroles by their plumes of condensed vapour, than from a ground thermal signature.

## 7.7 Case studies of remote sensing of low-temperature thermal anomalies

Some of the difficulties encountered in measuring ground surface temperatures using airborne thermal infrared surveys are discussed by Brivio et al. (1989), who noted considerable disparities between temperatures measured on the ground and those determined by an airborne thermal infared survey of Vulcano, and by Kieffer et al. (1981) in a broadly inconclusive attempt to locate new thermal anomalies that were precursor signs of the 18 May 1980 eruption of Mount St Helens. Nevertheless, ground surface temperature anomalies have been proposed as thermal precursors of eruptions that occurred on Mount Etna in 1983 (Bonneville & Kerr 1987), and 1986 (Bonneville & Gouze 1992). We will consider the latter study here because it incorporates a high-resolution image (Landsat TM) and because it illustrates quite well the processing techniques described earlier, and then consider briefly the special case of crater lakes.

### 7.7.1 Low-temperature ground surface anomalies

The TM image used by Bonneville & Gouze (1992) was recorded on 23 October 1986, 1 month after a summit eruption and 1 week before a flank eruption involving the eastern and northeastern flanks of the volcano between the altitudes of 2900 and 2200 m. During that eruption more than $60 \times 10^6 \, m^3$ of lava were produced.

Following the methodology outlined in Section 7.5.4, Bonneville & Gouze (1992) used the split window technique with low-resolution (1 km pixel) AVHRR channels 4 and 5 to derive a map of how atmospheric transmission varied across the image. They then used this to calculate the atmospherically corrected radiance in high-resolution (120-m pixel) TM band 6 data, and finally used a digital terrain model to remove the effect of adiabatic air cooling with altitude, derived as in Equation 7.15). The efficacy of this method is shown by the relative enhancement of thermal anomalies from their uncalibrated TM image (Fig. 7.8a) to the final image (Fig. 7.8b). Bonneville & Gouze (1992) identified five anomalous thermal fields labelled A1 to A5.

A1. In the summit zone, three anomalies of high amplitude but small spatial extent correspond to the northeast, central (Bocca Nova), and southeast craters, all of which were active. The lava flows extending from the northeast crater in a northwest direction (September 1986 eruption (Smithsonian Institution 1986a)) as well as those from the southeast crater are identified. The cones are particularly evident as hot areas around all the craters. The mean temperature anomaly reaches about 3°C with respect to the surrounding area. The total anomalous heat flow on the summit zone between altitude levels 3000 and 3300 m (an area of 1 km$^2$) can be estimated at 200 MW using the empirical relationship expressed in Equation 7.14. Bonneveille & Gouze (1992) proposed that a measurement such as this could be used as a guide to the overall state of Mount Etna's heat budget.

A2. This anomaly corresponds to the 1984 lava field extending through the higher part of Valle del Bove that Pieri et al. (1990) observed using TM bands 5 and 7. Evidently the flow was still perceptibly warmer than its surroundings 2 years later

A3. This spatially small anomaly (100–200 m) is on the northern edge of Valle del Bove, close to the 2000 m contour. This is on the site of the fissure-fed eruption that began on 30 October 1986, 1 week after the data acquisition (Smithsonian Institution 1986b).

A4. This anomaly in the southern edge of Valle del Bove does not correspond to any recent volcanic activity, although it is associated with fissure zones like anomalies A2 and A3.

A5. This large anomaly around Rifugio Sapienza may be linked to recent lava fields (eruptions of 1983), and numerous fracture zones.

Comparison of Figures 7.8a and 7.8b shows that, except for the summit anomalies A1, the anomalies are detectable only after the altitude correction described in Section 7.5.2, even though the amplitude of each is at least 2°C with respect to the local environment. The sparsity of vegetation within these zones allows them to be measured with a high degree of confidence, and they correspond to heat-flow anomalies of about 130 W m$^{-2}$. Anomaly A3 could be considered as a thermal precursor of the 30 October 1986 eruption, but it could not be expected to be recognized as such before the event, without repeated imaging capable of monitoring the development of the anomaly. The presence of the larger anomaly A4, which is not obviously linked either to eruptive activity or to solar warming, shows that there is still some way to go before remotely sensed low-temperature anomalies can be relied on as a guide to impending volcanic activity.

(a)

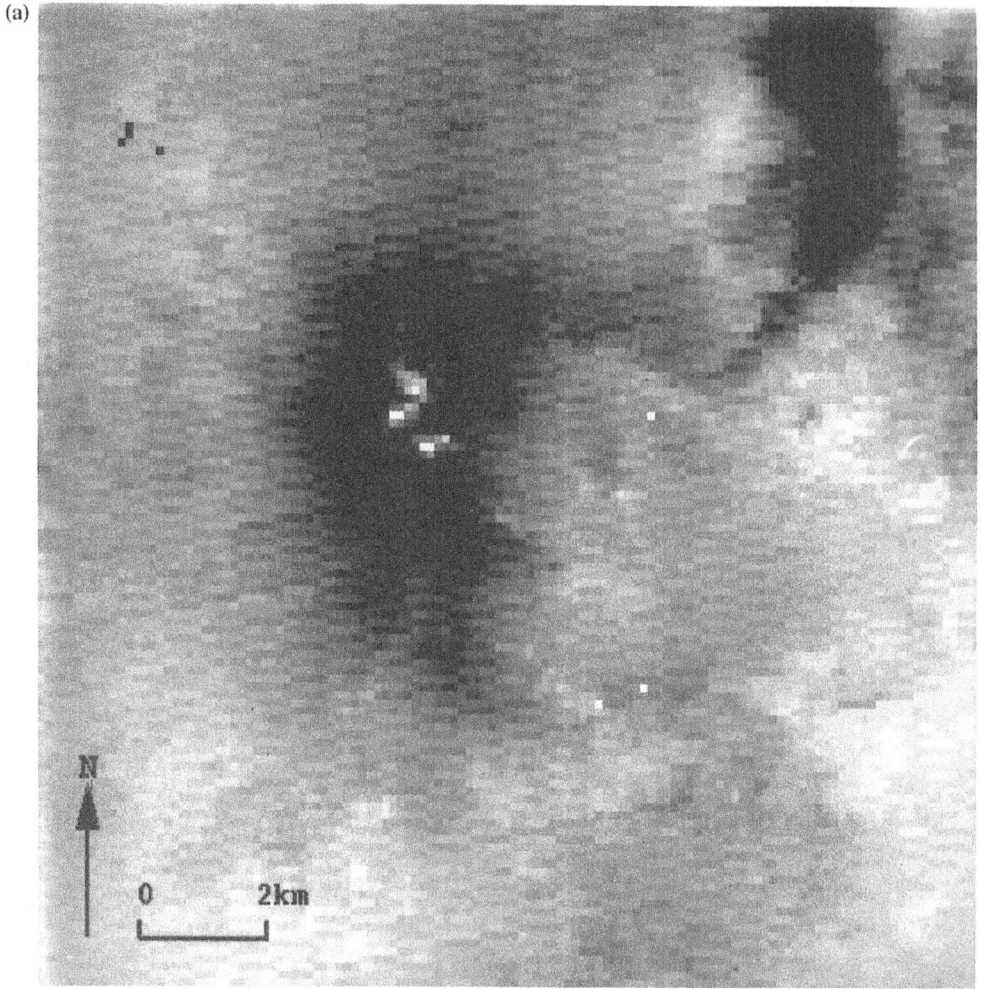

**Figure 7.8**  (a) Uncalibrated TM band-6 image recorded 22.00 h local time on 23 October 1986. Grey tones from black to white represent temperature ranging from cold to warm.

## 7.7.2 Thermal anomalies at crater lakes

Volcanic crater lakes also provide targets for thermal remote sensing. Increases in lake-water temperature prior to eruptions have been documented by ground-based measurements at several sites, such as Taal, Philippines (Moxham 1967), Ruapehu, New Zealand (Hurst & Dibble 1981) and Poàs, Costa Rica (Brown et al. 1989, 1991). In the latter example calorimetry of the crater lake was regarded as an important constraint on the energy budget of the volcano. Using AVHRR data and the split window algorithm, the temperature of a crater lake more than a few kilometres across could be measured to an accuracy of a fraction of a degree in the same way as the sea surface, but such targets are rare. At present, Landsat TM band 6 provides the highest resolution TIR satellite data in the civilian sector. In view of its 120-m

204

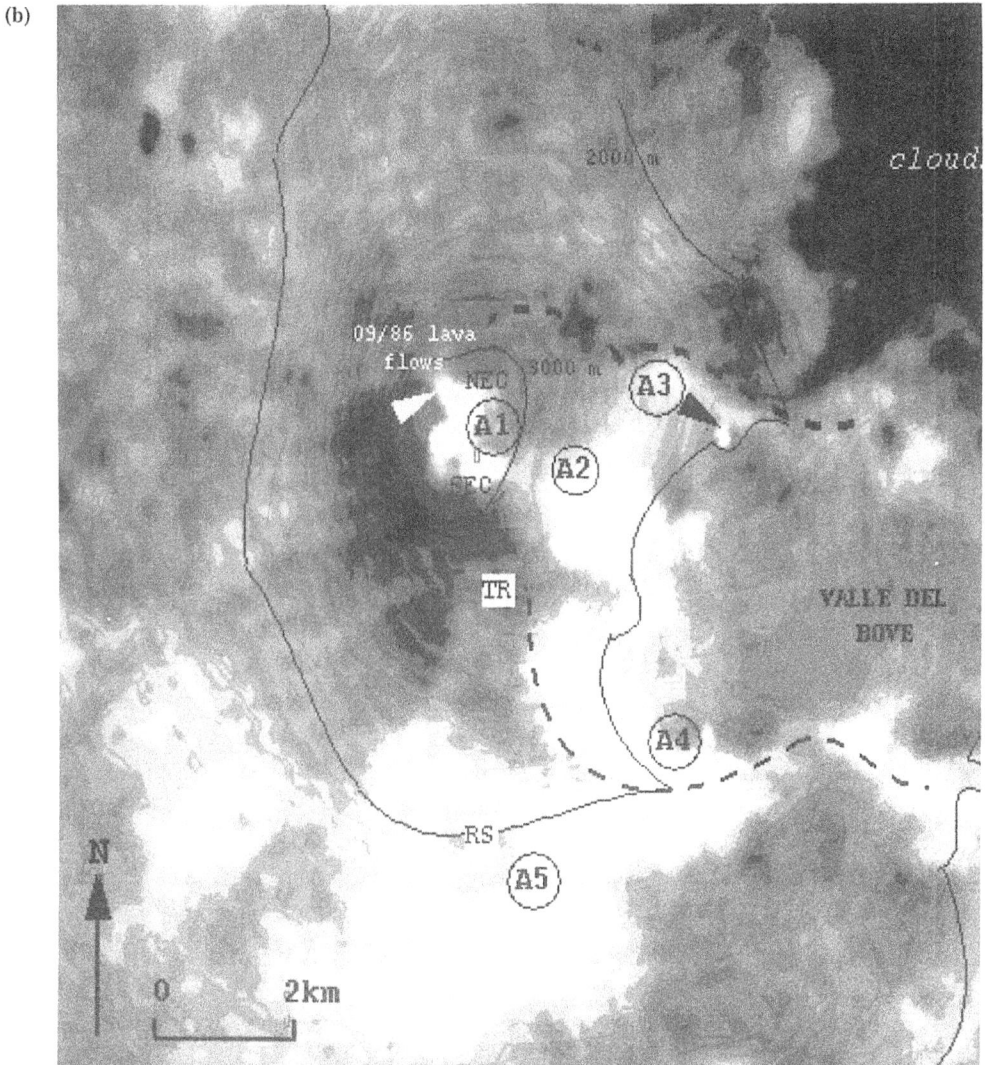

**(b)**

**Figure 7.8** (b) The same image corrected for atmospheric absorption and adiabatic air cooling with altitude. Each grey level now represents a 1 °C range in pixel-integrated temperature from 14 °C (black) to 27 °C (white). Five thermal anomalies A1–A5 are discussed in the text. Superimposed jagged lines mark the edge of Valle del Bove, and fine lines mark the 1000 m and 2000 m contours. RS, Rifugio Sapienza; TR, Torre del Filosofo; SEC, southeast crater; NEC, northeast crater.

pixel dimension, geometric considerations mean that these data can measure the temperature of a circular lake only if it exceeds 340 m in diameter (Oppenheimer 1993).

Oppenheimer (1993) presented an analysis of three TM thermal images of the crater lake at Poás, Costa Rica, finding increasing discrepancies between TM- and field-based measurements as the lake constricted in size. Without an adequate way to correct for the radiative properties of the atmosphere the accuracy of such measurements is uncertain, but in principle at least remotely sensed data offer a means of performing temperature and calorimetric monitoring.

## 7.8 Ground-based infrared measurements of hot phenomena

Many ground-based techniques involve temperature measurements of either gas or soil temperature by contact thermometers (see Chs 3 and 14) (Pinkerton 1993). Typically, ground-temperature probes are set up in a network so as to measure the temperature at more than one depth in thermally "normal" and "anomalous" areas. Meteorological parameters may also be measured and the data can be either stored locally or transmitted to remote observatories through very high frequency (VHF) links or satellite data relay systems such as ARGOS (Ch. 3). This approach has three main advantages:

(a) it is cheap, and easy to install (at least for simple networks without radio transmission);
(b) it allows continuous monitoring of temperature; and
(c) it provides reliable data on absolute temperatures and temperature gradients.

However, there are some concomitant disadvantages:

(a) data are obtained for only a few isolated points, making it difficult or impossible to discriminate between very local effects (such as water penetrating along fractures) and those of volcanic origin; and
(b) the lifetime of ground-based equipment may be severely limited by the hostile and corrosive environment in which it is sited.

We shall discuss here only non-contact methods using infrared devices that are the ground-based (and usually non-imaging) analogues of the remote-sensing devices providing the sort of data discussed in the rest of this chapter. Oppenheimer & Rothery (1991) describe the use of broad-band infrared thermometers exemplified by the Minolta Cyclops series. These are about the same size and weight as a "camcorder", cost about US$2000 and have proved durable and fairly reliable in the field (though they do not work well at ambient temperatures much below freezing) and can measure the brightness temperature of a field-of-view 1° of arc or less across. These instruments, mass produced for use in metallurgy and the ceramics industry, have effectively made optical pyrometers (as used, for example, by Le Guern et al. (1979)) redundant, because they do not rely on the surface to be measured being visibly radiant. In practice, one infrared thermometer of the new type measuring in the 8–14 μm region can measure temperatures from ambient up to several hundred degrees celcius, and a second one operating in a shorter waveband can measure an overlapping range extending up beyond feasible magmatic temperatures.

The main advantage of such thermometers is that they can be used to take readings of inaccessible surfaces at distances of up to several hundred metres, for example open vents on the floors of steep-walled craters. They can verify the presence of high-temperature sources when visual identification is prevented by fumes, or simply by daylight. Limitations arise in using temperature data quantitatively because of the difficulty in repeating measurements, contamination by reflected radiation (especially in visible or SWIR wavelengths) and solar heating/wind cooling (in the TIR), and uncertainties in surface emissivity and atmospheric transmission. In consequence, interpretation of measurements of dynamic phenomena such as rapidly moving or overturning lava bodies is not straightforward. Oppenheimer & Rothery (1991) and Oppenheimer et al. (1993b) have used such thermometry to record spatial/temporal temperature distributions associated with fumarole fields.

206

Other devices capable of non-contact measurements of temperature in the field include spectrometers, that measure radiance in a hundred or more narrow channels extending from the visible into the SWIR or beyond. An example of this type of instrument is the GER "IRIS" spectroradiometer used in Hawaii by Flynn et al. (1991 1993) and Flynn & Mouginis-Mark (1992). Unfortunately, this type of equipment is heavy and is field-portable only with care. A lighter weight 0.4–3.0 μm FTIR spectroradiometer for field use is currently under development at the University of Hawaii (P. Mouginis-Mark, personal communication). There are other FTIR field spectrometers that are lightweight and commercially available, but we are not aware of any such device having been used for thermal studies. The operation of a video camera capable of recording moving images in the 8–12 μm region (mentioned in Section 7.6.2, and suitable for use on the ground as well as from the air) is described by Rothery & Pieri (1993).

## 7.9 Satellite data: where to get them, what to ask for, and what they cost

This section is provided as a guide to volcanologists who are seriously considering using satellite-based remote sensing data for measuring thermal phenomena on volcanoes. Its intention is to offer practical advice on data sources, what formats of data to ask for, and the likely costs.

When buying remotely sensed data it is prudent, if possible, to make a visual inspection first. Although catalogues usually provide an estimate of the total amount of cloud cover within an image, only by checking visually can you be sure that cloud or fumes are not obscuring the part of the volcano that is of interest to you. It is cheaper to buy an image as a photograph than in digital form, and this price differential may be very significant for high-resolution satellites such as Landsat. Photographs can be interpreted visually, providing a simple way of recording changes between one eruption and the next, but only digital data can be used to derive quantitative radiance values necessary for thermal modelling. If you intend to measure radiances to derive temperatures by means of the dual-band method, then try to avoid using data in which the pixels have been resampled in any way, as this destroys the integrity of the original radiance information. This means never using any data that are sold as "geocoded", because such an image has been resampled to fit a map projection, and although this is useful for mapping purposes it introduces irreversible radiometric degradation within pixels.

We give below some information on where data can be ordered. Many countries have national remote-sensing centres (too numerous to list here) that will give further advice. If your research is funded by a national research council then this body may already have established procedures for enquiring after and buying satellite data. This is the case, for example, with the Natural Environmental Research Council in the UK. A general rule whenever buying data from which you intend to derive temperatures, especially if you intend to use the dual-band technique, and whichever data supplier you deal with, is to state clearly what you intend to do with the data and to seek advice on the most suitable format for your purpose. Bear in mind that the salespersons who handle most requests for data are unlikely to be familiar with the problems facing a remote sensing volcanologist, so you would be well advised to ask to speak to somebody able to give specific technical advice.

### 7.9.1 Landsat data

Data from the Landsat TM (and the less useful, but cheaper, Multispectral Scanner instrument (MSS)) are held at many centres worldwide. Some Landsat data are received directly at ground stations over a range of about 3000 km, and are available from local distribution centres. Directly transmitted data covering north America, and data covering other parts of the globe transmitted via relay satellites are held centrally in the USA. At present, you can expect to pay over US$3000 for a quarter of a full TM image, covering an area of approximately 90 km × 90 km. A black and white photographic transparency covering a full scene at 1:1 million scale costs about $2700, but photographic versions of simultanously recorded images from the MSS (80-m pixels) are much cheaper at US$120, and should be adequate for cloud and fume assessment. In addition, it is worth checking with your supplier to see if any kind of "quick-look" product is available for this purpose. Pre-1986 TM data are available as 180 km × 180 km full scenes on CCT, but only in resampled format, for US$300. Landsats-4 and -5 were operated on a commercial basis by the Eosat company. The replacement Landsat-6 may mark the beginning of de-commercialization of the Landsat programme. It will be operated jointly by the US Defense Department and Eosat, and this may offer the opportunity of significantly reduced prices for scientifc research, at least for projects funded from US Government sources.

Non-resampled data were formerly supplied by Eosat (see below) under the description of "A-format" data, as opposed to the more commonly used "P-format" or geometrically resampled varieties. A-format data did not have forward and reverse sweeps of the scan mirror lined up, so the untreated image has alternate sets of 16 lines displaced, but this was the only format in which the radiometric quality of the data were fully to be trusted (Rothery et al. 1988a, Glaze et al. 1989a). A-format was discontinued in 1991, but data continued to be available (though only as full scenes) in raw form under the description of "level 0" (zero) data, in which the visually obtrusive 16-line offsets were removed but the integrity of the raw pixels is unaltered. The analogous format supplied by European stations goes under the name "level 1".

An alternative strategy is to buy geometrically corrected image data in which the original pixels have been resampled by nearest-neighbour resampling only (avoiding images that have been processed by, say, cubic convolution resampling, which look less blocky but in which the DN of each output pixel is a linear or non-linear function of several neighbouring original pixels). A small fraction of pixels may have been duplicated or omitted in the nearest-neighbour resampling procedure but, in principle, the band-to-band radiometry of each pixel should be identical to that on the original image. Unfortunately, Glaze et al. (1989a) found that pixels had been replicated in different positions in TM bands 5 and 7 in nearest-neighbour resampled

TM data, so that these two bands were locally displaced from one another by 1 pixel, thereby undermining the basis of the dual-band technique at unpredictable locations. Eosat now claim that TM data supplied with a 30-m resampling interval and produced by nearest-neighbour resampling will not exhibit replicated pixels (D. Fischel, Eosat Chief Scientist, personal communication). The only drawback in using such data is that a small proportion (less than 1%) of the original pixels will not be represented in the resampled dataset, and in consequence there is a slight chance that a thermal anomaly confined to a single pixel could be missed.

Initial enquiries, requests for a free computer printout of all centrally held data over any area you specify, and special requests for the satellite to record an image (e.g. of an on-going eruption, or by night) should be directed to:

EOSAT,
c/o EROS Data Center,
Sioux Falls, SD 57198, USA

Tel. (1) 605 594 6511

or

EOSAT Customer Services,
4300 Forbes Boulevard,
Lanham, Md 20706, USA

Tel. 1-800-344-9933 or (1) 301 552 0537. Fax (1) 301 522 0507

A recently established means of interrogating a global data base of Landsat and other satellite data (including AVHRR) is provided by the Global Land Information System (GLIS), which can be reached using an alphanumeric terminal or terminal emulator package on a PC as follows:

from NSI/DECNET:   $SET HOST GLIS
                   USERNAME: GLIS
from INTERNET:     $TELNET glis.cr.usgs.gov
                   or $TELNET 152.61.192.54

Direct dial: set modem to 8 bits, no parity, 1 top bit.
Dial: (605) 594-688 or FTS 753-7888.

Direct dialing is not recommended from outside the USA, but the Internet connection works well. Assistance and information about GLIS can be obtained by phoning 1-800-252-4547 (from within the USA only) or (1) 605 594 6099

Enquiries for Landsat images covering Europe and north Africa, may also go to:

ESA-ESRIN,
Earthnet User Services,
Via Galileo Galilei,
00044 Frascati, Italy

Tel. (39) 69401360. Fax (39) 694180 361

From within the UK, enquiries about all varieties of satellite data can be directed to the following address (but seek also research council advice if you are so funded):

National Remote Sensing Centre Ltd (NRSCL),
Delta House,
Southwood Crescent,
Southwood,
Farnborough,
Hants GU14 0NL, UK

Tel. 01252 541464. Fax 01252 375016

### 7.9.2 JERS-1 (Fuyo-1) data

A Japanese satellite JERS-1 launched in February 1992 records images by means of its OPS instrument in four wavebands in the SWIR region, and has potential for thermal measurements on lava bodies. This satellite was officially renamed Fuyo-1 after launch, but both names seem to be in current use. Prices may be in line with those for Landsat TM.
  Enquiries about JERS-1 data should be directed to:

Remote Sensing Technology Center of Japan,
Uni-Roppongi Building 7-15-17,
Roppongi, Minato-Ku, Tokyo 106, Japan

Tel. (81) (0)3-3403-1761. Fax (81) (0)3-3403-1766. Telex 02426780 RESTEC J

### 7.9.3 AVHRR data

The satellites carrying the AVHRR are operated by the National Oceanographic and Atmospheric Adminstration (NOAA) of the USA on a non-commerical basis, as a result of which the data are cheap to buy, and some versions can even be received direct from the satellite. There are essentially two pixel sizes of data available: 1-km and 8 km (at nadir). The 1 km data are obtained either when the data are transmitted in real time to a ground station in high-quality form (known as High Resolution Picture Transmission (HRPT)) or when they are recorded on-board at 1-km resolution and transmitted to a NOAA ground station later (these are Local Area Coverage (LAC) data). For all volcanological purposes, the 1-km data are much to be preferred. The 8-km (global area coverage (GAC)) data result from on-board recording at reduced resolution. GAC data are continuously available, LAC data are usually obtained by special request only, and HRPT data are recorded routinely for specific regions only.
  AVHRR data are also transmitted to ground in analogue (non-digital) form by automated picture transmission (APT) in the form of a signal that can be received by a simple antenna (costing less than US$1000) and then displayed on a video screen. However, these data are

not suitable for quantitative radiometric studies.

For general worldwide imaging requiring data of AVHRR from a rolling archive (extending back over about 4 years for LAC data) contact GLIS (listed above), or:

National Climatic Data Center,
Satellite Services Division, Room 100,
World Weather Building,
Washington DC, 20233, USA

Tel. (1) 301 763 8111 (or, within the USA, FTS 763 8111).
Telex 248376 OBSWUR

To request scheduling of AVHRR LAC data contact:

Interactive Processing Branch,
NESDIS, Room 510,
World Weather Building,
Washington DC, 20233, USA

Tel. (1) 301 763 8142 (or, within the USA, FTS 763 8142). Telex 248376 OBSWUR

Further information, and advice on alternative sources of data, can be sought from national remote sensing centres and meteorological agencies. Anyone wishing to do quantitative thermal work should consult the *NOAA Polar Orbiter users' guide*, which can be obtained from the addresses in the World Weather Building quoted above. The cost of AVHRR data is very reasonable, you should expect to pay less than about US$100 for a tape that could contain extracts of several original images.

### 7.9.4 Future sensors

Future remote sensing instruments capable of recording thermal radiance from hot lava in more than two wavebands will enable cross-checking of the dual-band solutions to two thermal components, and also more refined modelling where more than two thermal components need to be considered. Multiple channel devices and imaging spectrometers will make it possible to attempt more realistic corrections for atmospheric effects. The main instrument packages in the Earth observing system (EOS), a series of satellites to be deployed by NASA and other space agencies in the late 1990s, are described in Mouginis-Mark et al. (1991a).

## 7.10 Conclusions

Weather satellite data useful for measuring low-temperature thermal anomalies are cheap and easily available. The cost of purchasing infrared satellite data in digital form, having sufficently high resolution to be useful in studies of high-temperature thermal anomalies, now outweighs the price of the simplest image processing facilities, but data costs to scientists may fall significantly as a consequence of decommercialization.

Satellite or airborne remotely sensed data offer a way to collect data covering the whole extent of a particular phenomenonon at one time. For example, Landsat TM or ETM data can be used to determine temperatures of both "crust" and "core" of an active lava flow, and of gathering this information at 30 m intervals along the whole length of a flow instantaneously. To achieve this sort of coverage depends on the good luck of cloud-free conditions and (for short-lived or rapidly changing events) of the satellite overpassing the target area at the appropriate time. It is important to beware the pitfalls posed by geometrically corrected image data, if comparisons between spectral channels are to be used as the basis of temperature determinations. Despite the many difficulties, remote sensing looks set to play an increasingly valuable role in studies of active volcanism in the future.

# Acknowledgements

DAR's contribution to this study was facilitated by support from the Natural Environmental Research Council (GR9/14 and GR3/8006) and the UK Overseas Development Administration, and CO was supported by the Natural Environmental Research Council (GR3/8006). We thank Lori Glaze and Harry Pinkerton for their thoughtful and useful reviews, and Peter Francis, David Stevenson and Dave Pieri for other discussions relevant to the content of this chapter.

# References

Abrams, M., L. Glaze, M. Sheridan 1991. Monitoring Colima Volcano, Mexico, using satellite data. *Bulletin of Volcanology* **53**, 571–4.

Andres, R. J. & W. I. Rose 1994. Description of thermal anomalies at two active Guatemalan volcanoes using Landsat Thematic Mapper imagery. *Photogrammetric Engineering & Remote Sensing* in press.

Bartollucci, L. A., M. Chang, P. E. Anuta, M. R. Graves 1988. Atmospheric effects on Landsat TM thermal IR data. *IEEE Transactions of Geoscience and Remote Sensing* **26**, 171–5.

Bonneville, A. & P. Gouze 1992. Thermal survey of Mount Etna volcano from space. *Geophysical Research Letters* **90**, 725–8.

Bonneville, A. & Y. Kerr 1987. A thermal forerunner of the 28th March 1983 Mount Etna eruption from satellite thermal infrared data. *Journal of Geodynamics* **7**, 1–31.

Bonneville, A., G. Vasseur, Y. Kerr 1985. Satellite thermal infrared observations of Mount Etna after the 17th March 1981 eruption. *Journal of Volcanology and Geothermal Research* **36**, 209–32.

Brivio, P. A., E. Lo Giudice, E. Zilioli 1989. Thermal infrared surveys at Vulcano Island: an experimental approach to the thermal monitoring of volcanoes. In *IAVCEI Proceedings in Volcanology vol. 1: Volcanic hazards – assessment and monitoring*, J. H. Latter (ed.), 357–71. Berlin: Springer.

Brown, G. C., H. Rymer, J. Dowden, P. Kapadia, D. Stevenson, J. Barquero, L. D. Morales 1989. Energy budget analysis for Poás crater lake: implications for predicting volcanic activity. *Nature* **339**, 370–3.

Brown, G. C., H. Rymer, D. Stevenson 1991. Volcano monitoring by microgravity and energy budget analysis. *Journal of the Geological Society, London* **148**, 585–93.

Crisp, J. & S. M. Baloga 1990. A model for lava flows with two thermal components. *Journal of Geophysi-

*cal Research* **91**, 9543–52.

Deschamps, P. Y. & T. Phulpin 1980. Atmospheric correction of infrared measurements of sea surface temperature using channels at 3.7, 11 and 12 microns. *Boundary-Layer Meteorology* **18**, 131–43.

Dragoni, M. 1989. A dynamical model of lava flows cooling by radiation. *Bulletin of Volcanology* **51**, 88–95.

Dragoni, M., S. Pondrelli, A. Tallarico 1992. Longitudinal deformation of a lava flow: the influence of Bingham rheology. *Journal of Volcanology and Geothermal Research*, **52**, 247–54.

Drury, S. A. 1993. *Image interpretation in geology*, 2nd edition. London: Chapman & Hall.

Flynn, L. P. & P. J. Mouginis-Mark 1992. Cooling rate of an active Hawaiian lava flow from nighttime spectroradiometer measurements. *Geophysical Research Letters* **19**, 1783–6.

Flynn, L. P., P. J. Mouginis-Mark, J. C. Gradie, P. G. Lucey 1991. Radiative temperature measurements taken at Kupianaha lava lake: final results and implications for satellite remote sensing. *EOS Transactions of the American Geophysical Union, AGU 1991 Fall Meeting Program & Abstracts*, 562. Washington DC: American Geophysical Union.

Flynn, L. P., P. J. Mouginis-Mark, J. C. Gradie, P. G. Lucey 1993. Radiative temperature measurements at Kupianaha lava lake Kilauea volcano, Hawaii, *Journal of Geophysical Research* **98**, 6461–76.

Francis, P. W. 1979. Infrared techniques for volcano monitoring and prediction – a review. *Journal of the Geological Society of London* **136**, 355–9.

Francis, P. W. & D. A. Rothery 1987. Using the Landsat Thematic Mapper to detect and monitor active volcanoes. *Geology* **15**, 614–7.

Friedman, J. D., D. Frank, H. G. Kieffer, D. L. Sawatzky 1981. *United States Geological Survey Professional Paper 1250, Thermal infrared surveys of the May 18 crater, subsequent lava domes and associated volcanic deposits. The 1980 eruptions of Mount St Helens, Washington*, P. W. Lipman, D. R. Mullineaux (eds), 557–67. Washington DC: United States Government Printing Office.

Glaze, L. S., P. W. Francis, D. A. Rothery 1989a. Measuring thermal budgets of active volcanoes by satellite remote sensing. *Nature* **338**, 144–6.

Glaze, L. S., P. W. Francis, S. Self, D. A. Rothery 1989b. The 16 September 1986 eruption of Lascar volcano, north Chile: satellite investigation. *Bulletin of Volcanology* **51**, 149–60.

Gupta, R. K. & K. V. S. Badarinath 1993. Volcano monitoring using remote sensing data. *International Journal of Remote Sensing* **14**, 2907–2918.

Hardee, H. C. 1982. Permeable convection above magma bodies. *Tectonophysics* **84**, 179–95.

Hardee, H. C. 1993. Convection heat transfer rates in molten lava. In *Monitoring active lavas*, C. J. Kilburn (ed.), 193–201. London: UCL Press.

Harris, A. J. L. 1992. Volcano detection and monitoring using AVHRR: the Krafla eruption, Iceland 1984. MSc thesis, University of Dundee, UK.

Hurst, A. W. & R. R. Dibble 1981. Bathymetry, heat output and convection in Ruapehu crater lake, New Zealand. *Journal of Volcanology and Geothermal Research* **9**, 215–36.

Kahle, A. B., A. R. Gillespie, E. A. Abbott, M. J. Abrams, R. E. Walker, G. Hoover, J. P. Lockwood 1988. Relative dating of Hawaiian lava flows using multispectral thermal infrared images: a new tool for geologic mapping of young volcanic terrains. *Journal of Geophysical Research* **93**, 15239–51.

Kieffer, H. H., D. Frank, J. D. Friedman 1981. *United States Geological Survey Professional Paper 1250, Thermal infrared surveys at Mount St Helens. Observations prior to the eruption of May 1 The 1980 eruptions of Mount St Helens, Washington*, 257–77. Washington DC: United States Government Printing Office.

Kneizys, F. X., E. P. Shettle, L. W. Abreu, J. H. Chetwynd, G. P. Anderson, W. O. Gallery, J. E. A. Selby, S. A. Clough 1988. *Air Force Geophysics Laboratory Environmental Research Paper 1010, Users' guide to LOWTRAN 7*. Hanscom, Massachusetts: Hanscom AFB.

Krueger, A. J., L. S. Walter, C. C. Schnetzler, S. D. Doiron 1990. TOMS measurement of the sulphur dioxide emitted during the 1985 Nevado del Ruíz eruptions. *Journal of Volcanology and Geothermal Research* **41**, 7–15.

Le Guern, F., J. Carbonelle, H. Tazieff 1979. Erta 'Ale lava lake: heat and gas transfer to the atmosphere. *Journal of Volcanology and Geothermal Research* **6**, 27–48.

Matson, M., G. Stephens, J. Robinson 1987. Fire detection using data from NOAA-N satellites. *International Journal of Remote Sensing* **8**, 961–70.

McClelland, L., T. Simkin, M. Summers, E. Nielsen, T. C. Stein (eds) 1989. *Global Volcanism 1975–1985*. Englewood Cliffs, New Jersey: Prentice-Hall.

Mouginis-Mark, P., S. Rowland, P. W. Francis, T. Freidman, H. Garbeil, J. Gradie, S. Self, L. Wilson, J. Crisp, L. Glaze, K. Jones, A. Kahle, D. Pieri, H. Zebker, A. Kreuger, L. Walter, C. Wood, W. Rose, J. Adams, R.Wolf 1991a. Analysis of active volcanoes from the Earth Observing System. *Remote Sensing of the Environment* **36**, 1–12.

Mouginis-Mark, P., S. Rowland, H. Garbeil, P. Flament 1991b. AVHRR observations of the Kupianaha eruption, Hawaii. *EOS Transactions of the American Geophysical Union, AGU 1991 Fall Meeting Program & Abstracts,* 562. Washington DC: American Geophysical Union.

Moxham, R. M. 1967. Changes in surface temperatures at Taal volcano, Philippines 1965–1966. *Bulletin Volcanique* **31**, 215–34.

Moxham, R. M. 1971. Thermal surveillance of volcanoes. In *The surveillance of volcanic activity*, 103-22. Paris: UNESCO.

Oppenheimer, C. 1991a. Volcanology from space: applications of infrared remote sensing. PhD thesis, Open University, UK.

Oppenheimer, C. 1991b. Lava flow cooling estimated from Landsat Thematic Mapper infrared data: the Lonquimay eruption, Chile 1989. *Journal of Geophysical Research* **96**, 21856–78.

Oppenheimer, C. 1993. Infrared surveillance of crater lakes using satellite data. *Journal of Volcanology and Geothermal Research* **55**, 117–28

Oppenheimer, C. M. M. & D. A. Rothery 1991. Infrared monitoring of volcanoes by satellite. *Journal of the Geological Society, London* **148**, 563–9.

Oppenheimer, C, D. Pieri, V. Carrere, M. Abrams, D. Rothery, P. W. Francis 1992. Volcanic thermal features observed by AVIRIS. In *Summaries of the Third Annual JPL Airborne Geoscience Workshop June 1–5 1992, vol. 1: AVIRIS Workshop*, R. O. Green (ed.), 41–43. Pasadena: Jet Propulsion Laboratory.

Oppenheimer, C., P. W. Francis, D.A. Rothery, R. W. Carlton, L. S. Glaze 1993a. Infrared image analysis of volcanic thermal features: Láscar Volcano, Chile 1984–1991. *Journal of Geophysical Research* **98**, 4269–86.

Oppenheimer, C, D. A. Rothery, P. W. Francis 1993b. Thermal distributions at fumarole fields: implications for infrared remote sensing of active volcanoes. *Journal of Volcanology and Geothermal Research* **55**, 97–115.

Oppenheimer, C., D. A. Rothery, D. C. Pieri, M. Abrams,V. Carrere 1993c. Analysis of Airborne Visible/Infrared Imaging Spectrometer (AVIRIS) data of volcanic hot spots. *International Journal of Remote Sensing* **14**, 2919–34.

Pieri, D. C., L. S. Glaze, M. J. Abrams 1990. Thermal radiance observations of an active lava flow during the June 1984 eruption of Mount Etna. *Geology* **18**, 1018-22.

Pinkerton, H. 1993. Measurements of the properties of flowing lavas. In *Monitoring active lavas*, C. J. Kilburn (ed.), 175–91. London: UCL Press.

Pollack, J. B., O. B. Troon, B. N. Khare 1973. Optical properties of some terrestrial rocks and glasses. *Icarus* **19**, 372–89.

Price, J. C. 1984. Land surface temperature measurements from the split window channels of the NOAA7 advanced very high resolution radiometer. *Journal of Geophysical Research* **89**, 7231–7.

Realmuto, V. J., K. Hon, A. B. Kahle, E. A. Abbot, D. Pieri 1992. Multispectral thermal infrared mapping of the 1 October 1988 Kupianaha flow field, Kilauea volcano, Hawaii. *Bulletin of Volcanology* **55**, 97–115.

Reddy, C. C., S. K. Srivastav, A. Bhattacharya 1992. Use of short wavelength infrared data for detection and monitoring of high temperature related geoenvironmental features. In *Remote sensing applications and geographic information systems*, I. V. Muralikrishna (ed.), 216–20. New Delhi: Tata McGraw-Hill.

Reddy, C. S. S., A. Battacharya, S. K. Srivastav 1993. Night time TM short wavelength infrared data analysis of Barren Island volcano, south Andaman, India. *International Journal of Remote Sensing* **14**, 783–7.

Rees, W. G. 1990. *Physical principles of remote sensing*. Cambridge: Cambridge University Press.

Rothery, D. A. & C. M. M. Oppenheimer 1991. Monitoring volcanoes using short wavelength infrared images. In *ESA SP-319, Proceedings of the 5th International Colloquium on Spectral Signatures of Objects in Remote Sensing, Courchevel, France, 14–18 January 1991*, 513–16. Noordwijk: European Space Agency.

Rothery, D. A. & C. M. M. Oppenheimer 1994. Monitoring Mount Erebus by remote sensing. In *Volcanic*

*Studies of Mount Erebus, Antarctica*, P. Kyle (ed). Antarctic Research Series, American Geophysical Union.

Rothery, D. A. & D. C. Pieri. 1993. Remote sensing of active lava. In *Monitoring active lavas*, C. J. Kilburn (ed.), 203–32. London: UCL Press.

Rothery, D. A., P. W. Francis, C. A. Wood 1988a. Volcano monitoring using short wavelength infrared data from satellites. *Journal of Geophysical Research* **93**, 7993–8008.

Rothery, D. A., P. W. Francis, C. A. Wood 1988b. Volcano monitoring by short wavelength infrared satellite remote sensing. In *Proceedings of the 6th Thematic Conference on Remote Sensing for Exploration Geology, Houston, Texas, May 16–19 1988*, 283–91. Ann Arbor: Environmental Research Institute of Michigan.

Rothery, D. A., A. Borgia, R. W. Carlton, C. Oppenheimer 1992. The 1992 Etna lava flow imaged by Landsat TM. *International Journal of Remote Sensing* **13**, 2759–63.

Sabins, F. F. 1986. *Remote sensing: principles and interpretation* 2nd edn. New York: W. H. Freeman.

Sawada, Y. 1989. The detection capability of explosive eruptions using GMS imagery, and the behaviour of dispersing eruption clouds. In *IAVCEI Proceedings in Volcanology vol. 1: Volcanic hazards – assessment and monitoring*, J. H. Latter (ed.), 233–45. Berlin: Springer.

Sekioka, M. & K. J. Yuhara 1974. Heat flux estimation in geothermal areas based on the heat balance of the ground surface. *Journal of Geophysical Research* **79**, 2053–8.

Smithsonian Institution 1986a. Mt Etna. *Scientific Event Alert Network (SEAN) Bulletin* **11**, 4–8.

Smithsonian Institution 1986b. Mt Etna. *Scientific Event Alert Network (SEAN) Bulletin* **11**, 7–9.

Smithsonian Institution 1989. Niyamuragira. *Scientific Event Alert Nework (SEAN) Bulletin* **14**, 21.

Smithsonian Institution 1992a. Erta 'Ale. *Bulletin of the Global Volcanism Network* **17**, 5.

Smithsonian Institution 1992b. Erta 'Ale. *Bulletin of the Global Volcanism Network* **17**, 4.

Smithsonian Institution 1992c. Erta 'Ale. *Bulletin of the Global Volcanism Network* **17**, 2.

Wan, Z. & J. Dozier 1989. Land-surface temperature measurement from space: physical principles and inverse modelling. *IEEE Transactions of Geoscience and Remote Sensing* **27**, 268–77.

Wiesnet, D. R. & J. D'Aguanno 1982. Thermal imagery of Mount Erebus from the NOAA-6 satellite. *Antarctic Journal of the United States* **17**, 32–4.

Woods, A. W. & S. Self 1992. Thermal disequilibrium at the top of volcanic clouds and its effect on estimates of eruption column height. *Nature* **355**, 628–30.

# Appendix: list of acronyms and related terms

ARGOS  a system for data collection and transmission developed jointly by CNES, NASA and NOAA

ATSR   Along-Track Scanning Radiometer (an instrument carried by ERS-1)

AVHRR   Advanced Very High Resolution Radiometer (the imaging instrument carried by NOAA TIROS-N satellites)

AVIRIS   Airborne Visible/Infrared Imaging Spectrometer

CCT   computer compatible tape (a standard medium for distributing image data)

CD   compact disc (the same format as used for audio equipment, capable also of storing large volumes of image data)

CNES   Centre National d'Etudes Spatial (the French national space agency)

DN   digital number (number assigned to the radiance detected in a pixel of an image)

DTM   digital terrain model (a computer file containing height information for specified localities, often gridded to match image data)

EOS   Earth Observing System (a series of NASA Earth observation satellites with "next-generation" remote sensing instruments, due for deployment in 1998)

Eosat   a commercial organization operating Landsat and marketing its data from

215

1986 onwards

ERS-1 European Remote Sensing satellite-1

ESA European Space Agency

ETM Enhanced Thematic Mapper (a more versatile version of the Thematic Mapper, carried by Landsat-6)

Eurimage an organization marketing Landsat (and other) image data recorded in Europe

FTIR Fourier transform infared

GAC Global Area Coverage (low-resolution, 8-km pixel, AVHRR data)

GER Geophysical Environmental Research Inc.

GLIS Global Land Information System (a computer data base of Landsat and other remotely sensed data)

HRPT High Resolution Picture Transmission (a format of 1-km pixel AVHRR data transmitted direct to ground)

IFOVS instantaneous field-of-view (the region of surface from which radiance is measured by a sensor, similar in size, but not necessarily identical, to the size of a pixel)

IRIS Infrared Intelligent Spectroradiometer (the name of a spectrometer marketed by GER)

JERS-1 Japanese Earth Resources Satellite-1 (also known as Fuyo-1)

LAC Local Area Coverage (1-km pixel AVHRR data recorded on board and transmitted when in range of a NOAA ground station)

Landsat not an acronym, but the name of a satellite series

LOWTRAN the name of a computer code designed to calculate atmospheric spectral transmittance and spectral radiance for a given atmospheric path

MSS Multispectral Scanner (the basic imaging instrument of the Landsat series, recording $80 \times 50$m pixels, its longest waveband is 0.8–1.1 μm so it is less suitable than the TM for detecting SWIR thermal radiance)

NASA National Aeronautics and Space Administration (USA)

NOAA National Oceanic & Atmospheric Administration (USA)

NS001 an airborne instrument, also known as the "Thematic Mapper Simulator", because it mimics the characteristics of the Landsat TM

OS Optical Scanner (the visible and SWIR imaging system carried by JERS-1)

PC personal computer

pixel picture element (discrete unit of which a digital image is composed)

SWIR short wavelength infrared (1–3 μm wavelength)

TIMS Thermal Infrared Multispectral Scanner (an airborne thermal infared imaging instrument)

TIR thermal infared (>3 μm wavelength)

TIROS-N Television & Infrared Observation Satellite (a family of meteorological satellites with a long pedigree, the "N" refers to the current series)

TM Thematic Mapper (the principal imaging instrument of Landsat-4 and Landsat-5, recording 30-m pixels)

# 8 Microgravity monitoring

H. Rymer

## 8.1 Introduction

Microgravity is becoming increasingly recognized as a valuable tool for mapping out the subsurface mass redistributions that are associated with volcanic activity. It is essential that elevation data are obtained at the same time as gravity data for unambiguous interpretation, and by combining these datasets, far more information is available than using either method alone. I consider here the instrumentation required for the various types of gravity survey that may be undertaken, and the suitability of microgravity surveys to a range of volcano settings.

The technique has been applied systematically to no more than about 20 volcanoes world-wide, but where good-quality data exist they are summarized here to illustrate the versatility and insight provided by combining gravity with other geophysical and geochemical monitoring methods. The units used in the literature range from SI ($ms^{-2}$) to gravity units (1 g.u. = $10^{-5} ms^{-2}$) to Gals (1 Gal (sometimes written gal) = $10^{-2} ms^{-2}$). Since Gals are becoming the most commonly used unit, we use them here, with gravity changes being typically of the order $10–1000 \mu Gal$ ($10–1000 \times 10^{-8} ms^{-2}$). The term "microgravity monitoring" follows from the use of microgals to measure the small gravity changes of interest here.

## 8.2 Background

The theory of gravity relevant for geophysics can be found in many textbooks (e.g. Tsuboi 1983), so in this chapter only the applications of the theory to microgravity techniques and practices relevant to the monitoring of active volcanoes are discussed. Basically there are two distinct types of field gravity survey.

The first is the conventional Bouguer survey in which observations of gravity are made at several field stations and at a base or reference station. A Bouguer anomaly map is then produced based on a comparison between the values of gravity at the field stations relative to the base or reference station. Before data can be expressed in this way, several corrections must be applied to allow for the differing latitude and height of the stations and the gravitational effect of material lying vertically between stations. Once these corrections have been made, the data are usually expressed either as relative gravity differences with respect to the local base station or some international gravity station. This is a common geophysical technique employed to determine subsurface density structures and has been widely used in volcanic areas to map out caldera infill thickness, magma feeder pipe dimensions and the extent of hydrothermal alteration (for reviews see Yokoyama (1972), Rymer & Brown (1986)). Typically the size of gravity anomalies of interest in these cases are tens to hundreds of milligals.

The second type of field gravity survey, the microgravity survey is the subject of this chapter. Once again data are expressed relative to a base or reference station, but in addition to information on the spatial variation of gravity, temporal variations are measured. Usually a set of gravity differences for field stations relative to a base station are measured and compared with the values obtained on subsequent surveys which may be made days, months or even years later. The changes in gravity observed as a function of time are much smaller than the Bouguer anomalies described above, being typically tens to hundreds of milligals. Since the variations of interest are so much smaller than for a Bouguer survey, high-precision techniques must be applied both to data acquisition and analysis (Rymer 1989). Although the basic mathematics of gravity is the same for a Bouguer survey as it is for a microgravity survey, when working at the limits of precision some factors become more important than when working at lower precision and these are highlighted in this chapter.

## 8.3 Technique

Despite the wide range of applications of microgravity to volcanology (see Section 8.5), the instrumentation and methodology remain essentially the same. The differences are largely limited to survey design which depends on the size and depth of the subsurface changes of interest.

### 8.3.1 Instrumentation

Both absolute and relative measurements of the Earth's gravity can now be made to accuracies of about $1\,\mu\text{Gal}$ ($10^{-8}\,\text{m}\,\text{s}^{-2}$). However, absolute measurements are usually made by using the falling mass technique and the instrument takes several hours to stablize at each location and then several more hours for a reliable high precision gravity value to be obtained. For microgravity and Bouguer field surveys, relative gravity measurements are made. In principle, any instrument that can measure relative gravity, that is the difference in the acceleration due to gravity at two points may be used for gravity surveying. Gravity meters (often abbreviated to "gravimeters") developed for this purpose fall into two main categories, stable (or static) and unstable (astatic) types. The stable meters consist of a system (usually a sensitive balance) that is displaced from its equilibrium position by a change in gravity. The displacement is very small and is magnified electronically, optically or mechanically before measurement. In contrast the unstable meter is also displaced from equilibrium by a change of gravity but the restoring force needed to recover equilibrium rather then the displacement is measured. Systems of gears mean that the restoring force can be measured very precisely. The Askania, Gulf or Hoyt, Norgaard and Bollden are examples of static gravimeters mostly developed for the oil industry and well suited for Bouguer surveys where the expected gravity variations are large. The magnitude of the gravity changes of interest for microgravity surveying require high-precision instrumentation, field practice and analysis techniques. Astatic instruments such as those manufactured by Lacoste and Romberg, Scintrex, Sodin and Worden are the only ones with the portability, ruggedness and precision suitable for microgravity surveys and as such they are also preferable for Bouguer surveys.

These astatic gravimeters all function like a long period seismograph. A mass on the end of a beam is held in place at one end by a supporting beam and is balanced by a stretched spring (Fig. 8.1). The spring is set up in such a way that its extension is equal to the distance between the points at which its ends are fixed. It behaves as a "zero-length" spring because its length, which is defined as its real (unstretched) length minus its extension is zero. In reality of course, it does not shrink away to nothing when no force is put onto the beam, because the spring is always under stress since it is coiled. When the torque on the mass (from the force of gravity) is perfectly balanced by the torque from the spring, the net torque on the mass becomes zero. In this case, the mass exhibits simple harmonic motion, but the period tends to infinity and the equilibrium is said to be unstable.

In the case of the Worden, Scintrex and Sodin instruments, the spring is made of fused quartz, but the Lacoste and Romberg springs are metallic. There are several instrument models available from each manufacturer depending on the type of gravity survey to be undertaken.

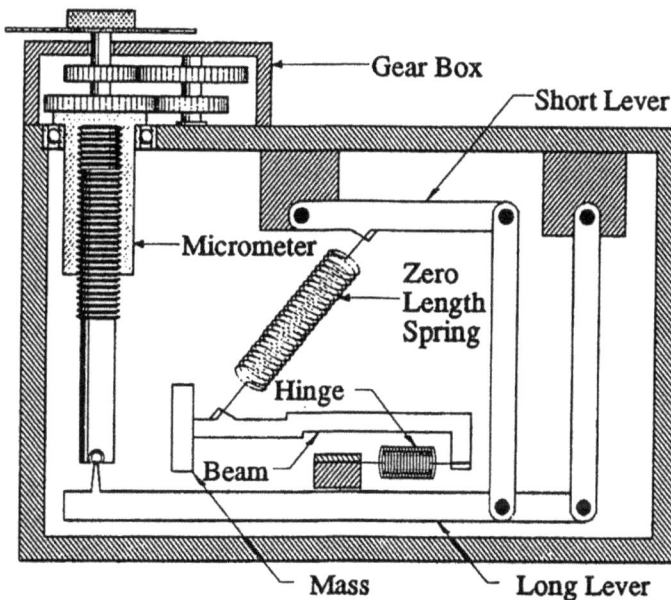

**Figure 8.1** Basic mechanism of a Lacoste and Romberg gravimeter (courtesy of Lacoste and Romberg).

Sodin gravimeters are designed for conventional Bouguer surveys and as such do not have the precision required for microgravity measurements. Scintrex instruments are easy for inexperienced operators to use and Worden instruments are competitively priced. The Lacoste instruments are usually more expensive, but experience has shown that they tend to suffer less from instrumental drift and calibration problems than the other types and therefore are recommended for microgravity surveying. Both Lacoste and Romberg model g and d meters are suitable for this type of work. If electrostatic feedback is fitted to the instrument (Harrison & Sato 1984) it becomes both easier and more accurate to read.

### 8.3.2 Sources of error

Before relative gravity variations can be measured with any degree of certainty, the various sources of error must first be considered and minimized. Errors may be simply classified into external, reader and instrumental effects (Rymer 1989) and are summarized in Table 8.1.

#### 8.3.2.1 External effects

There are two external effects on the measured value of gravity, the first is the Earth tide which is predictable and the second, noise, is not predictable. The gravitational effect at the Earth's surface due to the Earth's rotation and the relative movements of the Sun and Moon is up to $240\,\mu$Gal and it changes at a maximum rate of about $50\,\mu$Gal h$^{-1}$. The only foolproof method for making tidal corrections is to have a continuously recording gravity meter recording tides in the area of interest at the time of the surveys. Unless there are any gross errors in the calibration of the tidal meter, this will give the best estimate of the real tidal signal in the area. However, computer programs (e.g. Brouke et al. 1972) are usually used to predict the tidal effect on gravity for any time (usually Greenwhich Mean Time (GMT)) and place (defined by latitude, longitude and elevation to about 100 m horizontally and vertically). When precisions of better than say $10\,\mu$Gal are required, there are three complicating factors that must be considered.

(a) Tidal predication tables and programmes give the amplitude of the tide on a spherical solid Earth and this value must be multiplied by the gravimetric factor which is a constant that allows for the real elasticity of the Earth. The gravimetric factor ranges from 1.155–1.165 and on average is taken to be 1.16 (Melchior 1983, Baker 1984) but strictly it is a latitude and time-dependent variable. Although there have been suggestions that the range of the factor is much larger in active volcanic areas (Mason et al. 1975), use of the accepted average value is recommended unless direct observation of the factor can be made using a tidal gravimeter.

(b) The solid Earth responds to the tidal forces at a slightly different rate from the real elastic Earth, and so there is a $-6°$ to $+3°$ phase difference between the observed and predicted tide (Melchior 1983). The only way to quantify this effect in a region of interest is to make tidal observations over a period of at least 3 months. The effect will be greatest in anomalous regions such as those underlain by extensive volumes of melt (volcanically active areas) and coastal locations where the ocean responds differently from the land.

**Table 8.1** Summary of external, instrumental and reader errors using a Lacoste and Romberg model G meter (from Rymer 1989).

| Cause | Comments | Approx. size of error maximum | minimum |
|---|---|---|---|
| **External** | | | |
| Earth tide | | | |
| Gravimetric factor | Value ranges from 1.155 to 1.165 depending on Love numbers and latitude | $<2\,\mu$Gal | $<1\,\mu$Gal |
| Phase lag | Observed and predicted tides may be out of phase by $-6°$ to $+3°$ | Unknown but small | $(<1\,\mu$Gal$)$ |
| Ocean loading | Caused by tilting of plages and the gravitational attraction of mass of water in the oceans | $<10\,\mu$Gal | $<1\,\mu$Gal |
| Noise | Low frequency ($<1$ Hz) disturbances cause beam to swing; also produce tares | $<50\,\mu$Gal | $<1\,\mu$Gal |
| **Reader** | | | |
| Leg length | Height of meter is varied by changing leg lengths, gravity varies according to the free air gradient of $-3\,\mu$Gal cm$^{-1}$ | $<10\,\mu$Gal | $<1\,\mu$Gal |
| Sensitivity and levelling | Sensitivity can be varied manually, but it drifts with time. Failure to level, especially along the long level, effectively changes the reading line and changes the sensitivity | $<20\,\mu$Gal | |
| Dial movements | Slack in the gears will cause errors unless the reading is approached from the same side each time | $<40\,\mu$Gal | $<1\,\mu$Gal |
| Timing | Provided the reading is steady, there is no evidence that there is an advantage in waiting before making a reading | Negligible | – |
| **Instrumental** | | | |
| Evidence | The rms deviation about the mean reading when the gravity meter is moved and reset between readings is greater than if it is not moved between readings | | |
| Meter calibration | Polynomials and Fourier series can be used to model the calibration features. There are periodic terms due to the way the LCR is constructed, but over small ranges the effect can be kept down to a few microgals | $500\,\mu$Gal in $500$ mGal or $0.1\%$ | $<1\,\mu$Gal |
| Tares | | | |
| Thermally induced | Low battery power or a sudden change in external temperature may cause a thermal shock to the measuring system unless a secondary thermostat is fitted. If the internal temperature is allowed to fall to room temperature, the effect is much larger | ca. 10 mGal | |
| Mechanical shock | Hysteresis effects in the spring and physical jolting of the system can cause tares of almost any magnitude | ca. 10 mGal | $<1\,\mu$Gal |
| **Totals** | | **50 µGal to several mGal** | **$<10\,\mu$Gal** |

(c) The lithospheric plates tilt as the oceans move in response to the tides. In addition, there is a gravitational effect due to the moving mass of the oceans. These effects together may account for as much as 4% of the observed tide, or 10μGal, and is particularly important in coastal regions. The only way to guarantee that the real tidal correction is applied to data is to observe the tide continuously using the same type of gravimeter as is used for the survey work. However, with careful use of the prediction programs and fastidious noting of the timing of observations, errors for inland regions should be less than 1μGal. The tidal correction is made automatically by the Scintrex meters, the user types in space and time co-ordinates, and thus any change to the phase of the correction or to the gravimetric factor cannot be made without disabling the correction or modifying the software.

Noise is a much bigger problem than the Earth tides since it cannot be controlled or predicted. Since gravimeters are long period seismometers, urban traffic noise is an obvious problem when making gravity measurements. Other sources of noise are wind and the microseisms caused by the interaction of the atmosphere and the Earth's surface. For a microgravity survey in an active volcanic region, the problems are most likely to be due to a combination of tourists and volcanic tremor. The reference station away from the active region can usually be chosen to avoid noise. In some cases it is clear that noise is too great to obtain meaningful measurements as the reading will be unsteady (see later in this chapter), however if there is continuous low frequency ground vibration, stable but erroneous measurements can be made. For example, laboratory measurements on Lacoste and Romberg meter G 105 have shown that at 37.5Hz a reproducible and stable downwards displacement of 320μGal will occur (Hallinan 1991). This experiment was carried out after anomalous gravity changes were observed at geothermal well heads when production ceased (Hallinan et al. 1991). Thus although well heads provide easily identified and relocatable microgravity stations, they should be avoided because of the variable noise present. Volcanic tremor is a cause for concern here, and clearly in an ideal world, microgravity measurements should not be made while the noise is present. However, the source of volcanic tremor is more widely dispersed than the tremor source at a well head, and so if all stations are affected in the same way, then the effect should largely cancel out. The only way to be sure that there is no significant noise effect in data is to make numerous repeat observations at various times of day and to note variations not only between the field stations and the base, but also between individual field stations. The response of each gravimeter to a particular frequency will be different and so it is essential that the same instruments are used on all surveys.

### 8.3.2.2 Reader effects
Reader effects are errors put into the reading by the observer. Good field practice minimizes these errors. Before a gravity reading is made (whichever type of instrument is used), the meter must be set up carefully. This is done by placing a purpose built base plate onto the station and then putting the instrument onto this smooth surface. The meter is levelled carefully so that only the vertical component of gravity is measured, this is checked using mutually perpendicular spirit levels or pendulums. The level is adjusted by lengthening and shortening the three legs supporting the instrument. It is important to minimize the amount

of adjustment made to the leg lengths, since it can be found after several readings that the meter is a few centimetres higher or lower on the base plate than for the first observation. This would put an immediate error of up to $9\,\mu$Gal (for $3\,$cm) into the data. The simplest way to avoid the necessity of changing the leg lengths is to insure that the station is flat, and then to move the meter around gently on the concave base plate until it is almost perfectly level, then use the leg adjusters only for "fine tuning".

Manufacturers' instruction books give details on the adjustment of sensitivity, but it is worth pointing out here that, since reading line is a function of sensitivity, this must always be checked after adjustment. This is particularly important when an instrument is used for other purposes or by other operators in addition to the microgravity survey. Although there is no evidence that it is beneficial to wait for any length of time between unclamping the meter and making the first reading (Rymer 1989), good field practice suggests that several repeat readings should be made at each location, and that the readings should always be made in the same way. The gearing mechanism on the Lacoste and Romberg instruments means that an error of $40\,\mu$Gal on a model g meter and $4\,\mu$Gal for a model d meter will be produced by changing the direction from which the reading line is approached (see Section 8.3.3).

### 8.3.2.3 Instrumental effects

The combined effect of external and reader errors (Table 8.1) is about $10\,\mu$Gal (Rymer 1989). Instrumental effects are very important because they are difficult to quantify and yet they may be considerably larger than the effects already discussed. Errors in Bouguer gravity surveys have traditionally been lumped together and called "drift" which, for simplicity, is taken to be linear (Milsom 1990). In some surveys, even the effect of the Earth tides is removed by a linear correction. However, by making several repeat readings in a variety of conditions, it can be shown that instrumental drift for the Lacoste and Romberg instruments is extremely small (a few microgals per day). Nevertheless, even after the effects of the Earth tides have been removed, there are often differences as large as tens or even hundreds of microgals between reported base station or other frequently measured station readings during the course of a day. This is because the instrument suffers tares; all spring meters suffer this effect. There are basically two types of tare, thermal and mechanical (Rymer 1989). A tare is a sudden apparent jump in reading and may be positive or negative and may be only a few microgals (and, therefore, would not be noticed) or a few milligals. Detectable tares are usually greater than 10 or $20\,\mu$Gal and are found when, after tidal correction repeated readings at a particular station are different. Thermal tares occur when the thermostat and heater inside the instrument are unable to keep up with rapid temperature changes outside (usually this occurs when the meter is removed from the carrying case into a cold environment prior to making a measurement). Even a drop of a fraction of a degree in the temperature inside the instrument will cause the apparent reading to drift for a few hours, and a thermal tare may result so that the final stable reading is different from the reading that would have been obtained if the temperature had not been allowed to change. Thermal tares are also induced by poor battery contact and care must always be taken that the power cable has not been knocked out as the meter is removed from and replaced into its case. Mechanical tares occur when the instrument has been jolted either in the clamped or unclamped "reading" state. Although a tare might reasonably be expected to follow if an instrument falls out of the back of a landrover (clamped!), tares may also occur if the instrument is lightly knocked when unclamped, and even if the wind is very strong when a reading is being made, so that the beam swings from end to end banging on the stops. The important point about tares is that they tend to be step functions rather than linear drift. Thus if a tare has occurred (identified by a change in reading at a station after the tidal correction has been made), it should be removed from the data *where it occurs* as a step. It should be possible to detect where and when the tare occurred, so all subsequent data must be adjusted to allow for its effect (see Table 8.2, for example). As with all errors, the effect on the final data may be minimized by good field practice and clearly the more repeat readings that have been made during a survey day, the more likely it is that the tare(s) can be identified, quantified and removed from the affected readings.

There is some evidence that for gravity differences between the base station and a field station greater than about 20 mGal there is an hysteresis effect in the spring mechanism (Rymer 1989). It is almost impossible to separate the effects of hysteresis from tares and the only ways to minimize the effects are:

(a)  to make sure that the base station while being in a separate and stable region away from the active parts of the volcano is in a place where the value of gravity (and therefore the elevation) is about the same as the region of interest; and

(b)  to ensure that there are several repeat readings made between the base and field stations by going to and fro between stations. (See tares on repeat readings of large gravity differences in Table 8.2.)

Table 8.2 A gravity data set illustrating tares (differences in repeat readings) and their removal. The instrument used was Lacoste and Romberg G513 on the microgravity network at Mount Etna, Sicily, on 9 June 1992.

| Station | Tidally corrected reading (mGal) | Discrepancy (µGal) | Gravity difference relative to first SAP (mGal) | Gravity difference relative to second SAP (mGal) |
|---|---|---|---|---|
| SAP | 3168.945 | | 0 | −0.060 |
| TDF | 2914.434 | | −254.511 | −254.571 |
| Lowry | 2923.243 | | −245.702 | −245.762 |
| Thrushcross | 2918.452 | | −250.493 | −250.553 |
| Lowry | 2923.258 | +15 | −245.687 | −245.747 |
| Jesopina | 2939.743 | | −229.202 | −229.262 |
| TDF | 2914.450 | +16 | −254.495 | −254.555 |
| 06 Inter | 2972.377 | | −196.568 | −196.628 |
| Pic | 3020.402 | | −148.543 | −148.603 |
| SAP | 3169.005 | +60 | 0.060 | 0 |

* Only three repeat readings were made, at stations Lowry, TDF and SAP. There is evidence for a 15–16 µGal tare between the first two repeated stations, but this is so close to the expected error (see Table 8.1) that it would normally be ignored and the gravity difference between these stations and the reference station (SAP) averaged. There is, however, a tare of 60 µGal between the two SAP readings and so there are two estimates of the gravity difference between each station and the SAP reference depending on where the tare occurred. By repeating this survey several times and by having as many stations revisited on each survey day as possible, the time of the tare can be deduced. In fact, on this occasion, a 60-µGal tare occurred between the first SAP and TDF readings, so that the final column of data is correct. If, on the other hand, the tare had occurred between readings at stations Inter and Pic then column-4 values would have been valid to Inter, and the Pic value from the fifth column.

A smaller instrumental effect comes from the calibration. This can change with time and is particularly noticeable when observations are made over a period of several years. Provided that the same instrument(s) are always used for microgravity surveys, and data for each instrument compared only with data from that same instrument (the data are not averaged), errors in the calibration of the meter do not usually pose a large problem.

Small physical and mechanical imperfections in the screw, lever system and gears mean that the response of the gravimeter spring is not linear throughout its range. The calibration of the instrument (whereby meter readings are converted to milligal or microgal) is therefore not linear. At the boundaries between the scales there is a discontinuity, which for microgravity surveys must be smoothed out prior to data reduction. The effect is usually less than about 1 μGal (Table 8.1), but is simple to avoid. Since the cumulative effect of tares is to make the readings on the instrument gradually increase with time (although tares of either sign can occur), any discrepancies at the boundaries between the various scales may eventually produce apparent gravity changes (Rymer 1989). The problem may be avoided by modifying the manufacturers' conversion tables so that there is no discontinuity in corrected reading in going from one scale to another (this is not possible on the Scintrex meter since the conversion to microgals from meter units is automatic). However, for the Lacoste and Romberg model d meter where the reading can be pre-set by adjusting the coarse screw, this problem does not arise.

Strict field procedures (see Section 8.3.3) are essential for obtaining optimum precision from instrumentation. In summary, it can be seen from Table 8.1 that the combined effect of external, instrumental and reader errors on a single gravity reading is less than 10 μGal for a Lacoste and Romberg model g meter. Provided that the overall gravity range is no more than about 150 mGal, the error for a Lacoste and Romberg model d meter is rather less. For a single gravity reading using a Lacoste and Romberg g meter the error ranges from 7 μGal (best case) to about 23 μGal (worst case). For a gravity difference reading (as would normally be required) the errors range from 10 to 33 μGal and decrease as the number of repeat readings go up (Rymer 1989). Errors for other types of gravimeter are larger than for the Lacoste and Romberg instruments.

### 8.3.3 Survey procedure

Manufacturers' instruction booklets explain how to make a gravity observation, and this of course depends on the type of instrument being used. However, the following procedures should be followed to maximize the precision of the survey.

It is essential that the same instrument(s) are used throughout a survey and that for the repeat surveys the same instrument(s) are again used. Since there are slight calibration differences between instruments (even those of the same type and model), data from different instruments should not be averaged. Gravity changes with time should be deduced for each instrument. For both Lacoste and Romberg and Worden meters, even with electronic readout, there is a large degree of reader skill required and therefore it is important that the same operator should always read the same instrument.

Each reading should be made in exactly the same way. Stations should be clearly marked for easy relocation. If possible they should be large purpose built concrete blocks with foundations of tens of centimetres. Often an existing rock outcrop or boulder will suffice, and the point is marked with a nail or paint spot. The base plate must be put in the same position on the station each time it is read. To facilitate this, it is usual to paint or in some other way mark the positions of the base plate legs on the ground. The meter should be orientated in the same way on each occasion, and the operator should sit or kneel or crouch in the same way each time. All other external factors such as traffic, spectators and weather should be as constant as possible throughout a survey.

Once the instrument has been levelled (see manufacturer's booklet for details), a reading is made manually for all instruments except the Scintrex. The reading must always be approached from the same direction (i.e. either always from upscale or always from downscale) to avoid whiplash effects. Several repeat readings should be made, turning the reading dial off the reading line between each attempt. The exact time (for tidal correction) must be noted, and of course the station name. It is good field practice also to note down weather conditions and other factors such as state of activity of the volcano that may affect the observations. If the instrument is known to have suffered a jolt during transit to the station, the last station should be returned to and a repeat reading made, although a mechanical tare can usually be isolated if the time and location of the probable tare is noted and there is another instrument also being used. It is very bad luck for two instrument to suffer tares at the same time! However the use of two or more instruments simultaneously should not be adopted to avoid regular repeat readings at key stations.

## 8.4 Interpretation

Bouguer gravity anomalies are interpreted in terms of various subsurface density contrasts and 2-dimensional, 2.5-dimensional and 3-dimensional models are either iterated by computer or developed interactively on line. Microgravity data can, in principle, be treated in the same way, but because of the time involved in collecting high precision data, there are usually too few data points to justify detailed computer modelling. The way in which microgravity data are presented depends on the quality and amount of the data and on the type of interpretation required. We will therefore consider the various ways in which microgravity data are presented and interpreted in turn, although this often depends on the type of volcano on which the survey is being carried (see Section 8.5).

### 8.4.1 The free-air gradient

For a Bouguer gravity survey, data are made comparable with the base or reference station firstly by correcting for elevation differences using the free-air gradient (FAG). At the Earth's surface, $FAG = -308.6\mu Gal\,m^{-1}$. This means that for every 1 m of elevation increase, the value of gravity decreases by $308.6\mu Gal$. If the FAG is actually measured, it is often found

**Figure 8.2** Measurement of the free-air gradient (FAG). Observations are made on the ground and on a tripod (both instruments measure both sites), and the height difference is measured.

to differ from the theoretical value. FAG measurements are easily made using a tripod adapted to hold a gravimeter baseplate (Fig. 8.2). The difference in height between the meter on the ground and on the tripod is measured and the observed gravity difference then divided into this value. Each meter is read on the ground and on the tripod several times. On the crater floor of Poás volcano (Costa Rica) the FAG is $-420\,\mu\mathrm{Ga\,m}^{-1}$, at the summit of Mount Etna, Sicily, it is $-365\,\mu\mathrm{Gal\,m}^{-1}$ and at Breiddalur, Southeast Iceland, it is $-201\,\mu\mathrm{Gal\,m}^{-1}$ (Rymer 1994). There are two reasons for this deviation from the theoretical value. Firstly, there is a terrain component, such that a measurement of the FAG made close to the edge of a cliff or crater wall will be less negative than a measurement made a few metres away from it. Secondly, there is a Bouguer anomaly component, such that a region of negative Bouguer anomaly will have a FAG that is less negative than the theoretical value, while a region of positive Bouguer anomaly will have a more negative FAG than the theoretical value (Berrino et al. 1992). The relative sizes of the terrain and Bouguer components in the FAG will depend on the severity of the topography and the density contrasts beneath the ground.

Uplift or subsidence of a microgravity station will involve the local terrain (i.e. the cliff or crater wall) so that its contribution to the FAG becomes irrelevant. This means that although the effect of the terrain is to decrease the FAG, any elevation changes at a microgravity station occur also in the local topography and so the gradient along which gravity varies with height will be the theoretical FAG. The Bouguer component does not behave in the same way though, because its source is usually considerably deeper than the source of any elevation changes. So, for example, if the source of a Bouguer anomaly in a caldera is a magma chamber at a few km depth, and the source of gravity and elevation variations is above it then the rate of change of gravity with height will be influenced by the magma chamber. Thus although an observation of the FAG will include both the terrain and Bouguer components, the terrain component does not affect the rate of change of gravity that will occur during microgravity surveying but the Bouguer component may if the source of the Bouguer anomaly is not the same as the source of the gravity and elevation changes. Obviously it is extremely difficult to discriminate between the terrain and Bouguer components of the FAG, although where the topography is relatively quiet, such as within a large caldera, it has been shown that it is better to use the observed rather than the theoretical FAG for microgravity data analysis interpretation (Berrino et al. 1992). Thus the FAG should always be measured (ideally at each station) before microgravity data are interpreted.

### 8.4.2 Bouguer corrected free-air gradient

Since gravity stations do not rise and fall in isolation, the gravitational effect of the rise and fall of the surrounding ground must be taken into account when correcting gravity data for elevation changes. This correction is analogous to the Bouguer correction applied in a conventional "static" Bouguer survey. For a static survey, the Bouguer correction is an infinite flat plane of material with density $\rho$ and thickness $h$. Its gravitational effect $g$ is given by:

$$g = 2\pi G \rho h$$

where $G$ is the universal gravitational constant ($6.672 \times 10^{-11}$ $Nm^2kg^{-2}$). In a microgravity survey, small changes in gravity ($\Delta g$) and elevation ($\Delta h$) are monitored and the rate of change in gravity with height will be a combination of the FAG and the Bouguer components:

$$\Delta g/\Delta h = FAG + 2\pi G\rho$$

Thus on an infinite flat plane, assuming an average rock density of $2600\,kg\,m^{-3}$, an elevation increase of 1 m would produce a gravity decrease of

$$[-308.6 + (0.04191 \times 2600)] = -200\,\mu Gal$$

assuming the theoretical FAG. The Bouguer corrected FAG ranges from $-191\,\mu Gal\,m^{-1}$ for a density of $2800\,kg\,m^{-1}$ to $-225\,\mu Gal\,m^{-1}$ for a density of $2000\,kg\,m^{-1}$ (Fig. 8.3a). The source of the gravity changes is assumed to be within the infinite horizontal Bouguer slab.

A horizontal infinite slab is not always a good approximation to the source of gravity changes, especially in a volcanic area. A spherical point source is just as simple to handle mathematically and is often more realistic (especially when considering magma chambers). In this case, the Bouguer corrected FAG is given by:

$$\Delta g/\Delta h = (FAG + \tfrac{4}{3}\pi G\rho)$$

and assuming the theoretical FAG ranges from $-230\,\mu Gal\,m^{-1}$ for a density of $2800\,kg\,m^{-3}$ to

(a)                                    (b)                                    (c)

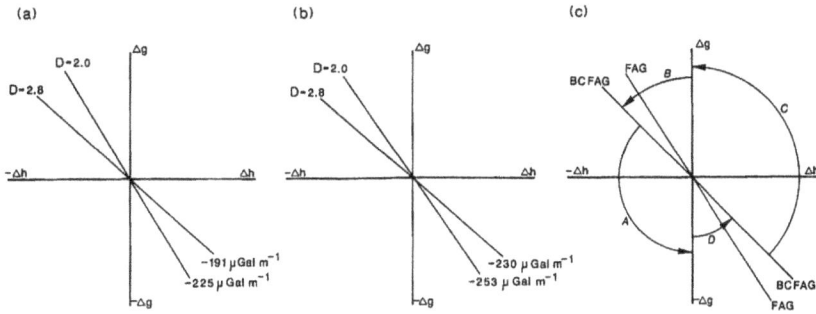

**Figure 8.3** Gravity–height relationships: (a) for a Bouguer slab; (b) for a spherical source; (c) data may fall into any of the regions A–D or along the lines shown in parts (a) and (b) (from Rymer 1994).

$-253\,\mu\mathrm{Gal\,m^{-1}}$ for a density of $2000\,\mathrm{kg\,m^{-3}}$. Thus the rate of change of gravity with elevation is numerically slightly larger for a spherical source than for the flat plane, but in both cases the relationship is linear (Fig. 8.3b).

The first stage in interpretation for a microgravity survey is to note the rate of change in gravity with elevation. Often this is averaged for each station, but strictly each station should be considered separately. Once the gradient has been deduced, it may be compared with the theoretical or observed FAG or the Bouguer corrected FAG. Departures from the FAG are interpreted in terms of changes in subsurface mass, while departures from the Bouguer corrected FAG are interpreted as density changes (Brown et al. 1991b). When data fall within region A in Figure 8.3c, a volcano has undergone deflation with gravity increasing rather less than predicted by the theory described above. This implies an overall gravity decrease and therefore subsurface mass and density decrease (Rymer 1994). This has been interpreted in terms of magma vesiculation (e.g. for Pacaya and Poás volcanoes see Eggers (1983), Rymer & Brown (1984)). Data falling in region B in Figure 8.3c again reflect deflation, but this time with a net gravity and therefore subsurface mass and density increase. This has been modelled in terms of upwards migration of relatively dense magma, degassing as it moves and regional deflation caused by reduced magma pressure at depth (Rymer & Brown 1989, Brown et al. 1991b).

It is worth pointing out here that gravity and height changes are not always caused by the same phenomenon, and may well occur on different time scales. A gravity–height gradient of $-171\,\mu\mathrm{Gal\,m^{-1}}$ was observed during a deflation event on Kilauea between November and December 1975 (Jachens & Eaton 1980), apparently associated with a magnitude 7.2 earthquake. These data, which fall in region A in Figure 8.3c, were interpreted in terms of magma migration from the summit area into the east rift zone. Between December 1975 and April 1977, further deflation occurred, but this time with a gravity–height gradient of $-607\,\mu\mathrm{Gal\,m^{-1}}$, falling in region B in Figure 8.3c. The data were thought to indicate intrusion of magma into the void region left by the earlier migration episode (Dzurisin et al. 1980). In both cases, deflation was probably on a more regional scale than the gravity changes and reflects magma drainage from a deeper source.

Data falling in regions C and D of Figure 8.3c can clearly be interpreted in the same way,

229

but this time for episodes of inflation. However, it is interesting to note that most published cases of gravity and height changes occur for deflation episodes, except at calderas (Table 8.3) where the gravity–height correlation tends to follow inflation along the spherical source gradient (Fig. 8.3b).

The surface deformation caused by volume changes within a deeply buried spherical magma chamber were first considered by Mogi (1958), and the relationships derived were later extended to include the Poisson ratio of the deforming rocks (Hagiwara 1977). The Mogi model assumes the crustal rocks to behave as a semi-infinite elastic body. Surface deformation is assumed to come from changes in hydrostatic pressure within the chamber and not to alter the bulk density of the magma chamber or the surrounding rock. More recent models for crustal behaviour (e.g. Wong & Walsh 1991) allow for the magma chamber to have distinct elastic properties and for the surrounding rock to have a finite permeability. Both the Mogi model and the more recent models predict that in volumetric terms surface ground deformation is less than deformation within the magma chamber. In both cases the gravity–height relationship is linear, and for realistic values of the rock physical properties, the Wong & Walsh model predicts gravity-- height gradients that fall between the slab and spherical models described above.

Microgravity changes on volcanoes are not always caused by variations in a slab (Fig. 8.4a), or spherical source (Fig. 8.4b). There may be fluctuations in the amount of steam within a hydrothermal system or small magma intrusions best modelled as cylindrical vertical bodies

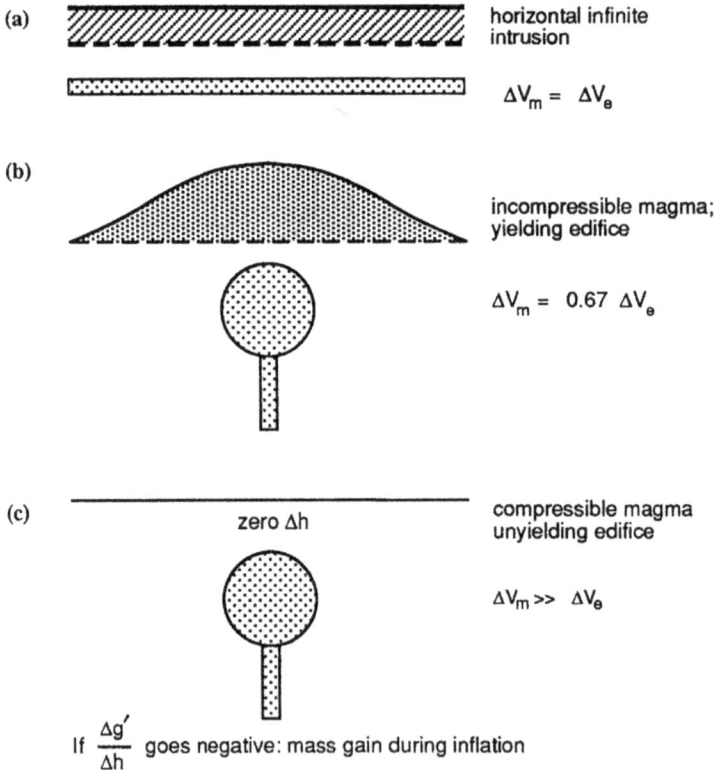

(a)    horizontal infinite intrusion

$\Delta V_m = \Delta V_e$

(b)    incompressible magma; yielding edifice

$\Delta V_m = 0.67 \ \Delta V_e$

(c)    zero $\Delta h$    compressible magma unyielding edifice

$\Delta V_m >> \Delta V_e$

If $\dfrac{\Delta g'}{\Delta h}$ goes negative: mass gain during inflation

Figure 8.4 Schematic diagrams showing sources of gravity and height changes (a) the Bouguer slab, (b) the spherical source, and (c) non-elastic behaviour.

**Table 8.3** Gravity–height correlations for a selection of active volcanoes with large silicic calderas, basaltic extensional rift zones or subduction-related andesitic stratocones (after Rymer 1994).

| Volcano | Period/No. sets of observations/ No. of stations | Δg (max) (μGal) | Δh(max) (m) | Δg/Δh (μGal m$^{-1}$) | Source | Comments |
|---|---|---|---|---|---|---|
| **Volcanoes with large silicic calderas** | | | | | | |
| Yellowstone Wyoming, USA | 1977–87 | −60 | 0.120 | −170 | Smith et al. (1989) | The increase is subsurface mass was thought to reflect some combination of magmatic (basaltic intrusion), tectonic and hydrothermal processes |
| Campi Flegrei, Italy | 1981–84 | −331 | 1.616 | −215 | Berrino et al. (1984) | Gradual inflation was observed centred close to Pozzuoli town during the Brady seismic crisis. The FAG was measured *in situ* (−290 μGal m$^{-1}$) and the *residual* gravity increase, using this FAG was −75 μGal m$^{-1}$. This was interpreted in terms of a mass increase of ca. $2 \times 10^{11}$ kg within a spherical magma chamber at ca. 3 km depth |
| Rabual, Papua New Guinea | 1971–85 | −410 | 1.800 | −216 | McKee et al. (1987) | An overall subsurface mass increase of ca. 108 kg was deduced within a "point source" magma chamber at ca. 1.8 km depth |
| **Volcanoes with basaltic extensional rift zones** | | | | | | |
| Krafla, Iceland | Jun. 1975–Aug. 1980 4 sets, ca. 16 stations | −160 | 0.7 | −112 | Kanngieser (1983) | Maximim changes were seen across the active rift zone. The existing voids were probably being filled as a precursor to the fissure eruption |
| Krafla, Iceland | Jan. 1978 2 sets, 40 stations | +150 | −0.90 | −166 | Johnsen et al. (1980) | The departure from the BUG was considered by Johnsen et al. to be due to groundwater movements. However, the gravity changes are quite large, and since Krafla magmas were transported by dykes from the magma chamber to the surface fissures during this time (Tryggvason 1986), magma movements are implicated. It seems more likely then that overall mass increases during inflation were due to magma being intruded inelastically. Subsequently, voids were created due to lateral magma drainage |
| | Jan.–Jun. 1978 2 sets, 40 stations | ca. −200 | ca. 0.8 | −250 | Johnsen et al. (1980) | |

**Volcanoes with basaltic extensional rift zones** (continued)

| Volcano | Period/No. sets of observations/ No. of stations | Δg (max) (μGal) | Δh(max) (m) | Δg/Δh (μGal m$^{-1}$) | Source | Comments |
|---|---|---|---|---|---|---|
| Krafla Iceland | Apr. 1977–Jul. 1978 2 sets, ca. 40 stations | +210 +100 | -1.000 -0.500 | -210 -200 | Johnsen et al. (1980) Johnsen et al. (1980) | The gravity–height correlation during these periods of caldera subsidence and inflation approximately follow the predicted BUG. This means that the volumes of |
| | Jul. 1975–Jul. 1979 6 sets, 18 stations | Cyclic 100–200 amplitude | 0.500–1.000 | -200 (average) | Torge (1981) Torge & Kanngieser (1980) | subsidence/inflation is the same as the volume of magma flowing out of/into the region. No overall density change occurs (since data fall on predicted BUG) but mass changes do occur (deviation from FAG) since magma moves in and out of the system |
| Kilauea Hawaii, USA | Nov.–Dec.1975 2 sets, 17 stations | +234 | -1.310 | -171 | Jachens & Eaton (1980) | Deflation was observed across the volcano summit after a small summit eruption. Overall mass deficit was interpreted in terms of the incomplete collapse of drained voids |
| | Dec. 1975 – Apr. 1977 2 sets, 16 stations | ca. +145 | ca. -0.80 | -607 | Dzurisin et al. (1980) | No surface activity during first period, but mass increase attributed to magma emplacement. Summit region continued to deflate and magma drained to east rift zone. |
| | Apr.–Oct. 1977 2 sets, 21 stations | ca. +140 | ca. -0.45 | -280 | Dzurisin et al. (1980) | The eruption in Sep.1977 lends support to the idea of lateral magma drainage from the (deflating) summit to the active rift zone |
| Etna Sicily, Italy | Sep. 1979–Jun. 1980 2 sets, 48 stations | -36 to +18 | -0.006 to +0.004 | Positive | Sanderson (1982) | These data are unique in that a positive ΔgΔh relationship is obtained. Observed changes are not much greater than the precision quoted by other workers |
| | Sep. 1980–Jul. 1981 3 sets, 25 stations | +63 | +0.017 | Positive | Sanderson et al. (1983) | (7–33 μGal; Rymer, 1989), but were interpreted in terms of changes in dyke magma pressure associated with explosive eruptions southeast of the summit in Aug.– Sep. 1981 and a major fissure eruption in Mar. 1981 |
| | Jun. 1990–Jun. 1991 2 sets, 27 stations | +408 | +0.021 | Large positive | Rymer et al. (1993) Rymer et al. (1994) | Large gravity increases in the summit region and along the line of fractures left open after the 1989 activity were associated with minimal ground deformation. Increases thought to be due to passive dyke intrusions into already opened crack system |

| Volcano | Period/No. sets of observations/ No. of stations | $\Delta g$ (max) ($\mu$Gal) | $\Delta h$(max) (m) | $\Delta g / \Delta h$ ($\mu$Gal m$^{-1}$) | Source | Comments |
|---|---|---|---|---|---|---|
| **Volcanoes with explosive, subduction-related andesitic stratocones** | | | | | | |
| Pacaya, Guatemala | Jan.–Jun. 1980 2 sets, ca. 48 stations | −231 | ≥0.019 | Large | Eggers (1983) | These observations immediately preceded phreato-magmatic strombolian activity and indicate that low density material was displacing high density material beneath the summit either by upwards stoping, or by *in situ* magma vesiculation |
| Usu, Japan | 1977–82 | >30 | <1000 | −160 | Yokoyama (1989) | Analysis of data from one station 2 km from crater indicated that deformation was caused by inflation of ground material induced by water permeation |
| Poás, Costa Rica | Mar. 1983–Apr. 1985 >25 sets collected every few days for periods of several weeks at a time: 13–16 crater stations, 6–8 control stations | Cyclic ca. 120 amplitude | Insignificant <0.05 | Large | Rymer & Brown (1984, 1987) | Cyclic gravity variations centred on the non-erupting crater (fumaroles and hot crater lake only) don't appear to be due to elevation changes. Models of the shallow structure of the summit regions (Rymer 1985, Brown et al. 1987) identify a cylindrical magma feeder pipe beneath the crater. Density changes within partially molten magma at depths exceeding 500 m could account for the observed changes. A 10% change in the degree of vesiculation (i.e. a density change of ±5 kg m$^{-3}$) which is possible under the presumed PT conditions was the preferred model |
| Poás, Costa Rica | Mar. 1985–Mar. 1989 8 sets, > 9 stations | +358 | −0.336 | Large | Rymer & Brown (1989) Brown et al. (1991b) | An ash eruption in Apr. 1989 was preceded by a gradual gravity increase and small surface deflation. The total mass increase (excess above the FAG) was $10^8$ kg and was considered to represent stringers of molten magma reaching up to within a few hundred metres of the surface. Magma devesiculation may also account for some of the changes, but water table movements cannot since this period was characterized by vaporization of the hydrothermal system at Poás. (Rymer & Brown 1989, Brown et al. 1991b) |

| Volcano | Period/No. sets of observations/ No. of stations | $\Delta g$ (max) ($\mu$Gal) | $\Delta h$(max) (m) | $\Delta g/\Delta h$ ($\mu$Gal m$^{-1}$) | Source | Comments |
|---|---|---|---|---|---|---|
| **Volcanoes with explosive, subduction-related andesitic stratocones (continued)** | | | | | | |
| Mt. St. Helens Washington, USA | March–Sept. 1980 | −56 | 0 or small | Large | Jachens et al. (1981) | Changes must be associated with the climactic eruption of 18 May 1980 but appear small, probably because of an unfortunate choice of timing and location for measurements (Eggers, 1987) |
| Sakurajima Japan | 1975–85 | ca. 120 | ca. 0.08 | Irregular | Yokoyama (1989) | The gravity/height profile across the volcanoes indicated that the same point source model was appropriate for both data sets. The source is at ca. 3 km depth and has a mass excess of $3 \times 10^{11}$ kg. Since $0.7 \times 10^{11}$ kg were erupted during the 1975–82 period, it was deduced that ca. $4 \times 10^{11}$ kg of magma was supplied to the volcano during this time |
| Mt. Baker Washington USA | May–Sept. 1975 9 sets, 2 stations | −550 | Insignificant | Large | Malone & Frank (1975) | Measurements were made on the Sherman crater. Gravity changes were interpreted in terms of the loss of water and gasses from funaroles and the melting of snow and ice. Such large gravity changes, however, represent a very large mass deficit and in view of the fact that funarole temperatures increased at this time it is likely that shallow magma vesiculation or migration also took place |
| Miyakejima Japan | 1980–1983 | ca. −30 to +20 | ca. +0.11 to −0.06 | −279 | Yokoyama (1989) | Most observations could be interpreted in terms of water table movements, but stations close to the fissure showed a net residual gravity increase above the best fit $\Delta g/\Delta h$ line of ca. 20 $\mu$Gal. This, and the fact that deflation was also observed led to the interpretation that $\Delta g/\Delta h$ changes near the fissure, had an additional component due to a dyke being filled with magma from the surface to a depth of 3 km even after eruption |

| Volcano | Period/No. sets of observations/ No. of stations | $\Delta g$ (max) ($\mu$Gal) | $\Delta h$(max) (m) | $\Delta g/\Delta h$ ($\mu$Gal m$^{-1}$) | Source | Comments |
|---|---|---|---|---|---|---|
| **Volcanoes with explosive, subduction-related andesitic stratocones (continued)** | | | | | | |
| Mihara Japan | Sep. 1950–Apr. 1951 3 sets, 62 stations | −400 | <0.01 | Large | Iida et al. (1952) | These data represent $\Delta g/\Delta h$ observations covering a period of inflation followed by deflation, with insignificant total change in height. Iida et al. interpreted the departure from the BUG in terms of a draining magma chamber. Eggers (1987) suggests that replacement of high density country rock by vesiculated magma is a more likely possibility. Alternatively, the deflationary period could have left voids which would also account for the mass deficit |
| Pacaya Guatemala | Jun. 1979–Jan. 1980 2 sets, 27 stations | +221 | ≥−0.196 | Large | Eggers (1983) | Measurements were made during a period of post-eruption deflation. Eggers (1987) suggests that the gravity increase can be accounted for by void filling. |
| Karkar Papua New Guinea | Apr.–May 1979 daily to weekly observations 10 stations | −60 | < tens of cm | >−300 | McKee et al. (1987) | These data were collected during a period of phreatic activity and Eggers (1987) suggests this departure from the BUG is due to ground water movements. This is probable, since $\Delta g$ is small and magma was not erupted, meaning that substantial magma movements at depths less than a few km are not implicated |

BUG, Bouguer corrected gradient (see text for details); FAG, free-air gradient; PT, presure and temperature; P, pressure, T, temperature

(Brown et al. 1991) or as inclined planes. In neither case will the resulting gravity–height relationship be linear as has been shown experimentally (Rymer and Brown 1989) and theoretically (Savage 1984). However, in most cases the errors in microgravity data are usually considered to be too large to detect deviations from the simple linear relationship and the average $\Delta g/\Delta h$ gradient is considered.

According to the Mogi model, the expected gravity–height gradient $\Delta g/\Delta h$ is given by:

$$\Delta g/\Delta h = \text{FAG} + \Delta V_m \, G r Z/\Delta h \, (X^2 + Z^2)^{3/2}$$

where $\Delta V_m$ is the change in magma volume, $G$ is the universal gravitational constant, $\rho$ is the magma density, and $Z$ and $X$ are the vertical and horizontal distances to the spherical point source.

From elastic theory (Johnson 1987), the change in edifice volume ($\Delta V_e$) during uplift or subsidence is given by:

$$\Delta V_e = 2\pi(X^2 + Z^2)^{3/2}\Delta h/Z$$

These terms may be combined to give the gravity–height gradient in terms of changes in magma and edifice volume changes assuming elastic behaviour;

$$\Delta g/\Delta h = \text{FAG} + 2\pi G\rho\Delta V_m/\Delta V_e$$

When the observed gradient is the same as the FAG, then $\Delta V_m = 0$ and there is no subsurface mass change. When $\Delta V_m = \Delta V_e$ the $\Delta g/\Delta h$ relationship simplifies to the Bouguer slab formula (Fig. 8.4a). For a Poisson ratio of 0.25, the Mogi model predicts $\Delta V_m/\Delta V_e = 2/3$ for a spherical source (Fig. 8.4b). However, in order to explain observed gravity/height gradients that differ greatly from these values (Table 8.3), almost infinite magma volume changes are required with negligible edifice volume changes (Fig. 8.4c). This can be achieved once the requirement for volcanic edifices to behave like an elastic half-space is relaxed, magma is considered to be compressible to some degree and pore spaces or fractures within the surrounding material are envisaged.

## 8.5 Applications

Most of the microgravity surveys that have been published to date involve only a few datasets often separated by intervals of a few years (see reviews by Rymer & Brown 1986, Eggers 1987, Rymer 1994). In many cases the reason for making repeated surveys is the occurrence of some sort of volcanic activity, and so interpretations of gravity changes are biased by the observed surface changes. Only limited information of the rate of change of ground deformation and microgravity is obtained in surveys of this kind, and it is important to realise that while the observed changes may be small or insignificant, the unobserved short-term changes may be considerably larger.

Since there are many different types of eruption, there will also be many different microgravity responses to eruption. The simplest way to classify volcanoes and their activity

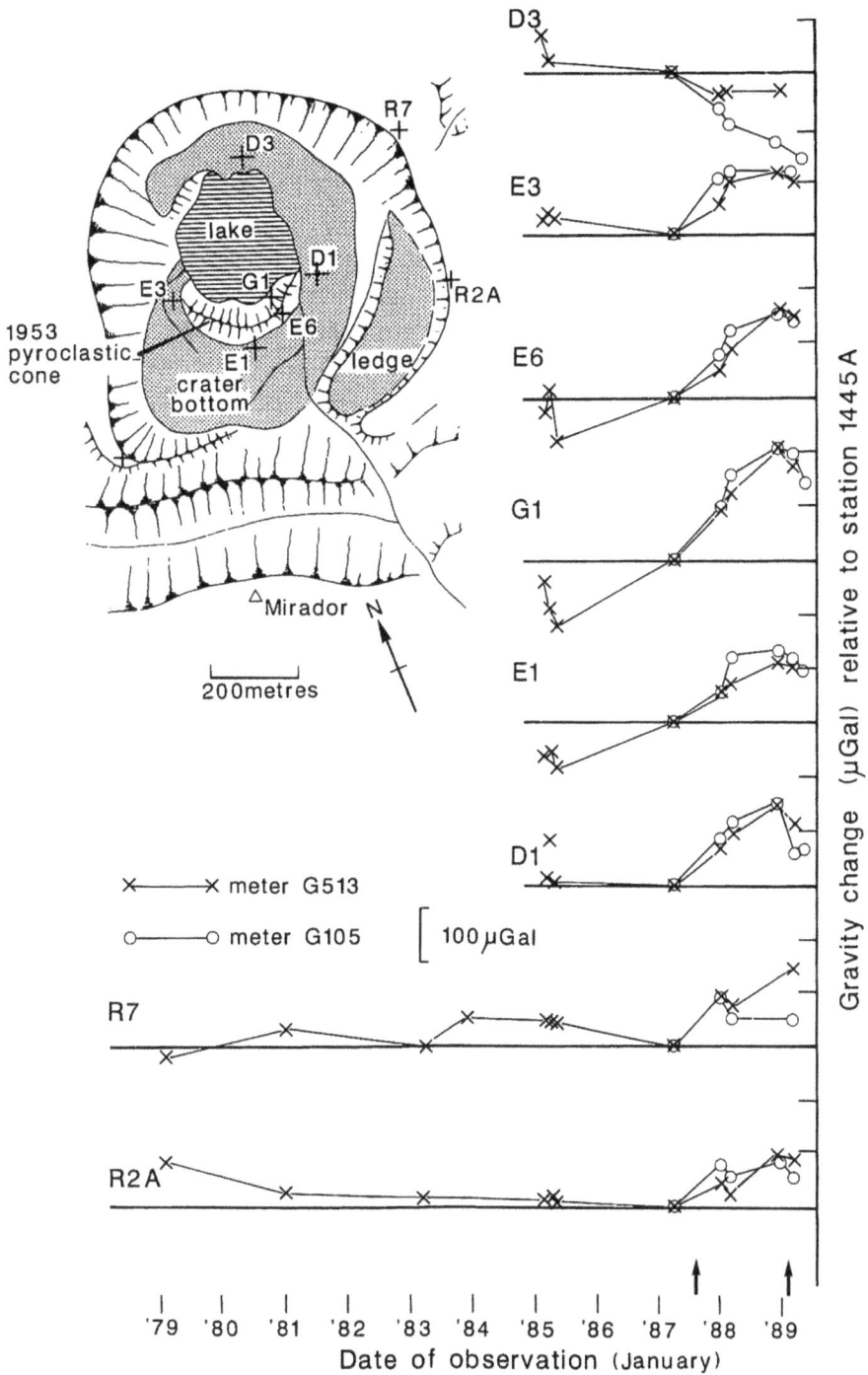

**Figure 8.5** Gravity changes at Poás volcano, Costa Rica for two Lacoste and Romberg gravity meters G105 and G513. Data (stations located on inset) are expressed with respect to a base station 2 km away on the southern flank of the volcano (from Rymer & Brown 1989).

is by eruptive products, so there are basically three groups at which microgravity changes have been identified (Rymer 1994) and published results from the best documented examples are summarized in Table 8.3.

## 8.5.1 Andesitic volcanoes

This explosive subduction related group constitutes the most important in terms of hazard and number on land. Microgravity and ground deformation surveys have been carried out at numerous circum-Pacific locations. Volcanoes in this group typically have relatively complex edifices and the magma feeding system is rather long and narrow (Brown et al. 1987). The summit region is usually characterized by low density and poorly consolidated material, and the intrusion of magma into such a region produces a dramatic increase in subsurface density and, therefore, in microgravity. A build up of magmatic gas would be expected to occur close to the surface in a volcano in this group, so that compared with volcanoes of the other groups (see below), it should be relatively easy to detect as a precursory gravity decrease.

It has been found that ground deformation at volcanoes in this group is limited, but that gravity changes are large, so that data typically plot close to the vertical axis of Figure 7.8.1 and not along the "usual" gravity/height correlation lines. Data and interpretations for well documented examples from this group are summarized in Table 8.3a. For example, more

**Figure 8.6**   Geysering in the crater lake at Poás volcano, Costa Rica, April 1988.

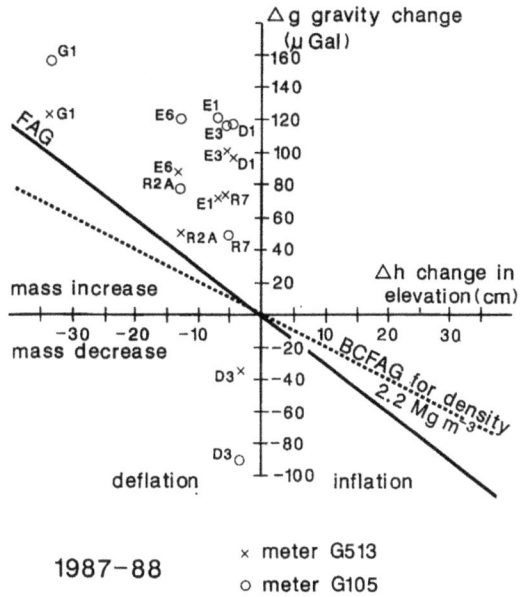

**Figure 8.7** Gravity/height changes observed between March 1987 and March 1988 at Poás volcano, Costa Rica, using Lacoste and Romberg gravity meters G513 and G105 (from Rymer & Brown 1989).

1987−88

× meter G513
o meter G105

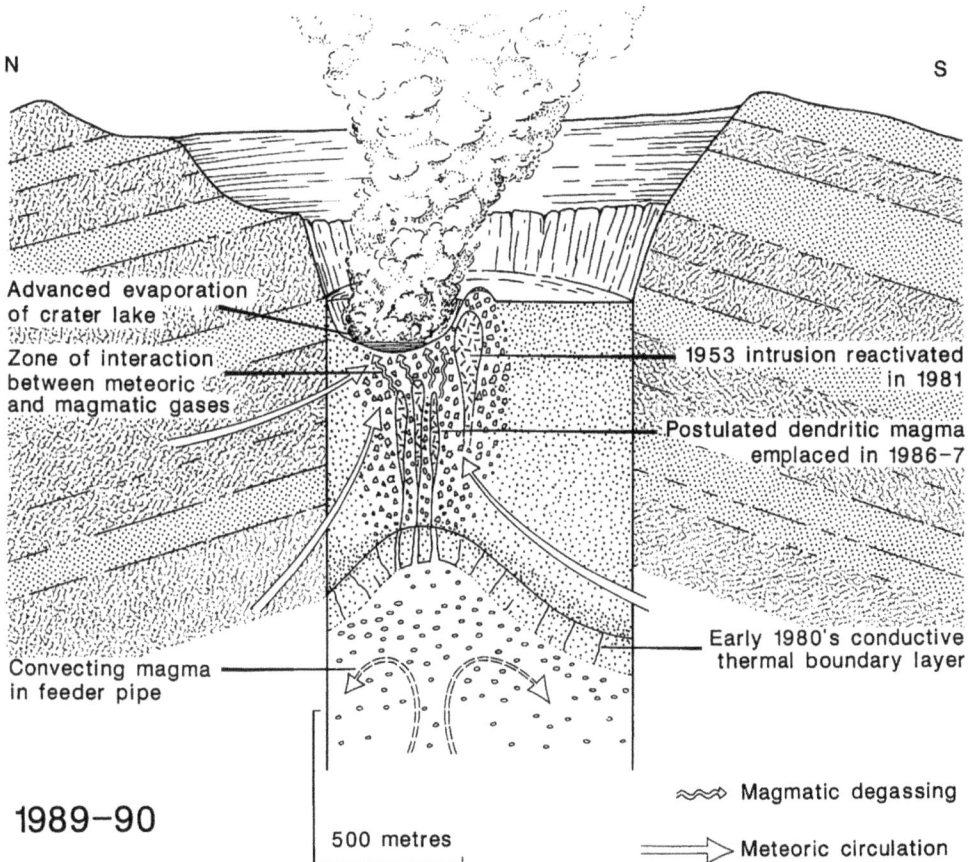

**Figure 8.8** Interpretation of gravity changes at Poás volcano, Costa Rica observed before the 1989 eruption. Mass increases of approximately $10^8$ kg with minimal ground deformation are thought to reflect dendritic intrusions of magma penetrating the poorly consolidated sediments and pyroclastics beneath the lake.

than 10 years of microgravity and ground deformation monitoring at Poás volcano, Costa Rica, combined with seismic and energy budget analysis as well as lake geochemistry and meteorological studies have provided a detailed understanding of the shallow structure and behaviour (Brown et al. 1991b). Microgravity observations have been made at least annually since 1979 at Poás using Lacoste and Romberg gravity meters G105 and G513. Gravity changes are expressed with respect to the base station 2 km south of the crater area which is invariant with respect to control stations up to 25 km away. The data show that, although there are no significant gravity changes (more than approximately 30 µGal) at crater rim stations (Fig. 8.5) during the period of observation, large gravity increases at crater bottom stations were detected between 1985 and 1989. From the mid-1960s, the crater lake at Poás was stable, fluctuating about a mean level with the seasons. Its temperature was about 40°C and pH 1. Fumarole temperatures on the cone (formed in the 1953 eruption) just to the south of the lake varied with time (Fig. 8.5). Although there was no surface eruptive activity between 1986 and 1988, the lake level began to fall and geysering began (Fig. 8.6), lake and fumarole temperatures rose, and gravity increased. There was limited deflation at crater bottom stations, which means that the data plot close to the vertical line in Figure 8.1, as is shown for a typical dataset from this period (Fig. 8.7). Since there is clearly no linear correlation between gravity and elevation changes at Poás, the usual elastic models involving slabs and spherical magma chambers (see Section 8.4) are obviously not appropriate in this case. We therefore interpret the gravity increases (and therefore subsurface mass increases) in terms of dendritic intrusions (Fig. 8.8) of magma beneath the highly permeable region beneath the crater lake (Rymer & Brown 1989, Brown et al. 1991). The mass of the intrusions may be quantified by integrating the magnitude of the gravity increases over the area of the increases. The average increase ($g$) is about 200 µGal ($2 \times 10^{-6}$ m s$^{-2}$) within an area ($S$) of about $10^4$ m$^2$. Thus using Guass' theorem, the increased subsurface mass ($\Delta M$) is given by:

$$\Delta M = \frac{1}{2\pi G} \rho' g S$$

where $\rho'$ is the ratio of the density of the intrusion to the difference between the densities of the intrusion (2670 kg m$^{-3}$) and the surroundings (2000 kg m$^{-3}$) and $G$ is the universal gravitational constant ($6.67 \times 10^{-11}$ N kg$^{-2}$ m$^2$). This evaluates to approximately $10^8$ kg and is the increase in calculated subsurface mass that occurred between 1986 and 1989 beneath the active crater of Poás (Rymer & Brown 1989). Interpretation in terms of dentritic intrusions of magma are consistent with the observation of increased power output at the same time (Rymer and Brown 1989, Dowden et al. 1991, Brown et al. 1991b, Fernandez et al. 1992). Although the simpler slab and spherical models considered earlier seem to be appropriate for some rhyolitic and basaltic group volcanoes (see Sections 8.4.2 & 8.4.3), this example shows that at least some volcanoes in this group are better modelled using more complex geometries (Fig. 8.6). Other published examples from volcanoes of this group are summarized in Table 8.3.

## 8.5.2 Basaltic volcanoes

Shield, extentional rift and island volcanoes such as Kilauea (Hawaii), Etna, (Sicily), Krafla and Askja (Iceland) undergo effusive basaltic eruptions as well as more evolved explosive activity. An important aspect of this group is the tendency of the structures to become gravitationally unstable and to collapse generating rift systems into which magma may intrude passively (Rymer et al. 1993). Large datasets for Kilauea and Etna in particular make this group one of the better studied, although it is of much reduced hazard compared with the rhyolitic and andesitic groups (see Sections 8.5.1 & 8.5.3). Periods of both inflation and deflation have been reported, and of overall subsurface mass increase and decrease. It is clear for this group at least that a single microgravity precursor to activity is difficult to identify, but important information on the mechanisms operating within active volcanoes have been discovered. For example, on Mount Etna, the eruption beginning 14 December 1991 was preceded by negligible ground deformation and seismicity, but in an elongated region running south-southeast from the summit craters gravity increases of over 100 μGal were observed

**Figure 8.9** Gravity changes observed using Lacoste and Romberg meter G513 between June 1990 and June 1991 in the summit region of Mount Etna, Sicily. Data are expressed with respect to a base station approximately 10 km away on the southern flank (from Rymer et al. 1993).

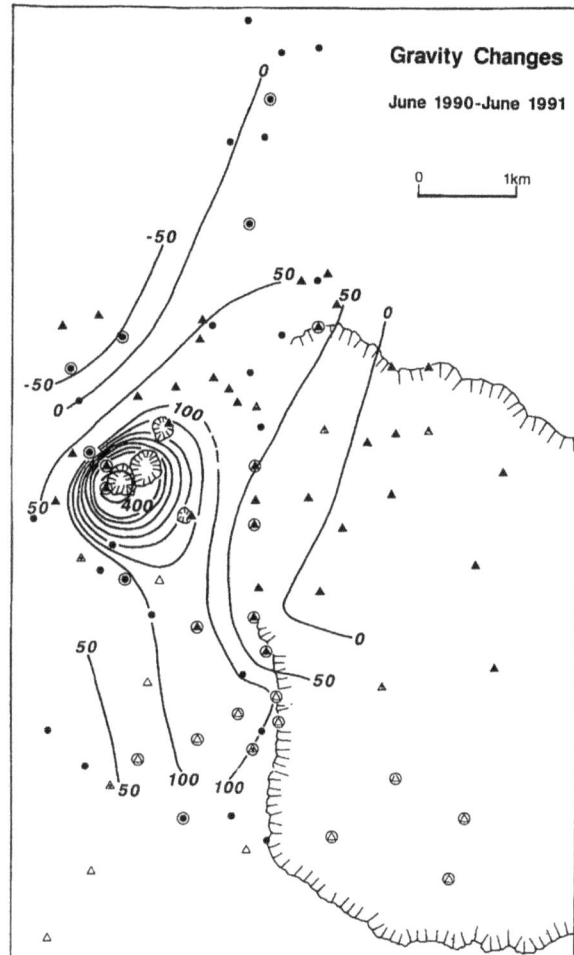

241

between June 1990 and June 1991 (Fig. 8.9). These increases are interpreted in terms of the passive intrusion of magma into a prestressed fracture that had remained "open" since the end of the 1989 activity (Rymer et al. 1993, 1994). The top of the fracture network cuts the surface in a line tending south-southeast for ca. 7 km from the summit. Since the intrusion was passive, there was no source of ground deformation or seismic signal. Published data from other volcanoes in this group are summarized in Table 8.3. Linear gravity/height correlations have been deduced for these datasets. Some data fall on the "expected" lines (Fig. 8.1), but in many cases the large gravity changes and small elevation changes suggest that inelastic processes are responsible.

### 8.5.3 Rhyolitic volcanoes

These tend to be silicic caldera structures where unrest has been observed, but where, although the long-term threat of a devastating eruption is present, no historic large scale explosive activity has been reported. Most data have been interpreted using the spherical model (see Section 8.4.2), and this reflects the large depth compared with lateral extent of the source of gravity change at calderas. A likely cause of eruption at calderas is the development of unstable temperature, chemical, density and viscosity gradients within a cooling magma chamber (Blake & Ivey 1986), such as fractional crystallization leading to oversaturation in volatiles. This would be expected to produce an overall density decrease observable at the surface as a gravity decrease, and so microgravity decreases might be a precursor to caldera eruption. The 1875 eruption of Askja was considered by Sigurdsson & Sparks (1981) to be triggered by a basaltic intrusion into a fractionating magma chamber, this would have produced a density and, therefore, microgravity increase before the eruption. Since no caldera eruptions have occurred since microgravity monitoring has been used to monitor activity, it is not clear which is the better model for caldera eruption triggering. It is likely that both scenarios occur and that microgravity changes of either sign should be considered seriously. Magma intrusions into deep spherical chambers have been identified at Campi Flegrei, Yellowstone and Rabaul, and drainage either from a magma chamber or from the overlying hydrothermal system at Campi Flegrei and Askja (Table 8.3).

Campi Flegrei inflated by up to 1.616 m between 1982 and 1984 (Fig. 8.10) and gravity decreased by up to 331 μGal. The FAG was measured at several stations in the region and an average value of $-290\,\mu\text{Gal}\,\text{m}^{-1}$ was deduced (Berrino et al. 1984). Using the spherical model (see Section 8.4.2), Berrino et al. (1984), Dvorak & Berrino (1991) and Berrino (1993) concluded that a new mass of $1.5 \times 10^{-11}\,\text{kg}$ was intruded into a magma chamber some 3 km (estimated from deformation data) beneath the surface during this period. Subsequent deflation followed a shallower gradient of $-120\,\mu\text{Gal}\,\text{m}^{-1}$ indicating that for each increment of height change more mass was lost during deflation than had been gained during the preceding inflation (Berrino et al. 1992). Thus a linear relationship was observed between gravity and height changes both during inflation and deflation, but the sources were not necessarily the same. The inflation may have resulted from heat transport in the overlying hydrothermal system (Bonafede 1991). There are other examples of periods of inflation at calderas such as Rabaul

**Figure 8.10** The ancient Roman market of Serapeo in the centre of Campi Flegrei caldera (Naples, Italy) has undergone both deflation (borings of molluscs indicate lengthy periods under the sea) and inflation (more than 1 m between 1982 and 1984). This is a classic example of the long period ground deformation characteristic of caldera unrest.

**Figure 8.11** View looking west across Askja caldera, Iceland. The explosion crater Viti formed during the eruption in 1875 as did the 5 km × 3 km caldera Oskjuvatn now filled by a lake. Oskjuvatn is located in the southeast corner of the main caldera that is 7 km × 8 km across. The caldera floor is covered with post-glacial basaltic lava flows and the walls largely comprise hyaloclastites.

and Yellowstone (see Table 8.3) but deflation has rarely been observed, although at these calderas characterized by unrest, it seems reasonable that both forms of deformation should be expected.

The only example of an overall subsurface mass increase accompanying deflation at a caldera comes from Askja in Iceland (Rymer & Tryggvason 1993). In this case the caldera is deflating and gravity increases have been observed only in one part of the caldera and are thought to be related to dyke intrusion, possibly as a result of drainage from the central magma chamber. However, although in 1875 Askja did undergo a caldera forming evolved explosive eruption, it more naturally fits into the basaltic group of volcanoes (Fig. 8.12) (Brown et al. 1991).

## 8.6 Summary and conclusions

Microgravity monitoring when combined with ground deformation and other geophysical surveillance techniques has been shown to be a valuable tool for understanding the processes occurring within a volcano before, during and after eruption. For field gravity surveys of the precision required for microgravity monitoring, Lacoste and Romberg instruments are recommended. Since the technique uses the instruments at the limits of their precision, at least two instruments should always be used, and the same instruments and operators should always be employed.

In some cases, gravity and height correlations are found to follow the simple elastic theory (see Section 8.4), but in other cases more complex but realistic models involving passive magma movements and pre-existing fractured regions are required. Detailed interpretations depend on the local situation, but generally it is clear that andesitic volcanoes tend to exhibit the largest microgravity changes prior to eruption (see Section 8.5.1), basaltic volcanoes undergo passive magma migration with only limited ground deformation (see Section 8.5.2) and rhyolitic centres tend to behave more in the classic elastic way (see Section 8.5.3). Thus although it has been shown that the technique is broadly applicable to the range of volcano types, interpretation of gravity changes is always ambiguous without elevation data, and should be used in conjunction with other methods.

## Acknowledgements

The methods described here, particularly the interpretation of microgravity data with the implications for volcano structure, evolution and eruption trigger mechanisms were developed in collaboration with my close friend and colleague Geoff Brown. Geoff was tragically killed while setting up a microgravity survey on 14 January 1993 in an eruption of Galeras volcano, Colombia. Thanks are due also to Gennaro Corrado who reviewed the manscript, to John Taylor and Andrew Lloyd for preparing the diagrams and Lynn Tilbury for preparing the text.

# References

Baker, T. F. 1984. Tidal deformations of the Earth. *Scientific Progress in Oxford* **69** 197–233.

Berrino, G. 1992. Gravity changes induced by height–mass variations at Campi Flegrei caldera. *Journal of Volcanology and Geothermal Research*.

Berrino, G., G. Corrado, G. Luongo, B. Toro 1984. Ground deformation and gravity changes accompanying the 1982 Pozzuoli uplift. *Bulletin of Volcanology* **47**, 188–200.

Berrino, G., H. Rymer, G. C. Brown, G. Corrado 1992. Gravity–height correlations for unrest at calderas. *Journal of Volcanology & Geothermal Research* **53**, 11–26.

Blake, S. & G. N. Ivey 1986. Density and viscosity gradients in zoned magam chambers, and their influence on withdrawal dynamics. *Journal of Volcanology & Geothermal Research* **30**, 201–30

Bonafede, M. 1991. Hot fluid migration: an efficient source of ground deformation: application to the 1982–1095 crisis at Campi Flegrei – Italy. *Journal of Volcanology and Geothermal Research* **48**, 187–98.

Brouke, R. A., W. E. Zurn, L. B. Slichter 1972. Lunar tidal acceleration on a rigid Earth. *Geophysics Monograph Series of the American Geophysical Union*. **16**, 319–24.

Brown, G. C., H. Rymer, R. S. Thorpe 1987. The evolution of andesite volcano structures: new evidence from gravity studies in Costa Rica. *Earth and Planetary Science Letters*. **82**, 323–34.

Brown, G. C., S.P. Everett, H. Rymer, D. W. McGarvie, I. Foster. 1991. New light on cladera evolution, Askja Iceland. *Geology* **19**, 352–5.

Brown, G. C., H. Rymer, D. Stevenson 1991. Volcano monitoring by microgravity and energy budget analysis. *Journal of the Geological Society of London* **148**, 585–93.

Dowden, J., P. Kapadia, G. C. Brown, H. Rymer 1991. The dynamics of a geysir eruption. *Journal of Geophysical Research* **96**, 18059–71.

Dvorak, J. J. & G. Berrino 1991. Recent ground movement and seismic activity in Campi Flegrei, Southern Italy: episodic growth of a resurgent dome. *Journal of Geophysical Research* **96**, 2309–23.

Dzurisn, D., A. Anderson, G. P. Eaton, R. J. Koyanagi, P. W. Lipman, J. P. Lockwood, R. T. Okamura, G. S. Puriwai, M. K. Sajo, K. M. Yamoshita 1980. Geophysical observations of Kilauea volcano, Hawaii, constraints on the magma supply during November 1975-September 1977. *Journal of Volcanology and Geothermal Research* **7**, 241–69.

Eggers, A. A. 1983. Temporal gravity and elevation changes at Pacaya volcano, Guatemala. *Journal of Volcanology and Geothermal Research* **19**, 223–37.

Eggers, A. A. 1987. Residual gravity changes and eruption magnitudes. Guatemala. *Journal of Volcanology and Geothermal Research* **33**, 201–16.

Fernandez, M., H. Rymer, G. C. Brown, E. Hernandez 1992. La desecacion del lago caliente del Volcano Poás (Costa Rica) a partir de 1986 y el ciclo eruptivo de cenizas en 1989: evidencias de un ascenso magmatico. In *Memoirs of the 7th Latinamerican Congress of Geology (Madrid)*.

Hagiwara, Y. 1977. Gravity changes associated with seismic activity. In *Advances in Earth and Planetary Sciences, vol. 2: Earthquake precursors*, C. Kisslinger & Z. Suzuki, Z (eds.) 137–46.

Hallinan, S. E. 1991. Gravity studies of the Guayabo caldera and the Miravalles geothermal field. PhD thesis, Open University, UK.

Hallinan, S. E., G. C. Brown, H. Rymer 1991. *Final Report ODA FNI 8788/541/093, Geophysical assistance in geothermal field assessment and monitoring, Miravalles, Costa Rica*. Open University, UK.

Harrison, J. C. & T. Sato 1984. Implementation of electrostatic feedback with a LaCoste-Romberg mode G gravity meter. *Journal of Geophysical Research* **89**, 7957–61

Iida, K., M. Hayakawa & K. Katayose 1952. *Geological Survey of Japan, Report No. 152, Gravity survey of Mihara volcano, Oosima Island, and changes in gravity caused by eruption*.

Jachens, K., C. & G. P. Eaton 1980. Geophysical observations of Kilauea volcano, Hawaii. I. Temporal gravity variations related to the 29th November earthquake and associated summit collapse. *Journal of Volcanology and Geothermal Research* **7**, 225–40.

Jachens, R., D. R. Spydell, G. S. Pitts, D. Dzurisn, C. W. Roberts 1984. Temporal gravity variations at Mount St Helens, March–May 1980. In *United States Geological Survey Professional Paper 1250, The 1980 eruptions of Mount St Helens, Washington*, P. W. Lipman & D. R. Mullineaux (eds), 175–82. Washington

DC: United States Government Printing Office.

Johnsen, G. V., A. Bjornsson, S. Sigurdson 1980. Gravity and elevation changes caused by magma movement beneath the Krafla caldera, Northwest Iceland. *Journal of Geophysical Research* **47**, 132–40.

Kanngieser, E. 1983. Vertical component of ground deformation in North Iceland. *Annales Geophysicae* **1**, 321–8.

McKee, C., J. Mori, B. Tali 1987. *Geological Survey of Papua New Guinea Report 87/29, Microgravity changes and ground deformation at Rabaul caldera 1973–1985.*

Mason, R. G., M. G. Bill, M. Muniruzzamann 1975. Microgravity and micro-earthquake studies. In *UK Research on Mount Etna 1974*, 43. London: The Royal Society.

Melchoir, P. 1983. *The tides of the planet Earth*. Oxford: Pergammon.

Milsom, J. 1990. *Field Geophysics*. Milton Keynes, UK: Open University Press/Halsted Press.

Mogi, K. 1958. Relationship between eruptions of various volcanoes and the deformation of the ground surfaces around them. *Bulletin of the Earthquake Research Institute* **36**, 99–134.

Rymer, H. 1985. Gravity studies of sub-surface structures and evolution of active volcanoes in Costa Rica. PhD thesis, Open University, UK.

Rymer, H. 1989. A contribution to precision microgravity data analysis using Lacoste and Romberg gravity meters. *Geophysical Journal* **97**, 311–22.

Rymer, H. 1994. Microgravity change as a precursor to volcanic activity. *Journal of Volcanology and Geothermal Research* **53**, 11–26.

Rymer, H. & G. C. Brown 1984. Periodic gravity changes at Poás volcano, Costa Rica. *Nature* **311**, 243–5.

Rymer, H. & G. C. Brown 1986. Gravity fields and the interpretation of volcanic structures: geological discrimination and temporal evolution. *Journal of Volcanology and Geothermal Research* **27**, 229–254.

Rymer, H. & G. C. Brown 1987. Causes of microgravity change at Poás volcano, Costa Rica: an active but non-erupting system. *Bulletin of Volcanology* **49**, 389–98.

Rymer, H. & G. C. Brown 1989. Gravity changes as a precursor to volcanic eruption at Poás volcano, Costa Rica. *Nature* **342**, 902–905.

Rymer, H. & E. Tryggvason 1993. Gravity and elevation changes at Askja, Iceland. *Bulletin of Volcanology* in press.

Rymer, H., J. B. Murray, G. C. Brown, F. Ferrucci, W. McGuire 1993. Magma eruption and emplacement mechanisms at Mt Etna 1989–1992. *Nature* **361**, 439–41.

Rymer, H., G. C. Brown, F. Ferrucci, J. B. Murray 1994a. Dyke intrusion mechanisms on Etna 1989–1992 and microgravity precursors to eruption. *Acta Vulcanologica* **4**, 109–114.

Sanderson, T. J. O. 1982. Direct gravimetric detection of magma movements at Mount Etna. *Nature* **297**, 487–96.

Sanderson, T. J. O., G. Berrino, G. Corrado, M. Grimaldi 1983. Ground deformation and gravity changes accompanying the March 1981 eruption of Mount Etna. *Journal of Volcanology and Geothermal Research* **16**, 299–315.

Savage, J. C. 1984. Local gravity anomalies produced by dislocation sources. *Journal of Geophysical Research* **89** 1945–52.

Smith, R. B., R. E. Reilinger, C. M. Meertens, J. R. Hollis, S. R. Holdahl, D. Dzurisin, W. K. Gross, E. E. Klingele 1989. What's moving at Yellowstone? *EOS Transactions of the American Geophysical Union* **70**, 113–25.

Sigurdsson, H. & R. S. J. Sparkes 1981. Petrology of rhyolitic and mixed magma ejecta from the 1875 eruption of Askja, Iceland. *Journal of Petrology* **22**, 41–84.

Torge, W. 1982. The present state of relative gravimetry. In *Proceedings of the General Meeting of International Association of Geophysics Toyko*, 319–24.

Torge, W. & E. Kanngieser 1980. Gravity and height variations during the present rifting episode in northern Iceland. *Journal of Geophysical Research* **47**, 125–31.

Tryggvason, E. 1986. Multiple magma reservoirs in a rift zone volcano: ground deformation and magma transport during the September 1984 eruption of Krafla, Iceland. *Journal of Volcanology and Geothermal Research* **28**, 1–44.

Tsuboi, C. 1983. *Gravity*. London: Allen & Unwin.

Wong, T-F. & J. B. Walsh 1991. Deformation induced gravity changes in volcanic regions. *Geophysical*

*Journal International* **104**, 513–20.

Yokoyama, I. 1972. Gravimetric, magnetic and electrical methods. In: *The surveillance and prediction of volcanic activity*. Paris: UNESCO.

Yokayama, I. 1989. Microgravity and height changes caused by volcanic activity, four Japanese examples. *Bulletin of Volcanology* **51**, 333–45.

# 9 Geoelectrical methods in volcano monitoring

J-F. Lénat

## 9.1 Introduction

Geoelectrical methods have been used extensively to study the structure of volcanoes. They provide an image of the internal form as a distribution of resistivities which must be interpreted in terms of volcanic layers and bodies. The essential rationale underpinning the application of geoelectrical methods to the monitoring of volcanic activity lies in the fact that the distribution of resistivities is often modified in response to internal volcanic processes. In the first part of this chapter, therefore, the factors which control rock resistivity and its modification in a volcanic environment are examined in some detail.

Although geoelectrical methods are promising tools for probing changes inside volcanoes, only a few experimental studies have been published to date, some of which will be examined here, with reference to case studies demonstrating how different geoelectrical techniques have been successfully applied to a number of volcanoes. Because the methods used for volcano monitoring differ little from those used for more classical geoelectrical surveys in geophysics, only the general principles and procedures will be described, unless an experiment requires a unusual methodology.

### 9.1.1 Resistivities of volcanic rocks

Generally speaking, the resistivity of rocks has little to do with their mineral composition, but depends strongly upon their porosity and upon the resistivity of the fluid that they contain. Indeed, with the exception of some ore-type minerals, most of the rock-forming minerals are very poor ohmic (electronic) conductors at normal temperatures. The dominant way in which electrical current is propagated in most rocks is electrolytic conduction, where the ions of the fluid (generally water) present in pores, cracks and along grain boundaries, conduct the electricity. Hence, the major factors determining the resistivity of a rock will be its effective, or interconnected, porosity, its degree of fluid saturation, and the resistivity of the fluid. Volcanic rocks exhibit a wide range of porosities, from low-porosity dense lava flows or intrusions to high-porosity vesiculated lava flows or poorly compacted pyroclastic deposits. As the rocks are progressively buried, during the growth of a volcano, their bulk porosity decreases by compression of cracks and pores (Ryan 1987), an effect which contributes to increasing resistivity with depth. In addition, in oceanic volcanoes, lavas erupted at depths of more than a few hundred metres below sea level are generally poorly vesiculated because the ambient hydrostatic pressure inhibits the degassing, therefore most submarine lavas are quite dense (Peterson & Moore 1987).

The degree of fluid saturation is generally controlled by the hydrogeological conditions. On most basaltic shield volcanoes, for example, a significant resistivity contrast will occur at the transition between the vadose zone, where rocks contain only moisture, and the zone beneath the water table, where rocks are saturated with water. On volcanoes that have significant amounts of ash within their structure, on the other hand, there may be no recognizable resistivity contrast between the vadose zone and rocks beneath the water table. At any individual volcano, the degree of water saturation will show a complex pattern which is dependent upon the specific characteristics of the volcano's hydrogeological system (e.g. the presence or absence of perched water tables, the nature of the rainfall regime, and the existence of large permeability and/or porosity contrasts within the edifice). An important and common feature, encountered mostly in basaltic shield volcanoes, is the presence of high-level water tables trapped by complexes of near-vertical dikes. This situation occurs particularly in summit regions and rift zones, and leads to the development of impermeable barriers which impede the lateral flow of ground water (Stearns 1942, Jackson & Kauahikaua 1987).

The resistivity of water depends upon the concentration of contained ions and on its temperature. For example, at normal temperatures, the resistivity of meteoric water ranges from about 30 to $10^3$ ohm.m, whereas the resistivity of seawater is about 0.2 ohm.m. If the temperature of a dilute solution of salt (e.g. NaCl) is raised (Quist & Marshall 1968) from a temperature of about 20°C, a decrease of resistivity is first observed with a minimum value reached at about 300°C where the resistivity is lowered by a factor of about 7. Beyond this temperature, the resistivity of the solution starts to increase due to an increased content of water vapour. The resistivity of a rock can, in principle, be estimated from its porosity, degree of saturation, and the resitivity of the contained fluid, using the empirical formula of Archie (1942). Experiments on Hawaiian and Icelandic basalts (Olhoeft 1977, Rai & Manghnani 1981, Flovenz et al. 1985), however, show departures from Archie's law when the temperature is above 80°C or when salinity of the water is low. The lower than expected resistivities are tentatively explained by interface-conduction phenomena, due to the presence of hydrated minerals such as clay minerals or zeolites. Indeed, although these minerals are not themselves conductors, they have high cation-exchange capacities with the fluids, which leads to the build up of a conductive double layer at their interface with the pore water. With a temperature increase, the number of ions in the mobile part of the double layer also increases and, as a result, the resistivity is lowered more than expected from a simple increase in the mobility of electrical charges in the water. In volcanoes, hydrothermally altered zones will, therefore, be characterized by low resistivities, due to the presence of a high proportion of hydrated minerals.

Finally, let us examine what happens at the high temperatures characteristically encountered in active volcanic regimes. Above about 300°C, the resistivity of a dilute solution of saline water, and therefore of rock, increases. This is attributed to the increased content of

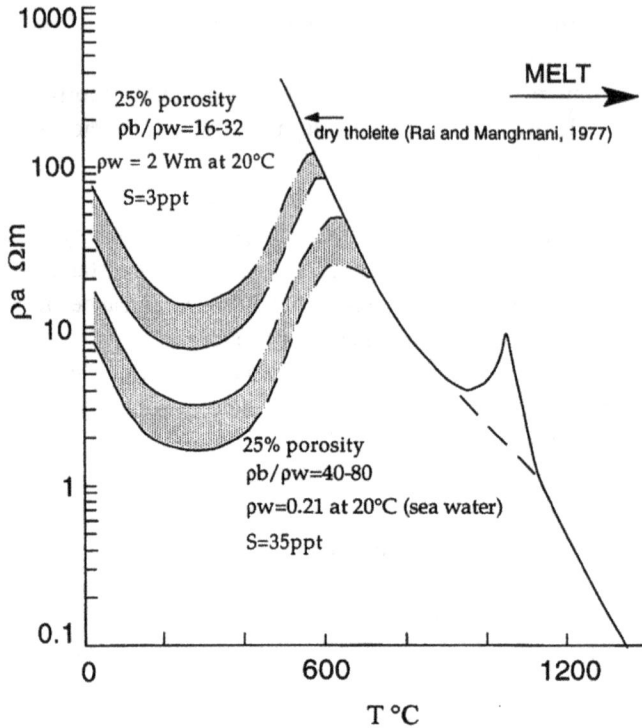

**Figure 9.1** Expected variations in basalt resistivity for temperatures from 20°C to 1400°C. The solid line shows the resistivity expected for completely dry rock; the two shaded areas show ranges of resistivities for typical basalts saturated with two different salinity fluids. The cusp in the curve for dry tholeiite is probably an effect of the speed of experimental heating and cooling (Rai & Manghnani 1977) and is not significant for field survey interpretation. The arrow labeled "melt" indicates the range in which basalts are molten. (Figure and caption from Kauahikaua et al. 1986.)

vapour in a two phase (liquid–vapour) fluid system. Such shallow vapour-dominated systems or layers have been probed by coupled geoelectrical surveys and test wells (Zhody et al. 1973, Anderson 1987). Kauahikaua et al. (1986) report (Fig. 9.1) that for Kilauea basalts, the curves of resistivities of saturated and dry rocks intersect at about 600°C. Above this temperature, "the conduction through the rock matrix exceeds conduction through the pore electrolyte", and the resistivities again decrease. The resistivity of molten lava is known, both from the field (Frischknecht 1967, Halbwachs 1983) and the laboratory (Murase & McBirney 1973, Rai & Manghnani 1977), to range from about 20 to less than 1 ohm.m. A direct measurement in a lava flow adjacent to an eruptive fissure was performed by Bartel et al. (1983), giving a value of 40 ohm.m, but, since the measurement was made in a medium where vesicles were forming because of ongoing degassing, this value is probably higher than for non-vesiculated body of magma. Some idea of the possible resistivity distribution within an active volcanic system is given by the illustration schematic in Figure 9.2

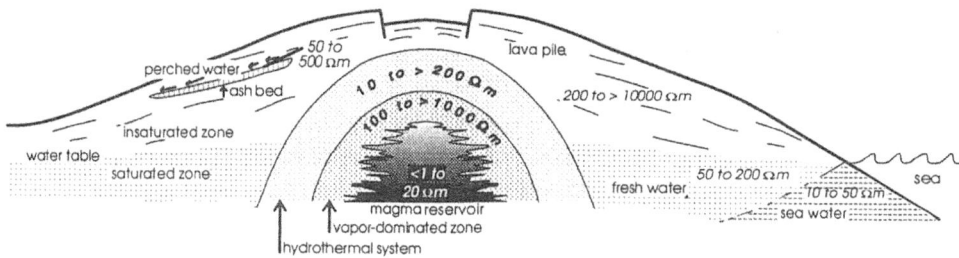

**Figure 9.2** Idealized geoelectrical model of a volcano. The ranges of resistivity values are given only to show the general trends in volcanic terrains. Values outside these ranges can be encountered according to the specific lithology and the hydrogeological conditions of volcanoes. Smaller scale features such as intrusions, fault zones, and dyke-trapped water tables are not shown in this general model.

### 9.1.2 Possible causes of resistivity changes in volcanoes

Having some knowledge of the factors that determine the resistivity of rocks, we can speculate on the kinds of volcanic phenomena that modify this resistivity at depth. Porosity variations may arise from volcano tectonic phenomena or from chemical activity. The well-established inflation and deflation cycles that are linked to magma accumulation or withdrawal within volcanic edifices (Decker 1987) involve modifications of internal stresses which are often expressed by opening or closing of cracks, thereby modifying the megascopic porosity of large volumes of rock. Similar effects may be expected from slumping of a volcano's flanks. Porosity may also be reduced by the formation of zeolites and clays minerals on the walls of intravolcanic fractures and within pores, during hydrothermal alteration. This results in the loss of resistivity by reduction of porosity, but this is offset by additional conductivity due to the presence of the hydrated minerals. Pore fluid resistivity may be lowered by the assimilation of an influx of magmatically derived, mineralized fluids which will increase the salt concentration of water. It is likely that this kind of event will also be associated with a temperature increase which would further contribute to a fall in pore fluid resistivity.

Besides modifications of the resistivities in existing volcanic structures, the development of new features, such as the emplacement of new intrusive bodies, may have a major impact on the resistivity distribution. The emplacement of an intrusion will replicate, at a different scale and depth and with a different geometry, some of the resistivity features found around and within a magma reservoir. The resistivity distribution of the host volume of rocks will be modified by fracturing, the elevation of temperature, the creation of an hydrothermal system around the cooling intrusion, and by the presence of molten magma.

## 9.2 Direct-current resistivity methods

Direct-current resistivity techniques have been employed in geoelectrical surveys for decades, and exhaustive descriptions of these methods may be found, therefore, in many geophysics textbooks (see e.g. Grant & West 1965, Telford et al. 1990). Such techniques are commonly described as resistivity or galvanic methods because they emit a direct measure of apparent ground resistivity and because they make use of electrical contact with the ground. The apparent resistivity of the ground is measured with an array of four electrodes. Direct current is injected into the soil through a pair of electrodes called current electrodes, thus establishing a stationary electric field. Potential difference, or potential gradient, is measured using a pair of potential electrodes. Application of Ohm's law leads to an expression of the resistivity as:

$$\rho = (\Delta V/I)2\pi k$$

where $\Delta V$ and $I$ are, respectively, the measured difference of potential and the input current, $k$ is a factor which depends upon the geometry of the array of electrodes, and $\rho$ is the apparent resistivity. In the ideal case of a totally homogeneous isotropic ground, $\rho$ and the true resistivity would be the same. In reality, the resistivity of the ground is not, as we have shown in the preceding section, homogeneous; the apparent resistivity is, therefore, a function of the resistivity distribution in the ground, and the geometry of the electrode array. It is worth noting here that, when the array is expanded, deeper levels of the ground are probed. Quantitative interpretation of the data is achieved by modelling, with the apparent resistivity signature of a given resistivity distribution being computed and compared with the observed data. Various methods have been proposed to achieve a satisfactory agreement between computed and oberved data (see, e.g., Koefoed 1979). One has to be aware, however, that such interpretations suffer from non-uniqueness arising from problems known as equivalence and suppression (Keller & Frischknecht 1966). Practically, realistic contraints on the number of layers and on their range of resistivity have to be introduced in the initial model which is then adjusted using computer software.

### 9.2.1 Equipment and procedures

Two approaches are used for resistivity monitoring: periodically repeated surveys and the use of a permanently installed network with data often sent by telemetry to a data collection centre or stored in the field by a recording device. In both cases, the necessary equipment is essentially the same and consists of insulated electric wires, a direct-current source, and devices to measure $\Delta V$ and $I$ or their ratio. It is important to note that a new generation of meters, using microprocessors, has appeared in the last decade which allow signal processing, including the stacking of signals, which permits data collection using less powerful sources. This may be of importance since, depending upon the resistivity of the terrrain and upon the range of depths to be probed, it may be necessary to work with very small signals because it is not always possible to carry large generators in the field. Repeating surveys in areas of interest will generally provide more data than using a permanent network which, for logistic and economic reasons, will usually consist of a limited number of electrodes. On the other hand, periodic surveys are time consuming and require more personnel. Prior to initiating a resistivity monitoring programme on a volcano, it is sensible, therefore, to first undertake a geoelectrical survey of the area of interest to study and establish its electrical structure, and afterwards to select the best array for the monitoring network, based on the results of the survey.

## 9.2.2 Case study: monitoring apparent resistivity at Izu-Oshima volcano, Japan

Apparent resistivity monitoring has been carried out at Izu-Oshima, an active basaltic volcano with a caldera and a nested central cone crowned by a summit crater, since 1975 (Yukutake et al. 1983, 1987) using an axial dipole-dipole arrangement. The electrodes were deployed in a line across the crater, with the current electrodes on one side and the potential electrodes on the other, as shown in Figure 9.3, allowing probing of the apparent resistivity in the zone beneath the crater. Geoelectrical soundings on the caldera floor and on the central cone show a general decrease in resistivity with depth. Beneath the central cone, a conductive layer (less than 200 ohm.m) at a depth of about 120 m, is overlain by more resistive rocks (ranging from about 1100 to 15000 ohm.m). Theoretical considerations (Yukutake et al. 1983) suggest that if magma intrudes beneath the crater, the apparent resistivity would probably decrease, at least for the longest dipole–dipole C (Fig. 9.3), because of the temperature rise. If the volume of the crater (about 230 m deep and 300 m wide at the time the experiment started) was reduced, due to the ascent of fresh magma, the apparent resistivity for A, and to a less extent for B, would increase. This is close to the pattern that was

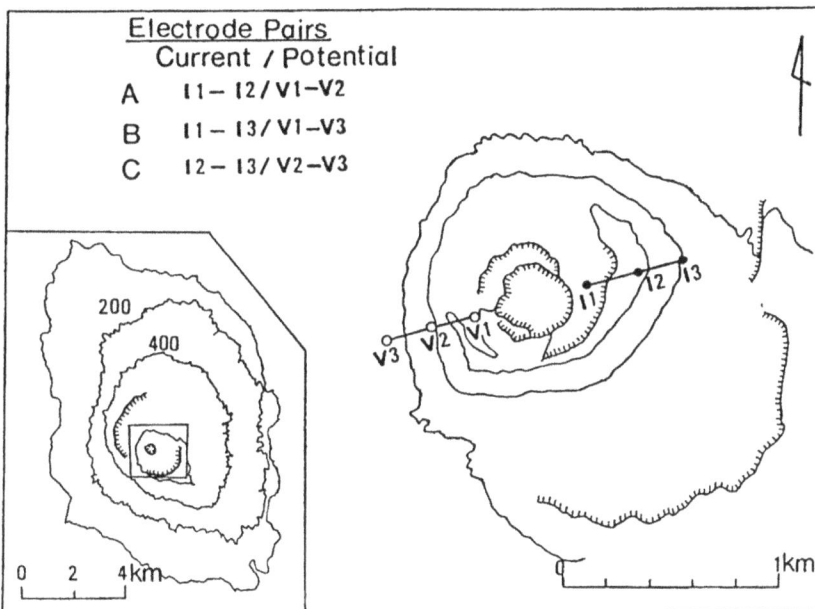

**Figure 9.3** Electrode arrangements across the summit crater of Izu-Oshima volcano for apparent resistivity monitoring. I1, I2 and I3 are the current electrodes and V1, V2 and V3 are the potential electrodes. A, B and C are the different arrays used for measuring the apparent resistivity. The inset is a generalized map of the island with a box showing the location of the main figure (from Yukutake et al. 1983).

253

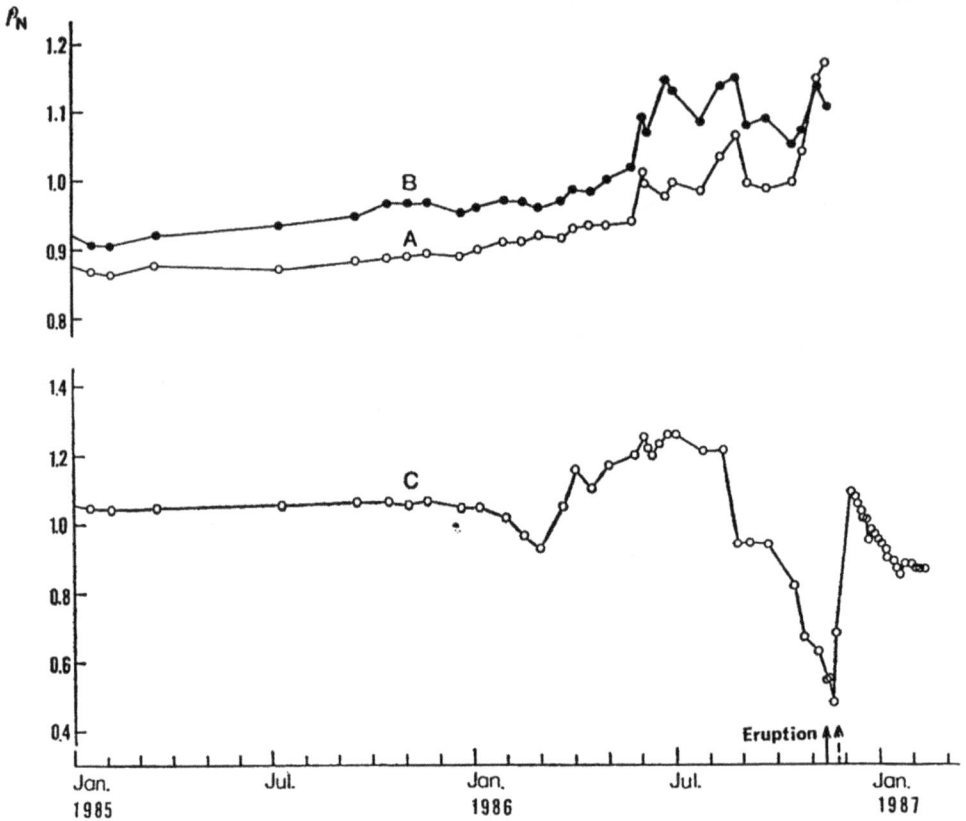

**Figure 9.4** Time variations of the apparent resistivity recorded at Izu-Oshima volcano at the network shown in Figure 9.3. The times of the eruptive outbreaks are shown by arrows. The observed variations are discussed in the text (from Yukutake et al. 1987).

observed during the 1986 crisis (Aramaki 1988) (Fig. 9.4). An eruption started on 15 November 1986 in the summit crater, following volcanic tremor that had been recorded since July 1986. The general decrease of apparent resistivity for C can be explained by intrusion of magma and the resulting temperature increase. For A and B, the last measurements were made on 16 November when the crater was filled with lava to within 100 m of the surface. The increase in apparent resistivity would thus be in agreement with the theoretical models of crater filling. Such models cannot explain, however, the changes observed before the onset of the eruption, unless a pervasive magma invasion of voids beneath the crater is inferred. Although interpretation of the apparent resistivity changes is not fully established in this case, the experiment clearly shows that significant and coherent resistivity variations are observed before volcanic eruptions.

## 9.3 Electromagnetic methods

Electromagnetic surveying methods have been developed largely during the last few decades (Ward 1980), and are applied to a wide range of studies, from observations of the deep interior of the Earth to near-surface surveys for groundwater or engineering. The principles underpinning the technique, the equipment required, accounts of operating procedures, and basic data interpretation are discussed in standard geophysics textbooks such as those by Telford et al. (1990), Grant & West (1965) and Keller & Frischknecht (1965). Electromagnetic methods are based on the response of conductive ground to the propagation of electromagnetic fields. A primary electromagnetic field may be artificially generated by passing an alternating current through an antenna, coil, or loop of electrical wire, or naturally produced by thunderstorms or solars emissions. In the presence of a conductive body, the magnetic component of the electromagnetic field induces currents in the conductor, these currents, in turn, give rise to a secondary electromagnetic field. A receiver will measure the resultant of the primary and the secondary fields, which will differ, in phase as well as in amplitude, from the primary field. These differences may be interpreted in terms of location, geometry and electrical properties of the conductor.

A host of electromagnetic methods have been developed based upon these general principles. One of their main common characteristics is that no physical contact with the ground is required, for either the transmitter of the primary field, or for the receiver. The latter normally consists of one or two coils which measure the magnetic field induced by the primary and secondary fields. This absence of the ground contact requirement probably constitutes the principal advantage of the electromagnetic technique over direct-current methods the use of which may be severely restricted in dry volcanic terrain by high contact resistances (which limit the amount of current that can be impressed into the ground).

### 9.3.1 Case study: electromagnetic monitoring of Kilauea volcano, Hawaii

The results of an experiment using electromagnetic monitoring at Kilauea volcano have been published by Jackson et al. (1985), and are summarized here. The Kilauea network (Fig. 9.5) was composed of a quasi-square loop transmitter (Tx) laid on the ground, and three sensors. Two of the sensors (OTL and KKK) were mu-metal cored coils and the third (PUH) was a square loop laid on the ground and composed of four turns of electric wire. All the sensors were sensitive only to the vertical component of the total magnetic field changes. The controlled-source electromagnetic (CSEM) system transmitter, operated at discrete frequencies between 0.01 and 8 Hz and transmitted a peak-to-peak sinusoidal current of 80 A. The transmitting and receiving electronic systems were time-synchronized to provide a phase reference. Measurements were repeated at intervals ranging from a few hours to 3 weeks. Although they were initially made at five different frequencies (between 0.1 and 8 Hz), they were later reduced to the measurement at 1 Hz and only the data for this frequency are reported and discussed by Jackson et al. (1985). Both the amplitude of the vertical component of the magnetic field and its phase angle relative to the source were recorded.

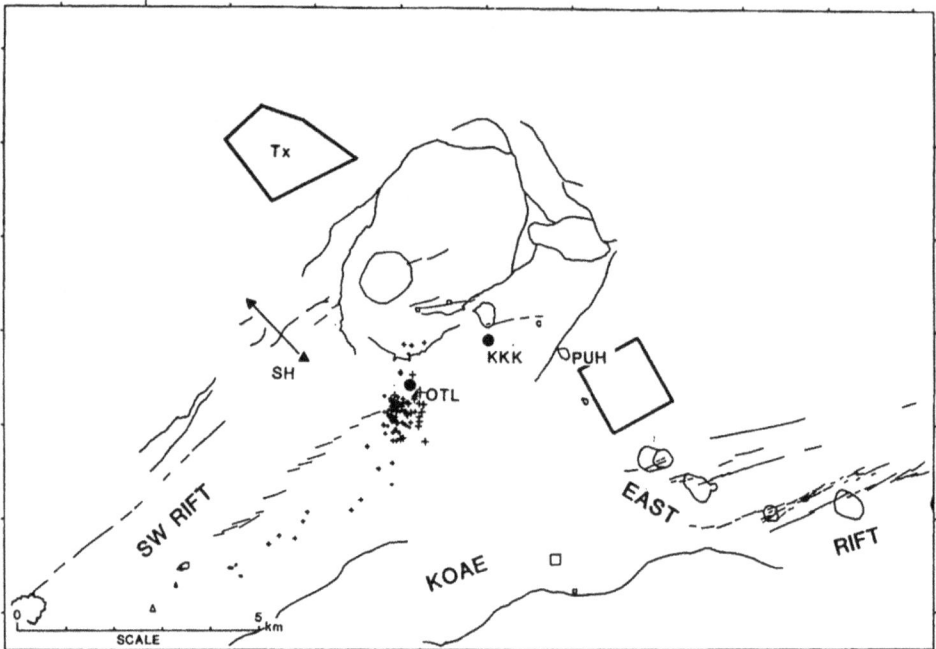

**Figure 9.5** Electromagnetic monitoring network of Kilauea. Tx, the transmitter loop; PUH, a receiver loop; and OTL and KKK, receiver coils. The main tectonic features of the summit of Kilauea are shown, including fractures and faults, and the positions of the caldera and pit craters. The crosses represent the epicentres associated with the January–February 1981 intusive events, and the arrow indicates the tilt change vector between 20 January and 6 February at station SH (Sand Hill). (From Jackson et al. 1985.)

The depth of penetration of an electromagnetic field is related to both the frequency and the ground resistivity, and increases when the frequency decreases (this is known as the "skin effect"). In this case, it had been shown by a previous electromagnetic survey (Kauahikaua 1982, Kauahikaua et al. 1986), using the same transmitting device, that the frequency of 1 Hz was sensitive to resistivity structures to a depth of about 2 km. During the periods of operation of the electromagnetic monitoring network (March to October 1979 and May 1980 to February 1981), six intrusive events were recorded by the seismic and deformation surveillance networks. The electromagnetic monitors showed no response or only moderate responses for some of the events, and remarkable signals for others, an example of the latter situation being illustrated in Figure 9.6. The sequence of events here was as follows. In early December 1980, seismicity and deformation patterns suggested that a slow inflation was developing in the south of the caldera, in the zone where OTL is located. On 21 January a minor intrusive event occurred beneath this area (see epicentre location in Fig. 9.5). On 24 January a second phase of slow intrusion developed up to 3 km down the southwest rift zone over a period of 3–4 days. Eventually, after an apparently aseismic movement of magma further downrift (Klein et al. 1987), a co-seismic intrusion was recorded at a distance of 15 km from the caldera. During these events, the OTL area inflated as shown by the Sand Hill tiltmeter record (Fig. 9.6) and the tilt vector (Fig. 9.5).

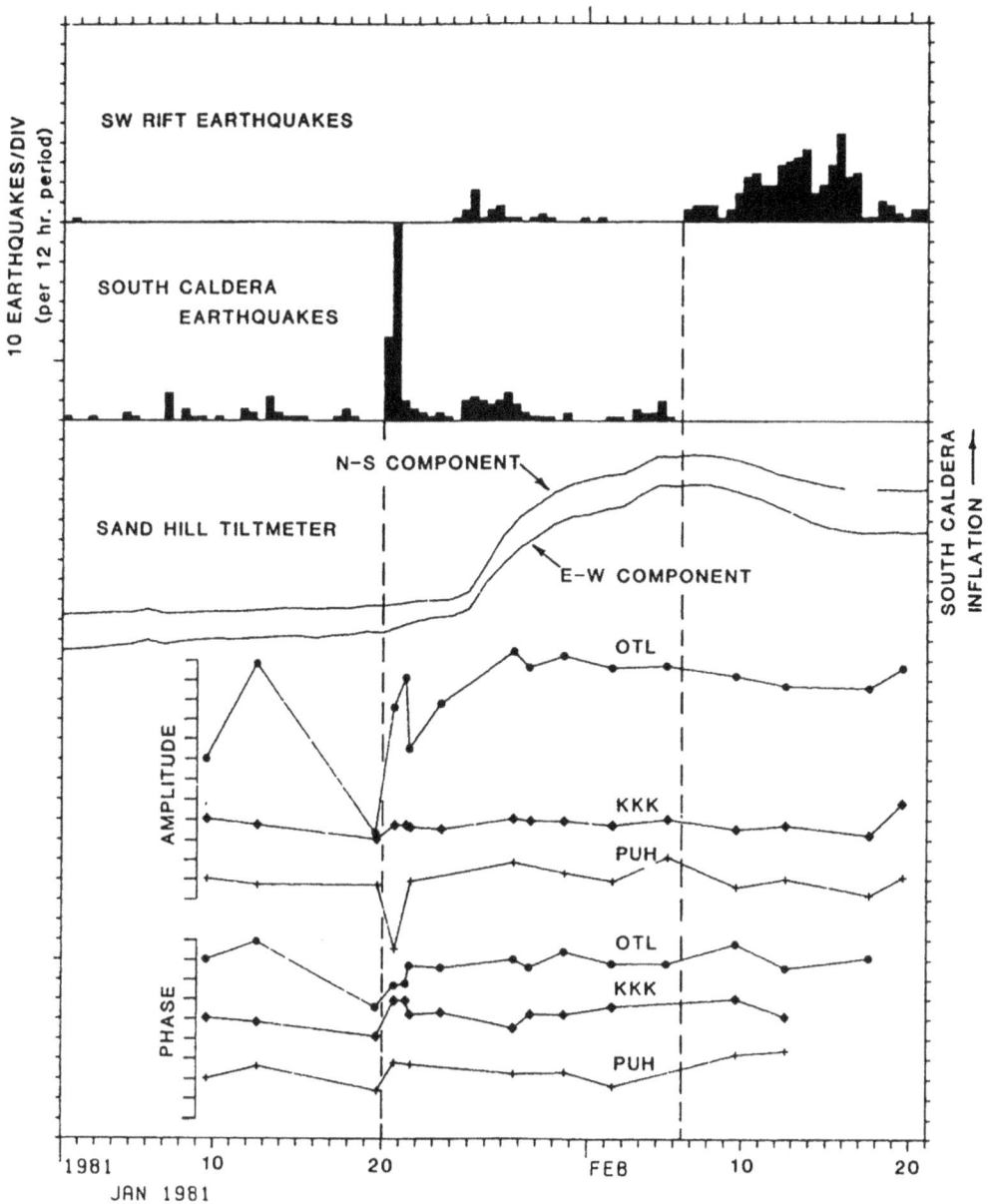

**Figure 9.6** Phase and amplitude changes at the Kilauea electromagnetic network during the January–February 1981 intrusive event. Seismic (earthquakes in the 0–5 km depth range) and deformation records are shown in the upper part of the figure. See Figure 9.5 for station and epicentre locations, and the tilt direction at Sand Hill. Amplitude is expressed as a percentage variation of the initial values (1 % per division), and phase in degrees of variation relative to the initial values (1° per division) (from Jackson et al. 1985).

Electromagnetic monitoring during these events reveal concurrent amplitude and phase variations at OTL which were dramatic and significant (noise level at the sensors was estimated to 0.3 % for the amplitude and to 0.2° for the phase), while much lower variations, or no variations at all, were observed at KKK and PUH which were located farther from the the zone of this complex intrusive sequence. Jackson et al. (1985) proposed that the changes observed at OTL were related to resistivity changes caused by the formation of a long sheet-like conductor (i.e. the emplacement of a low-resistivity dyke). The fact that large resistivity variations were recorded at OTL before the beginning of the main seismic and deformation events is important, suggesting perhaps that aseismic movements of magma which are not accompanied by ground deformation may be detected by resistivity monitoring. A significant response of the CSEM network was also recorded during an intrusion along the east rift zone in August 1980. From their observations, Jackson et al. (1985) concluded that the CSEM network was sensitive only to intrusive events that involve large movements of magma or to the shallow emplacement of magma near to the sensors.

### 9.3.2 Case study: very low frequency detection and monitoring of active lava tubes at Kilauea

The very low frequency (VLF) electromagnetic method is particularly well adapted to detecting shallow conductors in a more resistive medium. The technique has been successfully applied to the study of various subsurface volcanic processes and structures at Kilauea volcano. These applications are comprehensively reviewed by Zablocki (1978a), and include the detection of the molten cores of freezing lava lakes and the presence of active lava tubes, studies of recent eruptive and non-eruptive fissures, and observations of large fumarolic areas. As far as the monitoring of volcanic activity is concerned, the possibility of locating active lava tubes in flow fields appears to be the most important operational use of the VLF method. Since 1986, periodic VLF surveys undertaken over the Kupaianaha lava field at Kilauea have been successful in detecting and monitoring the concealed lava tube networks (Jackson et al. 1987, 1988), and the results of this study are outlined here. The VLF method uses radio transmitters as electromagnetic sources. Several powerful stations, located in different parts of the world, broadcast radiowaves in the VLF band (15–25 kHz) for long-range communications and radio positioning purposes. At large distances from the source, the radiated electromagnetic field is essentially planar with the magnetic component both horizontal and perpendicular to the azimuth of the station. Thus, if a conductor strikes in a direction close to that of the transmitter, it will be cut by the magnetic component producing both induction currents and a secondary magnetic field.

Paterson & Ronka (1971) have described the principles and the method of operation for ground VLF measurements. The commercially available instruments are very compact, lightweight, hand-held devices, consisting essentially of two orthogonal coils and a radio receiver which can be tuned to the desired station. The most commonly measured quantities are the tilt angle and the ellipticity. The tilt angle measures the departure from the horizontal of the magnetic field resulting from the vector addition of the primary (horizontal) and the secondary (spacially variable in direction) magnetic fields. The ellipticity refers to the fact that the resulting vector from the (alternating) primary and secondary fields describes an ellipse (Grant & West 1965, Smith & Ward 1974) whose shape depends upon the relative amplitude and phase difference of the two fields. The apparent resistivity can also be obtained with some VLF instruments which allow measurement of the electrical and magnetic fields simultaneously; in this case, the apparent resistivity is determined essentially as in the scalar magnetotelluric method (Telford et al. 1990), where an apparent resistivity is derived from the ratio of the amplitude of the horizontal electric field, in one direction only, over that of the magnetic field in the orthogonal horizontal direction (in contrast to the determination of a tensor apparent resistivity where more components of the fields are required.) Because of the relatively high frequencies (about 15–20 kHz) of the VLF method, the depth of investigation is usually restricted to near-surface terrain, between about 10 and 100 m, depending on the resistivity of the medium. Theoretically conductive structures with boundaries striking at right angle to magnetic field can be detected by the method. In practice, however, the azimuthal dependency is less severe and anomalies can still be observed when the strike of the structures is within ±45° of the optimal strike (Zablocki 1978a). Furthermore, this drawback can be avoided if two VLF transmitter stations with complementary bearings can be received.

For the detection of active lava tubes, the resistivity contrast between cooled lava (resitivity generally > 5000 ohm.m) and molten lava (resistivity of a few ohm.m or less) flowing in tubes, makes the VLF method ideally suited. In terms of geoelectrical models, an active lava tube would correspond to a 2-dimensional model of a semi-infinitely long conductor, with a more or less rectangular cross section, encased within a resistive medium, and observed data usually fit quite well with such models. Figure 9.7 (from D. B. Jackson & J. Kauahikaua

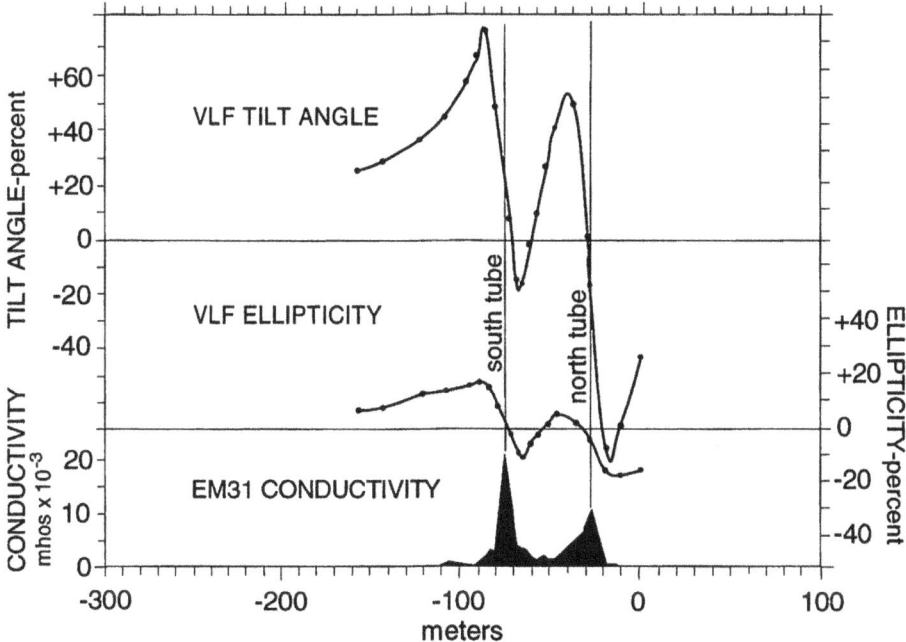

**Figure 9.7** Typical very low frequency (VLF) and shallow electromagnetic resistivity over two active lava tubes at Kilauea. The patterns shown are due to the presence of conductors (flowing lava) in a resistive medium (cooled lava flows). This example clearly demonstrates the capacity of the VLF method to detect concealed lava tubes (D. B. Jackson & J. Kauahikaua, personal communication).

personal communication) illustrates the typical shape of the VLF anomalies over observed lava tubes. Over a linear conductor which has a symetrical cross section but does not have a large lateral extension relative to its depth, the tilt angle and the ellipticity anomalies are antisymmetric, with the zero crossover above the center of the conductor. Over a broader conductor, the minima and maxima are near the edges of the structure, and amplitude and ellipticity are virtually flat over the central area. The profile in Figure 9.7 crosses two concealed active lava tubes at Kilauea and, although the two anomalies are partly coalescent, the typical and expected patterns for semi-infinitely long conductors of moderate lateral extension are clearly observed. The interpretation of these anomalies as being due to the presence of underlying active lava tubes is further confirmed by conductivity measurements (Fig. 9.7) made with different electromagnetic equipment (a two loop Min-Max system, model EM-31), and by visual obervation of one of the tubes through a "skylight" at short distance from the profile.

Figure 9.8a shows the VLF tilt angle along profiles located over a large, actively growing, flow field supplied by multiple active lava tubes in the east rift zone of Kilauea. The pattern of anomalies is complex because several lava tubes have formed close to one other. The lava field had not yet evolved into a stable system with only one or two tubes as is usually the case in more mature lava fields, and the individual anomalies are, therefore, more or less merged. Analysis of the profiles, however, and comparison with theoretical models allows the paths of both the main tubes and some of the secondary ones to be traced (Fig. 9.8b). Jackson et

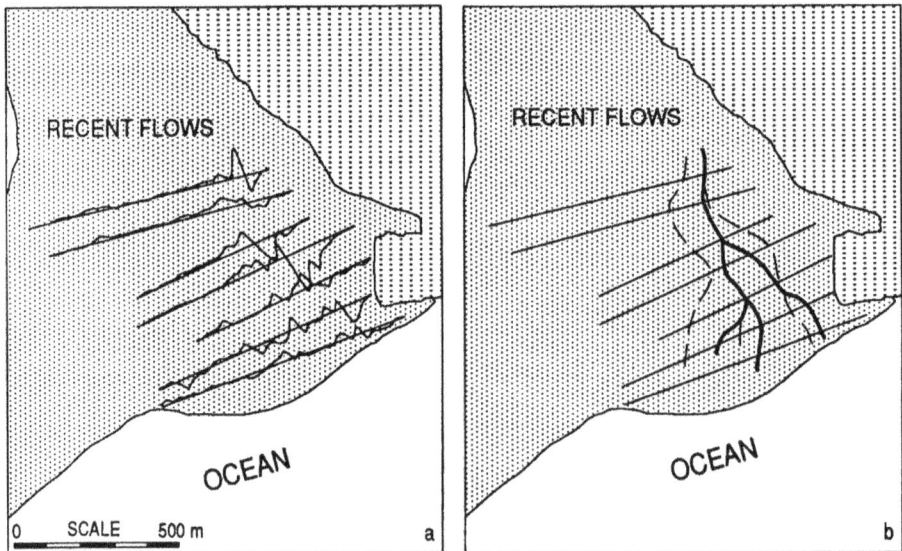

**Figure 9.8** Detection and mapping of active lava tubes in a lava flow field at Kilauea. (a) Very low frequency (VLF) tilt angle along profiles located over a large flow field fed by an eruption on the east rift zone of Kilauea. The pattern of anomalies appears complex because several lava tubes have formed close to one other. The individual anomalies are, therefore, more or less merged. (b) Analysis of the profiles and comparison with theoretical models allows the path of the main tubes and of some secondary ones to be traced (see text for explanation) (from D. B. Jackson & J. Kauahikaua, personal communication).

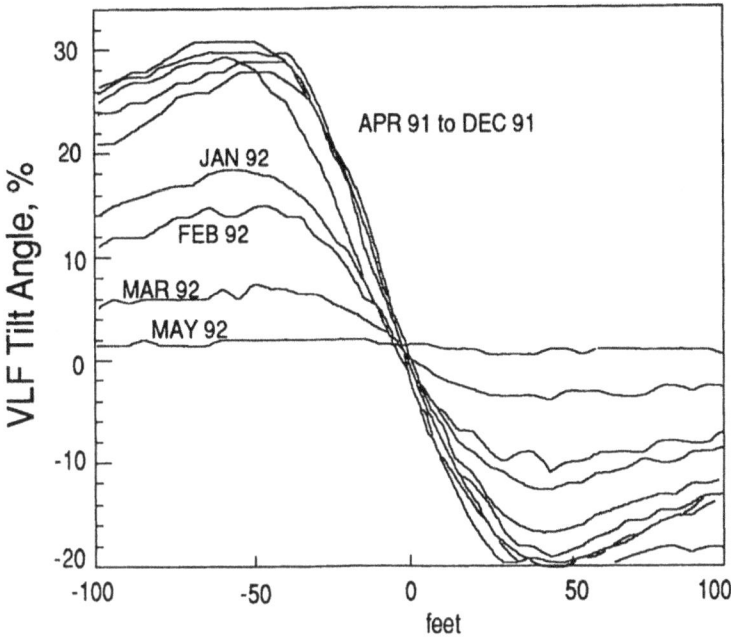

**Figure 9.9** Time evolution of the very low frequency (VLF) anomaly observed over a lava tube at Kilauea, from its period of maximum activity to its extinction (personal communication from J. Kauahikaua). Theoretical models indicate that changes in the cross-sectional area of a tube can be monitored using this method (Jackson et al. 1988; Kauahikaua et al. 1992).

al. (1988) have shown that more quantitative information than simply the location can be determined from the VLF monitoring of lava tubes. Figure 9.9 illustrates the time evolution of the VLF tilt angle over a lava tube, from the time of maximum activity to its extinction. Theoretical models indicate that the VLF anomaly of a tube, the depth of which is larger than its width and thickness, is practically affected only by the cross-sectional area or the conductance of the tube. Thus, if we assume that the resistivity of the lava flowing in the tube remains constant throughout the eruption, the cross-sectional area of the tube can be monitored and, if the flow velocity can be measured (usually by measuring the travel time of a burning wood stick or other object between skylights), the flux of lava can be estimated (Jackson et al. 1988).

## 9.4 Self-potential methods

Naturally occurring static electric-field potentials are present at the Earth's surface. They can have various electrochemical, mechanical or thermal origins (see Corwin & Hoover 1979). Although this method was developed in the nineteenth century, and used mostly for mineral exploration (Sato & Money 1960), it has always remained a marginal method in geophysics (except for well logging), largely due to difficulties in data interpretation. There has, however,

been a renewal of interest in the self-potential method in recent years for geothermal exploration and volcanological studies (e.g. Zohdy et al. 1973, Zablocki 1976, 1978b, Corwin & Hoover
1979, Jackson & Kauahikaua 1987) because prominent, and sometimes very high amplitude,
anomalies have been found over thermal anomalies and/or active volcanic structures.

Two main mechanisms are likely to generate self-potential anomalies on volcanoes (Corwin
& Hoover 1979), electrokinetic and thermoelectric couplings. The electrokinetic coupling
involves the generation of electric potential gradients (called "electrokinetic" or "streaming
potentials") by the flow of a fluid through a porous medium, a process also known as
"electrofiltration". These phenomena have been well described for rocks by, for example,
Ishido & Mizutani (1981) and Morgan et al. (1989). Briefly, potential gradients result from
a separation of electrical charges in a liquid. Ions of one sign (generally positive) are preferentially attracted and absorbed at the liquid-solid interface by electrostatic and chemical
absorption forces. When the liquid is in motion, under a pressure gradient, ions from the other
sign are preferentially transported, thus establishing a current. Simple models calculated by
Corwin & Hoover (1979) and field observations by Zablocki (1978b) and Jackson &
Kauahikaua (1987), strongly suggest that streaming potentials are the dominant source of high
amplitude self-potential anomalies at active volcanoes. Thermoelectric coupling involves the
generation of a voltage gradient when a temperature gradient is applied across a volume of
rock. This phenomena derives from a Soret effect in which a differential diffusion of ions in
the pore fluid is imposed by the thermal gradient. In volcanoes, thermoelectric and
electrokinetic phenomena can coexist, for example in hot convective systems surrounding
magma bodies.

Despite the publication of some important works during the past decade (e.g. Fitterman
1978, 1979a & b, Corwin et al. 1981, Fitterman & Corwin 1982, Ishido et al. 1983, Sill 1983,
Fitterman 1983, 1984, Fournier 1989), the quantitative interpretation of self-potential data
remains poorly developed. This is due to insufficient general understanding of the source
mechanisms and to difficulties in identifying the relative contributions of such mechanisms.
This weakness in the method is counterbalanced, however, by some remarkable qualitative
results involving the the detection of structures and thermal anomalies at active volcanoes (e.g.
Zablocki 1976, Jackson & Kauahikaua 1987).

### 9.4.1 Possible causes of temporal self-potential variations at volcanoes

Assuming that the electrokinetic and thermoelectric couplings are the main sources of self-
potential anomalies (Corwin & Hoover 1979), possible causes of self-potential signal variations can be anticipated. For streaming potentials, a coupling coefficient, $C_E$, is defined as
the ratio of the potential difference, $\Delta V$, over the pressure difference, $\Delta P$, along the flow
path of the pore liquid. $C_E$ is estimated by the Helmholtz–Smoluchowski equation:

$$C_E = - (\rho \varepsilon \zeta)/\eta$$

where $\rho$, $\varepsilon$, and $\zeta$ are the electrical resistivity, the dielectric constant, and the viscosity of the
pore fluid, respectively, and $\eta$ (the zeta potential) is the potential difference across the layer

of ions bounded to the solid–liquid interface (see Morgan et al. 1989). Within volcanoes, several of the quantities which determine $\Delta V$ may change in response to internal activity. For example, the intrusion of hot magma can generate or enhance the flow of fluids ($\Delta P$) in convective cells or hydrothermal systems. The resistivity of the fluids can be lowered by the increase in temperature and/or by an increase of salinity (assimilation of elements from volcanic gases or dissolution of rock minerals due to the temperature increase). The zeta potential can also vary in response to chemical and thermal variations (Ishido & Mizutani 1981, Morgan et al. 1989). In addition, the contribution of thermoelectric coupling will be increased. As might be expected, the reverse effects may be associated with a cooling magma body.

Stress changes in volcanoes can also modify or cause fluid motion and, therefore, generate changes in the streaming potential pattern, an effect which has received much attention in earthquake prediction studies (Mizutani et al. 1976, Fitterman 1979b). Additionally, experiments have also been undertaken to monitor subsurface flows in civil engineering applications by means of self-potential measurements (see Merkler et al. 1989). External phenomena, such as rain and atmospheric temperature variations can also have an effect on self-potential. For example, large self-potential variations have been observed at the permanent network installed at Mount Etna (see later in this chapter) at times of heavy rainfall. Following heavy rainfall on 12 July 1983, a temperature decrease of 7°C (77 to 70°C) accompanied by an increase of self-potential of about 100 mV was observed at a station installed in a fumarole area (whereas the self-potential reference electrode was in a thermally quiet zone). Self-potential levels returned to normal within a few days of the rainstorm. As with the direct-current resistivity and electromagnetic methods, self-potential monitoring can be undertaken by means of the periodic surveying of established stations or by using a permanently recording network. The equipment and procedures differ between these two approaches, both of which are discussed below with reference to studies at the Piton de la Fournaise and Mount Etna volcanoes.

### 9.4.2 Monitoring of a self-potential profile at Piton de la Fournaise volcano, Réunion Island

The equipment required for a self-potential survey is very simple, consisting of a pair of non-polarizing electrodes, a millivoltmeter, and insulated single-conductor wire. The non-polarizing electrodes used are generally the $Cu—CuSO_4$ type, while the millivoltmeter should have an input impedance large enough so that only negligible current will be drawn from the ground. When making measurements on volcanic terrain, such as dry, unweathered, lava flows, contact resistances between the ground and the electrodes may reach megaohms, it is essential, therefore, to use a very high input impedance voltmeter (typically $10^{10}$ ohm or more). Filters are also recommended to reject signals from long-period telluric currents, when long (several hundeds of metres) self-potential lines are used, and industrial pickup from powerlines. Readings are commonly taken with an accuracy of 1 mV, which is suitable for most purposes. During the initial survey, as well as for reoccupations for monitoring purposes, the data must be tied to a reference potential, since we can only measure potential differences or gradients. It is always desirable, although not always achievable, to have a reference potential located out-side the zone where spatial or temporal anomalies may exist. The installation of a network that will be monitored by periodic surveys requires stations in the field to be marked (with paint, nails or any other convenient means) so that the electrodes can be set up at precisely the same place at every reoccupation.

Only scarce data from the self-potential monitoring of volcanoes have been published so far, and these results are briefly discussed here. At Kilauea, Zablocki (1976), for example, twice remeasured a self-potential profile across an eruptive fissure, observing a positive anomaly which increased by more than 150% in amplitude after the eruption and decreased afterwards. Jackson (1988) gives an exhaustive report of the results from various types of geoelectrical monitoring (electromagnetic, VLF, and self-potential) undertaken during intru-sive activity in the east rift zone of Kilauea between September 1982 and January 1983. The self-potential surveys involved repeated measurements along profiles and permanent electrode arrays, which revealed self-potential changes accompanying some of the intrusions and a 68 mV transient excursion, lasting less than 1 h, that was observed about 1 day before an intrusion near the self-potential array. At Piton de la Fournaise volcano (Réunion Island), following a self-potential survey of the central zone in 1981 (Bonneville et al. 1984, Lénat 1987, Malengreau et al. 1994), a 3.8 km long profile was established around the summit area and reoccupied periodically since. The spacing between the stations varies from 10 to 50 m, with an average of 25 m. Some of the stations are marked in the field and the intermediate stations are located by measuring distances from the marked stations. The reference potential has been arbitrarily taken at a given station of the profile and, therefore, the variations observed be-tween two sets of measurements are relative to that reference. It would have been preferable to have a reference farther from the summit area, which is the most active zone, but, for logistical reasons, this proved impracticable. Figure 9.10 shows the differences between sur-veys undertaken at different times along the profile (remember that the profile is a closed loop around the summit craters). Figure 9.10a shows the initial profile in October 1981, while Figure 9.10b represents the observed changes between October 1981 and April 1983 (this later survey involved only a partial reoccupation of the profile). During this period, no vol-canic activity occurred and mostly low-amplitude variations were observed. In contrast, very large variations were observed between the two surveys that bracket the period July 1988 to April 1986 (Fig. 9.10c). This period was marked by a higher than normal frequency of vol-canic events (Delorme et al. 1989, Lénat et al. 1989a), including three eruptions and two intrusions in the summit area, and the collapse of a pit in the main summit crater (Dolomieu). A large (about +750 mV), general increase of the self-potential in the east-southeast portion of the profile is superimposed on shorter wavelength variations. In this area, the self-potential changes amount to about 50% of the amplitude of the (relative) negative anomaly (Fig. 9.10a). Subsequent datasets (Malengreau et al. 1994) show a trend to recovery of the initial negative amplitude. Since there has been no intermediate survey during this period, it is not possible to know if the self-potential variations are correlated with one of the events or if they repre-sent the cumulative effect of several events. Neither is it possible to determine how fast the self-potential changes occurred. Given the constraints imposed by the limited nature of the dataset, the simplest interpretation of these changes is in terms of a shallow hydrothermal

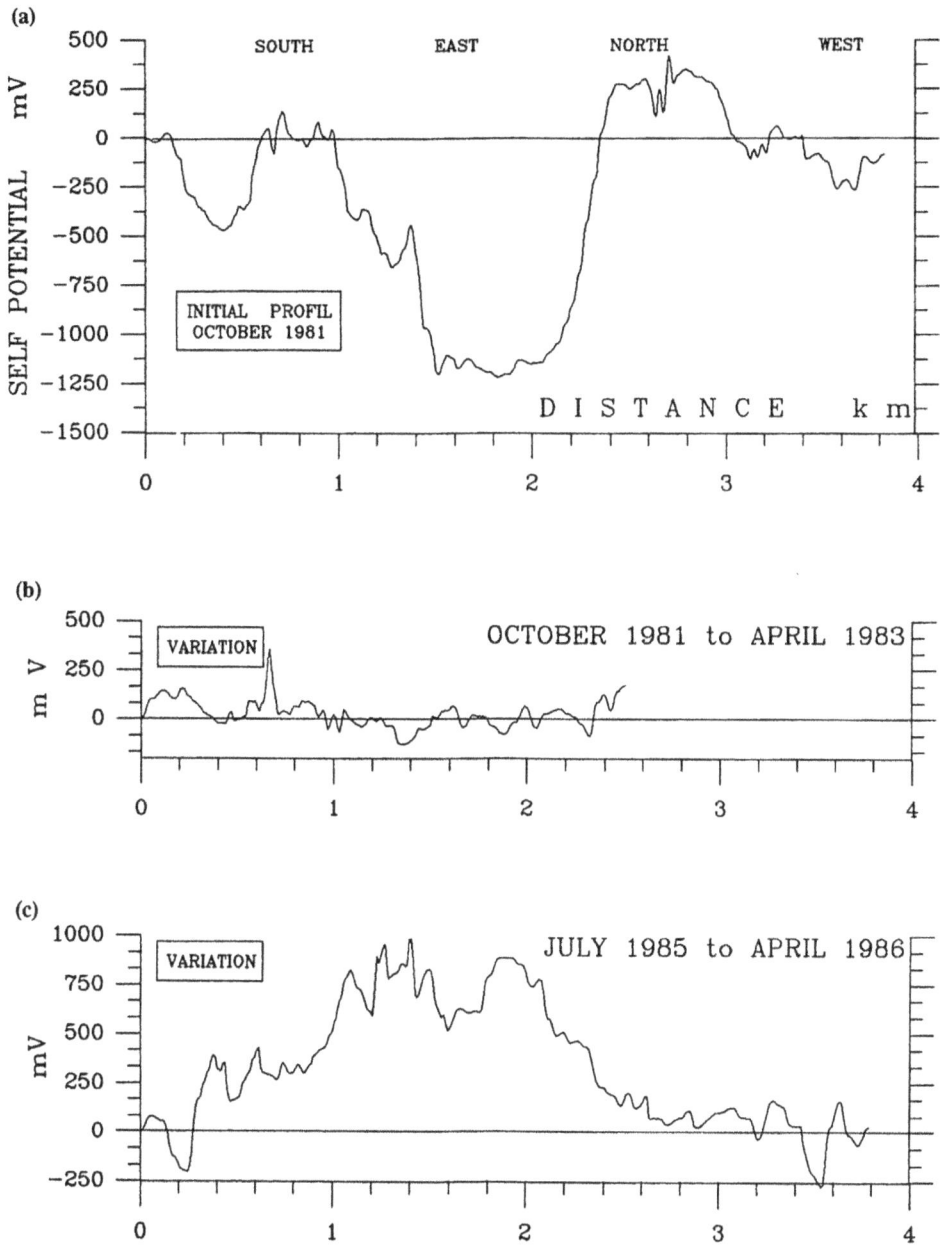

**Figure 9.10** Self-potential monitoring along a profile at the summit of Piton de la Fournaise. A 3.8 km long profile was established around the summit craters (see inset in Fig. 9.12). Self-potential was first measured along this profile in October 1981 (a), and measurements have been repeated periodically since. (b) The self-potential variation between October 1981 and April 1983, a period with no volcanic activity when only low-amplitude self-potential changes were observed. In contrast, large self-potential variations were recorded between July 1985 and April 1986 (c), a period marked by several volcanic events. For a more detailed discussion of these changes, see text (from Malengreau et al. 1993).

system which was disturbed by the emplacement of hot intrusions, and possibly by the opening and closing of cracks during one, or several, of the deformation episodes associated with the volcanic events (Delorme et al. 1989, Lénat et al. 1989a & b). It should be stressed here that any self-potential map of an active volcanic zone is only valid for a limited period of time, because large self-potential changes can occur in response to a number of events, such as the high-level emplacement of magma, resulting rapidly in a very different picture.

The Piton de la Fournaise study illustrates the potential the self-potential monitoring method and underlines some important points relating to field procedures. The self-potential anomalies, which are present over some active volcanic areas, may change with time in response to volcanic activity. Although the origin of the changes is not clearly identified, they probably reflect disturbances in fluid temperature and circulation as a consequence of the intrusion fresh magma or due to the cooling of established magma bodies. Furthermore, such changes may be large compared with the amplitude of the initial anomalies. Self-potential monitoring by repeated surveys has the advantage of clearly illustrating the spatial distribution of observed variations, and allows correlations to be made with events (such as dyke emplacement) at specific locations. When establishing a self-potential network, however, two important precautions should be taken to maximise the quality of the accumulated data. The reference potential selected should be located outside of the zone where the changes are expected to occur, in order to record the sign of the changes instead of (as in Fig. 9.10) the relative variations within the anomaly area. Secondly, the periodicity of the surveys should be adapted to the frequency of the volcanic events or the self-potential changes, in order to allow correlation of such changes with events and to effectively study the development of the changes over time. If these requirements are fulfilled, some useful insights into intravolcanic phenomena may be achieved using the self-potential method.

### 9.4.3 Continuous self-potential monitoring networks at Mount Etna and Piton de la Fournaise

Continuous monitoring of self-potential provides access to the timing and kinetics of self-potential variations. Two experiences of continuous monitoring are described briefly here to illustrate the type of information that can be obtained and to reveal some of the techniques used in such a monitoring strategy. A pioneering experiment in continuous self-potential monitoring was undertaken by Zablocki (1980) at Kilauea, but this suffered from two technical problems which hindered operation of the network, long-term drift of the potential electrodes ($Cu-CuSO_4$ and $Ag-AgCl$ non-polarizing electrodes were used) and lightning strikes on the cables.

In more recent experiments undertaken at Mount Etna and Piton de la Fournaise, we have introduced the use of $Pb-PbCl_2$ electrodes on the basis of work published by Petiau & Dupis (1980). These authors have tested various types of non-polarizing electrodes for low-frequency telluric observations. They concluded that $Pb-PbCl_2$ electrodes were far better than any other type, having an excellent long-term stability (in the range of a 1.3 mV drift per year, compared with $Ag-AgCl$ electrodes which are characterized by drift an order of magnitude larger), and also a small temperature coefficient ($-40~\mu v~°C^{-1}$) which is important in volcanological work where the ground temperature may change with time. The problem of lightning strikes is inherent in the self-potential monitoring of active volcanoes, involving, as it does, the laying of long electrical lines on the ground in areas of elevated relief which are particularly prone to lightning. It is possible to minimize the damage caused by lightning strikes by using protection devices similar to the ones used by telephone and electric companies to protect their lines. These devices are designed to protect equipment from surges and to bleed into the ground the very large currents induced by lightning. No protection system is perfect, however, and several breakdowns in the Etna and Piton de la Fournaise networks were the result of lightning strikes.

The self-potential network established at Mount Etna in September 1981 was included in a continuing project of ground temperature monitoring (Archambault et al. 1981, Archambault 1982). Three non-polarizing electrodes were laid at a depth of 1.2 m, their location selected on the base of preliminary self-potential mapping (Lénat 1987). One electrode, taken as the self-potential reference, was established in an area with no prominent self-potential anomaly (the Torre del Filosofo) (Fig. 9.11), another was installed on the southern edge of the central crater, and the third at Vulcarolo fumarole. The latter two sites were located in areas characterised by frequent volcanic events (explosions, intrusions, and eruptions) and typified by large self-potential anomalies. The summit electrode was later (June 1982) moved to the southeastern foot of the central cone. A weather station was also installed (providing data on air temperature, relative humidity, wind speed and direction, rainfall and atmospheric pressure) to allow consideration of meteorological parameters in data interpretation. The data were transmitted to France using the Argos satellite-based system (this allows the transmission of numerical data from beacons to data collection centres through National Oceanic & Atmospheric and Administration (NOAA) satellites, although the length of a message is limited to 32 byte). Signals from the different sensors were acquired about every 2 h and stored in a data-acquisition system. Data stored in the system memory were read out by an interface which generated a serial message every 210 s prior to message transmission to NOAA satellites. Data from between 6 and 10 acquisition cycles were obtained every day, and collected in France from the Argos service. Typical records of self-potential variations recorded between June 1982 and April 1983, a period including the beginning of the 1983 eruption, are illustrated and discussed below.

Figure 9.11 shows the self-potential curves recorded between September 1982 and April 1983. Between June 1982, when the network was repaired after a failure during the winter, and September, the self-potential variations were weak, the only noticeable changes being associated with the occurrence of heavy rainfalls (see earlier in this chapter). Self-potential events were recorded in November and January. In November the self-potential increased by 250 mV at station 3 and by 100 mV at station 1, while in January 1983 a larger increase (about 400 mV) occurred at both stations. At station 3, closer to the summit region, the latter increase was particularly sharp, occurring over a few days, while at station 1 the self-potential increase took place progressively over a period of about 2 weeks. This suggests strongly that the observed changes were not caused by variations at the reference electrode. Between late January and the eruption the self-potential stayed at a high level at station 1, with a tendency to decrease from mid-March followed by a sharp fall at the outbreak of the eruption on 28 March. At station 3, the self-potential steadily decreased between late January and mid-March when an

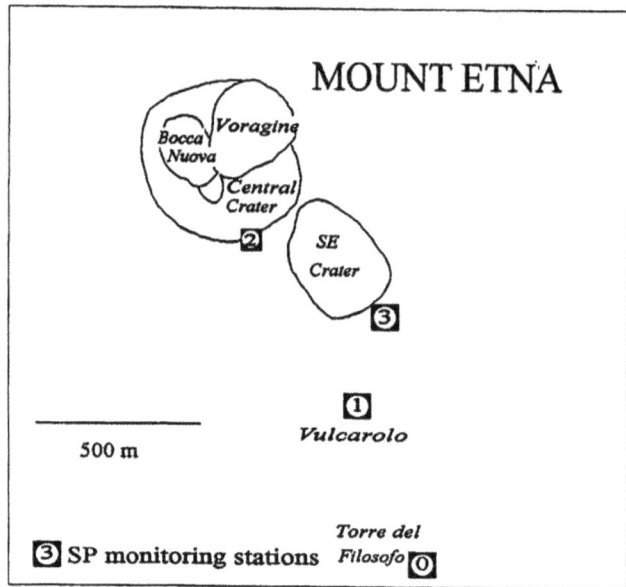

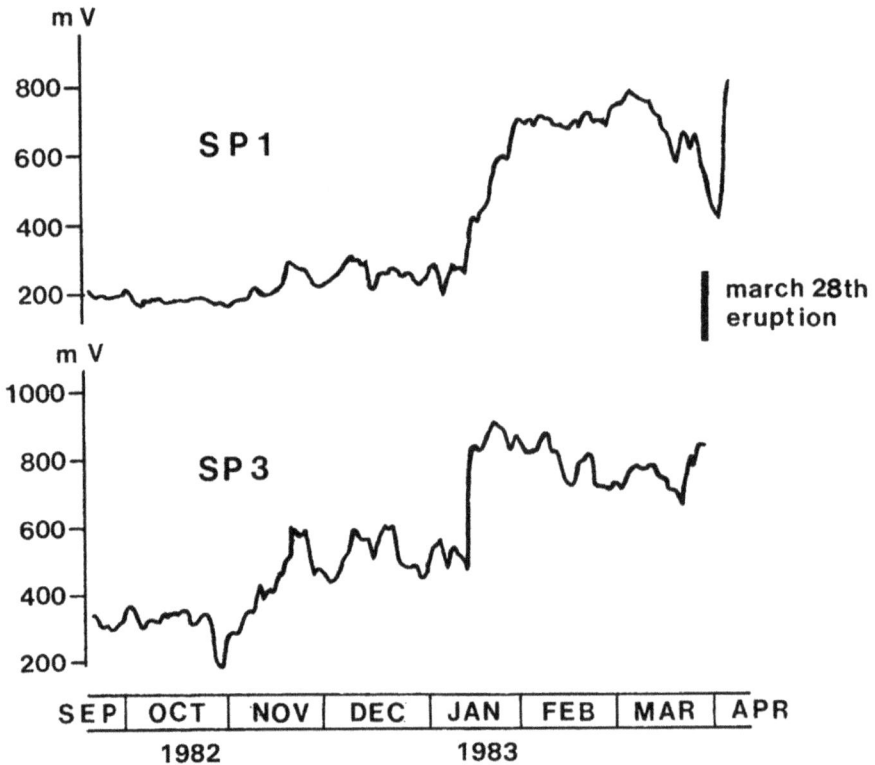

**Figure 9.11** Long-term self-potential monitoring at Mount Etna before the March 1983 eruption. The inset at the top shows the location of the self-potential stations to the south of the summit region. Two episodes of self-potential increase, which may be associated with magma intrusion, were recorded in the November and January preceeding the March 1983 eruption (records of stations 1 and 2 are shown, station 3 was not operating; station 0 is the reference). Ground tilt was also recorded during the January self-potential event (Grimaldi et al. 1984).

increase was recorded. Unfortunately, the station-3 cable was broken about 24h before the eruption, probably because of large ground movements associated with the intrusion of magma toward the surface, cutting of the flow of data. During the early days of April a sharp increase of 400 mV was recorded at station 1 before it also suffered the fate of a broken cable. The observed long-term self-potential variations are not recognized in the temperature records which show a large seasonal component, and cannot explain the changes over the monitoring period. The mid-January self-potential increase might be correlated with a significant ground-tilt episode reported by Grimaldi et al. (1984) who, at a permanent tilt station located about 4 km south of the summit, recorded continuous inflation of the upper part of the volcano from mid-January to about 12h before the eruption. This suggests a pressure increase in the feeding system during the 2½ months preceding the eruption and supports the hypothesis that the very large self-potential increase could originate from an increase of fluids and/or heat transferred in the edifice as magmatic pressures accumulated. It is regrettable that, due to equipment failures, this monitoring programme failed to record the complete year of data that would be necessary, at the very least, to assess the amplitude of any seasonal variations. The amplitude and duration of the recorded self-potential changes, however, and their correlation with the volcanic events, are large enough to suggest that they are associated with some kind of magma-related process. A self-potential time variation of 400 mV is of the same order of magnitude as the larger self-potential anomalies previously mapped, and it is highly likely, therefore that changes over the period of observation were indeed due to the formation of new self-potential anomalies.

A network for monitoring self-potential (five electrodes) changes was also installed at the summit of Piton de la Fournaise volcano in 1983 (Bonneville et al. 1984), together with sensors for monitoring air and ground temperature. The electrodes were linked to a central data acquisition system in the field, the voltages were transformed from analogue into digital form, multiplexed, transmitted by radio to the local volcanological observatory, and recorded on magnetic tapes. A set of data was acquired every hour during the operation of the network, from November 1983 to October 1984 and again from October 1987 to January 1988. Breaks in the records occurred due both to failures of parts of the field acquisition and transmission systems and to damage caused by lightning strikes. Nevertheless, some interesting features of self-potential changes were observed. The reference electrode was located about 1 km from the summit, beyond the main summit anomalies revealed by a survey undertaken in 1981 (Bonneville et al. 1984, Lénat 1987, Malengreau et al. 1994). A daily variation of 20–80 mV was observed at all the stations, which had a phase shift of about 12h relative to changes in ground temperature. A less well-defined seasonal variation, of about 150–200 mV, was also observed. Unlike at Mount Etna, no clear signals were found associated with periods of heavy rainfall. One explanation for this different behaviour may lie in the fact that at Mount Etna the measuring electrode but not the reference electrode was located on a fumarole, whereas at Piton de la Fournaise all the electrodes were established in similar terrain. Figure 9.12 shows the record of a station during the 4 December 1983 to 18 February 1984 eruptive crisis (Lénat et al. 1989b). Before the eruption, the pattern of self-potential was regular, showing only the daily variation. After the 4 December outbreak, we observed a decrease in self-potential and the data became less regular. During the inflation period that preceded the sec-

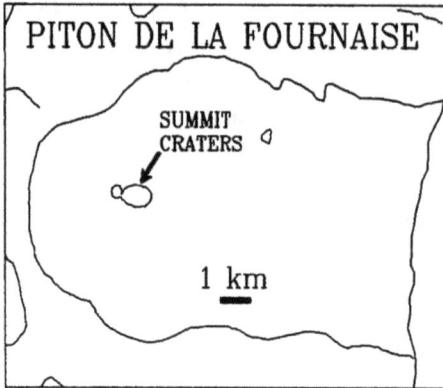

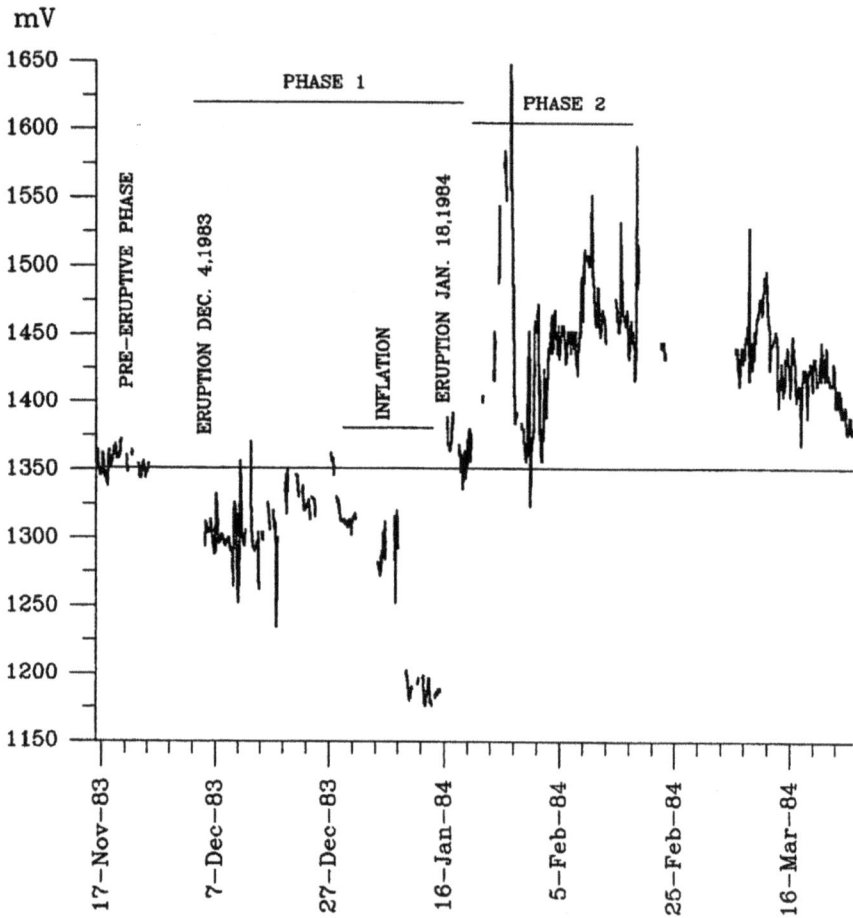

**Figure 9.12** Self-potential variations during the December 1983 to February 1984 crisis at Piton de la Fournaise. The permanent network was installed in the southwest area of the summit (see inset) For a more detailed explanation, see text.

ond outbreak of 18 January, a further decrease in the self-potential occurred. During the first part of the second eruptive phase, the self-potential dramatically increased and became jerky. The decreases and the jerks of the self-potential have been inferred to represent pertubations of the hydrothermal system due to the intrusion of magma at shallow depths (Malengreau et al. 1994). However, we cannot entirely rule out the possibility that the jerky signals were due to instrumentation problems, although the fact that they were observed only during the eruptive crisis, and not before and after, suggest that these were true self-potential signals. Although the general deformation pattern of the summit during these periods was one of inflation, some parts of the fissures which constitute the dense fracture network of the summit region (Bachèlery et al. 1983), underwent contraction during the first phase (Lénat et al. 1989b) of the eruption, a phenomenon which has the potential to temporarily disturb the circulation pattern of fluids within the edifice by closing established paths and creating an unstable regime of circulation, until the fissures become stabilized in a new equilibrium. The intrusions associated with the second eruptive phase may have opened paths to fluid circulation, thus allowing an increase in the circulation regime which might also have been enhanced by heat generated by the intrusions of fresh magma.

## 9.5 Conclusion

Geoelectrical methods have not as yet been used as extensively for volcano monitoring as the more established seismic and deformation techniques. As illustrated by the case studies presented here, however, surveillance programmes designed to reveal the subsurface resistivity and self-potential can provide unique and invaluable insights into the workings of internal volcanic processes, and may provide a useful complementary tool to forecasting eruptive events. The promising results achieved in the experimental geoelectrical studies undertaken to date, are likely to lead to their use on increasing numbers of active volcanoes. The further utilization of these techniques will also be engendered by technical developments which should result in more rugged and versatile equipment, not only for taking measurements, but also for improved data recording and transmission in the hostile volcanic environment. It is to be hoped that such improvements in equipment and field operations will also be accompanied by the development of improved models, which will be essential if the resistivity and self-potential modifications occurring in response to volcanic events, are to be better understood.

## Acknowledgements

The author thanks Dallas Jackson for sharing his experience in geoelectrical methods during field works in Hawaii and Réunion. The author is very grateful to Dallas Jackson and Jim Kauahikaua for providing some of their unpublished data, and for their constructive remarks on an earlier version of the manuscript.

# References

Anderson, L.A. 1987. Geoelectric character of Kilauea Iki lava lake crust. In *United States Geological Survey Professional Paper 1350, Volcanism in Hawaii*, R. W. Decker, T. L. Wright, P. H. Stauffer (eds), 1345-55. Washington DC: United States Government Printing Office.

Aramaki, S. (ed) 1988. *The 1986-1987 eruption of Izu-Oshima volcano*. Tokyo: Earthquake Research Institute, University of Tokyo.

Archambault, C. 1982. Remote monitoring of Etna. *ARGOS Newsletter* **15**, 1-5.

Archambault, C., J. Stoschek, G. Scarpinati, J-C. Tanguy 1981. *Bulletin PIRPSEV 96, Résultats des mesures de températures du sol effectuées dans la zone sud de l'Etna du 20 janvier 1980 au 4 avril 1981*, 27. France: CNRS-INSU.

Archie, G. E. 1942. The electrical resistivity log as an aid in determining some reservoir characteristics. *AIME Transactions* **146**, 54-62.

Bachèlery, P., L. Chevallier, J. P. Gratier 1983. Caractères structuraux des éruptions historiques du Piton de la Fournaise. *Comptes Rendus de l'Académie des Sciences, Paris, II*, **296**, 1345-50.

Bartel, L. C., H. C. Hardee, R. D. Jacobson 1983. An electrical resistivity measurement in molten basalt during the 1983 Kilauea eruption. *Bulletin Volcanologique*, **46**, 271-6.

Bonneville, A., F. X. Lalanne, J-F. Lénat 1984. *Bulletin PIRPSEV 95, Expérimentation des méthodes de polarisation spontanée, de surveillance thermique et de l'émanation du radon au Piton de la Fournaise. Présentation du réseau de la zone sommitale*, 95. France: CNRS-INSU.

Bousquet, J. C., G. Lanzafame, L. Villari 1984. Les ruptures de surface liées à l'éruption du 28 Mars 1983 d l'Etna (Sicile). *Bulletin Volcanologique* **47**, 895-907.

Corwin, R. F. & D.B. Hoover 1979. The self-potential method in geothermal exploration. *Geophysics* **44**, 226-45.

Corwin, R. F., G. T. DeMoully, R. S. Harding Jr, H. F. Morrison 1981. Interpretation of self-potential survey results from the East Mesa geothermal field, California. *Journal of Geophysical Research* **86**, 1841-8.

Decker, R.W. 1987. Dynamics of Hawaiian volcanoes: an overview. In *United States Geological Survey Professional Paper 1350, Volcanism in Hawaii*, R. W. Decker, T. L. Wright, P. H. Stauffer (eds), 997-1018. Washington DC: United States Government Printing Office.

Delorme, H., P. Bachèlery, P. A. Blum, J-L. Cheminée, J-F. Delarue, J-C. Delmond, A. Hirn, J-C. Lépine, P. Vincent, J. Zlotnicki 1989. March 1986 episodes at Piton de la Fournaise volcano (Réunion Island). *Journal of Volcanology and Geothermal Research* **36** 199-208.

Fitterman, D. V. 1978. Electrokinetic and magnetic anomalies associated with dilatant regions in a layered earth. *Journal of Geophysical Research* **83**, 5923-8.

Fitterman, D. V. 1979a. Calculation of self-potential anomalies near vertical contacts. *Geophysics* **44** 195-205.

Fitterman, D. V. 1979b. Theory of electrokinetic-magnetic anomalies in a faulted half-space. *Journal of Geophysical Research* **84**, 6031-40.

Fitterman, D. V. 1983. Modeling of self-potential anomalies near vertical dikes. *Geophysics* **48**, 171-80.

Fitterman, D. V. 1984. Thermoelectrical self-potential anomaliies and their relationship to the solid angle subtended by the source region. *Geophysics* **49**, 165-70.

Fitterman, D. V. & R. F. Corwin 1982. Inversion of self-potential data from the Cerro Prieto geothermal field, Mexico. *Geophysics* **47**, 938-45.

Flovenz, O. G., L. S. Georgsson, K. Arnason 1985. Resistivity structure of the upper crust in Iceland. *Journal of Geophysical Research*, **90**, 10136-50.

Fournier, C. 1989. Spontaneous potentials and resistivity surveys applied to hydrogeology in a volcanic area: case history of the Chaîne des Puys (Puy-de-Dôme, France). *Geophysical Prospecting* **37**, 647-68.

Frischknecht, F. C. 1967. Fields about an oscillating magnetic dipole over a two-layer earth and application to ground and airborne electromagnetic surveys. *Quarterly Journal of the Colorado School of Mines*, **72**, 326.

Grant, F. S. & G. F. West 1965. *Interpretation theory in applied geophysics*. New York: McGraw Hill.

Halbwachs, M. 1983. Electrical and electromagnetic methods. In *Forecasting volcanic events*, H. Tazieff & J-C. Sabroux (eds), 507-27. Amsterdam: Elsevier.

Ishido, T. & H. Mizutani 1981. Experimental and theoretical basis of electrokinetic phenomena in rock-wa-

272

ter systems and its applications to geophysics. *Journal of Geophysical Research* **86**, 1763–75.

Ishido, T., H. Mizutani, K. Bada 1983. Streaming potential observations, using geothermal wells and in situ electrokinetic coupling coefficients under high temperature. *Tectonophysics* **91**, 89–104.

Jackson, D. B. 1988. Geoelectric observations (including the September 1982 summit eruption). In *United States Geological Survey Professional Paper 1463, The Pu'u' 'O'o eruption of Kilauea volcano, Hawaii: episodes 1 through 20, January 1983 through June 8 1984*, E. W. Wolf (ed.), 237–51. Washington DC: United States Government Printing Office.

Jackson, D. B. & J. Kauahikaua 1987. Regional self-potential anomalies at Kilauea Volcano. In *United States Geological Survey Professional Paper 1350, Volcanism in Hawaii*, R. W. Decker, T. L. Wright, P. H. Stauffer (eds), 947–59. Washington DC: United States Government Printing Office.

Jackson, D. B., J. Kauahikaua, C. J. Zablocki 1985. Resistivity monitoring of an active volcano using the controlled-source electromagnetic technique : Kilauea volcano, Hawaii. *Journal of Geophysical Research* **90**, 12545–55.

Jackson, D. B., M. K. Hort, K. Hon, J. Kauahikaua 1987. Detection and mapping of active lava tubes using the VLF induction techniques, Kilauea volcano, Hawaii. *EOS* **68**, 1543.

Jackson, D. B., J. Kauahikaua, K. Hon, C. Heliker 1988. Rate and variation of magma supply to the active lava lake on the middle east rift zone of Kilauea volcano, Hawaii. *Geological Society of America Annual Meeting*, Denver, Colorado. Abstracts with program 20, 397.

Kauahikaua, J. 1982. The subsurface resistivity structure of Kilauea volcano, Hawaii. PhD dissertation, University of Hawaii, Honolulu, Hawaii.

Kauahikaua, J., M. Mangan, C. Heliker, T. Mattox 1992. The death of the Kupainaha vent, Kilauea volcano, Hawaii. *EOS* **73**, 629.

Kauahikaua, J., D. B. Jackson, C. J. Zablocki 1986. Resistivity structure to a depth of 5 km beneath Kilauea volcano, Hawaii from large-loop-source electromagnetic measurements (0.04–8 Hz). *Journal of Geophysical Research* **91**, 8267–83.

Keller, G. V. & F. C. Frischknecht 1966. *Electrical method in geophysical prospecting*. New York: Pergamon.

Klein, F.W., R. Y. Koyanagi, J. S. Nakata, W. I. Tanigawa 1987. The seismicity of Kilauea's magma system. In *United States Geological Survey Professional Paper 1350, Volcanism in Hawaii*, R. W. Decker, T. L. Wright, P. H. Stauffer (eds), 1019–1185. Washington DC: United States Government Printing Office.

Koefoed, O. 1979. *Geosounding principles, vol. 1: Resistivity sounding measurements*. Amsterdam: Elsevier.

Lénat, J-F. 1987. Structure et Dynamique internes d'un volcan basaltique intraplaque océanique: le Piton de la Fournaise (Ile de la Réunion). Doctorat d'Etat thesis, University of Clermont II, France.

Lénat, J-F., P. Bachèlery, A. Bonneville, A. Hirn 1989a. The beginning of the 1985–87 eruptive cycle at Piton de la Fournaise (La Réunion), new insights in the magmatic and volcano-tectonic systems. *Journal of Volcanology and Geothermal Research* **36**, 209–32.

Lénat, J-F., P. Bachèlery, A. Bonneville, P. Tarits, J-L. Cheminée, H. Delorme 1989b. The December 4, 1983 to February 18 1984 éruption of Piton de la Fournaise (La Réunion, Indian Ocean): description and interpretation. *Journal of Volcanology and Geothermal Research* **36**, 87–112.

Malengreau, B., J-F. Lénat, A. Bonneville 1994. Polarisation spontanée au Piton de la Fournaise: un outil potentiel de cartographie et de surveillance des structures volcaniques actives. *Bulletin de la Société Géologique de France* **165**, 221–32.

Merkler, G-P., H. Militzer, H. Hötzl, H. Armbruster, J. Brauns (eds) 1989. *Lectures notes in Earth sciences, vol. 27: Detection of subsurface flow phenomena*. Berlin: Springer.

Mizutani, H., T. Ishido, T. Yokokura, S. Ohnishi 1976. Electrokinetic phenomena associated with earthquakes. *Geophysical Research Letters* **3**, 365–8.

Morgan, F. D., E. R. Williams, T. R. Madden 1989. Streaming potential properties of werterly granite with applications. *Journal of Geophysical Research* **94**, 12449–61.

Murase, T. & A. R. McBirney 1973. Properties of some common igneous rocks and their melts at high temperatures. *Geological Society of America Bulletin* **84**, 3563–92.

Olhoeft, G. R. 1977. *United States Geological Survey Open File Report, Electrical properties of water-saturated basalt – preliminary results to 506K (233°C)*, 77–785. Washington DC: United States Government Printing Office.

Paterson, N. R. & V. Ronka 1971. Five years of surveying with the very low frequency electromagnetic method.

*Geoexploration* **9**, 7–26.

Peterson, D. & R. B. Moore 1987. Geologic history and evolution of geologic concepts, Island of Hawaii. In *United States Geological Survey Professional Paper 1350, Volcanism in Hawaii,* R. W. Decker, T. L. Wright, P. H. Stauffer (eds), 158–89. Washington DC: United States Government Printing Office.

Petiau, G. & A. Dupis 1980. Noise, temperature coeffecient, and long time stability of electrodes for telluric observations. *Geophysical Prospecting* **28**, 792–804

Quist, A. S. & M. H. Marshall 1968. Electrical conductances of aqueous sodium chloride solutions from O to 800°C and pressures to 400 bars. *Journal of Physical Chemistry* **72**, 684–703.

Rai, M. P. & M. H. Manghnani 1977. Electrical conductivity of basalts to 1550°C. Chapman conference on partial melting in the earth's upper mantle proceedings. *Oregon Department of Geology and Mineralogy Bulletin* **96**, 219–32

Rai, M. P. & M. H. Manghnani 1981. The effect of saturant salinity and pressure on the electrical resistivity of hawaiian basalts. *Geophysical Journal of the Royal Astronomical Society* **65**, 219–32.

Ryan, M.P. 1987. Elasticity and contractancy of Hawaiian olivine tholeiite and its role in the stability and structural evolution of subcaldera magma reservoirs and rift systems. In *United States Geological Survey Professional Paper 1350, Volcanism in Hawaii,* R. W. Decker, T. L. Wright, P. H. Stauffer (eds), 1395–1447. Washington DC: United States Government Printing Office.

Sato, M. & H. M. Money 1960. The electrochemical Mechanism of sulfide Self Potentials. *Geophysics* **25**, 226–49.

Sill, W. R. 1983. Self-potential modeling from primary flows. *Geophysics* **48**, 76–86.

Smith, B. D. & S. H. Ward 1974. On the computation of polarization ellipse parameters. *Geophysics* **39**, 867–9.

Stearns, H. T. 1942. Hydrology of lava-rock terranes. In *Hydrology,* O. E. Meinzer (ed.), 678–703. New York: McGraw-Hill.

Telford, W. M., L. P. Geldart, R. E. Sheriff 1990. *Applied geophysics,* 2nd edn. Cambridge: Cambridge University Press.

Ward, S. H. 1980. Electrical, electromagnetic, and magnetotelluric methods. *Geophysics* **45**, 1659–66.

Yukutake, Y., T. Yoshino, H. Utada, T. Shimomura, E. Kimoto 1983. Changes in the apparent electrical resistivity of Oshima volcano observed during a period of highly elevated tectonic activity. *Earthquake Prediction Research* **2**, 83–96.

Yukutake, Y., T. Yoshino, H. Utada, H. Watanabe, Y. Hamano, Y. Sasai, T. Shimomura 1987. Changes in the electrical resistivity of the central cone, Miharayama, of Izu-Oshima volcano, associated with its eruption in November 1986. *Proceedings of the Japanese Academy, Series B,* **63**, 55–8.

Zablocki, C.J. 1976. Mapping thermal anomalies on an active volcano by the self-potential method, Kilauea, Hawaii. In *Proceedings of the Second UN Symposium on the development and use of geothermal resources, San Francisco, California, May 1975,* Proceedings 2, 1199–209.

Zablocki, C.J. 1978a. Application of the VLF induction method for studying some volcanic processes of Kilauea volcano, Hawaii. *Journal of Volcanology and Geothermal Research* 3, 155–95.

Zablocki, C.J. 1978b. Streaming potentials resulting from the descent of meteoric water. A possible source mechanism for Kilauean self-potential anomalies. *Geothermal Resources Council, Transactions* 2, 747–8.

Zablocki, C.J. 1980. *United States Geological Survey Open-file report, Observations from self-potential monitoring studies on Kilauea volcano, Hawaii (1973–1975),* 80–99. Washington DC: United States Government Printing Office.

Zohdy, A. A., L. A. Anderson, L. P. Muffler 1973. Resistivity, self-potential and induced-polarization surveys of a vapor-dominated geothermal system. *Geophysics* **38**, 1130–44.

# 10 Geomagnetic surveying methods

J. Zlotnicki

## 10.1 Introduction

For a number of decades a range of geophysical and geochemical effects associated with volcanic activity have been monitored, including seismicity, ground deformation, gas emissions, and variations of temperature, gravity, electricity and magnetism. Among these, the study of variations in the local magnetic field are relatively young. From the beginning of this century it has been known that the pattern of the Earth's magnetic field can be disturbed locally by tectonic and volcanic activity. Initial studies focused on changes of the inclination and declination of the magnetic field, most notably on Mihara volcano on the Japanese island of Oshima (Rikitake 1951, Yokoyama 1957). However, with the development of proton precession magnetometers during the 1960s (Overhauser 1953, Sigurgeirsson 1970), continuous measurements of the absolute intensity of the Earth's magnetic field were made possible, and the study of the volcano-magnetic effect was extended to several other areas including Ruhapehu in New Zealand (Johnston & Stacey 1969) and Kilauea in Hawaii (Davis et al. 1973). Since then several workers have clearly demonstrated that volcano-magnetic signals of ten or more nanotesla (nT) can be induced by volcanic activity (Pozzi et al. 1979, Johnston et al. 1981, Davis et al. 1984, Zlotnicki 1986, Zlotnicki et al. 1987, Zlotnicki & Le Mouël 1988, Yukutake et al. 1990a, Sasai et al. 1990, Hamano et al. 1990, Tanaka 1993).

## 10.2 Magnetic fields

### 10.2.1 The Earth's magnetic field

Changes in the local magnetic field associated with volcanic activity, of the order of a few tens of nanotesla, must be measured within the total field which can include non-local changes of up to hundreds of nanotesla during magnetic storms, as well as the Earth's main field of 30000 to 50000 nT, depending on latitude. One can express the magnetic field as follows: at a point P of a volcano's surface the magnetic field $\mathbf{B}$ is, at any time $t$, the superimposition of three terms:

$$\mathbf{B}(P,t) = \mathbf{B}_p(P,t) + \mathbf{B}_t(P,t) + \mathbf{B}_v(P,t) \qquad (10.1)$$

$\mathbf{B}_p$ is the main field, $\mathbf{B}_t$ is the transient magnetic field, and $\mathbf{B}_v$ is the volcano-magnetic field.

$\mathbf{B}_p$ is the sum of a regular magnetic field generated by fluid currents within the Earth's core, and of a crustal magnetic field induced by magnetized rocks below their Curie temperature. The order of magnitude of the main field depends on the latitude and longitude of the volcano. For example, in the Lesser Antilles (Guadeloupe or Martinique) and at Réunion island, the order of magnitude of the main field is 38 000–40 000 nT. The main field is subjected to slow secular variation, with amplitudes of about –120 nT per year in the Lesser Antilles and +40 nT per year at Réunion. The spatial variation is 0.1 nT km$^{-1}$.

The transient magnetic field, $\mathbf{B}_t$, is composed of an external primary transient field and an internal secondary transient field. The primary transient field has its origin in the electric currents flowing in the ionosphere and the magnetosphere at altitudes of 100 km or more. Due to the altitude of the sources for this field, variations should be relatively homogeneous over conductive geological structures. The primary field induces a secondary transient magnetic field through the generation of electrical currents which circulate in the conducting rocks of a volcano. Temporal variations of this field range from magnetic pulses lasting hundredths of seconds to the 11-year solar cycle. Research on magnetic fields associated with volcanic activity has attempted to explain time variations ranging from seconds to years. The main variations are the daily solar variation, $S_R$, and magnetic storms, with the diurnal changes depending on a volcano's latitude and on the season, and varies between 20 and 100 nT. Magnetic storms induce rapid transient variations ranging from seconds to several hours and may last for a number of days, during which amplitudes can reach a few hundred nanotesla.

The volcano-magnetic field, $\mathbf{B}_v$, is most often generated by an overpressure in the magmatic reservoir or by a change of temperature within the volcano. Depending on the magnetization of rocks, the presence of heterogeneous conductive regions such as dykes, water tables, abnormal thermal areas and faults, thermomagnetic, piezomagnetic and electrokinetic effects and resistivity changes can occur in the edifice. Except for thermomagnetic signals that could reach several tens of nanoteslas (Yukatake et al. 1990a, Hamano et al . 1990, Tanaka 1993), other volcano-magnetic signals are less than 15 nT (Davis et al. 1979, 1984, Pozzi et al. 1979, Johnston et al. 1981, Zlotnicki & Le Mouël 1988).

## 10.2.2 Volcano-magnetic fields

Understanding of the different mechanisms which can generate volcano-magnetic signals has evolved considerably since first observations of the phenomenon. Prior to the 1960s, the single mechanism proposed to explain the observed changes was thermomagnetism (Rikitake 1952, Rikitake & Yokoyama 1955). However, more recent research by Yokoyama (1969) showed that piezomagnetic effect could also account for some of the observed changes, while Mitzutani et al. (1976) and Fitterman (1978) demonstrated that electrokinetic effects also constituted a possible mechanism for generating volcano-magnetic signals. Furthermore, a number of workers (Jackson & Keller 1972, Jackson et al. 1985, Yukutake et al. 1990b) have shown that changes in the resistivity of rocks at a volcano, due to the stress variations associated with volcanic activity, can induce a perturbation of the transient magnetic field (Zlotnicki 1986, Zlotnicki & Le Mouël 1988).

### 10.2.2.1 Thermomagnetism

Rocks submitted to temperature changes can be subject to demagnetization or remagnetization. When the temperature exceeds their Curie point, rocks lose their magnetization and so modify the static crustal magnetic field (Rikitake 1952, Rikitake & Yokoyama 1955). Conversely, emplaced lava flows or pyroclastic deposits that cool below their Curie temperature acquire a thermoremanent magnetization which is related to the intensity and direction of the Earth's magnetic field (see, e.g. Stacey & Barnejee 1974). Depending on the magnetic mineral content, these rocks can acquire a magnetization of up to $10\,A\,m^{-1}$. The local static magnetic field is slowly modified by the displacement of the Curie point isotherm with variations ranging in duration from several weeks to years.

Volcano-magnetic variations can be induced by thermal effects due to either convective or conductive heat exchange. Convective exchanges occur if fluids or gases are present to transport heat rapidly, for example, in fumarole fields, solfataras and hot water tables (Hurst & Christoffel 1973, Emeleus 1977, Tanaka 1993). On a larger scale, the emplacement of a dyke or the replenishment of a magma chamber can lead to extensive convective thermal exchanges. Where parts of the edifice are water saturated, both long-term and rapid volcano-magnetic variations are observed (Yukutake et al. 1990a & b, Sasai et al. 1990, Hamano et al. 1990). If thermal exchanges are essentially by conduction, only slow volcano-magnetic variations occur, such as the 5-year change preceding the the November 1986 eruption of Oshima volcano (Yukutake et al. 1990a & b, Sasai et al. 1990).

### 10.2.2.2 Piezomagnetism

Early experiments into magnetic susceptibility and remanent magnetization involved uniaxial loading, and determined that if a stress field is applied to a rock, its induced and remanent magnetic field is modified (Kalashnikov & Kapitsa 1952, Kapitsa 1955, Ohnaka & Kinoshita 1968, Pozzi 1973, Kean et al. 1976). For uniaxial stress fields, susceptibility and remanent magnetizations are proportional to the axial load applied to the rock when the maximum stress field remains below approximately $30\,MPa$. A mean coefficient may be within an order of magnitude of $1 \times 10^{-3}\,MPa^{-1}$ (Stacey & Johnston 1972, Hodych 1976, 1977). Variations in the intensity of the induced and remanent magnetizations depend broadly on the nature and the quantity of magnetic minerals in the rock (Stacey & Johnston 1972, Kean et al. 1976, Hamano et al. 1989), on the stress cycling, and on the size of magnetic domains (Revol et al. 1978).

To establish if piezomagnetism can account for observed volcano-magnetic signals it is necessary to establish the relationship between induced or remanent magnetization, or both, and the volcanic stress field. For example, the variation in the induced magnetic field, $\Delta J_i$, is related to the variation of the stress field, $\Delta s_{kn}$, by a four order tensorial relationship (Zlotnicki et al. 1981):

$$\Delta J_i = P_{ijkn}\,\Delta s_{kn}\,J_{oj} \tag{10.2}$$

where $\mathbf{J}_{oj}$ is the induced magnetization caused by the magnetic field, and $P_{ijkn}$ is a four-order tensor (the piezomagnetic tensor). For isotropic materials which have already been loaded close to their maximum bearing capacity the previous relationship can be expressed using two coefficients, $P_1$ and $P_2$:

$$\Delta \mathbf{J}_i = (P_1 \Delta s_{kk} \delta_{ij} + 2P_2 \Delta s_{ij}) \mathbf{J}_{oj} \qquad (10.3)$$

It is necessary to determine the coefficients $P_1$ and $P_2$ for each type of rock within the volcanic edifice. ($\delta_{ij}$ is the Kronecker symbol.)

Stacey (1964) pioneered the computation of the piezomagnetic field, and with colleagues (Stacey et al. 1965) he considered a spherical magma reservoir submitted to an hydrostatic pressure in an elastic domain in which the surrounding rocks were assumed to be uniformly magnetized. Later, Yukutake & Tachinaka (1967) computed the piezomagnetic field generated by an infinitely long cylinder in a semi-infinite medium. For a hydrostatic pressure of 10 MPa both models calculated a piezomagnetic field of the order of a few nanotesla. More recently, Davis (1976) considered viscoelastic mechanical behaviour as a possible mechanism for explaining the order of magnitude discrepancy between computed and observed magnetic fields at Kilauea volcano, while in Japan, Sasai (1979) and Sasai & Ishikawa (1991) determined analytical solutions inferred from the Mogi (1958) ground deformation model. Using finite-elements techniques, Pozzi et al. (1983) were able to estimate piezomagnetic field generated by a buried magma reservoir in semi-infinite elastic medium.

The piezomagnetic field ($\mathbf{B}_{Vp}$) at the ground surface of a volcano can be estimated (Zlotnicki & Le Mouël 1988) by:

$$\mathbf{B}_{V_p}(Q,t) = -\frac{\mu_0}{4\pi} \nabla \left[ \iiint_{V_p} \Delta \mathbf{J}(M,t) \cdot \frac{\mathbf{r}}{r^3} \cdot dv_m \right] \qquad (10.4)$$

Where $\mu_0$ is the magnetic permeability in the vacuum, $Q$ is the observation point, $M$ is a point in the rock volume $V_p$ submitted to a stress field, and $r$ is the distance between $M$ and $Q$.

Irrespective of the model used, the computed piezomagnetic field is of the order of 1 nT for an overpressure of about 10 MPa associated with a magma reservoir or similar body. From this work, several conclusions can be drawn: variations in the piezomagnetic field are related to changes in the stress field and range from years to seconds or less; if the mechanical behaviour of a volcano can be considered elastic, then the piezomagnetic field should be correlated with ground deformation; and if the magnetization of a volcano is initially uniform, the piezomagnetic field is at a maximum directly above the magma reservoir. The 1 nT field strength at 10 MPa does not alone seem able to account for all volcano-magnetic signals observed at volcanoes where amplitudes range from a few to tens of nanotesla or more. This is further supported by evidence for large volcano-magnetic signals at basaltic fissure volcanoes where high stress fields are not developed. In any case, inhomogeneous magnetized structures should be considered in piezomagnetic computations (Zlotnicki & Cornet 1986).

### 10.2.2.3 Electrokinetic phenomena

Electrokinetic phenomena have their origin in the existence of a double electric layer at the interface between rock and the pore fluid network. The double layer comprises a layer of ions exsolved from the liquid and firmly fastened to the rock wall, and a more diffuse layer in the liquid which is composed of opposite ions (Mitzutani et al. 1976, Fitterman 1978). For silicate minerals the exsolved ions are usually negative and an excess of positive ions can flow in the liquid. When an electric field is applied, it is mainly the free ions in the diffuse double layer and the liquid which move; an effect known as electro-osmosis. Conversely, when liquid is flowing under a pore-pressure gradient through relatively porous rock, an electric current is induced by the liquid, generating a phenomenon known as streaming electrokinetic effect.

Electrokinetic changes are irreversible phenomena which obey the laws of thermodynamics (Nourbehecht 1963). The relationship between the electric current density, $\mathbf{i}$, the fluid flow density, $\mathbf{j}$, the streaming potential gradient $\nabla E$ and the fluid pressure gradient (over hydrostatic pressure) $\nabla P$ is be expressed by:

$$\begin{bmatrix} -\mathbf{i} \\ -\mathbf{j} \end{bmatrix} = \begin{bmatrix} L_{11} & L_{12} \\ L_{21} & L_{22} \end{bmatrix} \begin{bmatrix} \nabla E \\ \nabla P \end{bmatrix} \tag{10.5}$$

Where $L_{ij}$ are generalized constant conductivities: $L_{11}$ is the electrical conductivity of the medium, $L_{22}$ is the permeability, $L_{21}/L_{22}$ is the streaming coefficient $C$ (or cross coupling coefficient), and $L_{21}/L_{22}$ is the electro-osmotic coefficient. The term $L_{11} \cdot \nabla E$ represents Ohm's law while $L_{22} \cdot \nabla P$ represents Darcy's law, and $L_{12} = L_{21}$ follows from Onsager's reciprocal relationship.

When a pore pressure gradient is applied, the maximum electric current density, $\mathbf{I}$, is obtained when $\nabla E$ is zero (Mitzutani et al. 1976, Fitterman 1978):

$$\mathbf{I} = -L_{12} \cdot \nabla P = \left( \frac{\varepsilon \zeta}{\eta} \right) \left( \frac{\sigma_r}{\sigma_f} \right) \nabla P \tag{10.6}$$

Where $\varepsilon$ is the dielectric constant of the fluid ($80/(36 \pi \times 10^9)$), $\zeta$ is the zeta potential, $\eta$ is the viscosity of the fluid, and $\sigma_r$ and $\sigma_f$ are the rock and fluid conductivity, respectively.

The parameters of Equation 10.6 are not well determined. The zeta potential changes between $10^{-2}$ V and $10^{-1}$ V, the order of magnitude of the viscosity is around $10^{-4}$ Pa s, and the mean ratio between medium and fluid conductivity is $10^{-2}$. If fluid conductivity is taken to equal $10^{-1}$ S m$^{-1}$, the streaming coefficient is found to be $10^{-6}$ V Pa$^{-1}$ (Mitzutani et al. 1976, Fitterman 1978, Morgan et al. 1989). All these parameters can be changed by orders of magnitude as a result of temperature variation, and pore pressure could vary by 1–10 MPa.

At any time $t$, the electrokinetic magnetic field $\mathbf{B}_{ve}(Q.t)$ at the surface of a volcano is given by the Biot and Savart law:

$$\mathbf{B}_{V_E}(Q,t) = \frac{\mu_0}{4\pi} \nabla \times \left[ \iiint_{V_E} \frac{\mathbf{i}(M,t)}{r} dv_M \right] \tag{10.7}$$

where $Q$ is a point at the ground surface, and $\mathbf{i}$ is the electric density current at point $M$ of the volume $V_E$ where the fluid is moving.

Several authors have calculated the electrokinetic magnetic field at the surface, but mainly in the context of seismo-tectonic effects (Mitzutani & Ishido 1976, Fitterman 1978, 1981, Murakami 1989), the order of magnitude of the electrokinetic magnetic field can reach around 10 nT or more for a pore pressure gradient of $1 MPa km^{-1}$.

Electrokinetic effects require the existence of fluids within the volcanic edifice. Ideally rain should be abundant, water tables should be present, and interconnected cracks and fissures should be more or less water-saturated. In addition, gas or magma flow may also play a role in the generation of electrokinetic phenomena. The temporal characteristics of electrokinetic signals are related to changes of the stress field (as for piezomagnetism) and also to thermal or chemical modifications of fluids within the interconnected cracks and fissures. Electrokinetic signals tend to be focused along channels where fluids are flowing, and may be associated with ground deformation and therefore with contemporaneous piezomagnetic effects. However, unlike piezomagnetism, electrokinetic effects are always associated with variations in the electric field.

### 10.2.2.4 Resistivity variations

In the Earth's crust, rock resistivity depends on the conductivity of the pore fluid and on the volume and geometry of interconnected pore networks. Laboratory studies conducted to establish how resistivity varies as a function of stress applied to rock (Brace et al. 1965, Nur & Walden 1990), have revealed that variations are strongly dependent on the level of water saturation in the interconnected pore networks (Nur 1972). Salinity, viscosity, permeability and porosity are all parameters that induce resistivity variations of up to one or two orders of magnitude. Electrical resistivity of rocks is also highly dependent on temperature (Keller & Frischknecht 1966). Resistivity changes in some parts of a volcano will, therefore, induce variations in the internal transient magnetic field. According to Davis et al. (1979) and Johnston et al. (1985), observed variations in the transient field may result at least in part to differences in the orientations of remanent and induced magnetization at nearby sites.

As determined by Zlotnicki & Le Mouël (1988), if uniform, tabular geological structures are larger than the heterogeneous structures, then the normal electric $E_n$ and magnetic $B_n$ fields are linked to the vector potential $A_n$ by:

$$E_n = -\frac{dA_n}{dt}$$

$$B_n = rot \ A_n$$

(10.8)

In a quasi-static approximation (Le Mouël & Menvielle 1982), the charge distribution $r$ is related to the normal transient field, and to the abnormal transient field $E_a$ which itself is induced by an abnormal conductivity heterogeneity $\sigma$:

$$\sigma = \varepsilon\sigma(E_n + E_a) \cdot \nabla(1/\sigma)$$

(10.9)

where $\varepsilon$ is the dielectric constant, and $\sigma$ is the sum of the normal conductivity $\sigma_n$ corresponding to stratified structures and the abnormal conductivity $\sigma_a$.

The abnormal resistivity induced magnetic field can be computed by:

$$\mathbf{B}_{ta}(Q,t) = -\frac{\mu_0}{4\pi}\left[\iiint_{V_{ta}} (\sigma\mathbf{E}_a + \sigma_a\mathbf{E}_n) \times \frac{\mathbf{r}}{r^3} dv_M\right] \tag{10.10}$$

where $r$ is the distance between the point $Q$ on the ground surface and the current point $M$ of abnormal resistivity structures, and $V_{ta}$ is the volume affected by changes in electrical resistivity.

The temporal characteristics of the abnormal resistivity induced magnetic field are governed by those of the stress field, although chemical and thermal modifications of rocks and fluids could also induce resistivity changes. The resistivity induced magnetic field can be strongly heterogeneous, and depend on the resistivity pattern of the volcano and the pattern of fluid movement. This phenomenon can occur contemporaneously with both piezomagnetic and electrokinetic effects, and can therefore be associated with ground deformation.

To date, no detailed analysis has been undertaken of induced resistivity changes, but examples based on field observations are documented. Variations in the strength of the resistivity field of up to 50% were observed on Oshima volcano during the November 1986 eruption (Yukutake et al. 1990b), when volcano-magnetic signals were also observed (Yukutake et al. 1990a, Sasai et al. 1990, Hamano et al. 1990). According to Zlotnicki & Le Mouël (1988), such resistivity changes can be determined directly from magnetic recordings.

## 10.3 Method

To date, volcano-magnetic monitoring has been used both to assist surveillance of active volcanoes and to develop a greater understanding of the physical processes which generate the magnetic signals. High frequency volcano-magnetic variations of more than 0.10 Hz and signals with amplitudes of less than about 1 nT are largely ignored because they are difficult to distinguish from the normal variations of the Earth's magnetic field (Yoshino & Tomizawa 1990). Stable, high-resolution magnetometers are required to identify volcano-magnetic variations of up to several nanotesla, and to eliminate the normal variation in the Earth's main and transient fields which are of the order of several tens of nanotesla.

Variations of the main and transient magnetic fields are largely eliminated by taking simultaneous simple differences at any time $t$ between the magnetic fields measured at two points $P_i$ and $P_j$ spaced several kilometres apart:

$$\Delta\mathbf{B}(P_i, P_j, t) = (\mathbf{B}_p(P_i, t) - \mathbf{B}_p(P_j, t)) + (\mathbf{B}_i(P_i, t) - \mathbf{B}_i(P_j, t)) + (\mathbf{B}_v(P_i, t) - \mathbf{B}_v(P_j, t)) \tag{10.11}$$

As explained previously, resistivity contrasts in the ground destroys the homogeneity of the transient magnetic field, and variations of up to 10 nT can appear in the difference $\Delta(P_i, P_j, t)$ irrespective of the state of volcanic activity. Several methods have been used tentatively to eliminate these disturbances. Stacey & Wescott (1965) have studied, without great success, second-order differences. Rikitake (1966) has proposed the use of weighted differences, but this method was found to be relatively ineffective. Although there have been improvements in the calculation of the difference $\Delta(P_i, P_j, t)$, where the volcano-magnetic effect is isolated from other transient magnetic variations, the methods are still not very sensitive. Subsequent work on tectonomagnetism by Beahm (1976), Ware (1979), and Davis et al. (1979, 1981) used transfer functions or multichannel predictive filtering to reduce the variations in the difference $\Delta(P_i, P_j, t)$ to 1 nT or less. These techniques are generally difficult to use in continuous volcano monitoring and require high stability, three-component magnetometers, which have not been widely used on volcanoes (Zlotnicki et al. 1986, Tanaka 1993).

Most research groups working on volcano-magnetic signals use simple simultaneous differences between the field intensities at a number of points on a volcano. These differences are averaged on 1-h (hourly mean), and 1-day (daily mean) values. Using daily mean values of the difference $\Delta(P_i, P_j, t)$ it is possible to reduce the abnormal transient magnetic variations to approximately 1–2 nT depending on the geological structure of the volcano. Although during periods of magnetic disturbance, such as magnetic storms, non-local variations still exceed the volcano-magnetic signal by tens of nanotesla. Most of the long-term studies of volcano-magnetic variations have shown that precursory signals prior to eruptions exceed the reduced value of the abnormal transient variations by up to one order of magnitude, and therefore this technique can be used with some degree of confidence. Furthermore, the calculation of the difference $\Delta(P_i, P_j, t)$ can be achieved without the need to access substantial computer processing facilities, and data are collected using relatively portable scalar magnetometers, enabling fieldwork to be carried out on poorly accessible volcanoes.

Magnetic monitoring of volcanic activity is usually based on the measurement of the field intensity at a number of points on the volcano. Generally the difference $\Delta(P_i, P_j, t)$ is calculated relative to a reference, or reduction, station. The location of this reference station is critical to the success of the monitoring, and should satisfy several criteria:
(a) the station should be far enough away from the volcano so that it is unaffected by the volcano-magnetic signal;
(b) the station should not be located close to features such as faults, cliffs, or shorelines, in order to minimize the possibility of local magnetic variations;
(c) the local static gradient should be low, less than about 10–20 nT m$^{-1}$; and
(d) the features of the variations at the reference station should be more or less identical to that at the stations located on the volcano in order to reduce the effect of abnormal transient variations.

Of course computations of any simultaneous difference $\Delta(P_i, P_j, t)$ can be done to study the different wave lengths of the volcano-magnetic variations. For example, a magnetic network of $N$ stations provides $N(N-1)/2$ possible differences.

## 10.3.1 Magnetic monitoring networks

Strategies for monitoring volcano-magnetic effects will depend on the principal aim of the research and on the resources available to the researchers. As discussed below, other factors, such as the ease of access to the volcano and the proximity of a specialized volcano observatory, will also dictate the strategy adopted.

## 10.3.1.1 Real-time magnetic networks

Where it is possible to use the facilities of a nearby volcano observatory, an extensive, real-time network can be installed. This has been done, for example, at Piton de la Fournaise volcano on Réunion island (Fig. 10.1). At such networks, a scalar magnetometer of 0.1 nT–0.01 nT resolution is used at each magnetic monitoring station and is controlled by a microcomputer via a serial line to the magnetometer. Generally the magnetic field is measured every minute, and readings stored in the computer's memory for later transmission to the observatory by radio or telephone line. If topography interferes with signal transmission it may be necessary to install relay stations to ensure that data reach the observatory. The energy requirements of each station are provided by batteries which are recharged either by solar panels or by small wind turbines. On Piton de la Fournaise, the batteries providing power for each station are rated at 38 Ah with solar panels generating around 48 W.

At the observatory, an on-line computer collects data from all parts of the magnetic network, including field stations and the reference station. The calculation of simultaneous differences relative to the reference station can then be automated. Graphical output of mean hourly values, mean daily values and simultaneous difference values allow immediate interpretation by observatory staff. One of the difficulties associated with real-time networks is the need to take simultaneous readings at each station. This can be achieved by providing a precision clock (about $10^{-8}$), at each station, or by using bi-directional transmissions between the observatory and magnetic stations and having the computer command all the field magnetometers. Within the next few years it will become more convenient to use the global positioning system (GPS)

**Figure 10.1** Station MGE (see Section 10.4.3) is a typical example of the type of magnetic monitoring station used at Piton de la Fournaise volcano. The station is autonomous and solar powered, and transmits data on a real-time basis to the local volcano observatory.

which will provide time signals with a precision of better than 0.01 s.

Once initial analysis of the data has been completed, further data interpretation can be accomplished using more advanced modelling. For this type of work data may require transmission to the observatory's host institution where appropriate computer facilities are more likely to be available. In the case of real-time monitoring, it is better if the data can be relayed to the host institution on a frequent basis. Data from the eight Piton de la Fournaise magnetic stations, for example, are relayed daily to France. With a measurement rate of one value per minute at each station, the information is compressed to a file of less than 15 kbyte and using a standard modem operating at 2400–9600 Bd, automatically sent in under a minute.

### 10.3.1.2 Long-term studies

It is necessary to make long-term measurements of magnetic variations in order to establish the baseline characteristic of each volcano, and evaluate any long period cyclicity of behaviour. For example, volcano-magnetic changes were recorded at Oshima volcano in 1981, (Yukutake et al. 1990a) 5 years before the November 1986 eruption. Therefore continuous self-recording magnetometers are often used prior to, and in addition to, a real-time network (Davis et al. 1984, Zlotnicki et al. 1987, Sasai et al. 1990). For long-term studies, stable, non-drift magnetometers must be used. For a number of years, magnetometers have been available with integral memory boards and these have been utilized in long-term studies, often stationed on a volcano for weeks or months at a time. However, the capacity of the memory board rarely exceeded 80000–90000 readings, requiring, therefore relatively frequent visits to the magnetometer to download data. With the reduction in cost of portable computers, it is now feasible to use their high storage capacity by linking one to the magnetometer by a serial cable allowing independent operation for several months.

A second type of monitoring network utilizes a number of relocatable benchmarks on the volcano, at which the magnetic field is regularly remeasured every few months (Pozzi et al. 1979, Zlotnicki 1986, Sasai et al. 1990). Individual measurements are made with a calibrated magnetometer at each benchmark, where the magnetometer sensor is held about 2.5 m high to attenuate any local static field gradient. Magnetic field readings are then subtracted from the values measured at the reference station. To reduce diurnal variation effects, the highest accuracy for this technique is achieved when the magnetometer is left at each benchmark for a number of hours or even days. Then abnormal transient magnetic fields, such as the diurnal variation, can be removed by calculating the difference $\Delta(P_i, P_j, t)$ relative to the reference station.

### 10.3.1.3 Mitigation and volcanic alert networks

Where a volcano is monitored solely as part of a hazard mitigation or alert network, the need for extensive real-time or long-term monitoring is obviously much reduced. In this case only a reference station and one or two other magnetometers need be installed at locations which reflect the activity of the volcano at any particular time. Radio transmission, perhaps by satellite, will relay readings to a central volcano observatory with minimal delay. With the onset of activity, more comprehensive monitoring equipment can be installed to assist eruption forecasting, and to aid interpretation of the volcano's eruptive mechanism. The range of parameters which can be relayed to the volcano observatory will depend on the transmission density of the medium being used. For instance, in lower latitude countries the Argos satellite may be used to relay information but, with a transmission density of 256 byte every 4 h, only mean daily values, standard deviation, and battery voltage can be sent to observatories.

## 10.4 Case studies

### 10.4.1 Mount St Helens

Mount St Helens is a conical andesitic strato-volcano (United States, 46°11'N, 122°10'W) which became reactivated during March 1980 after a number of years of quiescence. Expulsion of steam and ash accompanied increasing levels of seismic activity throughout March and April. Three self-recording magnetometers of 0.25 nT sensitivity and 10 min sampling rates were installed on the volcano 10 days before the catastrophic eruption of 18 May 1980. Two of the magnetometers were destroyed by the eruption, and Johnston et al. (1981) were obliged to use very remote reference stations to evaluate volcano-magnetic signals. The last operating station (Blue Lake), located 5 km west of the crater, revealed no precursory signal in the 10 days of recording prior to the eruption of 18 May, but did register a positive anomaly of approximately 8 nT at the time of eruption (Figs 10.2 & 10.3). This anomaly was observed even though an estimated 2.5 km³ of magnetized material was removed during the eruption. Johnston et al. (1981) calculated (from work by Yukutake & Tachinaka 1967) that a release of 100 MPa of overpressure acting on a spherical or cylindrical source of 1 km radius and 5 km depth, could have generated the piezomagnetic field necessary to induce the observed

**Figure 10.2** Comparative plots for the 280 km baseline on Mount St Helens of magnetic field differences between stations VIC and SHW and stations BLM (Blue Lake) and GDM. Occurrence times (in UTC) of the major eruptions of 18 May, 25 May and 12 June 1980 are indicated by arrows (after Johnston et al. 1981).

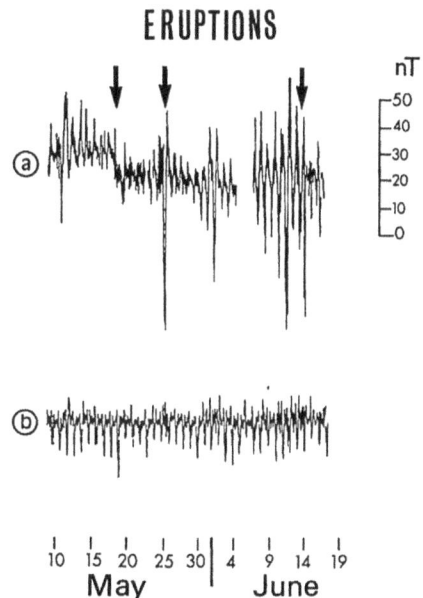

285

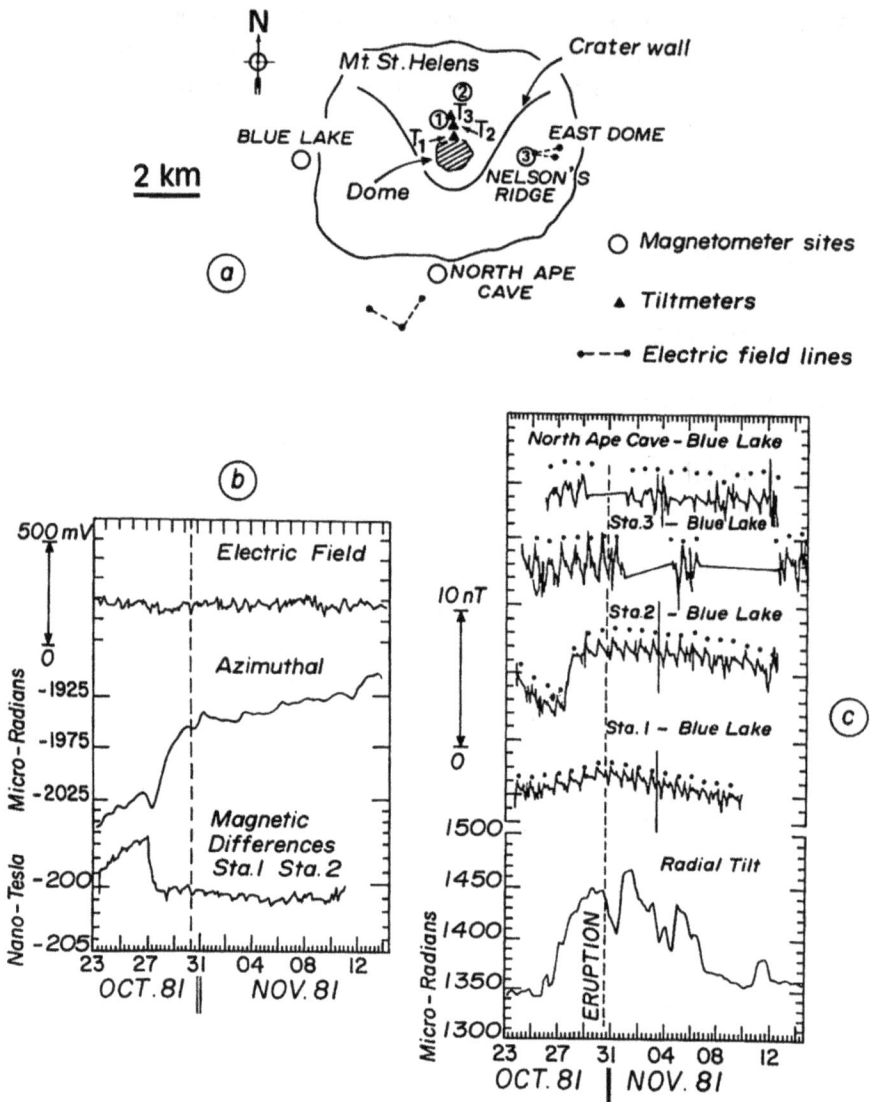

**Figure 10.3** (a) Location map of electric, magnetic and tiltmeter stations on Mount St Helens; (b) electric, magnetic and tilt fields; and (c) magnetic differences and radial tilt (after Davis et al. 1984).

volcano-magnetic signal. Transient magnetic variations were also associated with the eruptions of 25 May and 13 June 1980, with precursory transient volcano-magnetic signals being identified several hours prior to the former eruption. Ionospheric disturbances were induced by the atmospheric pressure wave associated with this event, and were observed up to 1000 km from the volcano (Johnston et al. 1981).

Later, four total-field magnetometers and two recording electric field lines were installed on the volcano (Davis et al. 1984), allowing recording of volcano-magnetic signals between 23 October and 11 November 1981 when a dacitic dome was extruded, and during an eruption on 30 October (Fig. 10.3), when magnetic variations of up to 4 nT were observed at two stations. On the north flank of the volcano an increase of 1 nT day$^{-1}$ was recorded until 27 October, and then followed by a decrease in the signal over 16 h. The reduction in the signal was accompanied by an azimuthal tilt of several tens of microradians. On the east flank of the volcano neither magnetic signals nor electric field variations were observed. From these observations, Davis et al. (1984) did not rule out electrokinetic effects, although they did not observe any electric-field variation associated with the tilt and magnetic variations and accompanying volcanic activity during 18 months of recording.

## 10.4.2 Oshima volcano

Oshima volcano is a basaltic strato-volcano (34°44'N, 139°21'E) located on the Japanese island of Oshima (Fig. 10.4). Large-scale eruptions occur every 100–150 years or so, interrupted by smaller scale events similar to those of 1974 and 1986. Yukutake et al. (1990c) assume that the activity which finally led to the 1986 eruption started around 1981, when large earthquakes of $M = 7.0$ and $M = 6.7$ occurred offshore, disturbing the regional stress pattern. At the time of the eruption the abnormal secular variation of the magnetic field on Oshima island returned to its normal trend, except to the south of the central cone where the total magnetic field began to decrease (Yukutake et al. 1990a) (Fig. 10.5). Small and slow resistivity variations seem to have started during the same period (Yukutake et al. 1990b). The magnetic field was decreasing at a rate of –0.44 nT per month which accelerated in April 1985 to –2.2 nT per month. The amplitude had reached –27 nT when the eruption occurred. At the same time, the rate of change of the resistivity variations increased from July 1985 with changes of up to 50% of the normal measurement. No unusual seismicity was recorded below Oshima volcano before the eruption, and neither did geodetic measurements provide precursory signals. Short-term magnetic (Sasai et al. 1990) and resistivity (Yukutake et al. 1990b) precursors were observed during and after October 1986 (Fig. 10.6). Steep magnetic variations of a few minutes duration were associated with the onset of the eruption on 15 November. Emission of electromagnetic waves were also observed at a frequency of 82 kHz by Yoshino & Tomizawa (1990). No noticeable magnetic signals had been observed outside the Oshima caldera during the few months preceding the November eruption, but from March 1987 large volcano-magnetic signals were again observed (Hamano et al. 1990), prior to a gas explosion on 16 November 1987.

Yukutake et al. (1990a & b) and Sasai et al. (1990) have proposed that the long-term magnetic and resistivity signals were triggered by perturbation of the regional stress pattern in 1981. High temperature gases and liquids escaped from a magma reservoir, filled pore networks and reheated meteoric waters which subsequently demagnetized the surrounding rocks. A magmatic body was then thought to have been intruded, leading to an increase in the rate of magnetic and resistivity change. Significant magnetic variations at the start of the activity were attributed to the eruption of the initial batch of magnetic materials, while post-eruption variations were assumed to be of piezomagnetic origin (Sasai et al. 1990).

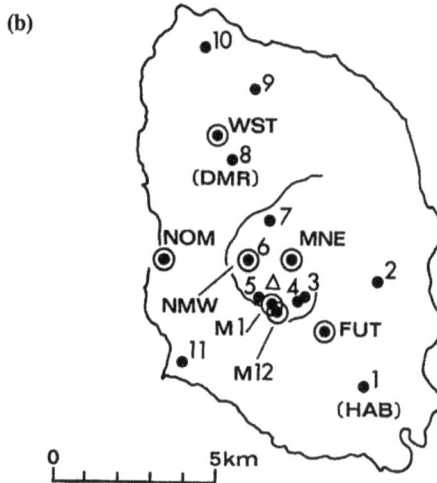

**Figure 10.4** (a) Map showing locations of Izu-Oshima island and the reference station KNZ; (b) The distribution of magnetometers (double circles with station names) and repeat survey points (closed circles with numbers) on Oshima island (after Sasai et al. 1990).

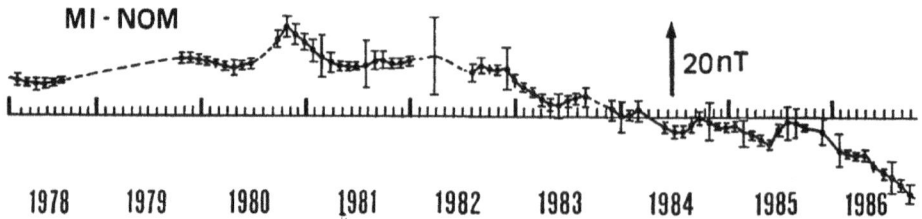

**Figure 10.5** Monthly mean differences of the geomagnetic total intensity between stations MI and NOM at Oshima island (after Yukutake et al. 1990a) See text for explanation.

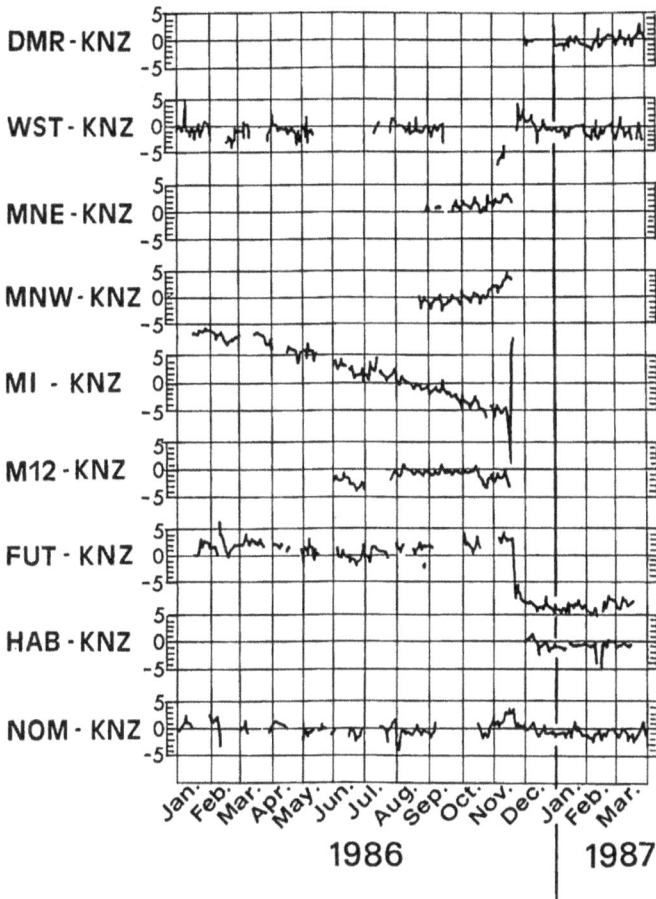

**Figure 10.6** Changes in the total magnetic field intensity on Oshima island relative to KNZ during the period January 1986 – March 1987. The plots show the difference between whole-day means and the value at KNZ (after Sasai et al. 1990).

## 10.4.3 Piton de la Fournaise

Piton de la Fournaise is a basaltic shield volcano (2632 m high) situated in the southeast part of Réunion island (21°20'S, 155°40'E). A number of calderas are situated on the volcanic edifice, the youngest of which (10 km × 15 km) is open on its east side through to the Indian Ocean and contains a 430 m high cone. Numerous fractures are found on the volcano, primarily in major zones which cut the whole volcano (Fig. 10.7). Eruptions occur mainly along fissures and discharge fluid basaltic flows; they occur on average every 18 months, and exception being the 1985–88 period during which 19 eruptions took place. Small magnitude (less than $M = 2$) earthquakes are located at depths either between 0.5 and 2.5 km, or below 4.5 km and are recorded before an eruption begins (Hirn et al. 1991). Ground deformation is mainly local to and centred around the volcanic cone, where tilt variations of several hundred

289

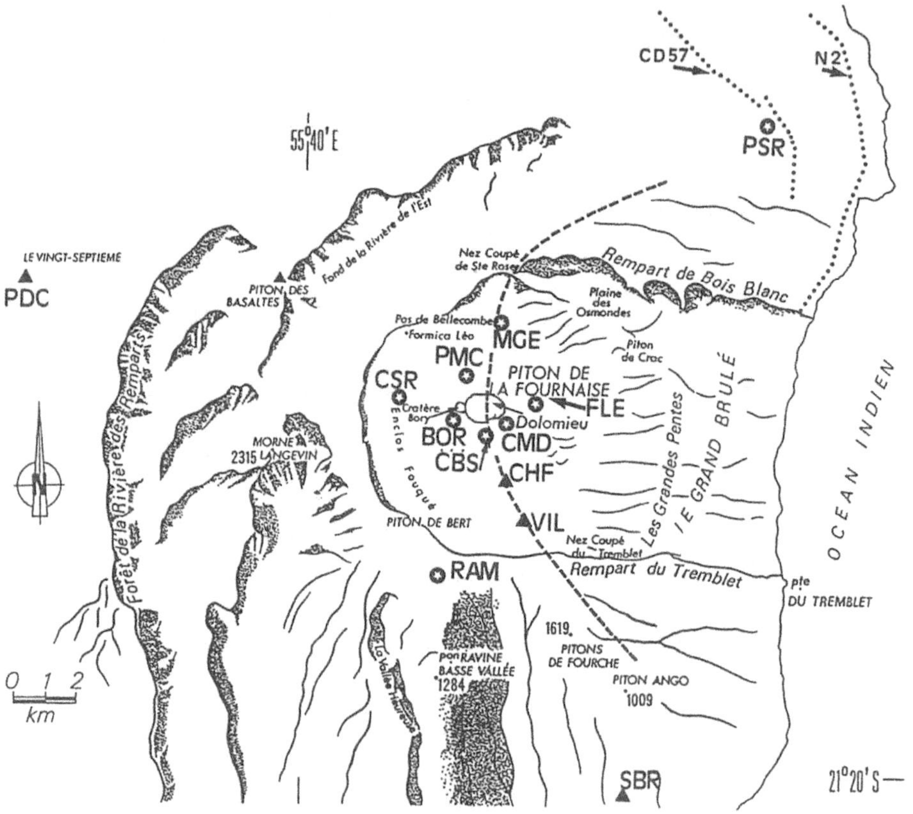

**Figure 10.7** Map of Piton de la Fournaise volcano. The dashed line represents the main fracture zone, while the stars are the telemetered magnetic stations, and the triangles are autonomous stations.

microradians (associated with magma migration) are observed on the east side of the major fractures zone (Delorme et al. 1989). Magma reaching the surface is usually channelled along pre-existing fractures (Lénat et al. 1989, Zlotnicki et al. 1990).

The magnetic network now operating is composed of eight autonomous radio-linked stations (see Section 10.3.1.1). Recording total field magnetometers (of 0.01 nT resolution and capable of storing 87 000 readings), horizontal conponent magnetometers (H and D components with 0.1 nT resolution) and associated electric field meters (resolution on the order of 0.02 mV) are sometimes installed in the field to support the real-time system. The nature of the data obtained is illustrated by observations made between 1985 and 1988(Zlotnicki & Le Mouël 1988, 1990a, Zlotnicki et al. 1992).

Mean daily values of the differences between total magnetic fields (noted $(P_i - \mathrm{CSR})$ are shown in Figures 10.8 to 10.10; caution must be exercised in data interpretation as misssing values can sometimes distort the daily mean values. The sharp, reversible peaks observed on some days are mostly induced by magnetic storms, while high frequencies variations in the (CBS–CSR) curve are generated by large abnormal transient variations. The other main results are summarized below.

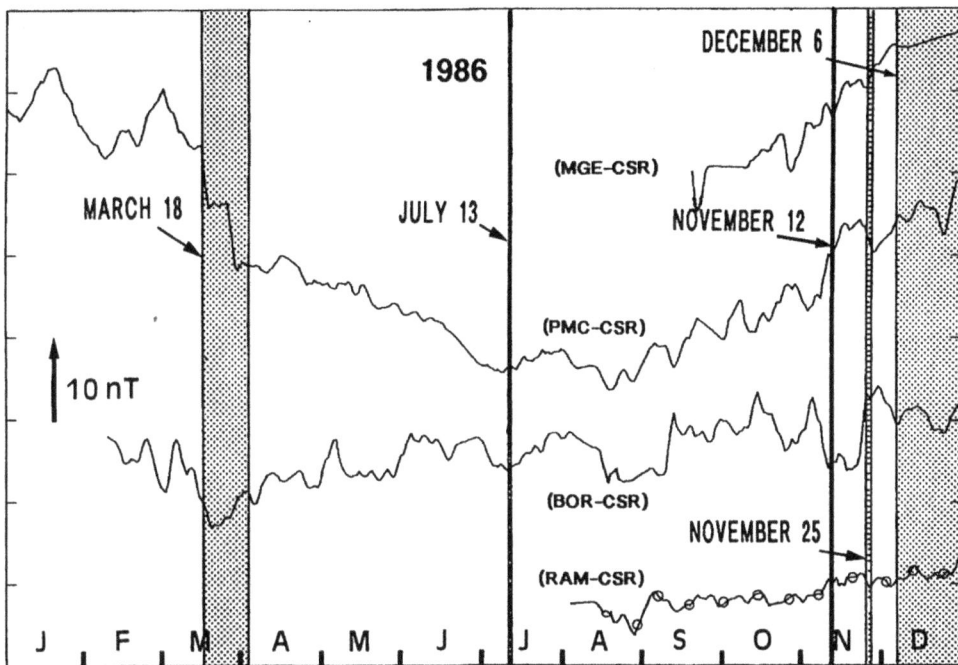

**Figure 10.8** Mean daily values of differences between total magnetic fields at Piton de la Fournaise during 1986.

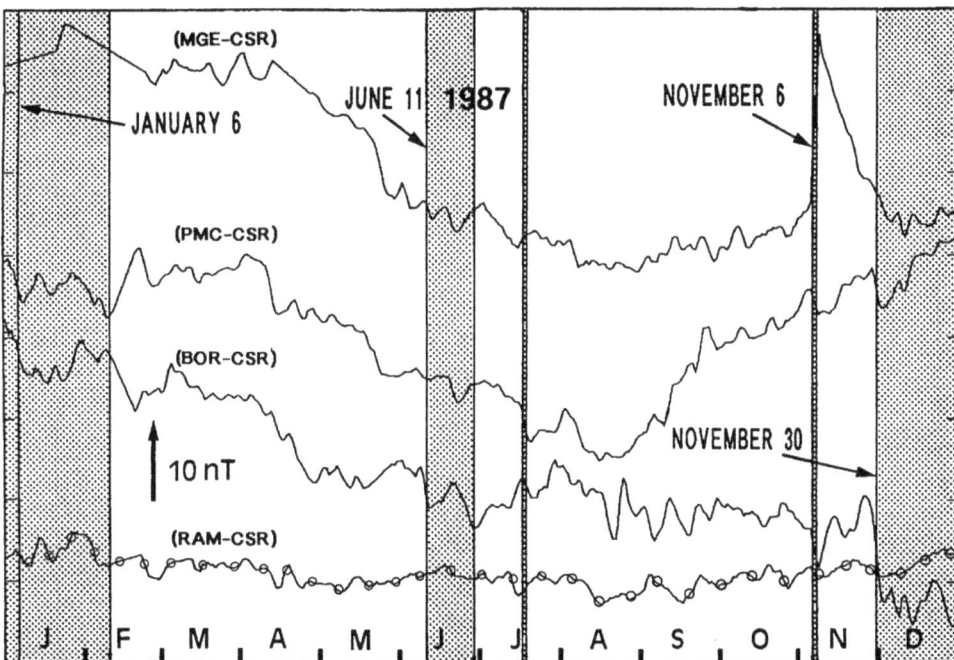

**Figure 10.9** Mean daily values of differences between total magnetic fields at Piton de la Fournaise during 1987.

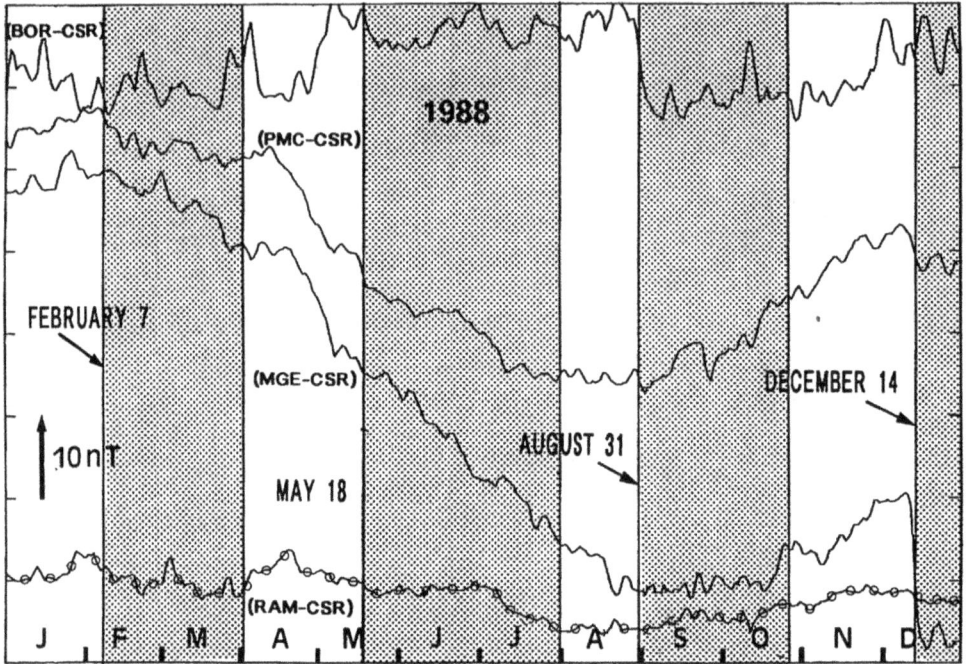

**Figure 10.10** Mean daily values of differences between total magnetic fields at Piton de la Fournaise during 1988.

### 10.4.3.1 Precursory signals

Most of the eruptions (March 1986, November 1986, January 1987, June 1987, November 1987, May 1988, and December 1988) were clearly preceded by magnetic signals, lasting typically for several days. Signals frequently appeared progressively in the days before eruption (e.g. November 1986, January 1987, November 1987, and December 1988), although some were observed a few weeks before activity (e.g. March 1986, June 1987, and May 1988). These signals are less than 5 or 6 nT and are observed essentially at the PMC, MGE, and BOR stations; rarely are signals recorded between remote stations (such as RAM-CSR), and thus the spatial extent of volcano-magnetic signals is of the order of 6 km.

### 10.4.3.2 Intrusive crises and short-term signals

When seismic swarms associated with magma movement and ground deformation are recorded in the few hours, or occasionally days, preceding eruptions, they correlate well with volcano-magnetic variations. These seismic crises, associated with the opening of fissures, can last for a few minutes (15 min for the December 1988 eruption) or up to several hours (18 March 1986 eruption). During this period volcano-magnetic signals are enhanced as the dyke migrates towards the surface. Rapid and substantial magnetic signals seem to be closely correlated with magma migration since the largest signals, whose amplitudes can reach around 15 nT, are observed in the vicinity of the eruptive vents (e.g. 6 November 1987) (Fig. 10.11). A major feature of these rapid signals has been their tendency to rapidly diminish: for example, the sharp peak at MGE during the 6 November 1987 eruption disappeared in less than a month.

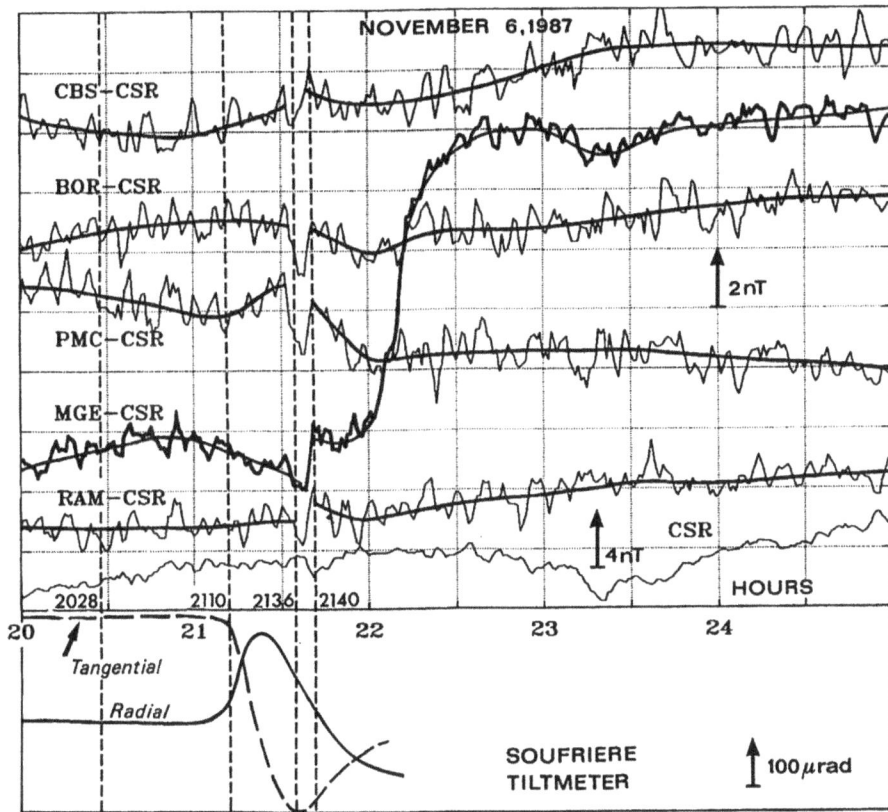

**Figure 10.11**  Differences between total magnetic fields at Piton de la Fournaise, associated with the intrusive crisis related to the 6 November 1987 eruption. Measurements were made at 1-min intervals.

## 10.4.3.3 Magnetic signals during eruptions

These variations remain difficult to estimate. It seems that volcano-magnetic variations are correlated with large variations in tremor activity (linked to variations in lava flow rates, explosive activity, etc.), as seen during the December 1986, February 1987, and May 1987 eruptions. Signals associated with the end of eruptions are not generally observed on magnetic recordings.

## 10.4.3.4 Disturbances of the diurnal solar variation

When simultaneous magnetic differences are calculated between two stations, homogeneous variations such as the primary secular transient variations of the ionosphere are removed. On volcanoes, however, not all the transient field components can be removed, and local abnormal transient variations (due to electrical conductivity contrasts) remain, with amplitudes up to 25% of the magnetic field intensity (Pozzi et al. 1979, Zlotnicki 1986, Zlotnicki & Le Mouël 1988). The main component of the transient field is the diurnal solar variation, $S_R$. Figure 10.12 presents an example of the mean hourly values of differences $(P_j - CSR)$. It can be seen from the 7 years of continuous recordings on Piton de la Fournaise volcano that the diurnal

293

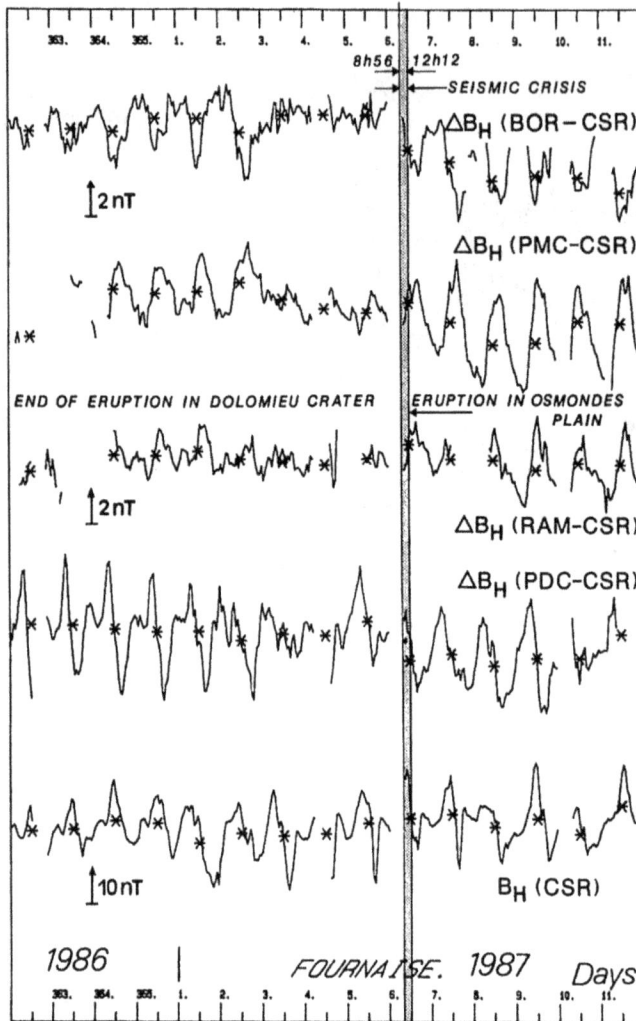

**Figure 10.12**   Mean hourly values of the magnetic field at the CSR station (Piton de la Fournaise), and of some total magnetic field differences between December 1986 and January 1987 (after Zlotnicki & Le Mouël 1988).

solar variation is most strongly disturbed during volcanic activity. In some cases the diurnal variation almost disappears (November 1986, February 1986, December 1986, and January 1987 eruptions). This abnormal behaviour of the diurnal variation lasts for periods of up to 10 days.

### 10.4.3.5 Long-term magnetic variations

Long-term variations are shown by the monthly running averages of the differences (Fig. 10.13). It can be seen that irreversible variations (on the time-scale of these recordings) are observed at the BOR and MGE stations with amplitudes of 10 and 50 nT respectively. When

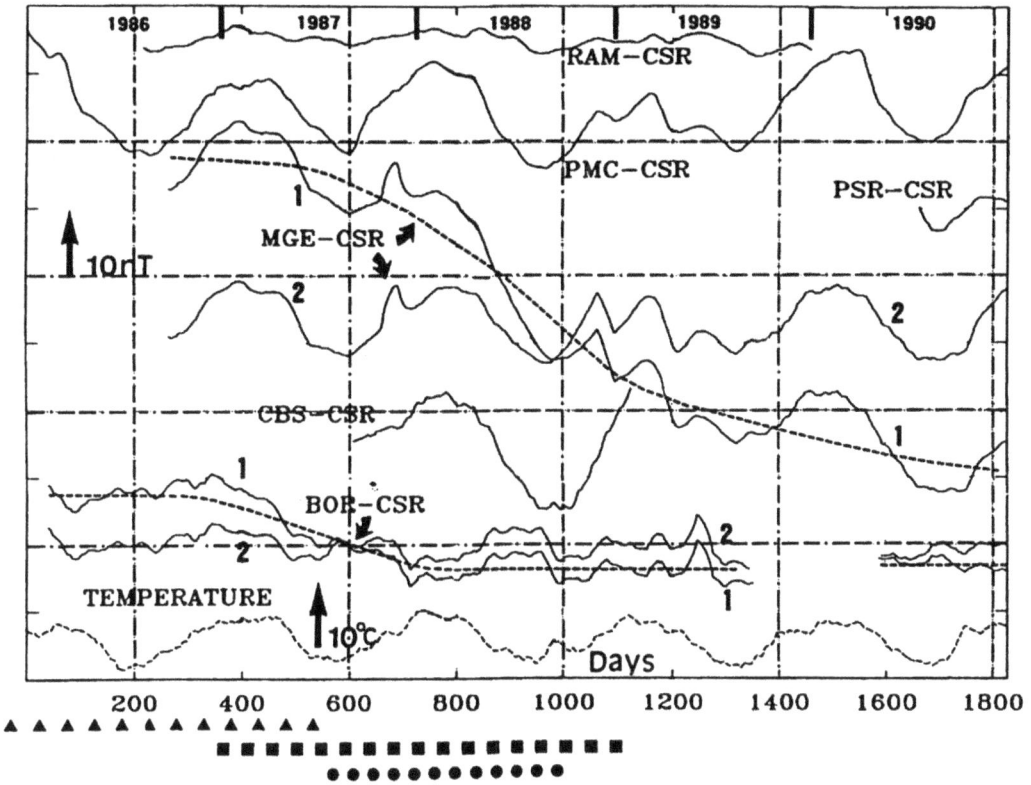

**Figure 10.13**  Mean monthly values of total magnetic field differences at Piton de la Fournaise between some stations (MGE, PMC, BOR, CBS, RAM, and PSR) and CSR station. Two curves are drawn for both the MGE and BOR stations, those labelled (1) including slow drift. Sites of volcanic activity: (▲) inside the Dolomieu crater; (■) on the north flank; (●) on the south flank.

effusive activity remains confined to the summit area of the volcano, a drift in the long-term magnetic signal is recorded at the summit station (BOR). This trend is reversed within 2 months when the eruption shifts to the volcano's flanks. Slow magnetic variations also appear on the north part of the cone when eruptions are situated on the north flank. The maximum rate of decrease in the (MGE-CSR) difference occurred in 1988, reaching −22 nT per year at its peak, forming part of the 1986–1991 eruptive period. The decrease at MGE becomes slower as north flank activity reduces.

Large annual variations of tens of nanoteslas were observed in 1990 and 1991 at several stations, including the autonomous stations (SBR, VIL, and CHF). The spatial distribution of these signals is strongly inhomogeneous. The amplitude is at its maximum at the summit station (CBS, 17 nT) and decreases with distance from the cone, but remains relatively higher along the main fracture zones. The annual variation is larger on the north flank, with an amplitude of the order of 15 nT at PMC, compared with the amplitude at station CHF of less than 10 nT. This annual variation becomes more homogeneous outside the Enclos Fouqué caldera, at some distance from the volcano (PSR and SBR stations).

*10.4.3.6 Origin of the volcano-magnetic signals*

No large stress field variation takes place within Piton de la Fournaise volcano. Small dykes are issued from a magma complex located at 2.5–4.5 km depth. Ground deformation is mainly recorded when the magma migrates towards the surface (Delorme et al. 1989) and affects a relatively small area (approximately 1000 m) around the dykes and vents (Zlotnicki et al. 1990). Larger volcano-magnetic signals sometimes appear far from the pressure source (magma reservoir). No widespread thermomagnetic mechanism can explain both the long term decrease on the north flank at MGE and an annual magnetic variation correlated with seasonal temperature changes. Therefore electrokinetic effects, rather than piezomagnetism or thermomagnetism, should be considered to explain the volcano-magnetic signals on Piton de la Fournaise. The volcanic edifice is the site of water circulation due to the high seasonal rainfall of up to 6 m per year. This movement is preferentially directed along the major fractures to the ocean, and is the cause of high positive self-potential anomalies inside the Enclos Fouqué caldera and along the main fracture zones, as well as negative self-potential anomalies outside the caldera close to sea level. Fresh water has also been found in ocean areas where the main fracture zones are extended offshore. Any modification to this quasi-stationary water circulation regime induces local disturbances of the magnetic field. Variations in the conductivity pattern of pore networks also contribute to these signals. The perturbations are enhanced along the main fracture zones where water circulation is a major factor, and could explain why large and inhomogeneous magnetic signals are seen close to the fracture zones. Signals may appear a long time before onset of eruptions and are related to changes in the pattern of water circulation. When seismic activity increases, these signals are emphasized and the abnormal diurnal variations can be obscured by the magnetic differences. Long-term drift of magnetic signals can also be generated by slow modifications to water circulation when dykes fill the fractures previously used for water movement. Therefore, although these signals are not caused directly by volcanic activity, they do follow the volcanic activity of Piton de la Fournaise.

## 10.5 Conclusion

The study of variations in the magnetic fields in the vicinity of active volcanoes is relatively young, and when used in isolation needs careful interpretation if it is to yield useful information on the state of volcanic activity. Improvements in the technology used to measure magnetic fields, and advances in techniques for isolating volcano-magnetic signals from other, non-local, variations, now present opportunities for employing this type of study in a range of situations. As this chapter has shown, magnetic studies may be used, even if tentatively, for forecasting eruptions, as well as for gaining a better understanding of volcanic systems and how they operate. When used in conjunction with other geophysical methods, the monitoring of volcano-magnetic signals clearly has an important role to play in observing both long-term changes of volcanic behaviour and in short-term forecasting.

# References

Beahm, T. J. 1976. Geomagnetic field gradient measurements and noise reduction techniques in Colorado. *Journal of Geophysical Research* **81**, 6276-80.

Brace, W. F., A. S. Orange, T. R. Madden 1965. The effect of pressure on the electrical resistivity of water saturated crystalline rocks. *Journal of Geophysical Research* **70**, 5669-79.

Davis, P. M. 1976. The computed piezomagnetic anomaly field for Kilauea volcano, Hawaii. *Journal of Geomagnetism and Geoelectricity* **28**, 113-22.

Davis, P. M., D. B. Jackson, J. Field, F. D. Stacey 1973. Kilauea volcano, Hawaii: A search for the volcano-magnetic effect. *Science* **180**, 73-4.

Davis, P. M., F. D. Stacey, C. J. Zablocki, J. V. Olson 1979. Improved signal discrimination in tectonomagnetism: Discovery of a volcano-magnetic effect at Kilauea, Hawaii. *Physics of Earth and Planetary Interior* **191**, 331-6.

Davis, P. M., D. D. Jackson, C. A. Searls, R. L. McPhernon 1981. Detection of tectonomagnetic events using multichannel predictive filtering. *Journal of Geophysical Research* **86**, 1731-7.

Davis, P. M., D. R. Pierce, R. L. McPhernon, D. Dzursin 1984. A volcano-magnetic observation on Mount St Helens, Washington. *Geophysical Research Letters* **11**, 233-6.

Delorme, H., P. Bachèlery, P. A. Blum, J. L. Cheminée, J. F. Delarue, J. C. Delmond, A. Hirn, J. C. Lépine, P. M. Vincent, J. Zlotnicki 1989. March 1986 eruptive episodes at Piton de la Fournaise volcano (Réunion island). *Journal of Volcanology and Geothermal Research* **36**,199-208.

Emeleus, T. G. 1977. Thermomagnetic measurements as a possible tool in prediction of volcano activity in the volcanoes of the Rabane caldera, Papua New Guinea. *Journal of Volcanology and Geothermal Research* **2**, 343-59.

Fitterman, D. V. 1978. Electrokinetic and magnetic anomalies associated with dilatant regions in a layered earth. *Journal of Geophysical Research* **83**, 5923-8.

Fitterman, D. V. 1981. Correction to Theory of electrokinetic magnetic anomalies in a faulted half space. *Journal of Geophysical Research* **86**, 9585-8.

Hamano, Y., R. Boyd, M. Fuller, M. Lanham 1989. Induced susceptibility anisotropy of igneous rocks caused by uniaxial compression. *Journal of Geomagnetism and Geoelectricity* **41**, 203-20.

Hamano, Y., H. Utada, T. Shimomura, Y. Tanaka, Y. Sasai, I. Nakagawa, Y. Yokoyama, M. Ohno, T. Yoshino, S. Koyama, T. Yukutake, H. Watanabe 1990. Geomagnetic variations observed after the 1986 eruption of Izu-Oshima volcano *Journal of Geomagnetism and Geoelectricity* **42**, 319-35.

Hirn, A., J. C. Lépine, A. Nercessian 1991. Episodes of Pit-crater collapse documented by seismology at Piton de la Fournaise. *Journal of Volcanology and Geothermal Research* **47**, 89-105.

Hodych, J. P. 1976. Single domain theory for the reversible effect of small uniaxial stress upon the initial susceptibility of rock. *Canadian Journal of Earth Science* **13**, 1186-200.

Hodych, J. P. 1977. Single domain theory for the reversible effect of small uniaxial stress upon the remanent magnetization of rock. *Canadian Journal of Earth Science* **14**, 2047-61.

Hurst, A. W. & D. A. Christoffel 1973. Surveillance of White island volcano 1968-72. Thermomagnetic effects due to volcanic activity. *New Zealand.Journal of Geology and Geography* **16**, 965-72.

Jackson, D. B. & G. V. Keller 1972. An electromagnetic sounding survey of the summit of Kilauea volcano, Hawaii. *Journal of Geophysical Research* **77**, 4957-65.

Jackson, D. B., J. Kauahikaua, C. J. Zablocki 1985. Resistivity monitoring of an active volcano using the controlled-source electromagnetic technique: Kilauea, Hawaii. *Journal of Geophysical Research* **90**, 12545-55.

Johnston, M. J. S. & F. D. Stacey 1969. Volcano-magnetic effect observed on Mt Ruapehu, New Zealand. *Journal of Geophysical Research* **74**, 6541-44.

Johnston, M. J. S., R. J. Mueller, J. Dvorak 1981. Volcano-magnetic observations during eruptions May--August 1980. In *United States Geological Survey Professional Paper 1250, The 1980 eruptions of Mount St Helens, Washington*, P. W. Lipman & D. R. Mullineaux (eds), 183-9. Washington DC: United States Government Printing Office.

Johnston, M. J. S., S. A. Silverman, R. J. Mueller, K. S. Breckenridge 1985. Secular variation, crustal contributions and tectonic activity in California 1976-1984. *Journal of Geophysical Research* **90**, 8707-17.

Kalashnikov, A. G. & S. P. Kapitsa 1952. Magnetic susceptibility of rocks under elastic stresses. *Izvestia Akademii Nauk USSR* 86, 521–3.

Kapista, S. P. 1955. Magnetic properties of eruptive rocks exposed to mechanical stresses. *Izvestia Akademii Nauk USSR Serija Geofiziceskaja*, 6, 489–504.

Kean, W., R. Day, M. Fuller, V. Schmidt 1976. The effect of uniaxial compression on the initial susceptibility of rocks as a function of grain size and composition of their constituent titanomagnetites. *Journal of Geophysical Research* 81, 861–72.

Keller, G. V. & F. C. Frischknecht 1966. *Electrical methods in geophysical prospecting.* Oxford: Pergamon.

Le Mouël, J. L. & M. Menvielle 1982. Geomagnetic variations anomalies and deflection of telluric events. *Geophysical Journal of the Royal Astronomical Society* 68, 575–87.

Lénat, J-F., P. Bachèlery, A. Bonneville, P. Tarits, J.L. Cheminée, H. Delorme 1989. The December 4 1983 to February 18 1984, eruption of Piton de la Fournaise (La Réunion, Indian Ocean): description and interpretation. *Journal of Volcanology and Geothermal Research* 36, 87–112.

Mitzutani, H. & T. Ishido 1976. A new interpretation of magnetic field variation associated with the Matsushiro earthquakes. *Journal of Geomagnetism and Geoelectricity* 28, 179–88.

Mitzutani, H., T. Ishido, T. Yokokura, S. Ohnishi 1976. Electrokinetic phenomena associated with earthquakes. *Geophysical Research Letters* 3, 365–8

Mogi, K. 1958. Relations between the eruptions of various volcanoes and the deformations of the ground surfaces around them. *Bulletin of Earthquake Research Institute, University of Tokyo* 36, 99–134.

Morgan, F. D., E. R. Williams, T. R. Madden 1989. Streaming potential properties of Nesterly Granite with applications. *Journal of Geophysical Research* 94, 12449–61.

Murakami, H. 1989. Geomagnetic fields produced by electrokinetic sources. *Journal of Geomagnetism and Geoelectricity* 41, 221–47.

Nourbehecht, B. 1963. Irreversible thermodynamic effects in inhomogeneous media and their applications in certain geoelectric problems. Thesis, Massachussets Institute of Technology, Cambridge, Massachussets.

Nur, A. 1972. Dilatancy, pore fluids, and premonitory variations of ts/tp travel times. *Bulletin of Seismological Society of America* 62, 1217–22.

Nur, A. & J. Walden 1980. *Time-dependent hydraulics of the Earth's crust, in "The role of fluids in crustal processes".* Washington DC: National Academy Press.

Ohnaka, M. & H. Kinoshita 1968. Effect of uniaxial compression on remanent magnetization. *Journal of Geomagnetism and Geoelectricity* 20, 93–9.

Overhauser, A. W. 1953. Parametric relaxation in metals. *Physics Review* 89, 689–700.

Pozzi, J. P. 1973. Effets de pression en magnétisme des roches. Thesis, University of Paris 6, France.

Pozzi, J. P., J. L. Le Mouël, J. C. Rossignol, J. Zlotnicki 1979. Magnetic observations made on La Soufrière Volcano (Guadeloupe) during the 1976-1977 crisis. *Journal of Volcanology and Geothermal Research* 5, 217–37

Pozzi, J. P., C. Philippe, J. L. Le Mouël, J. Zlotnicki, J. C. Rossignol 1983. *Bulletin du Programme Interdisciplinaire de Recherche sur la Prévision et la Surveillance des Eruptions Volcaniques 79, Anomalies magnétiques et déformations associées à l'activité de volcans andésitiques: exemple de la Soufrière de Guadeloupe.* Paris, France: PIRPSEV.

Revol, J., R. Day, M. Fuller 1978. Effect of uniaxial stress upon remanent magnetization: stress cycling and domain state dependence. *Journal of Geomagnetism and Geoelectricity* 30, 593–605.

Rikitake, T. 1951. Changes in magnetic dip that accompanied the activities of volcano Mihara. *Bulletin of Earthquake Research Institute, University of Tokyo* 29, 499–502.

Rikitake, T. 1952. On magnetization of volcanoes. *Bulletin of Earthquake Research Institute, University of Tokyo* 30, 71–82.

Rikitake, T. 1966. Elimination of non local changes from total intensity values of the geomagnetic field. *Bulletin of Earthquake Research Institute, University of Tokyo* 44, 1041–70.

Rikitake, T. & I. Yokoyama 1955. Volcanic activity and changes in geomagnetism. *Journal of Geophysical Research* 60, 165–72.

Sasai, Y. 1979. The piezomagnetic field associated with the Mogi model. *Bulletin of Earthquake Research Institute, University of Tokyo* 54, 1–29.

Sasai, Y. & Y. Ishikawa 1991. Tectonomagnetic signals related to the seismovolcanic activity in the Izu-Peninsula. *Journal of the Physics of the Earth* 39, 299–319.

Sasai, Y., T. Shimomura, Y. Hamano, H. Utada, T. Yoshino, S. Koyama, Y. Ishikawa, I. Nakagawa, Y. Yokoyama, M. Ohno, H. Watanabe, T. Yukutake, Y. Tanaka, Y. Yamamoto, K. Nakaya, S. Tsunomura, F. Muromatsu, R. Murakami 1990. Volcano-magnetic effect observed during the 1986 eruption of Izu-Oshima volcano. *Journal of Geomagnetism and Geoelectricity* **42**, 291–317.

Sirgurgeirsson, T. 1970. A continuously operating proton precession magnetometer for geomagnetic measurements. *Scienta Islanda* **2**, 64–7.

Stacey, F. D. 1964. The seismomagnetic effect. *Pure and Applied Geophysics* **58**, 5–22.

Stacey, F. D. & S. K. Barnejee 1974. *The physical principles of rock magnetism*. Amsterdam: Elsevier.

Stacey, F. D. & M. J. S. Johnston 1972. Theory of piezomagnetic effect in titanomagnetite-bearing rocks. *Pure and Applied Geophysics* **97**, 146–55.

Stacey, F. D. & P. Wescott 1965. Seismomagnetic effect- Limit of observability imposed by local variations in geomagnetic disturbances. *Nature* **206**, 1209–11.

Stacey, F. D., K. G. Barr, G. R. Robson 1965. The volcano-magnetic effect. *Pure and Applied Geophysics* **62**, 96–104.

Tanaka, Y. 1993. Eruption mechanism on inferred from geomagnetic changes with special attention to the 1989–1990 activity of Aso volcano. *Journal of Volcanology and Geothermal Research* **56**, 319–38.

Ware, R. H. 1979. High-accuracy magnetic field difference measurements and improved noise reduction techniques for use in tectonomagnetic studies. *Journal of Geophysical Research* **84**, 6291–5.

Yokoyama, I. 1957. Geomagnetic studies of Volcano Mihara. The 8th paper (continuous observations in geomagnetic declination during the period 1955–56). *Bulletin of Earthquake Research Institute, University of Tokyo* **35**, 567–72.

Yokoyama, T. 1969. Anomalous changes in the geomagnetic field on Oosima volcano related with its activities in the decade 1950. *Journal of the Physics of the Earth* **17**, 69–76.

Yoshino, T. & I. Tomizawa 1990. Observation results of low frequency electromagnetic emissions as precursors of volcanic eruption at Mihara in November 1986. *Journal of Geomagnetism and Geoelectricity* **42**, 225–35.

Yukutake, T. & H. Tachinaka 1967. Geomagnetic variation associated with stress change within a semi-infinite elastic earth caused by a cylindrical force source. *Bulletin of Earthquake Research Institute, University of Tokyo* **45**, 785–98.

Yukutake, T., H. Utada, T. Yoshino, H. Watanabe, Y. Hamano, Y. Sasai, E. Kimoto, K. Otani, T. Shimomura 1990a. Changes in the geomagnetic total intensity observed before the eruption of Oshima volcano in 1986. *Journal of Geomagnetism and Geoelectricity* **42**, 277–90.

Yukutake, T., T. Yoshino, H. Utada, H. Watanabe, Y. Hamano, T. Shimomura 1990b. Changes in the electrical resistivity of the central cone, Mihara-yama, of Oshima volcano observed by a direct current method. *Journal of Geomagnetism and Geoelectricity* **42**, 151–68.

Yukutake, T., Utada, H., T. Yoshino, E. Kimoto, K. Otani, T. Shimomura 1990c. Regional secular change in the geomagnetic field in the Oshima island area during a tectonically active period. *Journal of Geomagnetism and Geoelectricity* **42**, 257–75.

Zlotnicki, J. 1986. Magnetic measurements on La Soufriäre Volcano, Guadeloupe (Lesser Antilles) 1976–1984: A re-examination of the volcano-magnetic effects observed during the volcanic crisis of 1976–1977. *Journal of Volcanology and Geothermal Research* **91**, 709–18.

Zlotnicki, J. & F. H. Cornet 1986. A numerical model of earthquake induced piezomagnetic anomalies. *Journal of Geophysical Research* **91**, 709–18.

Zlotnicki, J. & J. L. Le Mouël 1988. Volcano-magnetic effects observed on Piton de la Fournaise Volcano (Réunion island): 1985-1987. *Journal of Geophysical Research* **93**, 9157–71.

Zlotnicki, J. & J. L. Le Mouël 1990. Possible electrokinetic origin of large magnetic variations at La Fournaise volcano. *Nature* **343**, 633–5.

Zlotnicki, J., J. P. Pozzi, F. H. Cornet 1981. Investigation on induced magnetization variations caused by triaxial stresses. *Journal of Geophysical Research* **86**, 11899–909.

Zlotnicki, J., R. Verhille, J. P. Viodé, J. C. Delmond, G. Simon, P. Guillement 1986. Magnetic network on Montagne Pelée volcano (Martinique, Lesser Antilles): A trial to discriminate volcano-magnetic signal. *Journal of Geomagnetism and Geoelectricity* **38**, 151–64.

Zlotnicki, J., M.G. Moreau, J.P. Viodé 1987. Volcano-magnetic variations related to the seismic crisis that occurred from December 1985 through May 1986 on the Montagne Pelçe, Martinique (Lesser Antilles). *Journal of Geomagnetism and Geoelectricity* **39**, 487–500.

Zlotnicki, J., J. C. Ruegg, P. Bachèlery, P. A. Blum 1990. Eruptive mechanism on Piton de la Fournaise volcano associated with the December 4 1983, and January, 18 1984 eruptions from ground deformation monitoring and photogrammetric surveys. *Journal of Volcanology and Geothermal Research* **40**, 197–217.

Zlotnicki, J., J. L. Le Mouël, C. Pambrun 1992. Variations anormales lentes du champ magnétique terrestre sur le Piton de la Fournaise. *Compte Rendus de l'Académie des Sciences, Paris* **314**, 661–9.

# 11 Remote sensing spectroscopy of volcanic plumes and clouds

R. J. Andres and W. I. Rose

## 11.1 Introduction

Volcanic plumes represent an extraordinarily diverse group of chemical phenomena. At first consisting largely of reactive gaseous species at high temperature (Table 11.1), the plumes mix with a much colder and strongly oxidizing atmosphere which results in numerous reactions, forming solid and liquid aerosols (Symonds et al. 1992). The elemental composition of plumes encompasses virtually the entire periodic table (Table 11.2). The understanding of the phenomena involved requires both broad thermodynamic description and the application of principles of heterogeneous chemistry (e.g. Turco et al. 1983).

Direct sampling of volcanic gases or plumes is a rare event. Gases can be sampled directly only when uncontaminated gas vents can be visited safely (e.g. Symonds et al. 1990),

Table 11.1 The ten most common volcanic gas species.

| 1<br>Gas species | 2<br>Estimated volcanic contribution to highly diluted plume | 3<br>Ambient tropospheric concentration | 4<br>Mixed plume concentration |
|---|---|---|---|
| $H_2O$ | 10–20 ppm | 40–40 000 ppm | 40–40 000 ppm |
| $CO_2$ | 0.5–0 ppm | around 300 ppm | 301–310 ppm |
| $SO_2$ | 1–2 ppm | 0.1–70 ppb | 1–2 ppm |
| HCl | 0.1–2 ppm | around 1 ppb | 0.1–2 ppm |
| $H_2S$ | 100–500 ppb | 0.08–24 ppb | 100—500 ppb |
| $S_2$ | 10–80 ppb | ? | ? |
| $H_2$ | 5–40 ppb | 540–810 ppb | 545–850 ppb |
| HF | 5–40 ppb | ? | ? |
| CO | 1–20 ppb | 0.05–0.2 ppm | 51–220 ppb |
| $SiF_4$ | 1–5 ppb | ? | ? |

The data are based upon homogeneous gas calculations (Symonds & Reed 992) and direct gas sampling at high temperature fumaroles. Dilution factors are constrained by direct data (Lazrus et al. 1979, Rose et al. 1980, Friend et al. 1982, Casadevall et al. 1984a, Rose et al. 1985b, 1988) which demonstrate factors of $10^4$ to $10^5$. Near-vent concentrations mat be higher due to lower dilution ratios. Plumes are assumed to have an $SO_2$ concentration of 1–2 ppm (Casadevall et al. 1981, Harris et al. 1981, Rose et al.,1985b) for comparison with other species. Species ambient atmospheric levels are from Graedel (1978) and Schidlowski (1986). Column 4 = column 2 + column 3.

**Table 11.2** Part of the calculated homogeneous equilibrium distribution of gas species for volcanic gas sampled from the dome of Augustine volcano (Alaska) on August 28th,1987 at 870°C and 1 atmosphere (adapted from Symonds et al. 1992). Data are mole fractions listed in order of abundance from top to bottom and left to right. Only gas species with concentrations of $>1 \times 10^{-11}$ are shown. Nearly 400 others were determined with abundances down to $1 \times 10^{-90}$ or less.

| Species | Value | Species | Value | Species | Value | Species | Value |
|---|---|---|---|---|---|---|---|
| $H_2O$ | $8.4 \times 10^{-1}$ | $ClSiF_3$ | $2.8 \times 10^{-7}$ | $SiF_3$ | $7.3 \times 10^{-9}$ | $S_5$ | $2.7 \times 10^{-10}$ |
| $HCl$ | $6.0 \times 10^{-2}$ | $AlF_2O$ | $2.7 \times 10^{-7}$ | $TiCl_4$ | $6.3 \times 10^{-9}$ | $SCl$ | $2.3 \times 10^{-10}$ |
| $SO_2$ | $5.7 \times 10^{-2}$ | $PbCl_2$ | $1.2 \times 10^{-7}$ | $AsCl_3$ | $6.1 \times 10^{-9}$ | $MgClF$ | $1.6 \times 10^{-10}$ |
| $CO_2$ | $2.4 \times 10^{-2}$ | $MoO_2Cl_2$ | $1.1 \times 10^{-7}$ | $S4$ | $6.0 \times 10^{-9}$ | $TiCl_3$ | $1.5 \times 10^{-10}$ |
| $H_2S$ | $1.0 \times 10^{-3}$ | $PbCl_4$ | $8.6 \times 10^{-8}$ | $H$ | $5.1 \times 10^{-9}$ | $KOH$ | $1.3 \times 10^{-10}$ |
| $H_2$ | $6.2 \times 10^{-3}$ | $CuCl$ | $6.9 \times 10^{-8}$ | $SiO_2$ | $4.1 \times 10^{-9}$ | $NaF$ | $1.2 \times 10^{-10}$ |
| $S_2$ | $1.9 \times 10^{-3}$ | $LiCl$ | $6.7 \times 10^{-8}$ | $OH$ | $4.1 \times 10^{-9}$ | $SCl_2$ | $1.0 \times 10^{-10}$ |
| $HF$ | $5.4 \times 10^{-4}$ | $M_2MoO_4$ | $6.5 \times 10^{-8}$ | $Br$ | $3.3 \times 10^{-9}$ | $CuBr$ | $9.6 \times 10^{-11}$ |
| $CO$ | $2.0 \times 10^{-4}$ | $MgCl_2$ | $4.8 \times 10^{-8}$ | $NiCl_2$ | $2.2 \times 10^{-9}$ | $Bi$ | $9.1 \times 10^{-11}$ |
| $SiF_4$ | $8.0 \times 10^{-5}$ | $Cl$ | $4.3 \times 10^{-8}$ | $FeCl_3$ | $1.6 \times 10^{-9}$ | $CrCl_4$ | $8.9 \times 10^{-11}$ |
| $S_2O$ | $2.3 \times 10^{-5}$ | $SiS2$ | $3.4 \times 10^{-8}$ | $BiCl$ | $1.6 \times 10^{-9}$ | $Pb$ | $8.7 \times 10^{-11}$ |
| $SO$ | $2.1 \times 10^{-5}$ | $FSiCl_3$ | $3.3 \times 10^{-8}$ | $VOCl_3$ | $1.5 \times 10^{-9}$ | $KF$ | $5.4 \times 10^{-11}$ |
| $H_2S_2$ | $1.6 \times 10^{-5}$ | $SbCl_3$ | $2.6 \times 10^{-8}$ | $SiCl_3$ | $1.2 \times 10^{-9}$ | $AlClF_2$ | $5.4 \times 10^{-11}$ |
| $COS$ | $1.4 \times 10^{-5}$ | $PbS$ | $2.5 \times 10^{-8}$ | $As_2$ | $1.1 \times 10^{-9}$ | $As$ | $4.7 \times 10^{-11}$ |
| $NaCl$ | $1.4 \times 10^{-5}$ | $S2Cl$ | $2.4 \times 10^{-8}$ | $SiHCl_3$ | $1.1 \times 10^{-9}$ | $HClO$ | $4.0 \times 10^{-11}$ |
| $HS$ | $8.9 \times 10^{-6}$ | $Cd$ | $2.1 \times 10^{-8}$ | $CrO_2Cl_2$ | $1.1 \times 10^{-9}$ | $LiBr$ | $3.3 \times 10^{-11}$ |
| $KCl$ | $8.7 \times 10^{-6}$ | $MnCl_2$ | $1.7 \times 10^{-8}$ | $AlCl_3$ | $1.0 \times 10^{-9}$ | $SiCl_2$ | $2.9 \times 10^{-11}$ |
| $HBr$ | $6.8 \times 10^{-6}$ | $(KCl)_2$ | $1.7 \times 10^{-8}$ | $CS_2$ | $1.0 \times 10^{-9}$ | $TiOCl_2$ | $2.9 \times 10^{-11}$ |
| $S_3$ | $4.1 \times 10^{-6}$ | $SiHF_3$ | $1.6 \times 10^{-8}$ | $BiS$ | $7.8 \times 10^{-10}$ | $TiF3$ | $2.5 \times 10^{-11}$ |
| $SiOF_2$ | $2.1 \times 10^{-6}$ | $NaBr$ | $1.6 \times 10^{-8}$ | $Zn$ | $7.2 \times 10^{-10}$ | $Sb$ | $2.4 \times 10^{-11}$ |
| $FeCl_2$ | $1.6 \times 10^{-6}$ | $SiO$ | $1.3 \times 10^{-8}$ | $Fe(OH)_2$ | $7.0 \times 10^{-10}$ | $LiF$ | $1.3 \times 10^{-11}$ |
| $AsS$ | $9.8 \times 10^{-7}$ | $SbS$ | $1.3 \times 10^{-8}$ | $SrCl_2$ | $5.3 \times 10^{-10}$ | $Mg(OH)_2$ | $1.2 \times 10^{-11}$ |
| $ZnCl_2$ | $6.4 \times 10^{-7}$ | $SO_3$ | $1.2 \times 10^{-8}$ | $SiS$ | $4.6 \times 10^{-10}$ | $H_2SO_4$ | $1.2 \times 10^{-11}$ |
| $SiCl_4$ | $5.4 \times 10^{-7}$ | $KBr$ | $9.3 \times 10^{-9}$ | $AlCl_2$ | $4.0 \times 10^{-10}$ | | |
| $CaCl_2$ | $4.6 \times 10^{-7}$ | $PbCl$ | $9.0 \times 10^{-9}$ | $NaOH$ | $3.3 \times 10^{-10}$ | | |
| $(NaCl)_2$ | $2.9 \times 10^{-7}$ | $S$ | $7.9 \times 10^{-9}$ | $Cl_2$ | $3.0 \times 10{-10}$ | | |

and direct sampling of active plumes requires airborne sampling missions (e.g. Rose et al. 1980, Chuan et al. 1987). Suitable instrumentation that can withstand the turbulence, abrasiveness and corrosive effects of direct contact with the plume requires special consideration, and the sampling itself can be risky. Furthermore, direct sampling only addresses a small fraction of the entire plume; it is not uncommon for a sampling device to have an intake of 1–10cm in diameter to sample a plume that could have a diameter of more than 1 km.

Because of the limitations of direct sampling mentioned above, remote sensing is an attractive tool for plume studies. The operator and the instrument are removed from the plume itself, and the synoptic perspective allows the entire plume to be studied. Several types of instruments, originally designed for other purposes, have been adapted to remotely sense volcanic plumes: cameras, spectrometers, radars and lidars can all be operated from the ground, from aircraft and from satellite platforms.

This chapter focuses on the most widely used volcanic plume remote sensing instruments. These include: a portable spectrometer; the correlation spectrometer (COSPEC); and satellite-

borne instruments, total ozone mapping spectrometer (TOMS) and advanced very high resolution radar (AVHRR). Works published from 1983 to 1991on the use of remote sensing
instruments for volcanic plume monitoring are given in the list of references at the end of
this chapter.

## 11.2 COSPEC

Very good descriptions of the design and field use of the COSPEC are given by Stoiber et al.
(1983), and these aspects will thus only be summarized here.

The COSPEC is a portable spectrometer which measures the absorption of solar ultraviolet
light by means of the sulphur dioxide ($SO_2$) molecule (Millán & Hoff 1978). The amount of
absorption is related to the burden of $SO_2$ in the field of view of the COSPEC. Cross-sectional
profiles of $SO_2$ burden in the volcanic plume are obtained by scanning the COSPEC through
the plume. These profiles are then multiplied by the plume width and plume speed to obtain
a flux, or mass of $SO_2$ emitted per unit time, often reported in metric tonnes per day or their
Système International (SI) equivalent (megagrams per day).

The COSPEC is the most widely used, non-satellite instrument for the remote sensing of
volcanic plume compositions. Comparable instruments for the monitoring of water vapour
($H_2O$) and carbon dioxide ($CO_2$), which usually represent the most abundant magmatic contributions in volcanic plumes, have not been developed or made widely available. This is
partially due to the high background of $H_2O$ and $CO_2$ occurring naturally in the troposphere
(Table 11.1). Separation of the signal generated by the volcanic flux of these species from
the signal generated by the natural background is a technological challenge. Similarly,
instruments for the detection of species less abundant than $SO_2$ have not been employed partly
because of their low concentration in volcanic plumes and/or because of the difficulty in
isolating a usable spectroscopic signature of the species (Table 11.1). These challenges may
be met by new applications using methods such as Fourier transform infrared (FTIR)
spectroscopy, laser Raman spectroscopy or by infrared (IR) or ultraviolet (IR) spectroscopy.
For example, Notsu et al. (1993) have demonstrated that IR spectral radiometers are useful
for measuring $SO_2$ and perhaps other volcanic gases.

The reliance upon the COSPEC for studies of volcanic plumes is substantial and has expanded in the last 10 years. Stoiber et al., (1983) gave a listing of COSPEC obtained $SO_2$ fluxes,
which is updated in Table 11.3. Additional fluxes reported in the *Scientific Event Alert Network (SEAN) Bulletin* and *Bulletin of the Global Volcanism Network (GVN)* of the Smithsonian
Institution are not included in this table because of the preliminary nature of that data.

Since the Stoiber et al. (1983) listing, at least 6300 measurements have been taken at a
minimum of 17 volcanoes in a wide variety of eruptive states (Table 11.3). Because of the
continuum of eruptive states, different sulphur contents of different magmas and generally
small datasets for each volcano, the database (which may initially appear large) is really quite
small, representing only a minuscule fraction of variable emission rates.

**Table 11.3** Compilation of published SO$_2$ fluxes at volcanoes from 1983 to 1991. Consultation of the reference will give more details about the measurements, i.e. COSPEC methodology, wind speed, average, standard deviation, dates and style of eruptive or non-eruptive activity. Volcanoes are listed in alphabetical order and fluxes under each volcano in chronological order.

| Volcano | Number of measurements | Activity | SO$_2$ flux (Mg/d) | Reference |
|---|---|---|---|---|
| Arenal | 5 | fuming, lava | 210 ± 30 | Casadevall al. (1984b) |
| | 3 | fuming, lava | 160 ± 35 | Casadevall et al. (1984b) |
| Augustine | 5 | eruptive | 23 820 ± 3300 | Rose et al. (1988) |
| | 5 | fuming | 380 ± 45 | Symonds et al. (1990) |
| | 5 | fuming | 45 ± 7 | Symonds et al. (1990) |
| | 15 | fuming | 27 ± 6 | Symonds et al. (1990) |
| Colima | 1 | fuming, lava | 230 | Casadevall et al. (1984b) |
| | 8 | fuming, lava | 320 ± 50 | Casadevall et al. (1984b) |
| Erebus | 27 | fuming | 230 ± 90 | Rose et al. (1985a) |
| | 41 | eruptive, fuming | 25 ± 10 | Kyle et al. (1990, 1993) |
| | 67 | fuming | 15 ± 7 | Kyle et al. (1990, 1993) |
| | 195 | eruptive, fuming | 21 ± 11 | Meeker et al. (1989), Kyle et al. (1990, 1993) |
| | 327 | fuming | 44 ± 27 | Kyle et al. (1990, 1993) |
| | 777 | fuming | 27 ± 9 | Kyle et al. (1993) |
| | 1170 | fuming | 52 ± 21 | Kyle et al. (1993) |
| | 513 | fuming | 71 ± 20 | Kyle et al. (1993) |
| Etna | >7 | fuming | 2300–5000 | Martin et al. (1986) |
| | 126 | fuming | 1056–921 | Andres et al. (1989a) |
| | 106 | fuming | 1120 ± 987 | Andres et al. (1993a) |
| | 374 | various | 1000–25 000 | Allard et al. (1991) |
| Fuego | >131 | eruptive, fuming | 160 | Andres et al. (1993b) |
| | 11 | fuming | 160 ± 70 | Andres et al. (1993b) |
| Kilauea* | 302 | fuming | 171 ± 52 | Connor et al. (1988) |
| | 233 | fuming | 170 ± 50 | Casadevall et al. (1987) |
| | 172 | fuming | 260 ± 90 | Casadevall et al. (1987) |
| | n.r. | fuming | 220 | Greenland et al. (1985) |
| | n.r. | eruptive | > 10 000 | Greenland et al. (1985) |
| | 906 | fuming | 167 ± 83 | Chartier et al. (1988) |
| | 37 | fuming | 1170 ± 400 | Andres et al. (1989b) |
| Lascar | 99 | fuming | 2300 ± 1120 | Andres et al. (1991) |
| Lonquimay | 54 | eruptive | 2380 ± 2720 | Andres et al. (1991) |
| Masaya | n.r. | fuming | <100–6000 | Stoiber et al. (1986) |
| Mount St Helens | n.r. | eruptive, fuming | 0–2600 | Casadevall et al. (1983) |
| Pacaya | >201 | eruptive, fuming | 260 | Andres et al. (1993b) |
| | 78 | eruptive | 30 ± 30 | Andres et al. (1993b) |
| Poás | n.r. | fuming | 600 | R.E. Stoiber (as cited in Prosser & Carr 1987) |
| | 5 | fuming | 810 ± 420 | Casadevall et al. (1984b) |
| | 5 | fuming | 700 ± 180 | Casadevall et al. (1984b) |
| | n.r. | n.r. | 500 | Stoiber et al. (1986) |
| | 11 | fuming | 90 ± 30 | Andres et al. (1991) |
| Redoubt | n.r. | fuming | 2000 | Hobbs et al. (1991) |
| Ruiz | >54 | fuming | 40–10 000 | Williams et al. (1986, 1990) |
| Santiaguito | >130 | eruptive, fuming | 80 | Andres et al. (1993b) |
| | 24 | fuming, lava | 50 ± 20 | Andres et al. (1993b) |
| White | 7 | fuming | 1230 ± 300 | Rose et al. (1986) |
| Island | 7 | fuming | 320 ± 120 | Rose et al. (1986) |
| | 5 | fuming | 350 ± 150 | Rose et al. (1986) |

* Fluxes from the Kilauea summit and the East Rift Zone are reported here. n.r., not reported.

## 11.3 Activity and SO₂ emission

One goal of COSPEC studies has been to use the $SO_2$ flux data to constrain the masses of magma that are degassing at or near the surface. That a general correlation between the $SO_2$ flux and volcanic activity exists has been clear since the beginning of volcanic $SO_2$ measurements (Stoiber et al. 1983). Data that demonstrate an exponential decrease in emissions correlating with decreasing volcanic activity have recently been accumulated at Mount St Helens (McGee 1992). Less robust datasets demonstrate the same thing at other volcanoes such as Augustine (Symonds et al. 1990) and Galunggung (Baddrudin et al. 1983). There are also numerous examples of data that suggest a direct relationship between activity and $SO_2$ (Casadevall et al. 1983, Greenland et al. 1985, Martin et al. 1986, Casadevall et al. 1987, Chartier et al. 1988, Andres et al. 1989b, Kyle et al. 1990, Symonds et al. 1990, Kyle et al. 1993).

However, not all COSPEC-obtained datasets show this positive correlation of $SO_2$ flux with eruptive activity. It has been suggested that tidal influences raise the level of $SO_2$ emissions without changing the state of eruptive activity (Stoiber et al. 1986, Connor et al. 1988). Other datasets show $SO_2$ emission variations of hundreds of per cent without any visible changes in eruptive activity (Andres et al. 1989a, 1991). Apparently, subsurface phenomena are responsible for these variations.

Several uncertainties may contribute to the variable $SO_2$ flux, and the relative contribution of each of these is unknown. We do not fully understand exsolution mechanisms in the magma (Turner et al. 1983, Whitney 1984, Carroll & Rutherford,1985, Gerlach 1986, Metrich et al. 1991) or bubble formation, growth and rise (Sparks 1978, Head & Wilson 1987, Toramaru 1989). These, in turn, control the style of eruptive activity (Blackburn et al. 1976, Kokelaar 1983, Vergniolle & Jaupart 1986, Greenland et al. 1988, Sahagian & Anderson 1991). Atmospheric processes such as attenuation of the solar UV light (Rose et al. 1985c), $SO_2$ to sulphate conversion (Martin et al. 1986, Gallagher et al. 1990) and wet and dry deposition (Finlayson-Pitts & Pitts 1986, Hicks et al. 1989, Kodosky & Keskinen 1990) also play a role in interpreting the data.

Investigators have tried to compare the volumes inferred from gas fluxes with observed magma volumes by using a melt inclusion based "petrologic" method (Anderson 1982, Rose et al. 1983, Devine et al. 1984, Palais & Sigurdsson 1989). Kilauea represents the best case of where these comparisons have volumes which match (Andres et al. 1989b). At Kilauea some $SO_2$ is lost from the summit, where magma is generally not vented to the surface, and the remainder is lost from vents along the East Rift, at or near the point where the lava itself emerges (Greenland et al. 1985, Gerlach & Graeber 1985). The gas flux measurements at Kilauea can be used, along with a wide variety of geochemical and geophysical data, to refine degassing processes and dynamics (Head & Wilson 1987, Greenland et al. 1988, Chartier et al. 1988), because there seems to be a general correlation between shallow magma volumes inferred from the amounts of gas emitted and the volumes known by other data. At a growing number of other volcanoes, chiefly of the circum Pacific, convergent plate boundary variety, such correlations do not work. There are now numerous examples of volcanoes in which the gas flux inferred volumes are much greater than the volumes observed at or near the surface (Casadevall et al. 1981, Rose et al. 1983, Williams et al. 1990, Andres et al. 1991,

Rose et al. 1991, Westrich & Gerlach 1992). The gas data require revision of our current understanding of degassing processes, and the details of this are not yet clear. At many, and perhaps most, circum Pacific volcanoes, $SO_2$ is lost from magma that is much deeper than the levels suggested by solubility experiments. While this discovery presently inhibits our models for inferring magmatic processes and possibly for eruptions, it may mean that eventually $SO_2$ measurements will give us insight into magma bodies at greater depths than we previously believed approachable.

## 11. 4 Improvements in COSPEC methodology

Improvements in COSPEC methodology are ultimately tied to a fundamental change in the way COSPEC data are viewed. In the early 1970s when the first COSPEC measurements were made (Moffat et al. 1972, Stoiber et al. 1983), the instrument was taken to as many volcanoes as possible and used to make spot measurements of the $SO_2$ emissions in order to characterize these emissions in terms of their contribution to the atmosphere. As time progressed, the characterization of $SO_2$ emissions began to show their variability. This variability was seen both between volcanoes and at one volcano over time-scales from minutes to years. COSPEC data then began to be viewed in terms of magmatic and volcanic processes. To use the data in this light requires larger datasets.

Large datasets have been created by either collecting small amounts of data every day for long time periods or by collecting large amounts of data over relatively short time periods. Examples of the former include COSPEC data collected at Kilauea, Hawaii (Casadevall et al. 1987) and Mount St Helens, Washington (Casadevall et al. 1983), which show long-term trends in $SO_2$ emissions that can be related to magmatic and volcanic processes occurring at these volcanoes. Examples of the latter have shown relationships between $SO_2$ emission rates and shallow magmatic processes such as convection rates and puffing (Andres 1988, Chartier et al. 1988, Andres et al. 1989a), seismicity (Connor et al. 1988) and magma migration (Andres 1988, Andres et al. 1989a, 1991). However, these manually collected datasets still do not, however, contain enough detail to investigate adequately the role of magmatic and volcanic processes in relationship to $SO_2$ emission rates.

Automation of the manual COSPEC technique has recently been accomplished (Kyle & McIntosh 1989, Kyle et al. 1993). This is a critical step in obtaining larger datasets that contain sufficient detail for better understanding of the contribution from various magmatic and volcanic processes which, in turn, contribute to the COSPEC signal. Automation involves replacing the tripod assembly, used to scan manually the volcanic plume, with an electric scanner assembly. Automation also includes electronically recording the output of the COSPEC into the memory of a portable computer.

There are three major advantages to automated COSPEC measurements over manual COSPEC measurements. Firstly, through longer measurement periods and more temporally closely spaced measurements larger datasets on $SO_2$ emission rates can be routinely collected. Secondly, error introduced by human operation is reduced by more consistent scanning of the COSPEC through the plume and more precise reduction of the data generated by the COSPEC. Thirdly, the standard deviation between consecutive measurements is reduced. This results from more closely spaced measurements, less human-induced error in the measurements and better plume speed determinations.

Plume speed is the greatest source of error in most COSPEC measurements. This error, which arises from using ground-obtained wind-speed measurements, visual observations of the moving plume or distant radiosonde data for the monitoring site, can be significantly reduced by simultaneous video recording of the plume being measured. Using the video recording and geometrical constraints, plume speeds can be determined for each COSPEC measurement with much greater accuracy.

The combination of COSPEC automation and video recording has resulted in the most detailed and statistically robust datasets ever recorded at a volcano (Kyle et al. 1993). These datasets provide for both proper use of simple statistics and for analysis with more sophisticated statistical techniques, which require larger sample populations. Continued acquisition of these datasets will allow for critical study of the relationships between $SO_2$ emission rates and magmatic or volcanic processes.

## 11.5 Remote sensing of other gases

Thermodynamic studies of magma degassing suggest which gases should evolve from magmas (Table 11.2). There are very few data that directly measure emission rates of nearly all of these gases. After $SO_2$, $CO_2$ is the only species that has been repeatedly and specifically measured (Harris et al. 1981). Because the method used for $CO_2$ measurements is more hazardous and less sensitive and because there is a high atmospheric background of $CO_2$, this has been done quite infrequently. There are data for Mount St Helens (Harris et al. 1981, McGee 1992) and for Kilauea (Greenland et al. 1985) which show the possible value of measurements of more than one gas species. Because one species (e.g. $CO_2$) may be less soluble and, therefore, be degassed prior to another, a potential exists for detecting intrusions and forecasting activity (see Gerlach and Graeber 1985). If, however, $CO_2$ routinely escapes after intrusion and prior to eruption, there may be problems detecting its release through faults and fractures scattered over the volcanic edifice (Allard et al. 1991).

Techniques designed to measure directly fluxes of other species are even more rare. One common approach has been to combine COSPEC measurements of the $SO_2$ flux with direct gas samples to obtain the ratios of $SO_2$ to other species, so that fluxes of other species can be calculated (Rose et al. 1986). This has been applied at many volcanoes and data on the flux of many major and minor species in plumes has been obtained (Casadevall et al. 1984b, Martin et al. 1986, Olmez et al. 1986, Stoiber et al. 1986, Le Guern et al. 1988, Rose et al. 1988, Andres et al. 1989a, Meeker et al. 1989, Kyle et al. 1990, Marty & Giggenbach 1990, Allard et al. 1991, Hobbs et al. 1991, Kodosky et al. 1991). Using similar approaches, global volcanic flux estimates for halogen gases (Symonds et al. 1988), $CO_2$ (e.g. Williams et al. 1992), lead (Patterson & Settle 1987, 1988), gold (Meeker et al. 1991), mercury (Siegal & Siegal

1984) and of trace gases and metals (Gemmell 1987, Symonds et al. 1987, 1990) have been published.

Using an indirect, COSPEC-based method is fundamentally inferior to actual measurement of the individual species. Development of a method to simultaneously measure fluxes of several species is necessary (Rose et al. 1989). Future development of new remote sensing flux methods, necessary so that fluxes of chemically important species can be better quantified and their variablity understood, may be the thrust for the next decade for gas researchers. Ultimately, accurate flux measurements of multiple species can be vital inputs to volcanic eruption forcasting and to volcano-climate studies.

## 11.6 Satellite methods

A revolutionary and promising new capability for remote flux measurements is satellite-based remote sensing. Satellite platforms are superior to ground based ones when the eruption is large, because the geometry of measuring a large opaque plume is difficult from the ground (Fig. 11.1). Since it is desirable to measure gas fluxes during eruptions when flux rates are highest and when material is most likely to reach the upper atmosphere and be widely dispersed, the satellite approach offers a capability not possible from the ground. During the past decade, a lot of satellite data concerning eruptions have been assembled. Meteorological satellites can map the dispersal of eruption clouds and can also estimate the altitude of the cloud by its equilibrated temperature (Sawada 1983, Matson 1984, Malingreau & Kaswanda 1986, Sawada 1987, 1989, Matson 1989). It is also possible to discriminate between eruption clouds and meteorological clouds with the AVHRR instrument on polar-orbiting meteorological satellites (Prata 1989, Holasek & Rose 1991, Schneider et al. 1994). The most important satellite for volcanic gas measurements is the Nimbus 7, which carries the total ozone mapping spectrometer (TOMS) instrument, a UV spectrometer designed for ozone mapping in the stratosphere, but which also can map volcanic $SO_2$ release of many recent eruptions (Krueger et al. 1990). These TOMS data are the first measurements of eruptive gas releases during explosive activity that reaches the upper troposphere or stratosphere. Other satellite detectors that have been used for volcanic eruption clouds are the solar backscattered UV (SBUV) instrument (McPeters et al. 1984), the solar mesosphere explorer (SME) (Barth et al. 1983, Clancy 1986, Naudet & Thomas 1987), the landsat thematic mapper (TM) (Glaze et al. 1989, Andres & Rose 1983) and the stratospheric aerosol measurement (SAM) (McCormick & Trepte 1987), which is particularly good at determining optical depth. A broad new satellite capability for volcanic clouds is planned for the National Aeronautics and Space Administration (NASA) Earth observing system (EOS) (Mouginis-Mark et al. 1991). An airborne multispectral thermal IR spectrometer (TIMS) has been used to estimate the $SO_2$ flux at Etna by Realmuto et al. (1993). This instrument is a prototype of a satellite instrument to be used in the EOS mission, so its success shows that IR spectrometry may have wide applicability in monitoring volcanic gases.

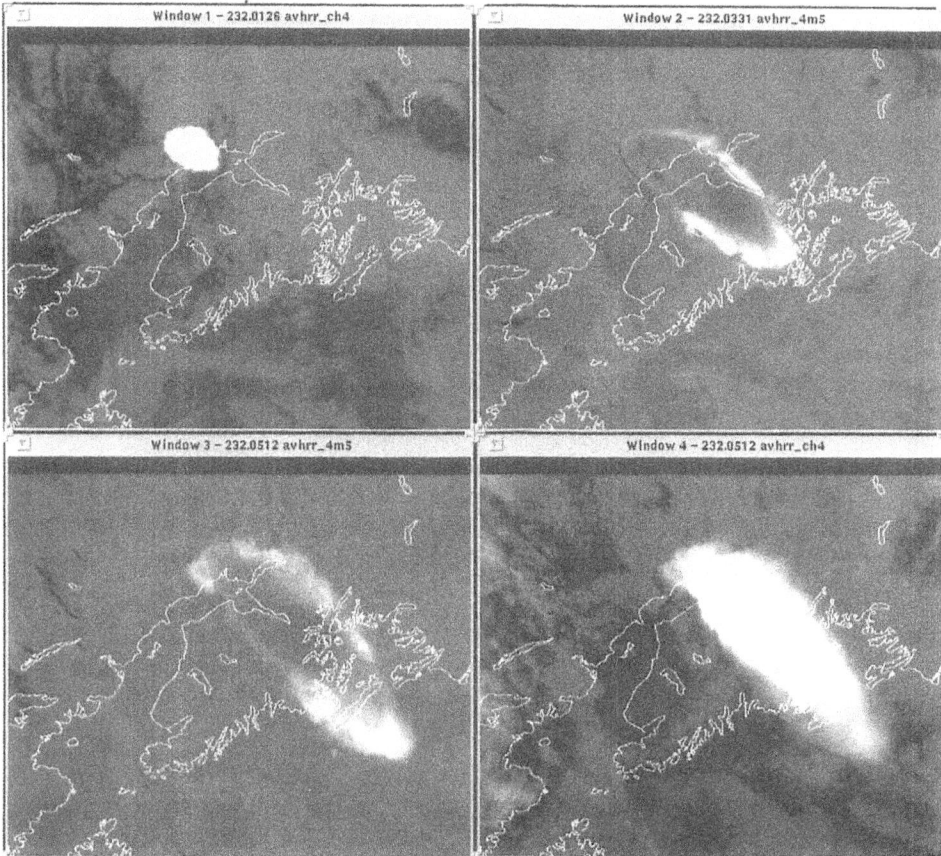

**Figure 11.1** Four satellite images of the 19 August 1992 eruption cloud from Crater Peak/Spurr volcano in Alaska (Schneider et al. 1994). The satellite imagery provides a synoptic view of the eruption and allows the volcanic cloud to be tracked and outlined. The four images shown are taken at 01.26, 03.31, 05.12 and 05.12 Greenwich mean time (GMT), respectively, on 19 August. The images shown in windows 1 and 4 (top left and bottom right, respecitvely) are thermal infrared channel-4 images from the AVHRR sensor, while windows 2 and 3 (top right and bottom left, respectively) show the cloud with band 4 minus band 5.

# References

Allard, P., J. Carbonnelle, D. Daajlevic, J. Le Bronec, P. Morel, M. C. Robe, J. M. Maurenas, R. Faivre-Pierret, D. Martin, J-C. Sabroux, P. Zettwoog 1991. Eruptive and diffuse emissions of $CO_2$ from Mount Etna. *Nature* **351**, 387–91.

Anderson, A. T. 1982. Parental basalts in subduction zones: inplications for continental evolution. *Journal of Geophysical Research* **87**, 7047–60.

Andres, R. J. 1988. Sulphur dioxide and particle emissions from Mount Etna, Italy. MSc thesis, New Mexico Institute of Mining and Technology. Socorro, New Mexico.

Andres, R. J. & W. I. Rose 1993. Detection of thermal anomalies associated with three active Guatemalan volcanoes with Landsat Thematic Mapper imagery. *Photogrammetric Engineering and Remote Sensing* in press.

Andres, R. J., P. R. Kyle, R. L. Chuan 1989a. Sulphur dioxide and particle emissions of Mount Etna, Italy from June to August 1987. *New Mexico Bureau of Mines and Mineral Resources Bulletin* **131**, 7.

Andres, R. J., P. R. Kyle, J. B. Stokes, W. I. Rose 1989b. Sulphur dioxide emissions from episode 48A, East Rift Zone eruption of Kilauea volcano, Hawaii. *Bulletin of Volcanology* **52**, 113–7.

Andres, R. J., W. I. Rose, P. R. Kyle, S. deSilva, P. Francis, M. Gardeweg, H. Moreno Roa 1991. Excessive sulphur dioxide emissions from Chilean volcanoes. *Journal of Volcanology and Geothermal Research* **46**, 323–9.

Andres, R. J., J. Barquero, W. I. Rose 1992. New measurements of $SO_2$ flux at Poás Volcano, Costa Rica. *Journal of Volcanology and Geothermal Research* **49**, 175–7.

Andres, R. J., P. R. Kyle, R. L. Chuan 1993a. Sulphur dioxide, particle and elemental emissions from Mount Etna, Italy during July 1987. *Geologishe Rundschau* in press.

Andres, R. J., W. I. Rose, R. E. Stoiber, S. N. Williams, O. Matiás, R. Morales 1993b. A history of sulphur dioxide emission rate measurements from Guatemalan volcanoes. *Bulletin of Volcanology* **55**, 379–88.

Baddrudin, M. 1986. Pancaran gas pada Letusan G. Galunggung 1982. In *Letusan Galunggung 1982–82*. J. A. Katili, A. Sudradjat, K. Kumumadinata (eds), 285–301. Direktorat Vulkanologi, Direktorat Geologi Dan Sumberdaya Mineral, Departmen Pertambangan Dan Energi, Bandung.

Barth, C. A., R. W. Sanders, R. J. Thomas, G. E. Thomas, B. M. Jakosky, R. A. West 1983. Formation of the El Chichón aerosol cloud. *Geophysical Research Letters* **10**, 993–6.

Blackburn, E. A., L. Wilson, R. S. J. Sparks 1976. Mechanisms and dynamics of strombolian activity. *Quarterly Journal of the Geological Society of London* **132**, 429–40.

Carroll, M. R. & M. J. Rutherford 1985. Sulfide and sulfate saturation in hydrous silicate melts. *Journal of Geophysical Research* **90**, C601–12.

Casadevall, T. J., D. A. Johnston, D. M. Harris, W. I. Rose, L. L. Malinconico, R. E. Stoiber, T. J. Bornhorst, S. N. Williams 1981. $SO_2$ emission rates at Mount. St Helens from March 29 through December 1980. In *United States Geological Survey Professional Paper 1250, The 1980 eruptions of Mount St Helens, Washington*, P. W. Lipman & D. R. Mullineaux (eds), 193–200. Washington DC: United States Government Printing Office.

Casadevall, T., W. Rose, T. Gerlach, L. P. Greenland, J. Ewert, R. Wunderman, R. Symonds 1983. Gas emissions and the eruptions of Mount St Helens through 1982. *Science* **221**, 1383–85.

Casadevall, T., S. de la Cruz, W. I. Rose, S. Bagley, D. Finnegan, W. Zoller 1984a. Volcano El Chichon, Mexico: the crater lake and thermal activity. *Journal of Volcanology and Geothermal Research* **23**, 169–91.

Casadevall, T. J., W. I. Rose, Jr, W. H. Fuller, W. H. Hunt, M. A. Hart, J. L. Moyers, D. C. Woods, R. L. Chuan, J. P. Friend 1984b. Sulfur dioxide and particles in quiescent volcanic plumes from Poás, Arenal, and Colima volcanos, Costa Rica and Mexico. *Journal of Geophysical Research* **89**, 9633–41.

Casadevall, T. J., J. B. Stokes, L. P. Greenland, L. L. Malinconico, J. R. Casadevall, B. T. Furukawa 1987. $SO_2$ and $CO_2$ emission rates at Kilauea volcano 1979–1984. In *United States Geological Survey Professional Paper 1350, Volcanism in Hawaii*, R. W. Decker, T. L. Wright, P. H. Stauffer (eds), 771–80. Washington DC: United States Government Printing Office.

Chartier, T. A., W. I. Rose, J. B. Stokes 1988. Detailed record of $SO_2$ emissions from Pu'u 'O'o between episodes 33 and 34 of the 1983-86 ERZ eruption, Kilauea, Hawaii. *Bulletin of Volcanology* **50**, 215–28.

Chuan, R. L., J. Palais, W. I. Rose, P. R. Kyle 1987. Particle sizes and fluxes of the Mt. Erebus volcanic plume. *Journal of Atmospheric Chemistry* **4**, 467–77.

Clancy, R. T. 1986. El Chichón and "mystery cloud" aerosols between 30 and 55 km: global observations from the SME visible spectometer. *Geophysical Research Letters* **13**, 937–40.

Connor, C. B., R. E. Stoiber, L. L. Malinconico, Jr 1988. Variation in sulphur dioxide emissions related to earth tides, Halemaumau crater, Kilauea volcano, Hawaii. *Journal of Geophysical Research* **93**, 867–71.

Deluisi, J. J., B. G. Mendonca, E. G. Dutton, M. A. Box, B. M. Herman 1983b. Radiative propeties of the stratospheric dust cloud from the May 18 1980, eruption of Mount St Helens. *Journal of Geophysical Research* **88**, 5290–8.

Devine, J. D. H. Sigurdsson, A. N. Davis, S. Self 1984. Estimates of sulfur and chlorine yield to the atmosphere from volcanic eruptions and potential climate effects. *Journal of Geophysical Research* **89**, 6309–25.

Finlayson-Pitts B. J. & J. N. Pitts, Jr 1986. *Atmospheric chemistry: Fundamentals and experimental techniques*. New York: John Wiley.

Friend, J. P., A. R. Bandy, J. L. Moyers, W. H. Zoller, R. E. Stoiber, A. L. Torres, W. I. Rose, M. P. McCormick, D. C. Woods 1982. Research on active volcanic emissions: an overview. *Geophysical Research Letters* **9**, 1101–4.

Gallagher, M. W., R. M. Downer, T. W. Choularton, M. J. Gay, I. Stromberg, C. S. Mill, M. Radojevic, B. J. Tyler, B. J. Bandy, S. A. Penkett, T. J. Davies, G. J. Dollard, B. M. R. Jones 1990. Case studies of the oxidation of sulphur dioxide in a hill cap cloud using ground and aircraft based measurements. *Journal of Geophysical Research* **95**, 517–37.

Gemmell, J. B. 1987. Geochemistry of metallic trace elements in fumarolic condensates from Nicaraguan and Costa Rican volcanoes. *Journal of Volcanology and Geothermal Research* **33**, 161–81.

Gerlach, T. M. 1986. Exsolution of $H_2O$, $CO_2$, and S during eruptive episodes at Kilauea volcano, Hawaii. *Journal of Geophysical Research* **91**, 177–85.

Gerlach, T. M. & E. J. Graeber 1985. Volatile budget of Kilauea volcano. *Nature* **313**, 273–7.

Glaze, L. S., P. W. Francis, S. Self, D. A. Rothery 1989. The 16 September 1986 eruption of Lascar volcano, north Chile: satellite investigations. *Bulletin of Volcanology* **51**, 149–60.

Graedel, T. E. 1978. *Chemical compounds in the atmosphere*. New York: Academic Press.

Greenland, L. P., W. I. Rose, J. B. Stokes 1985. An estimate of gas emissions and magmatic gas content from Kilauea volcano. *Geochimica et Cosmochimica Acta* **49**, 125–9.

Greenland, L. P., A. T. Okamura, J. B. Stokes 1988. Constraints on the mechanics of the eruption. In *United States Geological Survey Professional Paper 1463*, 155–64. Washington DC: United States Government Printing Office.

Harris, D. M., M. Sato, T. J. Casadevall, W. I. Rose Jr, T. J. Bornhorst 1981. Emission rates of $CO_2$ from plume measurements. In *United States Geological Survey Professional Paper 1250, The 1980 eruptions of Mount St Helens, Washington*, P. W. Lipman & D. R. Mullineaux (eds), 201–7. Washington DC: United States Government Printing Office.

Head III, J. W. & L. Wilson 1987. Lava fountain heights at Pu'u 'O'o, Kilauea, Hawaii: indicators of amount and variations of exsolved magma volatiles. *Journal of Geophysical Research* **92**, 715–9.

Hicks, B. B., T. P. Meyers, C. W. Fairall, V. A. Mohnen, D. A. Dolske 1989. Ratios of dry to wet deposition of sulphur as derived from preliminary field data. *Global Biogeochemical Cycles* **3**, 155–62.

Hobbs, P. V., L. F. Radke, J. H. Lyons, R. J. Ferek, D. J. Coffman, T. J. Casadevall 1991. Airborne measurements of particle and gas emissions from the 1990 volcanic eruptions of Mount Redoubt. *Journal of Geophysical Research* **96**, 735–52.

Holasek, R. E. & W. I. Rose 1991. Anatomy of 1986 Augustine volcano eruptions as recorded by multispectral image processing of digital AVHRR weather satellite data. *Bulletin of Volcanology* **53**, 420–35.

Kodosky, L. & M. Keskinen 1990. Fumarole distribution, morphology, and encrustation mineralogy associated with the 1986 eruptive deposits of Mount St. Augustine, Alaska. *Bulletin of Volcanology* **52**, 175–85.

Kodosky, L. G., R. J. Motyka, R. B. Symonds 1991. Fumarolic emissions from Mount St. Augustine, Alaska: 1979–1984 degassing trends, volatile sources and their possible role in eruptive style. *Bulletin of Volcanology* **53**, 381–94.

Kokelaar, B. P. 1983. The mechanism of Surtseyan volcanism. *Quarterly Journal of the Geological Society of London* **140**, 939–44.

Krueger, A. J., L. S. Walter, C. C. Schnetzler, S. D. Doiron 1990. TOMS measurement of the sulphur dioxide emitted during the 1985 Nevado del Ruiz eruptions. *Journal of Volcanology and Geothermal Research* **41**, 7–15.

Kyle, P. R. & W. C. McIntosh 1989. Automation of a correlation spectrometer for measuring volcanic $SO_2$ emissions. *New Mexico Bureau of Mines and Mineral Resources Bulletin* **131**, 158.

Kyle, P. R., K. Meeker, D. Finnegan 1990. Emission rates of sulphur dioxide, trace gases and metals from Mount Erebus, Antarctica. *Geophysical Research Letters* **17**, 2125–8.

Kyle, P. R., L. M. Sybeldon, W. C. McIntosh, K. Meeker, R. Symonds 1993. Sulphur dioxide emission rates from Mount Erebus, Antarctica. In *Volcanological studies of Mount Erebus, Antarctica, Antarctic research series*, P. Kyle (ed.), in press. Washington DC: American Geophysical Union.

Lazrus, A. L., R. D. Cadle, B. W. Gandrud, J. P. Greenberg, B. J. Huebert, W. I. Rose 1979. Trace chemistry of the stratosphere and of volcanic eruption plumes. *Journal of Geophysical Research* **84**, 7869–75.

Le Guern, F., R. X. Favre-Pierret, J. P. Garrec 1988. Atmospheric contribution of volcanic sulphur vapor and its influence on the surrounding vegetation. *Journal of Volcanology and Geothermal Research* **35**, 173–8.

Malingreau, J-P. & Kaswanda 1986. Monitoring volcanic eruptions in Indonesia using weather satellite data: the Colo eruption of July 28 1983. *Journal of Volcanology and Geothermal Research* **27**, 179–94.

Malone, S. D. & D. Frank 1975. Increased heat emission from Mount Baker, Washington. *Transactions of the American Geophysical Union* **56**, 679–85.

Martin, D., B. Ardouin, G. Bergametti, J. Carbonnelle, R. Faivre-Pierret, G. Lambert, M. F. Le Cloarec, G. Sennequier 1986. Geochemistry of sulphur in Mount Etna plume. *Journal of Geophysical Research* **91**, 249–54.

Marty, B. & W. F. Giggenbach 1990. Major and rare gases at White Island volcano, New Zealand: origin and flux of volatiles. *Geophysical Research Letters* **17**, 247–50.

Matson, M. 1984. The 1982 El Chichón volcano eruptions – a satellite perspective. *Journal of Volcanology and Geothermal Research* **23**, 1–10.

Matson, M. 1989. Monitoring volcanic eruptions using meteorological satellite data. *New Mexico Bureau of Mines and Mineral Resources Bulletin* **131**, 178.

McCormick, M. P. & C. R. Trepte 1987. Polar stratospheric optical depth observed between 1978 and 1985. *Journal of Geophysical Research* **92**, 4297–306.

McGee, K. 1992. The structure, dynamics and chemical composition of non-eruptive plumes from Mount St Helens. *Journal of Volcanology and Geothermal Research* **51**, 269–82.

McPeters, R. D., D. F. Heath, B. M. Schlesinger 1984. Satellite observation of $SO_2$ from El Chichón: identification and measurement. *Geophysical Research Letters* **11**, 1203–6.

Meeker, K. A., P. R. Kyle, D. Finnegan, R. Chuan 1989. Chlorine and trace element emissions from Mount Erebus, Antarctica. *New Mexico Bureau of Mines and Mineral Resources Bulletin* **131**, 184.

Meeker, K. A., R. L. Chuan, P. R. Kyle, J. M. Palais 1991. Emission of elemental gold particles from Mount Erebus, Ross Island, Antarctica. *Geophysical Research Letters* **18**, 1405–8.

Metrich, N., H. Sigurdsson, P. S. Meyer, J. D. Devine 1991. The 1783 Lakagigar eruption in Iceland: geochemistry, $CO_2$ and sulphur degassing. *Contributions to Mineralogy and Petrology* **107**, 435–47.

Millán, M. M. & R. M. Hoff 1978. Remote sensing of air pollutant by correlation spectroscopy – instrumental response characteristics. *Atmospheric Environment* **12**, 853–64.

Moffat, A. J., T. Kakara, T. Akitomo, L. Langan 1972. *Air note. Environmental measurements*. San Francisco.

Mouginis-Mark, P., S. Rowland, P. Francis, T. Friedman, H. Garbeil, J. Gradie, S. Self, L. Wilson, J. Crisp, L. Glaze, K. Jones, A. Kahle, D. Pieri, H. Zebker, A. Krueger, L. Walter, C. Wood, W. Rose, J. Adams & R. Wolff 1991. Analysis of active volcanoes from the Earth Observing System. *Remote Sensing of the Environment*, **36**, 1–12.

Naudet, J. P. & G. E. Thomas 1987. Aerosol optical depth and planetary albedo in the visible from the solar mesosphere explorer. *Journal of Geophysical Research* **92**, 8373–81.

Notsu, K., T. Mori, G. Igarashi, Y. Tohjima, H. Wakita 1993. Infrared spectral radiometer: a new tool for remote measurement of $SO_2$ of volcanic or fumarolic gas. *Geochemical Journal* in press.

Olmez, I., D. L. Finnegan, W. H. Zoller 1986. Iridium emissions from Kilauea volcano. *Journal of Geophysical Research* **91**, 653–63.

Palais, J. M. & H. Sigurdsson 1989, Petrologic evidence of volatile emissions from major historic and prehistoric volcanic eruptions. In *American Geophysical Union Monograph 52, Understanding Climate Change*, A. Berger, R. E. Dickinson, J. W. Kidson (eds), 31–53. Washington DC: American Geophysical Union.

Patterson, C. C. & D. M. Settle 1987. Magnitude of lead flux to the atmosphere from volcanoes. *Geochimica et Cosmochimica Acta* **51**, 675–81.

Patterson, C. & D. Settle 1988. Erratum. *Geochimica et Cosmochimica Acta* **52**, 245.

Prata, A. J. 1989. Infrared radiative transfer calculations for volcanic ash clouds. *Geophysical Research Letters* **16**, 1293–6.

Prosser, J. T. & M. J. Carr 1987. Poás Volcano, Costa Rica: Geology of the summit region and spatial and temporal variations among the most recent lavas. *Journal of Volcanology and Geothermal Research* **33**, 131–46.

Realmuto, V. J., M. J. Abrams, M. F. Buongiomo, D. C. Pieri 1993. The use of multispectral thermal infrared image data to estimate the sulphur dioxide flux from volcanoes: a case study from Mount Etna, Sicily, 29 July 1986. *Journal of Geophysical Research* in press.

Rose, W. I., R. L. Chuan, R. D. Cadle, D. C. Woods 1980. Small particles in volcanic eruption clouds.

*American Journal of Science* **280**, 671–96.

Rose, W. I., R. E. Stoiber, L. L. Malinconico 1981. Eruptive gas compositions and fluxes of explosive volcanoes: problems, techniques and initial data. In *Orogenic andesites and related rocks*. R. S. Thorpe (ed.). New York: John Wiley.

Rose, W. I., R. L. Wunderman, M. F. Hoffman, L. Gale 1983. A volcanologist's review of atmospheric hazards of volcanic activity: Fuego and Mount St Helens. *Journal of Volcanology and Geothermal Research* **17**, 133–57.

Rose, W. I., R. L. Chuan, P. R. Kyle 1985. Rate of sulphur dioxide emission from Erebus volcano, Antarctica, December 1983. *Nature* **316**, 710–2.

Rose, W. I., R. B. Symonds, T. Chartier, J. B. Stokes, S. Brantley 1985b. Simultaneous experiments with two correlation spectrometers at Kilauea and Mount St Helens. *Transactions, American Geophysical Union* **66**, 1142.

Rose, W. I., R. L. Chuan, W. F. Giggenbach, P. R. Kyle, R. B. Symonds 1986. Rates of sulphur dioxide and particle emissions from White Island volcano, New Zealand, and an estimate of the total flux of major gaseous species. *Bulletin of Volcanology* **48**, 181–8.

Rose, W. I., G. Heiken, K. Wohletz, D. Eppler, S. Barr, T. Miller, R. L. Chuan, R. B. Symonds 1988. Direct rate measurements of eruption plumes at Augustine volcano: a problem of scaling and uncontrolled variables. *Journal of Geophysical Research* **93**, 4485–99.

Rose, W. I., R. B. Symonds, A. Bernard, C. A. Chesner, R. J. Andres 1989. Volcanic gas releases to the earth's atmosphere: some important problems and ideas for solutions. *New Mexico Bureau of Mines and Mineral Resources Bulletin* **131**, 227.

Sahagian, D. L. & A. T. Anderson 1991. Classification of Hawaiian eruption styles on the basis of mechanisms of volatile release. *Transactions, American Geophysical Union* **72**, 296.

Sawada, Y. 1983. Attempt on surveillance of volcanic activity by eruption cloud image from artificial satellite. *Bulletin of the Volcanological Society of Japan* **28**, 357–73.

Sawada, Y. 1987. Study on analysis of volcanic eruptions based on eruption cloud image data obtained by the Geostationary Meterological Satellite (GMS). *Technical Report of the Meterological Research Institue (Japan)* **22**, 230.

Sawada, Y. 1989. Simultaneous observations of eruption cloud of the November 21 1986 Izu-Oshima eruption with images of GMS and NOAA, and weather radar. *New Mexico Bureau of Mines and Mineral Resources Bulletin* **131**, 233.

Schneider, D. J., W. I. Rose, L. Kelley 1994. Tracking of 1992 Crater Peak/Spurr eruption clouds using AVHRR. *US Geological Survey Bulletin*, in press.

Schidlowski, M. 1986. The Atmosphere. In *The natural environment and the biogeochemical cycles*, 2, O. Hutzinger (ed.). Heidelberg: Springer.

Siegel, S. M. & B. Z. Siegel 1984. First estimate of annual mercury flux at the Kilauea main vent. *Nature* **309**, 146–7.

Sparks, R. S. J. 1978. The dynamics of bubble formation and growth in magmas: a review and analysis 3, 1–37.

Stoiber, R. E., L. L. Malinconico, Jr, S. N. Williams 1983. Use of the Correlation Spectrometer at Volcanoes. In *Forecasting volcanic events*, H. Tazieff & J-C. Sabroux (eds), 425–44. Amsterdam: Elsevier.

Stoiber, R. E., S. N. Williams, B. J. Huebert 1986. Sulphur and halogen gases at Masaya Caldera complex, Nicaragua: total flux and variations with time. *Journal of Geophysical Research* **91**, 215–31.

Symonds R. B. & M. H. Reed 1993. Calculation of multicomponent chemical equilibria in gassolid-liquid systems: calculation methods, thermochemical data, and applications to studies of hightemperature volcanic gases with examples from Mount St Helens. *American Journal of Science* **293**, in press.

Symonds, R. B., W. I. Rose, M. H. Reed, F. E. Lichte, D. L. Finnegan 1987. Volatilization, transport and sublimation of metallic and non-metallic elements in high temperature gases at Merapi volcano, Indonesia. *Geochimica et Cosmochimica Acta* **51**, 2083–101.

Symonds, R. B., W. I. Rose, M. H. Reed 1988. Contribution of Cl- and F-bearing gases to the atmosphere by volcanoes. *Nature* **334**, 415–8.

Symonds, R. B., W. I. Rose, T. M. Gerlach, P. H. Briggs, R. S. Harmon 1990. Evaluation of gases, condensates, and $SO_2$ emissions from Augustine volcano, Alaska: the degassing of a Cl-rich volcanic system. *Bulletin of Volcanology* **52**, 355–74.

Toramaru, A. 1989. Vesiculation process and bubble size distributions in ascending magmas with constant velocities. *Journal of Geophysical Research* **94**, 523–42.

Turco, R. P., O. B. Toon, R. C. Whitten, P. Hamill, R. G. Keesee 1983. The 1980 eruptions of Mount St Helens: physical and chemical processes in the stratospheric clouds. *Journal of Geophysical Research* **88**, 5299–319.

Turner, J. S., H. E. Huppert, R. S. J. Sparks 1983. An experimental investigation of volatile exsolution in evolving magma chambers. *Journal of Volcanology and Geothermal Research* **16**, 263–77.

Vergniolle, S. & C. Jaupart 1986. Separated two-phase flow and basaltic eruptions. *Journal of Geophysical Research* **91**, 842–60.

Westrich, H. R. & T. M. Gerlach 1992, Magmatic gas source for the stratospheric SO2 cloud from the June 15 1991 eruption of Mount Pinatubo. *Geology* **20**, 867–870.

Whitney, J. A. 1984. Fugacities of sulphurous gases in pyrrhotite-bearing silicic magmas. *American Mineralogist* **69**, 69–78.

Williams, S. N., R. E. Stoiber, N. Garcia P., A. Londoño C., J. B. Gemmell, D. R. Lowe, C. B. Connor 1986. Eruption of the Nevado del Ruiz volcano, Colombia, on 13 November 1985: gas flux and fluid geochemistry. *Science* **233**, 964–7.

Williams, S. N., N. C. Sturchio, M. L. Calvache V., R. Mendez F., A. Londoño C., N. García P. 1990. Sulphur dioxide from Nevado del Ruiz volcano, Colombia: total flux and isotopic constraints on its origin. *Journal of Volcanology and Geothermal Research* **42**, 53–68.

Williams, S. N., S. J. Schaefer, M. L. Calvache V., D. Lopez 1992. Global carbon dioxide emission to the atmosphere by volcanoes. *Geochimica et Cosmochimica Acta* **56**, 1765–70.

# 12 Monitoring fluids and gases at active volcanoes

D. Tedesco

## 12.1 Introduction

The monitoring of active volcanic areas and the prediction of volcanic eruptions falls within the scope of a number of major international earth science programmes operating during the final decade of this century. Within these programmes, learning more about the geochemistry of volcanic fluids and gases will play an important role in increasing our understanding of how volcanoes function. At the present time the modelling of active volcanoes using geochemical methods is still in its infancy, and to date there have been few opportunities to monitor the geochemical changes of fluids and gases through a volcanic crisis which has culminated in eruption. Nevertheless, a number of useful studies of fumarolic fluids, soil gases and thermal waters have been undertaken in several countries, with the result that a sound body of basic knowledge is available. It is now possible, in consequence, to develop and extrapolate this knowledge in order to monitor the geochemical behaviour of volcanic systems, and to attempt to predict eruptions through variations of geochemical parameters. As in all types of volcano monitoring, it should be stressed, it is imperative that extensive baseline data be gathered, revealing the "normal" geochemical variations associated with quiescence, so that these can be filtered out from a future geochemical signal which might indicate a forthcoming eruption. Volcanic gases are one of the most important sources of information on contemporary degassing from the Earth's interior. The chemical nature of such gases is highly variable, and differs not only from volcano to volcano, but also from fumarole to fumarole. In addition, there is considerable evidence to indicate that the chemical composition of volcanic gases from a specific vent of a volcano changes over time (Saint-Claire Deville & Leblanc 1858, Mizutani & Matsuo 1959).

Modern research on fumarolic fluids started in France in the second half of the 18th century with the work of Saint-Claire Deville et al. (1863), who classified fumarolic manifestations into three types:

(a) high temperature (dry fumarole), T = 1000°C;
(b) medium temperature (acid fumarole), T = 300°C; and
(c) low temperature (neutral water fumarole), T = 100°C.

Silvestri (1862) was also an early worker in the field, and was the first to recognize the abnormal ammonium chloride content of medium temperature, acid fumaroles, as well as to establish a relationship between acidity and fumarolic temperatures. In 1913 Jaggar (Jaggar 1913), and a few years later Shepherd (1921), sampled the first high-temperature gases from the Halemaumau lava lake at Kilauea volcano, leading to the first complete analyses of fumarolic fluids. Modern analytical techniques were first used by Sicardi (1955) who worked on high- and low-temperature fumaroles at Vulcano (Aeolian Islands), achieving remarkable results comparable with data aquired using contemporary techniques. In 1961, Matsuo published the results of a study of the thermodynamic equilibrium of volcanic gases on Showashinzan volcano (Hokkaido), in which he calculated equilibria temperatures at depth. The work of Matsuo (1961) is considered by many to form the basis of modern volcanic gas geochemistry. Little new work in this field was forthcoming, however, until the 1970s, when Giggenbach (1975) laid down some basic rules on sampling and analytical procedures for volcanic fluids. It was not until the early 1980s that results of long-term geochemical monitoring at a single site became available. This important work was undertaken in the Bay of Naples, and the incentive was provided by the Campi Flegrei caldera bradyseismic crisis that occurred between 1982 and 1985.

Several active volcanic areas have since been intensively studied from a geochemical point of view, including the Campi Flegrei caldera, Vulcano island, Kilauea, White island (New Zealand), and Kusatsu Shirane (Japan). However, the number of samples collected for each area is often insufficient to yield the complete spectrum of chemical and isotopic data desirable. In the past, geochemical studies of volcanoes have been largely discontinuous, involving the collection of large amounts of data over relatively short time periods, interspersed with years during which no geochemical analyses are undertaken. In most cases these data represent only brief moments in the life of any volcano, and are insufficient to provide a complete, working model of geochemical behaviour. Although these data are useful, it is also necessary to collect routinely hundreds of samples for each fumarolic area over a period of years. Such a sampling strategy is required in order to construct a baseline standard for each geochemical parameter and to quantify the influence on chemical and isotopic species of factors, such as the contribution of marine and meteoric waters, earthquakes, tides and atmospheric pressure variations. This type of geochemical monitoring is essential to minimize the likelihood, particularly in densely populated volcanic areas, of "false alarms" as soon as some geochemical parameter starts to vary, and has recently been successfully undertaken at Vulcano.

Experience suggests that, whether sampling low- or high-temperature fumarolic gases, it is almost impossible to access a pure, juvenile fluid. All samples should be considered to represent mixtures of deep and surficial fluids or, for example, magmatic and hydrothermal fluids. Only in cases of gases trapped in rocks as fluid inclusions is it possible to analyze the actual composition of gases before eruption. However, quantifying the degree of mixing and the compositional variations/oscillations occurring over time, can give, the exact proportion of each different end member participating in the mixing process. Variations in fumarolic fluids, resulting from mixing, usually occur over a very short time, and can only be detected and recorded by collecting a large number of samples during the period of the monitoring programme. Such rapid geochemical variations have recently been recorded at Solfatara (Pozzuoli, Bay of Naples) during and after the bradyseismic crisis of the 1980s (Carapezza et al. 1984, Cioni et al. 1984, Tedesco & Sabroux 1987), at the Halemaumau fumaroles

(Kilauea) (B. J. Stokes, personal. communication), and in the active fumaroles at Vulcano (see later in this chapter). It appears that fumaroles at all volcanoes which have been sufficiently closely studied display such short-term variations in chemical and isotopic parameters, often independently of what is strictly defined as "volcanic activity", and such behaviour may well be typical of all active volcanic systems. In this chapter I will look at geochemical variations in fumaroles in more detail, with particular reference to recent studies at Campi Flegrei and Vulcano island. I will also discuss, more briefly, geochemical changes in both thermal waters and soil gases at a number of volcanic localities.

## 12.2 Sampling and analytical techniques

Volcanic fluids and gases can be analyzed using a number of very different methods, although some attempt has been made in recent years to standardize analytical techniques. When sampling fumaroles, fluids are conveyed from the vents to the sampling flask through special quartz dewar tubes (for temperatures up to 500°C) or titanium tubes (for temperatures over 500°C). These are inert, preventing chemical reaction and condensation of water vapour during the

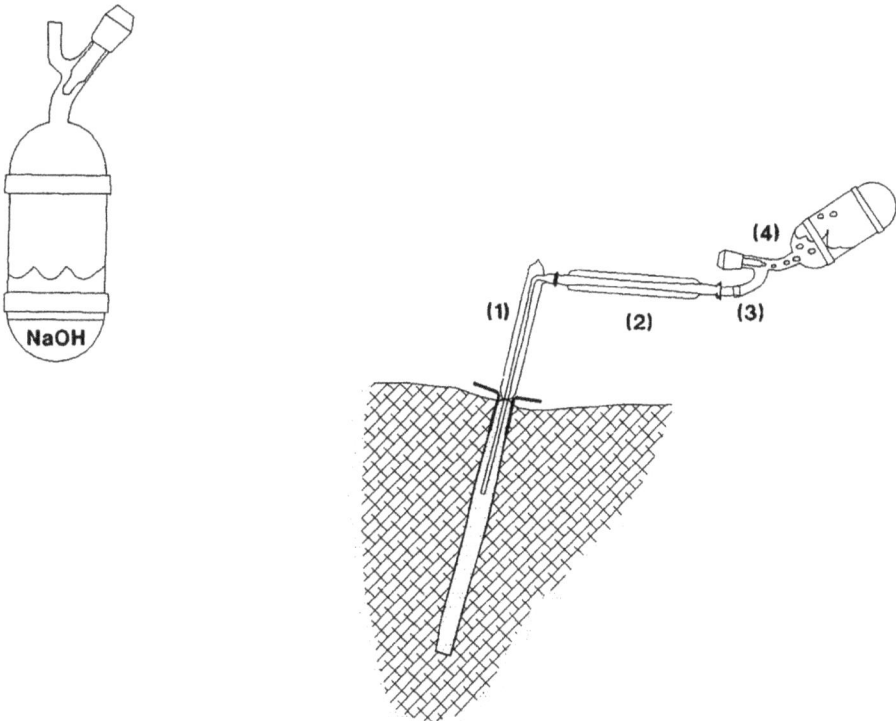

**Figure 12.1** Sampling device used at Campi Flegrei caldera and Vulcano. (1,2) Dewar quartz tubes; (3) pyrex and rubber connection (for temperatures below 300°C at the pipe exit); (4) pyrex flask.

sampling. The gas is usually collected for chemical analysis in a pre-evacuated flask partly filled with a 4 N sodium hydroxide absorbing solution (Giggenbach 1975), which may concentrate 100 times or so the original content of the so-called "uncondensable gases" (Fig. 12.1). Typically more than 1 mol of fumarolic fluids (dry gas + water vapour) is introduced into the sampling volume of around $250 \, cm^3$. The enrichment occurs because water vapour ($H_2O$) and acid species ($CO_2$, $SO_2$, $H_2S$, HCl and HF), known as the "condensable species", accounting for around 99% of the total gases, are trapped and dissolved in the alkaline solution until the solution saturated. At the same time the uncondensable species (He, $H_2$, $O_2$, Ar, $N_2$, CO and $CH_4$) remain out of alkaline solution. To analyze the gas compositions, a flame-ionization detector (FID) is used downstream from a methanizing oven, in series with a 3 m Porapak-R column at room temperature. Other gas species are analyzed with a thermal conductivity detector (TCD) downstream from a 3 m molecular sieve 5A column, at 65°C, with helium as a carrier gas (for analyzing He and $H_2$). After determining the volume ratio between the liquid phase and the gas phase in the sampling bottle, the condensable $CO_3^{2-}$ and $S^{2-}$ ions in the alkaline solution are analyzed by conventional methods, neutralization with HCl and titration by $Pb(ClO_4)_2$ using a pH electrode and sulphide ion–sulphide electrode, respectively (Tedesco 1987). $SO_2$ is oxidized by peroxide in an oven at 100°C, analyzed as sulphate ($SO_4^{2-}$) by means of liquid chromatography, and determined by the difference between the total sulphur ($SO_4^{2-}$) and $H_2S$. HF and HCl analyses are carried out by liquid chromatography in the same analytical sequence as $SO_4^{2-}$. The mass of water vapour can then be determined as it equals the difference between the total mass of fluid collected and the mass of the so-called "dry gases" (Table 12.1).

The helium isotope ratios are measured after the separation and purification of the gases, following a procedure described by Sano & Wakita (1985, 1988). About $0.3 \, cm^3$ standard temperature and pressure (STP) of each gas sample are introduced into a high-vacuum metallic line in which helium and neon are separated from the other components, using hot Ti–Zr getters and two charcoal traps held at liquid nitrogen temperature. The $^4He/^{20}Ne$ ratio is measured on-line with a Balzers MG 112 quadrupole mass spectrometer, and interference of

**Table 12.1** Typical analysis (% volumes) from one of the La Fossa crater fumaroles (FA) at Vulcano (Aeolian Islands) (see Section 12.4).

| Gas species | Uncondensable gases | Dry gases | Overall composition | Accuracy (%) |
|---|---|---|---|---|
| $H_2O$ | | | 85.59 | 2 |
| $CO_2$ | | 81.05 | 11.68 | 2 |
| $SO_2$ | | 8.52 | 1.23 | 4 |
| $H_2S$ | | 4.61 | 0.66 | 3 |
| HCl | | 3.87 | 0.56 | 4 |
| HF | | 0.54 | 0.077 | 4 |
| $H_2$ | 63.32 | 0.90 | 0.13 | 5 |
| $N_2$ | 34.79 | 0.49 | 0.071 | 10 |
| CO | 1.65 | 0.027 | 0.0034 | 2 |
| He | 0.04 | 0.0007 | 0.00 | 10 |
| $O_2$ + Ar | 0.20 | 0.0032 | 0.0004 | 15 |
| $CH_4$ | 0.03 | 0.0005 | 0.000073 | 2 |

doubly charged ions of $^{40}$Ar with $^{20}$Ne is kept negligible. Helium is then completely separated from neon by trapping the latter on activated charcoal at 40 K. Isotopic analyses are determined with a dual-collector modified VG 5400 mass spectrometer (Sano & Wakita 1988). Complete separation of $^{3}$He from H$_3$ and HD is ensured by a resolving power of 600 at the 5% peak height, the experimental blanks being negligibly small compared with actual sample size. Both $^{3}$He/$^{4}$He and $^{4}$He/$^{20}$Ne ratios are determined by comparison with a calibrated air standard and the accuracy of their measurements reaches 3% and 10%, respectively.

## 12.3 Interpretation of chemical and isotope data at Campi Flegrei caldera

The Campi Flegrei caldera (Fig. 12.2) has been affected by two different bradyseismic crisis (seismic activity and ground deformation) during the last two decades, the first occurring in 1970–72 and the second between 1982 and 1984. Since 1982, the Campi Flegrei volcanic district has experienced ground uplift (Fig. 12.3) of up to 180cm, and sustained seismicity involving several hundred felt earthquakes (Barberi et al. 1984) with a maximum magnitude of $M = 4.0$. The vents are monitored by a permanent surveillance network, which, in common with similar systems deployed elsewhere, such as the Long Valley–Mono Basin volcanic complex (Hermance 1983) and the Rabaul caldera (McKee et al. 1985), relies predominantly

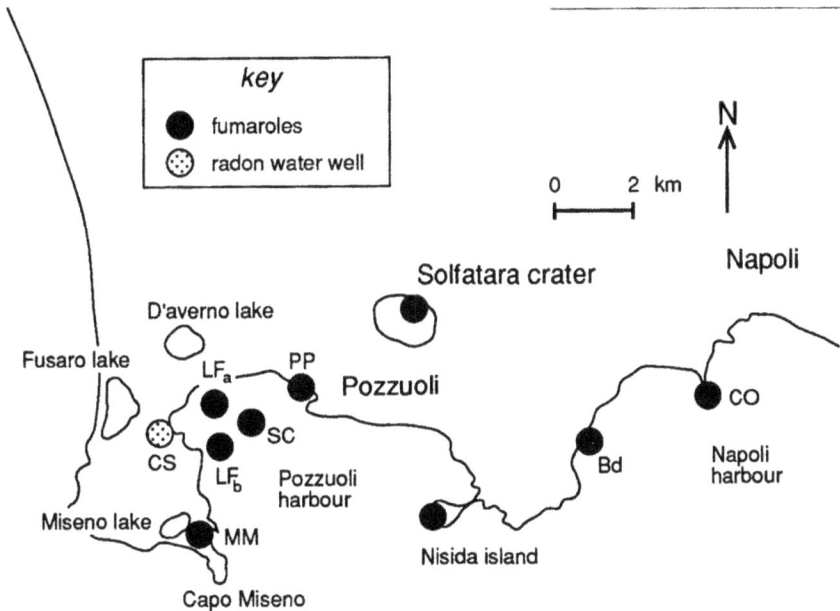

**Figure 12.2** Location map of the Campi Flegrei caldera (Bay of Naples) showing the position of the Solfatara Crater. Solid circles mark the positions of fumaroles, and open circles indicate water wells where radon sampling was undertaken. PP, Porto Pozzuoli; SC, Serra Caruso; LF, Le Fumose; MM, Mare Morto; MF, Mofete; Bd, Badessa; CO, Costagliola.

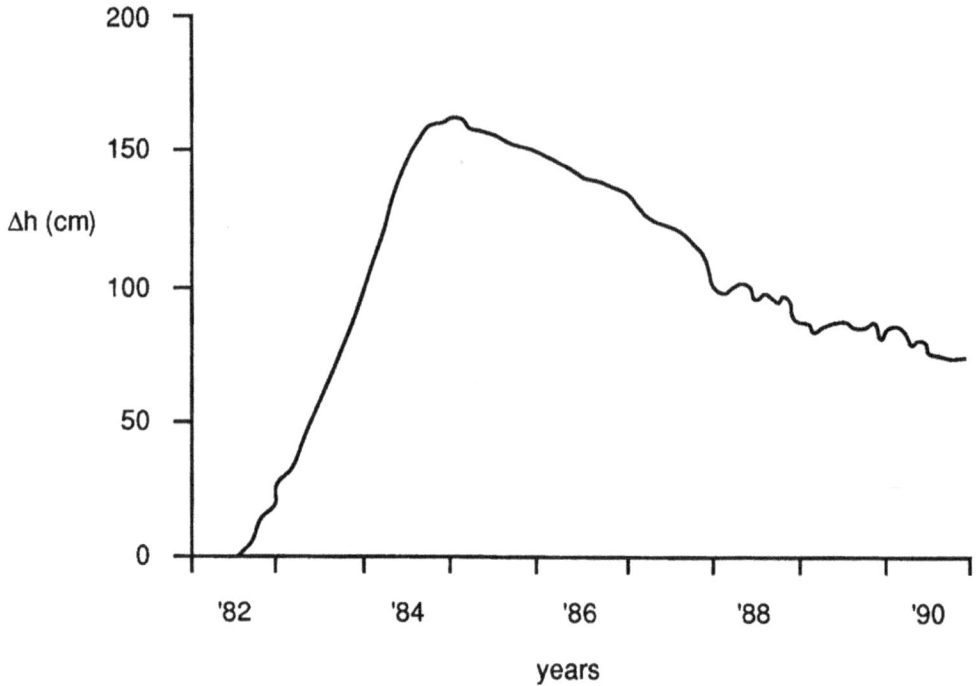

**Figure 12.3** Vertical ground deformation record at Pozzuoli harbour for the period 1982 to 1990. Uplift started in March 1982 and ceased in December 1984. Since then there has been a slow deflation, totalling around 80 cm.

on geophysical methods (e.g. seismometry, microgravimetry and precise levelling). At Campi Flegrei, however, it was also possible to undertake geochemical monitoring of fumarolic gases, given the conspicuous volumes of gas escaping from the Solfatara crater (the site of the largest superficial thermal anomaly in the area), and from several submarine fumaroles located in the Gulf of Pozzuoli and the Gulf of Naples. Accurate analysis of such gases provided information on the temperatures and pressures in the hydrothermal or magmatic reservoirs feeding the fumaroles, and on the physicochemical evolution of such reservoirs, and on any connections between these and an underlying body of hot magma. At the same time to better constrain the genesis of these fluids and the extent of these reservoirs (shallow or deep), several isotope analyses were carried out in order to establish the relationships between deep and shallow systems. A series of volcanic gas samples from the Solfatara crater and from several submarine fumaroles in the Pozzuoli and Napoli harbours, were collected at intervals after October 1983, and subsequently analyzed. Data were evaluated to examine the potential use of sampled fluids as geothermometers and geobarometers. Variations in the compositions of fumarolic fluids at Solfatara have been reported by other workers, both during the decline of seismic activity, and also before the probable start of the bradyseismic crisis (Cioni et al. 1984).

## 12.3.1 Data presentation and discussion

At Solfatara the proportion of water vapour ($H_2O$) increased from 1981 to 1982, while hydrogen sulphide ($H_2S$) content and the carbon/sulphur (C/S) ratio increased from, respectively, the end of 1980 and the beginning of 1981. Data on methane ($CH_4$) contents appear less reliable due to its low concentration in the fumaroles, but it also seems to have varied since 1981 (Cioni et al. 1984). All other species ($H_2$, $N_2$, Ar, CO and He) except carbon dioxide ($CO_2$), which decreased, exhibited practically constant values (Martini 1986).

A first approach to understanding the causes of these chemical variations requires examination of a possible correlation between the changes and an increase in the heat flux from

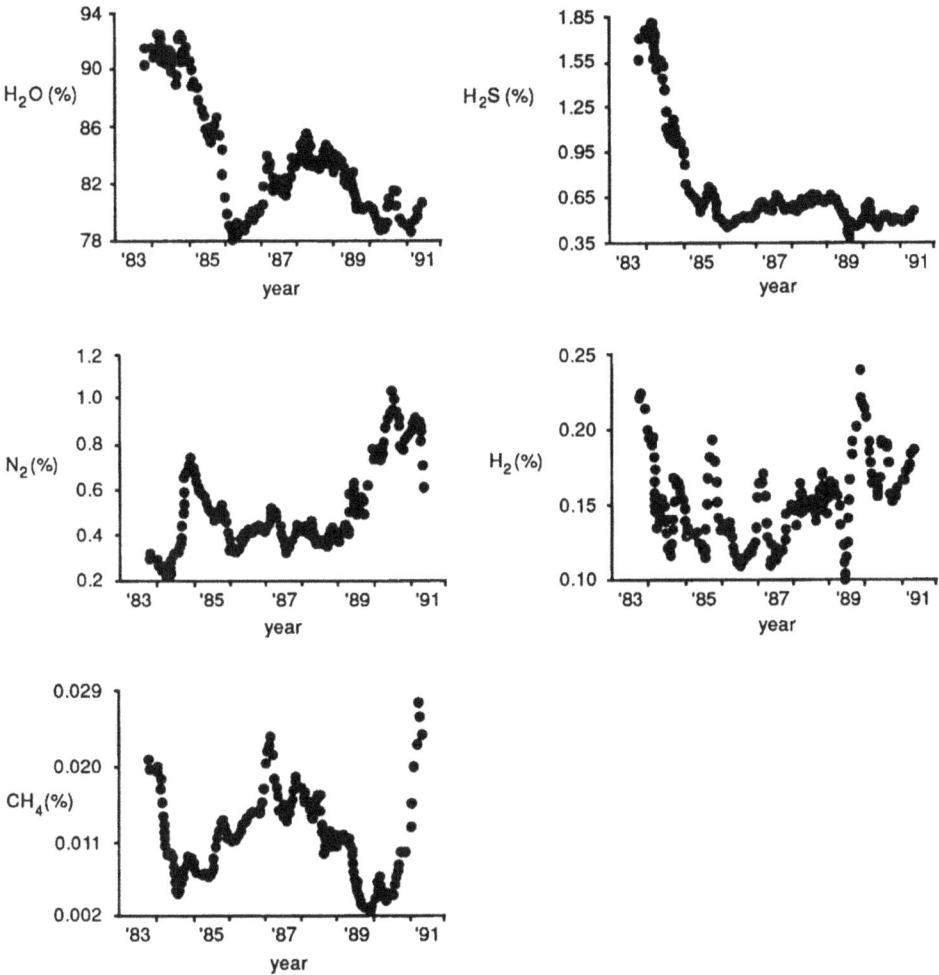

**Figure 12.4**   Geochemical changes at Solfatara crater, Campi Flegrei. Starting in late 1983/early 1984, dramatic changes were observed in the concentrations of several chemical species. $H_2S$, $N_2$, $H_2$ and $CH_4$ all show decreases at this time, before the cessation of both seismicity and ground deformation, followed by a fall in $H_2O$ starting in 1985. Concentrations start to increase again, in 1984 for $N_2$, 1985 for $H_2$ and $CH_4$, and 1986 for $H_2O$, and have remained variable since. See text for explanation.

Situation after 1970-72 bradyseismic crisis

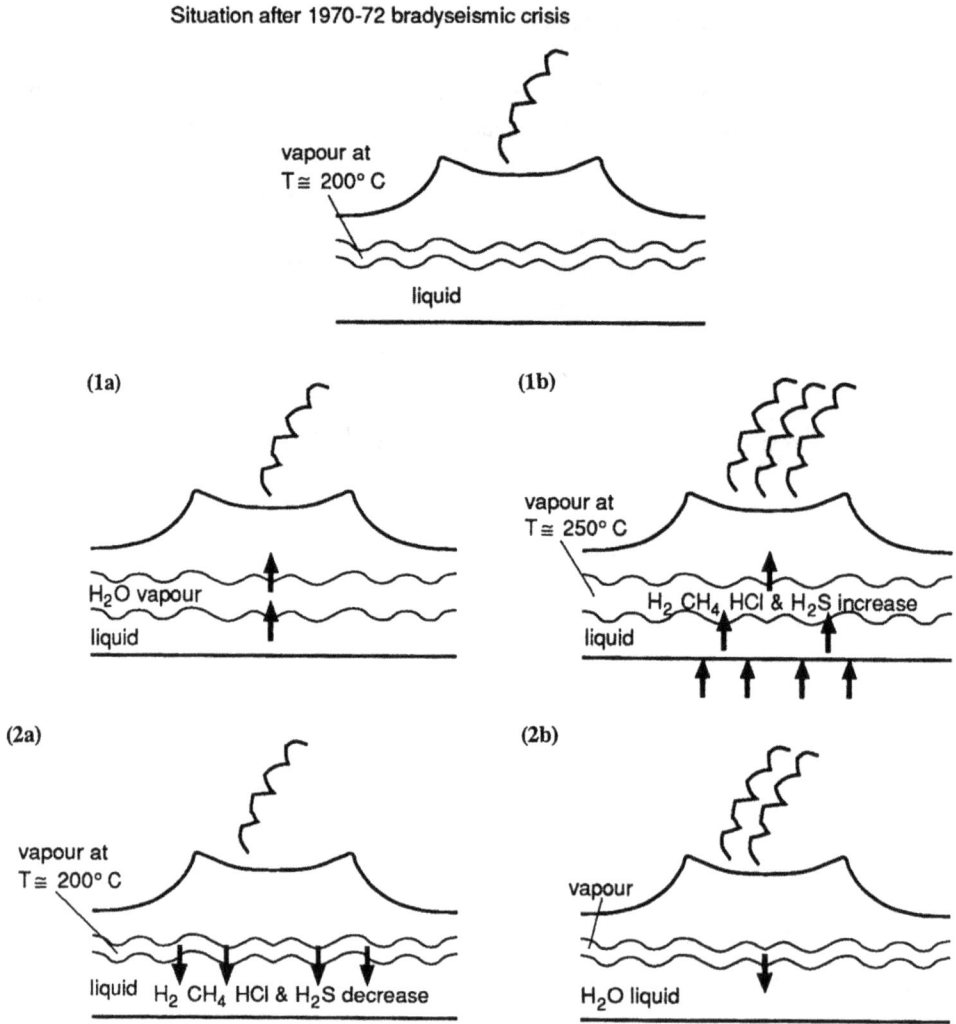

vapour at
T ≅ 200° C

liquid

(1a)

H₂O vapour

liquid

(1b)

vapour at
T ≅ 250° C

H₂ CH₄ HCl & H₂S increase

liquid

(2a)

vapour at
T ≅ 200° C

liquid  H₂ CH₄ HCl & H₂S decrease

(2b)

vapour

H₂O liquid

**Figure 12.5** Model explaining physical variations in the hydrothermal reservoir of Solfatara crater, during the recent bradyseismic crises, in response to changing heat flow. After the 1970–72 bradyseismic crisis, the temperature of the hydrothermal reservoir is estimated at 200°C. During the 1982–84 bradyseismic crisis, increased heat flow raised the temperature of the reservoir to 250°C, increased the water vapour content (1a), and introduced higher concentrations of $H_2$, $CH_4$, HCl, and $H_2S$ (1b). After the crisis, reduced heat flow resulted in a return of reservoir temperatures to 200°C, lower concentrations of $H_2$, $CH_4$, HCl, and $H_2S$ (2a) and a fall in the water vapour content (2b).

deeper reservoirs to the near-surface hydrothermal systems that feed the Solfatara crater (Tedesco et al. 1988a & b, 1990). Another possibility, which needs to be considered, involves establishing whether the geochemical variations can be explained by the direct injection of magmatic fluids into the superficial hydrothermal systems. Given the steady concentration of the so-called "high-temperature species" such as carbon monoxide (0.1 ppm of the total fluid), and the constancy of the helium isotope ratio, the latter is considered unlikely, with no new magmatic fluids involved either before or during the bradyseismic crisis (Tedesco et al. 1988b). At the beginning of 1984, a new episode of compositional variation was recognized in the fumarolic fluids (Fig. 12.4). Concentrations of $H_2S$, $CH_4$, $N_2$ and $H_2$ decreased strongly, followed, at the beginning of 1985, by a fall in the $H_2O$ content. During this period, however, the concentration of He and CO remained practically constant. These variations may be interpreted in terms of a decrease in the energy transfer (as heat flux) from deeper reservoirs. This simple model, illustrated in Figure 12.5, is also supported by thermodynamic calculations obtained using the water–gas shift reaction:

$$CO + H_2O \leftrightarrow CO_2 + H_2$$

This reaction allows the calculation of an equilibrium temperature for the Solfatara crater system, which is estimated at about 250°C for the 1983–84 period (Fig. 12.6) (Barin & Knacke 1973, Barin et al. 1977). From 1985, the equilibrium temperature dropped in two stages, by about 50°C. The first stage occurred between the end of the bradyseism in 1985 and the beginning of 1986, followed by a subsequent increase, and the second phase of falling temperature occurred during the 2 years following the beginning of 1989. These two events may also have been preceeded, before the seismic activity started in 1982, by a similar increase in the temperature of the hydrothermal reservoir, probably of the same magnitude. This would agree with Cioni et al. (1984), who estimated a positive $\Delta T$ of at least 20°C from the variation estimates of the $CO_2$/water ratio inferred before the bradyseismic crisis.

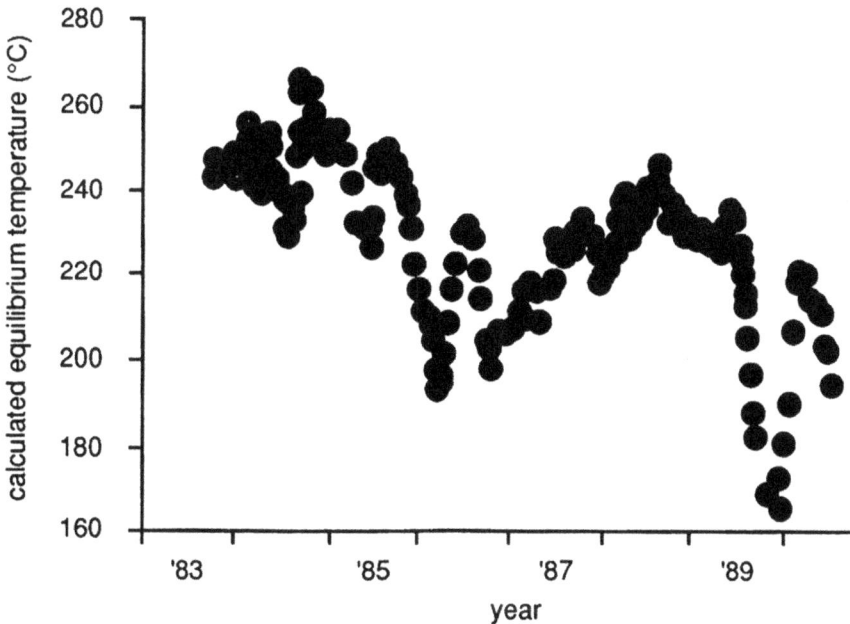

**Figure 12.6** Calculated equilibrium temperatures for the Solfatara hydrothermal system based on the water ¾ gas shift reaction $CO + H_2O \leftrightarrow CO_2 + H_2$ Although determined temperatures vary considerably, three distinct trends are recognizable: from 250 to 200°C between 1983 and 1986; from 200 to 240°C between 1987 and 1988; and from 240 to 200°C between 1989 and 1991. During this latter period, minimum calculated temperatures were around 160°C, this is close to the temperatures of the surface fumaroles.

## 12.3.2 Origin and variations of $\delta^{13}C$

Apart from CO and $CH_4$, which are present in the Solfatara fumaroles in concentrations of up to only a few parts per million, only $CO_2$ affects $\delta^{13}C$ values. The data obtained at the Solfatara fumaroles since 1970 indicate $\delta^{13}C$ values of between –1.9‰ and –0.8‰ before the bradyseismic crisis of 1982–84. Data obtained after the crisis seem more stable (Allard et al. 1991a) as they range between –1.9‰ and –1.6‰. Other values, very close to the previous ones were obtained by Carapezza et al. (1984) in 1983, and range between –1.73‰ and –1.49‰. These results show that, despite the strong variation of $CO_2$ in the gas phase, the $\delta^{13}C$ remained quite stable over more than 20 years. Allard et al. (1991a) discuss the origin of carbon in the Solfatara and Campi Flegrei fluids, and argue that the constant $\delta^{13}C$ values over the period 1970–88 is evidence for a stable source of carbon. This source must be both large and deep in order to explain the absence of isotopic balance effects with, respectively, the $CO_2$ output from the crater (170–350 t day$^{-1}$) (Carapezza et al. 1984), and the extensive mechanical and hydrological disturbances which affected the lower few kilometres of the caldera's basement during the bradyseismic crisis of 1982–84 (Barberi et al. 1984). As in the case of the helium isotope data, $\delta^{13}C$ values obtained for the Campi Flegrei caldera, are much higher than those normally obtained for mantle-magmatic carbon ($-6 \pm 2$‰) (Allard 1983, Javoy et al. 1984, Allard 1986), and are similar to those in normal marine carbonates which display a range of 0‰ $\pm$ 2‰ (Degens 1969). Moreover, fumaroles located in the eastern part of the caldera (Mare Morto, Secca Caruso, Le Fumose A and B, and Porto Pozzuoli; see Fig. 12.2), as well as geothermal fluids from wells also show values in the range of marine carbonates.

## 12.3.3 Temporal $^3He$ variations

In 1978, 4 years before the onset of the 1982–84 bradyseismic crisis, Poliak & Tolstikin (1980) measured a $R/R_a$ value of 2.7 at the Soffione fumarole at Solfatara. ($R/R_a$ is the enrichment of the helium isotope ratio in the sample compared to the helium isotope ratio in the atmosphere $(^3He/^4He)_s/(^3He/^4He)_a$). Results for the same fumarole in October and December 1983, at the height of the crisis, are very close ($2.8 \pm 0.1$) and show no evidence of increasing $^3He/^4He$ related to the ground deformation and seismic events. Over the same period, the Bocca Grande fumarole at Solfatara displayed somewhat lower $R/R_a$ values (2.4–2.5) despite its close proximity to Soffione (a few tens of metres). However, a value of 2.9, typical of the Soffione fumarole, was measured at this fumarole in June 1988, nearly 4 years after the end of the crisis, and between 1988 and 1991 $R/R_a$ values ranged from 2.9 to 3.2 (Fig. 12.7). It is noteworthy that in December 1984, as the bradyseism was subsiding, the Pozzuoli Porto submarine fumarole (Fig. 12.2), the closest to Solfatara, displayed a $^3He/^4He$ ratio similar to that of the Bocca Grande fumarole during the crisis. This similarity lends support to the idea that the helium isotopic composition of Campi Flegrei fumaroles had remained steady throughout the bradyseismic period. In contrast, the results obtained in 1986 show a tendency toward increasing $^3He/^4He$ in the Pozzuoli Porto, Secca Caruso, and Le Fumose (B) fumaroles (Fig.

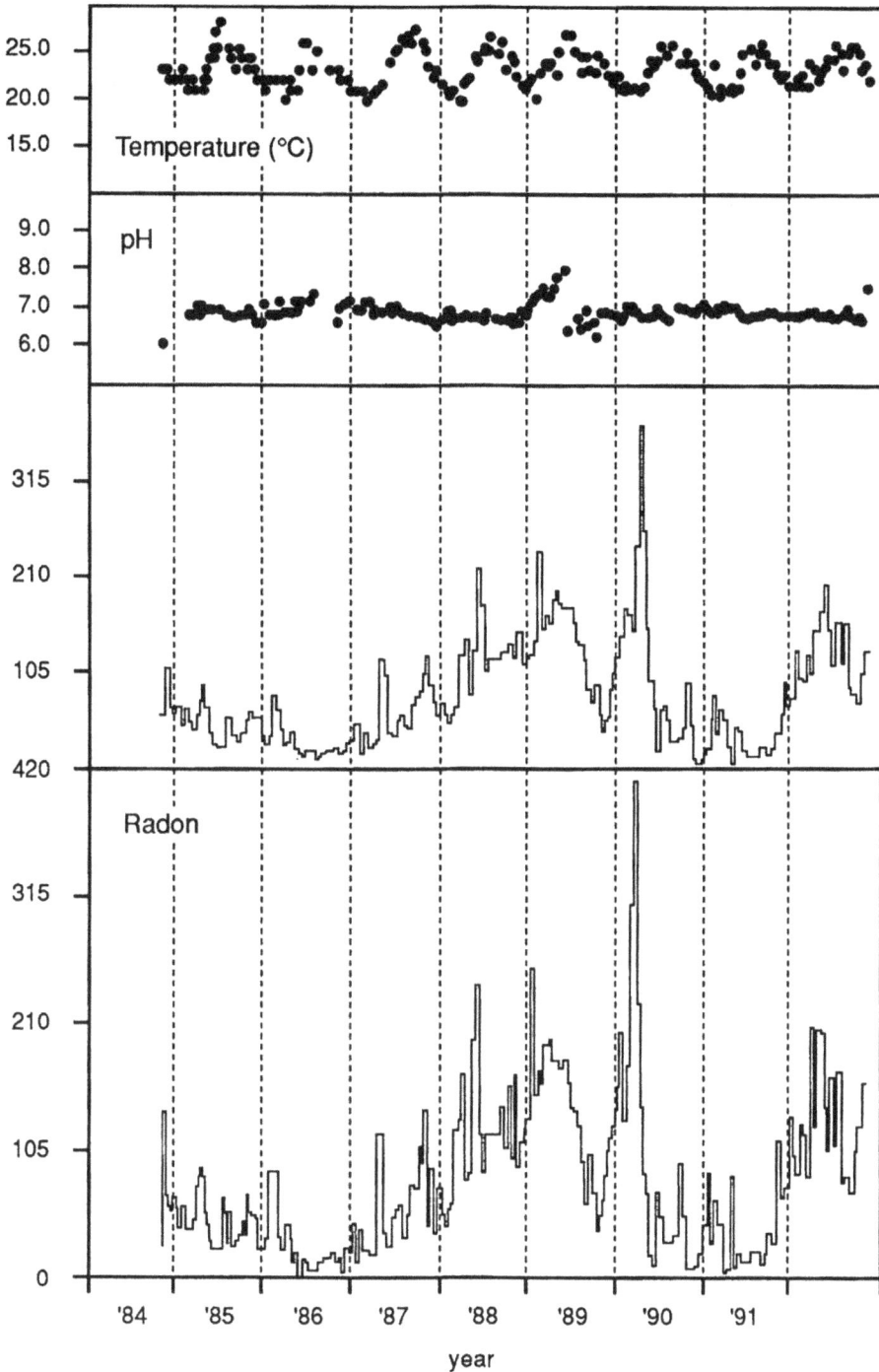

**Figure 12.7**  Helium isotope ratios for the Bocca Grande fumarole at Solfatara. Ratios determined for the 1983–84 period are similar to those obtained by Poliak & Tolstikin (1982). Ratios since 1988, 4 years after the end of the latest bradyseismic crisis, are much higher, ranging from 2.9 to around 3.2 $R/R_a$ (see text for explanation).

12.2) (Tedesco et al. 1990, De Natale et al. 1991). The variations exceed the analytical uncertainty and are thus considered significant. They represent a true [3]He increase in these
fluids after the crisis, and correspond to a temporary variation that can be clearly established
from the large number of data and the long duration of investigations. These data are in good
agreement with the chemical variations recorded at the Bocca Grande fumarole over the same
period. The results of radon monitoring in the caldera (e.g. Tedesco et al. 1988a) provide
additional information on geochemical changes at Campi Flegrei (Fig. 12.8). The variations
of this radiogenic rare gas in monitored hot-water wells follow a seasonal pattern, with radon activity increasing during the rainy winter season and then decreasing during the drier
May–September period. Such a pattern, suggestive of a rapid input of meteoric water into
the hydrogeological system, is attributed to enhanced or reduced stripping of radon by circu-
lating groundwaters from larger or smaller volumes of rocks, depending on the extent of water
circulation. If radiogenic [4]He accumulated in the same rocks is also extracted, then we might
expect a lower supply of this isotope to hot waters and fumarolic gases during the dry season
than in winter. Observed seasonal variations of helium in the Bocca Grande fumarole (Tedesco
1987, Tedesco et al. 1988b) are consistent with this hypothesis. At the same time, an increasing
radon trend has been recorded in different thermal wells of the Campi Flegrei region that
mirrors perfectly the chemical and isotopic variations registered at the Bocca Grande fumarole
(R. Pece, unpublished data). Such a mechanism could thus explain the trend of an increasing
[3]He/[4]He ratio in some of the submarine gases collected in May–June 1986, but not the higher
[3]He/[4]He ratio of the Bocca Grande fumarole from June 1988 until the present (Fig. 12.7). It
would also not account for the high value measured at the Soffione fumarole in December

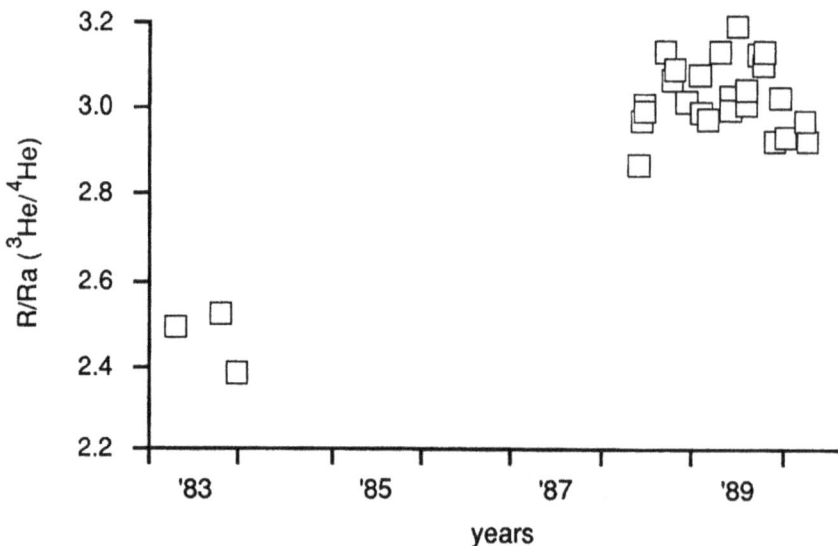

**Figure 12.8** Radon gas activity at Costagliola water well, Solfatara. Variations of radon in
hot-water wells at Solfatara follow a seasonal pattern, with activity increasing during the wet
Winter season, and decreasing during the drier May–September period. See text for explanation. (R. Pece, unpublished data.)

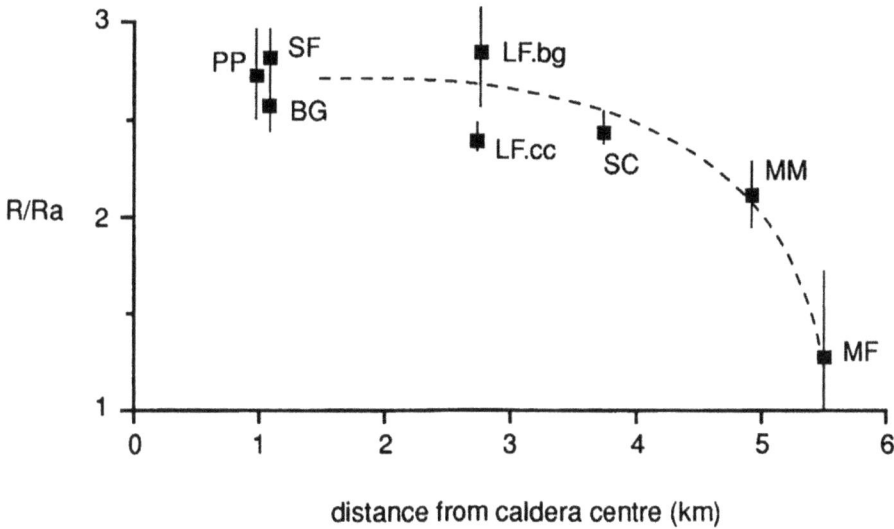

**Figure 12.9** Spatial variation of the $^3$He/$^4$He ratio in the Campi Flegrei area. Higher He ratios at the more centrally located fumaroles (with respect to Campi Flegrei), e.g. at Porto Pozzuoli (PP), and the Solfatara fumaroles (SF, Soffione; BG, Bocca Grande), compared with those encountered in the more peripheral fumaroles (LFbg & LFcc, Le Fumose; SC, Serra Caruso; MM, Mare Morte fumaroles), was originally explained (Tedesco et al. 1990) in terms of higher heat flux and greater input of magmatic fluids in the central part of the caldera. New data (see text and Fig. 12.10) seem to dispute this. (From Tedesco 1994.)

1983, during the period of high seismicity and ground deformation. Nor can it preclude the possibility of a true $^3$He/$^4$He and radon increase at Campi Flegrei after the 1982–84 events.

Using these $^3$He/$^4$He data, it has been possible to calculate that a maximum of 35% of the helium has a magmatic origin. Preliminary data suggested that the occurrence, in the central part of the caldera, of the hottest fluids (155°C at Solfatara) with the highest helium isotope ratios (compared to the lower values obtained in the more peripheral fumarolic fluids at Secca Caruso, Le Fumose and Mare Morto fumaroles), indicated a greater leakage of magmatic fluids and higher heat flux from the central part of the caldera (Tedesco et al. 1990). However, further data obtained from the same fumaroles, as well as those from new fumaroles, do not support this model (Fig. 12.9). New helium data indicate that the $^3$He/$^4$He ratio is similar inside and outside the caldera, ranging from $R/R_a$ = 2.5–2.9, with values of 2.8 and 2.7 (close to the maximum, $R/R_a$ = 2.9) up to 5 and 14 km, respectively, from the central part of the caldera (Fig. 12.10). These values do not support the notion that leaking magmatic fluids and heat flow are mostly concentrated in the central part of the caldera. It appears, instead, that an homogeneous gas phase is present both in the Campi Flegrei caldera and its surroundings, as suggested by Tedesco (1987). The small differences in the $^3$He/$^4$He ratio being attributed to the different paths followed by fluids to reach the surface, and to the different depths of the local, shallow hydrological systems present within and outside the inner caldera, through which the gases permeate (Rosi & Sbrana 1987). In this model, the origin of the gas is considered to be the same for all the fumaroles in the area, including those located in Naples

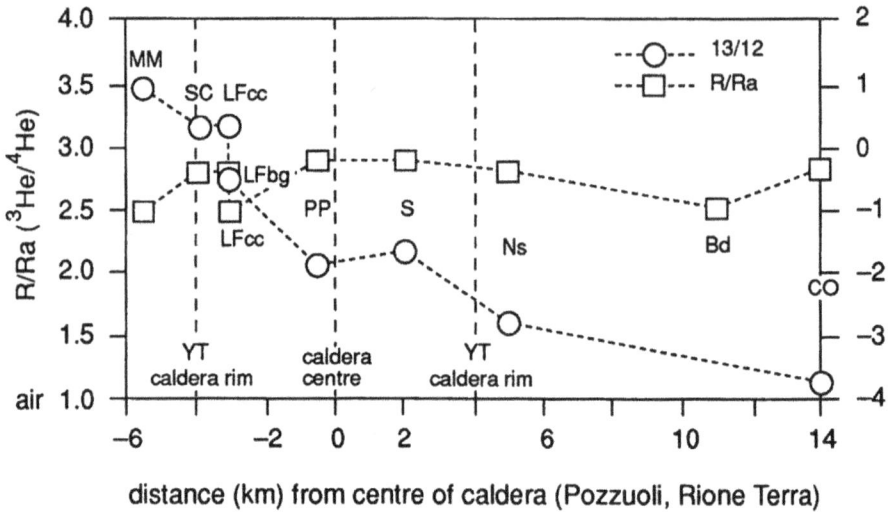

**Figure 12.10** Geographical distribution of $^3$He/$^4$He and $^{13}$C/$^{12}$C ratios in fumaroles at Campi Flegrei. In contrast to C ratios, He ratios are similar inside and outside the Campi Flegrei caldera (see text for explanation). Data points represent averages determined from several measurements at each fumarole. MM, Mare Morte; SC, Serra Caruso; LFbg & LFcc, Le Fumose; PP, Porto Pozzuoli; S, Solfatara, Ns, Nisida; Bd, Badessa; CO, Costagliola; Cal.bor., caldera border; Cal. cen., caldera centre; YT, yellow tuff caldera; AIR, concentration in atmosphere.

harbour. Small isotopic changes occur as the gas approaches the surface and passes through different layers and acquifers, where small amounts of crustal or atmospheric helium may be added. Accordingly, the gases from each fumarole are characterized by their own chemical and isotopic signatures, with even fluids collected only a few metres apart (as at the Le Fumose A and B fumaroles and in the Solfatara crater at the Soffione and Bocca Grande fumaroles) (Tedesco 1987), showing different $R/R_a$ values close to the minimum and the maximum values for the whole area ($R/R_a = 2.4$ and $2.9$, respectively). In conclusion, although the $^3$He/$^4$He ratios in this volcanic area are not high compared to other active volcanic areas (Tedesco et al. 1990) (see also Section 12.4) they are uniform over a very wide area. Most of the conclusions relating to He variations at Campi Flegrei in Tedesco et al. (1990) remain valid, and may be summarized as follows.

(a) Measurements of He isotopes in the Campi Flegrei fumaroles confirm that mantle-derived helium escapes from the caldera (Poliak and Tolstikin 1980). The $^3$He/$^4$He ratios ranges between $R/Ra = 2.5$ and $R/Ra = 2.9$ with no noteworthy difference between the ratios at subaerial and submarine fumaroles indicating that both are fed by a common helium source.

(b) The $^3$He/$^4$He ratio of the Campi Flegrei fumaroles is significantly lower than that of upper mantle volatiles and volcanic gases from other areas. Such a feature may reflect either a low mantle and/or magmatic He supply, a strong dilution of mantle gas by $^4$He from the crust, the radioactive ageing of an isolated magma chamber, and/or a true characteristic of the local mantle. Further isotopic analysis of helium in both volcanic rocks and mantle-derived xenoliths from this area could help to distinguish these alternatives. A shallow dilution by $^4$He-rich fluids, coupled with a decrease of the magmatic helium flux, may be responsible for some of the lower $^3$He/$^4$He ratios, although the low $^3$He/$^4$He of most volcanic gases and hydrothermal fluids in western Italy does coincide with a strong enrichment of the associated volcanic rocks in radiogenic Sr, LIL-elements, and $^{18}$O, a feature primarily attributed to anomalous mantle composition.

(c) Data indicate that the proportion of $^3$He in the emitted helium did not increase during the 1982–84 bradyseismic crisis. This observation has an important bearing on the origin of the crisis and, in particular, it excludes the emplacement of a shallow magma intrusion, associated with an increasing release of $^3$He-rich mantle and/or magmatic gas. Its significance, however, could be strongly weakened if the magma chamber, for any of the reasons mentioned above, does have a low $^3$He/$^4$He ratio, closer to that of the maximum helium isotope ratio of the caldera than of typical mantle helium.

## 12.4 Geochemical monitoring at Vulcano (Aeolian Islands)

The Aeolian Island archipelago (southern Italy) consists of seven volcanic islands and numerous seamounts. It is interpreted as a typical volcanic arc, generated by subduction beneath the Tyrrhenian Sea (e.g. Barberi et al. 1973, 1974, Beccaluva et al. 1985). This broad model is supported by recent petrological investigations and K/Ar dating studies of the volcanics. Vulcano, the southermost island, is one of the active volcanoes of the archipelago. It lies a few tens of kilometres north of the coast of Sicily (Fig. 12.11), and its active cone (La Fossa), 391 m high above sea level, last erupted in 1888–90 (Keller 1980). Since then, intense fumarolic degassing has persisted in La Fossa crater with a peak in 1926 when fumarole temperatures reached 600°C (Sicardi 1955). More recently, after a regional seismic event ($M = 5.5$) on 15 April 1978 (the Patti earthquake), and succeeding volcanic seismic swarm (Del Pezzo & Martini 1981, Falsaperla et al. 1989), the fumaroles underwent a temperature increase of more than 100°C, and showed variations in their chemical composition (Martini et al. 1980, Carapezza et al. 1981, Martini et al. 1989). A temperature decrease followed a few years later, but in 1987 the temperatures of some crater-rim fumaroles again started to increase with, in particular, the temperature of the F5 fumarole (Fig 12.11) increasing from about 200 to 330°C. Since 1988, temperature rises have been recorded at other fumaroles, with 470°C being measured in mid-1988, 550°C in mid-1989 and about 630–650°C in 1991–92, inside the crater at the same site at which Sicardi (1955) measured a temperature of 600°C in 1926. This recent thermal change is associated with an increasing gas flow, a spatial extension of the fumarolic field, and the opening of new fractures accross the rim of the crater (Smithsonian Institution 1988, Tedesco et al. 1991) and inside the crater. Because of the highly explosive potential of Vulcano eruptions (Sheridan & Malin 1983, Frazzetta et al. 1984), and due to the presence of a dense population on the island during the Summer months (up to 15000 people on any single day), these events provoked some concern about increasing volcanic hazard. Accordingly, in order to determine their cause, we have performed an intensive monitoring of the F5 and FA crater fumaroles and several other beach fumaroles since

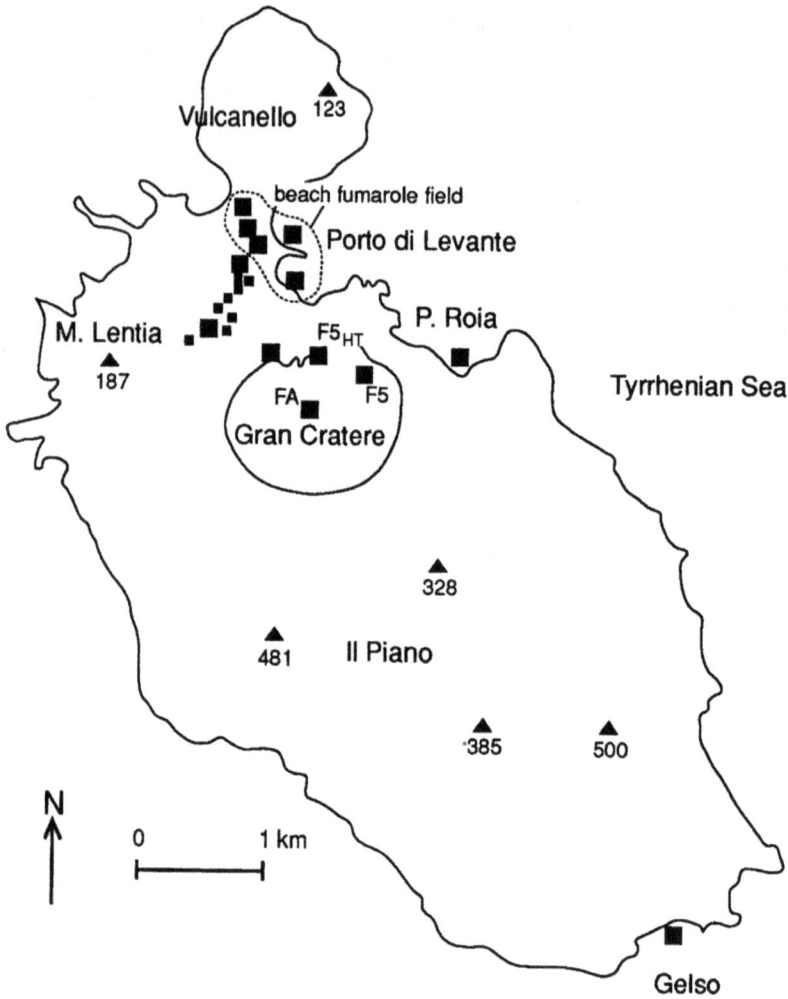

**Figure 12.11** Location map of Vulcano island (Aeolian Islands, Italy), showing the positions of sampled fumaroles and water wells.

mid-1987) (Tedesco et al. 1991). The F5 fumarole, located on the eastern rim of the Fossa crater, has been preferentially studied for a number of years by various authors (e.g. Tonani 1971, Allard 1978, Martini et al. 1980, Carapezza et al. 1981, Cioni and Corazza 1981, Le Guern and Faivre Pierret 1982, Cioni & D'Amore 1984, Martini et al. 1984, Mazor et al. 1988), due to its steady temperature and easy access.

## 12.4.1 Fluid compositions

Figure 12.12a shows that the amount of water vapour in the F5 fumarolic fluid oscillated during the sampling period (Tedesco et al. 1990), increasing from 81% in July 1987 to 95% in May

1988, and then falling to about 85% in December 1988. Since then a further change has occurred demonstrating a slightly different trend, an increase from January 1989 at the same rate as the previous rise, followed, once the maximum 97% value was achieved, by 3 months of stable water content and a subsequent gentle fall. The trend is still a decreasing one, as demonstrated by the September 1992 value of between 88 and 90%. Hydrogen has followed a similar pattern, its content varying by a factor 3 (Fig. 12.12b). In contrast, $SO_2$, $CO_2$, $N_2$, He (Fig. 12.12c–f) and HCl, exhibit exactly the opposite pattern, reflecting their complementary response to the changes in water vapour content which represents the dominant constituent. Only $H_2S$ (Fig. 12.12c) shows comparatively limited oscillations.

In order to eliminate the influence of water content variation, dry gas proportions and ratios were also considered. This revealed that hydrogen is the compound which fluctuates most obviously, by about one order of magnitude. A relative enrichment of sulphur with respect to carbon is also observed, and the S/C atomic ratio shows significant variation (50%), matching that of $H_2O$, and peaking at the time of maximum $H_2O$ content. At the same time, total sulphur ($H_2S + SO_2$) increased to about 18% of the dry gases, and then returned close to the

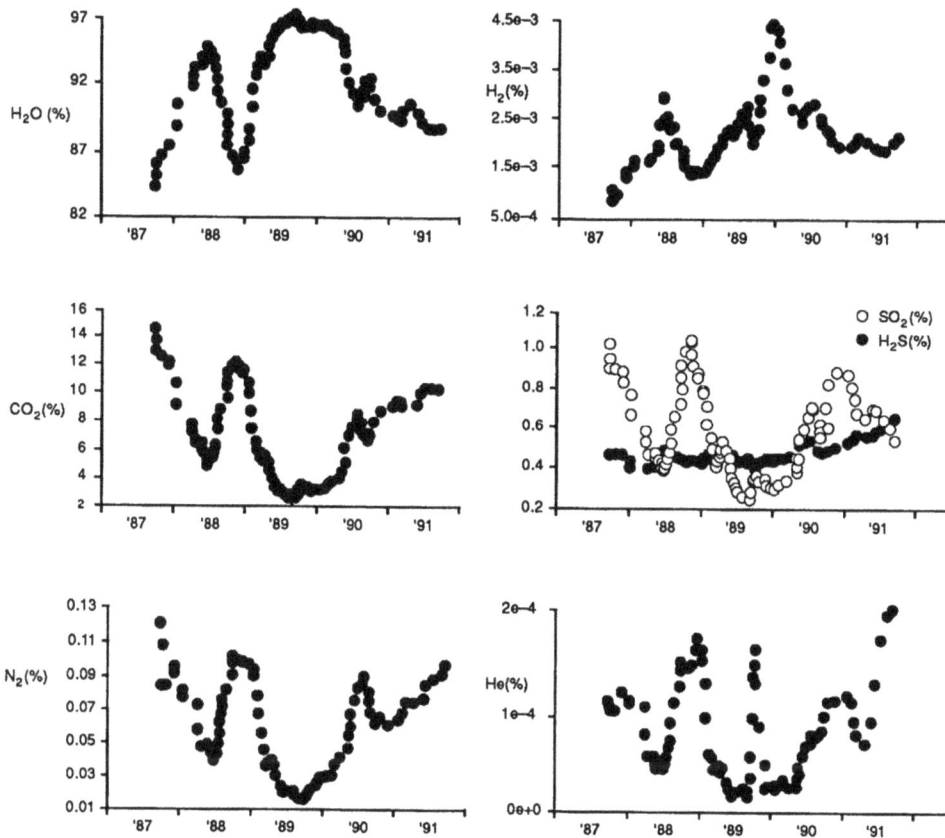

**Figure 12.12**   Temporal variations in the concentrations of the major gas species in the F5 fumarole at Vulcano. Complementary trends are displayed by $H_2O$ and $H_2$ on the one hand, and $CO_2$, $SO_2$, $H_2S$, $N_2$ and He on the other. See text for explanation.

initial value of around 8%. These changes were followed by a new phase of variation which started in 1989 with a similar pattern. Over the period of study, both nitrogen and helium displayed comparable trends (see Fig. 12.12).

### 12.4.2 Discussion

Recent investigations at Vulcano have shown that intensive gas emanations occur throughout the island. These are particularly significant in the crater and in the beach area at the northern foot of La Fossa cone (Fig. 12.11). Emissions from the two areas are characterized by distinctive chemical and carbon isotopic signatures (Martini et al. 1980, Carapezza et al. 1981, Le Guern & Faivre Pierret 1982, Cioni et al. 1984, Baubron 1987–90, Mazor et al. 1988, Baubron et al. 1990, Tedesco et al. 1991). In particular, gases in the crater area contain high proportions of acid compounds (such as $SO_2$, HCl, HF), and are rich in helium (around 10 ppm of dry gases), while the beach emissions contain no acid components and are relatively low in helium (around 2–3 ppm of dry gases). The difference in acid content has been related to the distribution of near-surface aquifers which are common in the beach area and may act as traps for the acid components (Carapezza et al. 1981, Cioni et al. 1984, Martini et al. 1984). This interpretation, however, cannot explain the observed differences in helium content (Baubron et al. 1990, Tedesco et al. 1991), because if the original gases feeding the two areas were the same, the beach gases would be expected to have higher proportions of helium (and other uncondensable phases) after removal of the acid species. The different helium contents may thus reflect the fact that, in addition to different near-surface conditions, the two areas are underlain at an intermediate depth by distinct and independent magmatic systems. Geochemical studies conducted during the past 15 years on the fumaroles of both crater and beach systems suggest that deep magmatic fluids ascend through a shallower acquifer system via a bi-phase (Tonani 1971, Allard 1978, Le Guern 1980, Carapezza et al. 1981, Mazor et al. 1988), or mono-phase magmatic reservoir (Cioni and D'Amore 1984) below La Fossa cone, and then feed the crater fumaroles. However, none of the cited authors provide supporting evidence for a second, deeper, hydrothermal system below the near-surface aquifer. The clearest evidence of a magmatic contribution is provided by the isotopic ratio of helium, which ranges from $R/R_a = 5$ to 6 at the crater fumaroles to around $R/R_a = 5$ at the beach fumaroles (Poliak & Tolstikin 1980, Shinohara & Matsuo 1984, Hooker et al. 1985, Tedesco 1987, Sano et al. 1989, Tedesco et al. 1994) contrary to the findings of Mazor et al., (1988). These values, which are slightly lower than typical arc volcanism (e.g. Poreda & Craig 1989), indicate that at least 65–80% of the helium in the fumaroles is mantle derived. The probable presence of a shallow aquifer (meteoric or marine water) adjacent to La Fossa cone and closely related to it, is also supported by results which indicate continuous rapid variation of the water vapour content of the fumaroles which seems to follow a seasonal pattern (Fig. 12.12a), which may be somewhat modified by other factors.

   A number of possible mechanisms operating in the fluid reservoir(s) may explain the observed chemical and isotopic variations of the F5, FA and other fumaroles since 1987. These include the following.

(a) Seasonal dilution of a deep gas reservoir with meteoric water.

(b) Variable injection(s) of hot gas from a magmatic reservoir. The existence of a magma intrusion at 2–4 km depth below La Fossa is suggested by gravimetric (Barberi et al. 1973), and seismic data (Ferrucci et al. 1991, Vilardo et al. 1991).

(c) Pressure variations in the fluid reservoir feeding the fumaroles, as a result of either tectonic or hydrodynamic events.

(d) Mixing between different magmatic and hydrothermal systems.

### 12.4.2.1 Seasonal dilution of a deep gas reservoir with meteoric water

According to the first hypothesis, the hydrothermal fluid may be variably diluted by superficial water through two processes: (a) addition of meteoric water (supported by the $\delta D$ of fumarolic condensates; Cheminée et al. (1969), Allard (1983)) which feeds the system on a seasonal basis; or (b) infiltration of meteoric and/or marine water as a result of stress field variations due to earth tides. A dilution effect could indeed explain the annual variations of $H_2O$ and the consequent decrease of all other species but $H_2$. The more soluble species, such as $SO_2$ and $HCl$, can be removed by a greater increase in water content; and $H_2/HCl$ and $H_2/SO_2$ ratios higher than $H_2/CO_2$, $H_2/N_2$ or $H_2/CO_2$ ratios are obeyed at the times of the two water peaks in the data.

### 12.4.2.2 Injections of hot gas from a deep magma reservoir

New injections of magmatic gas into a shallower reservoir (either an hydrothermal system, or an old magmatic intrusion) should be followed by an increase in the reservoir temperature, a higher steam/liquid water ratio in the ascending fluid, an increase in the concentration of typical magmatic species and then, by a rise in the equilibrium temperature. The outlet temperatures range from 335°C to 275°C, 415°C to 476°C and 535°C to 661°C fumaroles at F5, $F5_{HT}$ and FA, respectively (Fig. 12.11). Apparent equilibrium temperatures for the fumarolic gases have previously been calculated by Sabroux (1979, 1983), and equilibrium temperatures based on our results from the F5, $F5_{HT}$ and FA fumaroles have also been computed from thermodynamic data for both the pressure independent water-gas-shift reaction

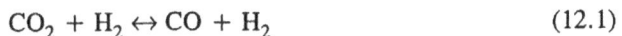

$$CO_2 + H_2 \leftrightarrow CO + H_2 \tag{12.1}$$

and the pressure-dependent $H_2S/SO_2$ equilibrium

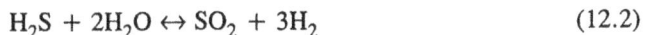

$$H_2S + 2H_2O \leftrightarrow SO_2 + 3H_2 \tag{12.2}$$

Temperatures calculated from reaction 12.1 range between 315 and 420°C for F5 (Fig. 12.13) and up to 440°C for the $F5_{HT}$ fumarole. Thus they are close to the higher outlet temperature, and do not provide evidence of higher temperature at depth. Those calculated from reaction 12.2 are slightly higher, ranging from 360 to 440°C, with some isolated higher values for both fumaroles. The differences between the outlet and the equilibrium temperatures, given by the pressure-independent reaction 12.1, indicate that cooling of the fluid during its transit from the reservoir to the surface is limited to 80–100°C at most and Carapezza et al. (1981) independently calculated a $\Delta T$ of only 20–40°C for cooling of the Vulcano crater fluids duriing and earlier period. Such values suggest, therefore, that the fumarolic fluid feeding the F5

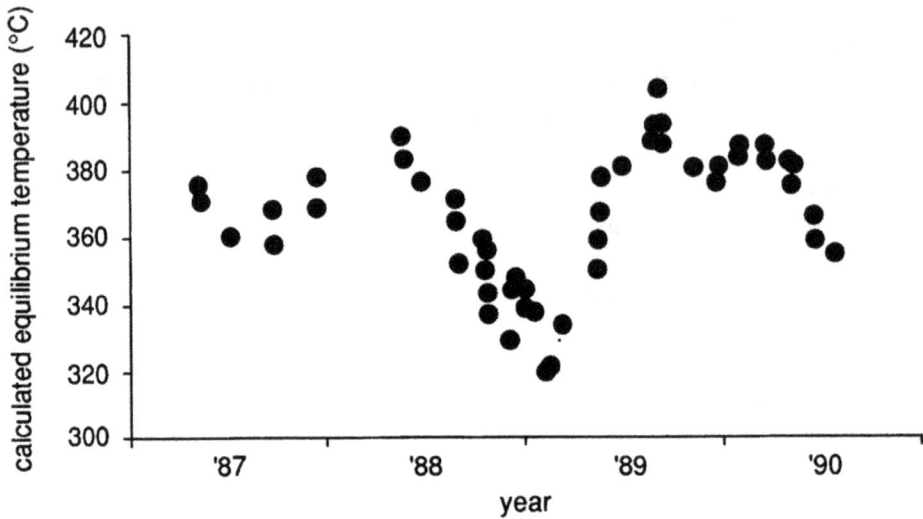

**Figure 12.13** Calculated equilibrium temperatures for the F5 fumarole. Using the water-gas shift reaction $CO_2 + H_2 \leftrightarrow CO + H_2$ these are determined to lie between 315 and 420°C. Equilibrium temperatures for the hotter $F5_{HT}$ fumarole are up to 440°C. See text for explanation. Temporal variations in the observed temperature trends are similar to those displayed by changes in the $H_2O$ concentration of the F5 fumarole, suggesting that the water vapour content may be a controlling factor.

and $F5_{HT}$ fumaroles equilibrates at a maximum temperature of around 410–440°C in a reservoir which must be shallow enough to limit gas re-equilibration during ascent. At the fumarole FA fluids reached an outlet temperature of between 535 and 661°C, with the temperature variations appearing to be dependent on the sampling point and the highest temperature site slowly migrating from place to place within the fumarolic field making relocation for sampling purposes difficult. Calculated equilibrium temperatures from reactions 12.1 and 12.2 indicate different results. Reaction 12.1 always gives temperatures lower, or at most similar to the emission point, indicating that chemical compositions, as for the F5 fumarole, re-equilibrate very rapidly under shallow conditions. In contrast, reaction 12.2 indicates chemical equilibrium at higher temperatures, with the calculated equilibrium temperature range of between 670°C and 900°C suggesting a trend of increasing temperature with time.

The possibility of an increasing input of magmatic gas since 1982 is not supported by a number of observations, the first of which concerns the carbon monoxide content of the fumarolic gases. At the F5 fumarole in particular, CO (which forms under high temperature and low oxygen fugacity conditions) occurred only in trace amounts and showed no recent increase compared to previous years. If an increasing magmatic gas input was the cause of the recent thermal increase at the crater fumaroles, then a significant increase in both CO and in the $CO/CO_2$ ratio would have been expected in all crater fumaroles, including at the F5 and $F5_{HT}$ vents. This was not observed during the 5-year period of investigation, the CO content remaining between 0.5 and 1.5ppm of the anhydrous phase (Tedesco et al. 1991). On the other hand, the FA fumarole, at a temperature of 600–650°C, shows a higher CO content compatible with its higher outlet temperature. Nevertheless, as already shown by thermodynamical calculation using reaction 12.1, it seems that fluids, in this case, also always reach chemical equilibria close to the outlet temperature. The greater CO concentrations in higher temperature fumaroles show a slow variation with time. Unfortunately it is not possible to determine whether the FA fumarole, inside the crater, and with the highest temperature, appeared during the last 5 years, or whether it already existed with a temperature higher than the rim crater fumarole and was simply not identified by gas geochemists.

An increasing input of magmatic gas is also made further unlikely by an examination of changes in helium ratios. The $^3He/^4He$ ratio continuously oscillated at the crater fumaroles over the period of study, with high values of $R/R_a$ ranging from 5.0 to more than 6.0 (Sano et al. 1989), similar to those previously obtained by other workers. In contrast, at the beach fumarolic field, the helium ratio remained at a constant, with a value of $R_a = 5.0$. The higher values at the crater fumaroles can be explained by a direct connection between the magmatic reservoir, at about 4 km depth or more (Vilardo et al. 1991), and a shallower reservoir (the magmatic intrusion of the 1888–90 eruption), with only very slight interactions with near-surface aquifiers. At the same time, mixing during ascent with superficial reservoirs, and marine or meteoric waters (Tedesco et al. 1991), all of which are poor in $^3He$, could produce the lower $^3He/^4He$ values (Tedesco et al. 1994). The variations of the $^3He/^4He$ ratios recorded at the F5 and FA crater fumaroles (Fig. 12.14) are explained by this model as a consequence of a continuous mixing between deep magmatic fluids ($^3He$ rich) and higher level marine waters ($^4He$ rich). The highest $^3He/^4He$ value at the crater fumaroles may be representative of uncontaminated juvenile fluid coming directly from the deepest magmatic reservoir, while the lowest values of the $^3He/^4He$ ratio (always recorded at the same fumaroles) may indicate increased mixing between the two reservoirs, when the addition of the superficial $^4He$ is at a

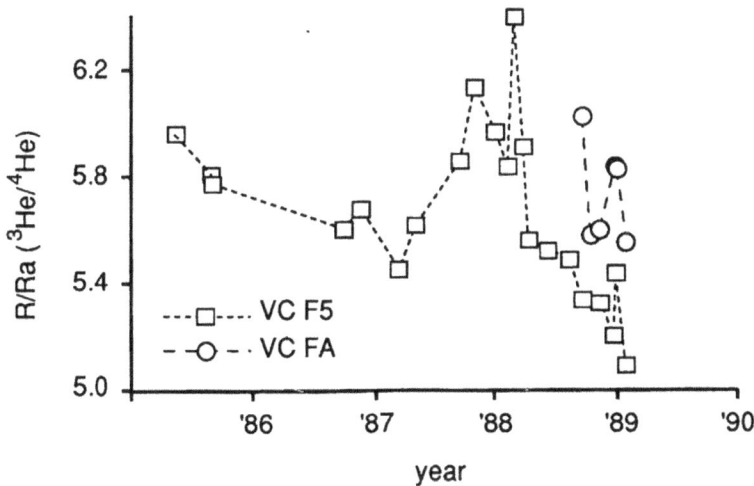

**Figure 12.14** Temporal variations in the $^3He/^4He$ ratios at the F5 and FA crater fumaroles. The higher ratios, compared to the beach fumaroles, are attributed to a direct connection with a deep magmatic reservoir. Oscillations in the ratios are the result of continuous mixing (to various degrees) between deep magma-derived fluids ($^3He$ rich), crustal fluids and superficial or deep-marine waters ($^4He$ rich).

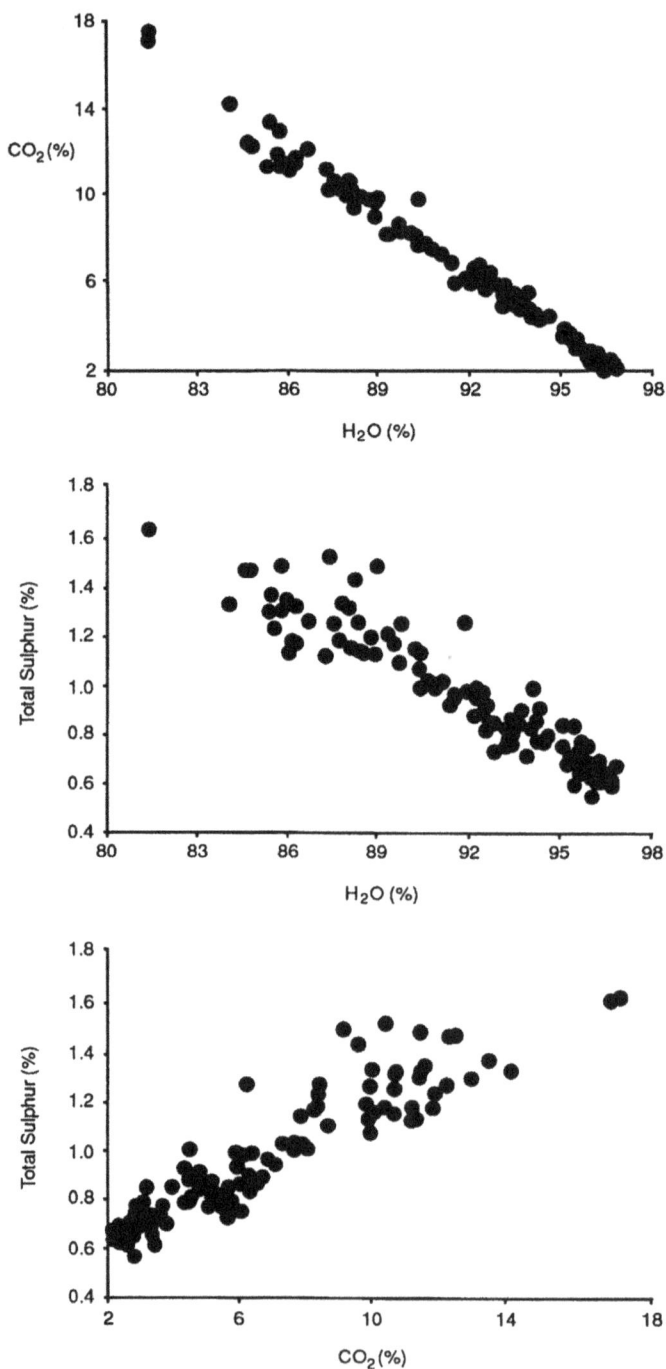

**Figure 12.15**  Two-component variation diagrams for the Vulcano fumaroles. The trends illustrate the different contributions from the shallow and deep reservoirs. End members define the chemical composition of the sampled fluids at two different times: when maximum mixing between the surficial and deep reservoirs occurs; and when the gas reaches the surface directly from the deep reservoir, with little or no addition of superficial fluids. All intermediate points on the graphs represent different degrees of mixing between the two end members. See text for further explanation.

maximum. It is worth noting that the minimum $^3$He/$^4$He value at the crater fumaroles is the same as the ratio repeatedly obtained at the beach fumaroles, further suggesting that the two systems are influenced by the same phenomena. The very high $^4$He/$^{20}$Ne ratios observed at nearly all sampling sites, imply that the superficial helium source is crustal and not atmospheric. These results are in good agreement with the interpretation that an older superficial eruptive fracture could exist, located close to the beach fumarolic system. This reservoir should be slightly depleted in $^3$He compared to that which feeds the crater fumaroles, and comparatively richer in $^4$He, without addition of atmospheric helium trapped and carried in the superficial meteoric waters but probably buffered by deep marine waters.

It is possible to distinguish the different contributions from the crustal, shallow and deep reservoirs with reference to Figure 12.15. End members of the illustrated trends define the chemical compositions of the sampled fluids at two different times: (a) when maximum mixing between the surficial and deep reservoirs occurs; and (b) when the gas reaches the surface directly from the deep reservoir, with little or no addition of superficial fluids. All intermediate points on the graphs represent different degrees of mixing between the two end members. The inverse correlation between $CO_2$ and total sulphur ($SO_2 + H_2S$) versus $H_2O$ (Fig. 12.15a & b) suggests that if water vapour has its origin in a shallow reservoir (Cheminée et al. 1969, Allard 1983), the other gas species might all be derived together from a deeper reservoir, without any further addition en route to the surface. Verification of such an interpretation must await further and more comprehensive study of the chemical and the isotopic oscillations recorded at the Vulcano crater fumaroles.

$H_2$ variations at Vulcano over the monitoring period might argue for the injection of deep, magma-derived $H_2$ and $CO_2$, and indeed Oskarsson (1984) showed that a non-equilibrated actively degassing magma can produce hydrogen pulses through diffusion processes, $H_2$ being the most mobile compound due to its low molecular weight. Although $H_2$ actually increased in the first period of our sampling at Vulcano, its variation appears to be related primarily to that of water, and the $H_2/H_2O$ ratio suggests rather steady redox conditions in the fluid equilibration zone. An increase in $P(H_2)$ can in fact be explained simply by an increase of $P(H_2O)$, provided that $P(O_2)$ remains constant (Gerlach 1980) (see the equation below) without the need to invoke a fresh input of magmatic $H_2$.

$$H_2 + \tfrac{1}{2}O_2 \leftrightarrow H_2O$$

Increasing vaporization of the underground reservoir would be also be expected to generate increasing concentrations of the less soluble gas species ($H_2S$ and $CO_2$) compared to others ($HCl$, $HF$ and $SO_2$), together with higher $H_2S/SO_2$ and $CO_2/H_2$ ratios. The fact that neither characteristic is observed at Vulcano also argues against a fresh input of magma-derived gas.

### 12.4.2.3 Influence of tectonic or hydrodynamic events
It is difficult to attribute the geochemical variations at Vulcano to a disturbance of the hydrothermal system due to increasing regional or local geophysical activity. No abnormal ground deformation was observed in the area during the five years of the geochemical survey, and neither were any significant tectonic earthquakes recorded over the same period (Ferrucci et al. 1991).

### 12.4.2.4 Mixing between different magmatic and hydrothermal systems

The coastal fumaroles of Vulcano are generally interpreted as resulting from the percolation of a crater-type fluid through superficial water tables, leading to a preferential loss of soluble species such as $SO_2$, HCl and HF (Martini et al. 1980, 1984, Cioni & D'Amore 1984, Mazor et al. 1988). If compared to the crater fumaroles, however, the coastal fumaroles not only have different chemical characteristics (high $H_2$ and $CH_4$, low He, $N_2$ and $He/CO_2$ ratios), but also different carbon and helium isotope ratios (Allard 1978, Cannata et al. 1988), which, according to Baubron et al. (1990) and Tedesco et al. (1991), may reflect their feeding from a separate hydrothermal system, distinct from the shallow reservoir feeding the crater. This hydrothermal system may be fed by the late-stage degassing of a cooling magma body, perhaps related to the Vulcanello complex which was active from BC 183 through to the 16th century (Keller 1980).

Interactions between the two systems are feasible, particularly considering the proximity of the respective fumarolic fields (less than one kilometre apart), and some of the chemical variations in the F5 fumarole are compatible with such a mixing process. For example, Tedesco et al. (1991) showed an inverse relationship between He and $H_2O$, such that He decreases as $H_2O$ increases. This can be interpreted simply as being due to water dilution or to mixing between an He-rich/$H_2O$-poor crater fluid and an He-poor/$H_2O$-rich beach fluid. The latter is supported by the inverse correlation between the calculated equilibrium temperature and the $H_2S/SO_2$ ratio (Tedesco et al. 1991) which points to mixing between a hot (high-$SO_2$) crater fluid and a colder (high-$H_2S$) beach fluid. Further indication of such mixing between two different reservoirs is illustrated by the trend of the $R/Ra$ values at the F5 and $F5_{HT}$ fumaroles (Fig. 12.14). Here, the minimum value obtained at the crater fluid ($R/R_a = 5$) corresponds to the normal value obtained for the beach fluids. This trend can be explained by mixing between a $^3$He-poor reservoir (beach) and a $^3$He-rich reservoir (crater).

### 12.4.2.5 Some conclusions about gas variations at Vulcano

The bulk of the chemical and isotopic data gathered at Vulcano strongly suggest that a fresh injection of deep magmatic fluid was not the source of the observed geochemical variations or the recent increase in fumarolic activity. In particular, this model is not supported by the constant and low CO contents of fumaroles F5 and $F5_{HT}$, the continuous oscillation of the $^3He/^4He$ ratio without a steadily increasing trend, and the rather low calculated equilibrium temperatures which are close to the emission temperatures and comparable to those observed during previous "variation cycles". Nevertheless, the trend of increasing vent temperature at the FA fumarole (660°C max.), and the high CO content could be explained by an injection of new magmatic fluid. Even if this is the case, however, it is likely to represent the more effective transport of a deep magma-derived fluid to the surface, rather than being due to a new rising magma body. This model may explain both the increase of the outlet temperature, and the chemical equilibrium (reaction 12.2, see Section 12.4.2.2) and composition thermodynamically derived from higher temperatures. At the same time, if reaction 12.1 (see Section 12.4.2.2) always points to shallow equilibrium conditions at low temperature, this may be attributed to partial re-equilibration of CO as compared to $H_2$ (or vice versa) as the deeper fluids travel towards the surface. The model presented here for Vulcano constitutes the first step towards developing a better understanding of the dynamic evolution of the volcano's geochemical system. A more constrained model will require increased chemical and isotopic study of all the hydrothermal systems, and a sampling strategy designed to cover a wider area of the island.

## 12.5 Thermal waters

Thermal water surveys are often carried out in volcanically active areas to determine the extent of a volcano's influence on the local hydrological system and to detect chemical and temperature anomalies: they may also be conducted simply because an active crater is inaccessible and it may be safer to undertake geochemical sampling at some distance from potentially dangerous zones. In particular, studies of both hot and cold springs often form an important component of any geothermal assessment and can provide a useful tool in monitoring active and dormant volcanoes.

Major component (Na, K, Ca, Mg, Cl, F, Li, $SO_4$, $HCO_3$ and Si) and isotope ($^{18}O/^{16}O$, deuterium and $^{13}C/^{12}C$) data obtained from thermal waters can initially be used to characterize the area of interest, chemically and isotopically, and to distinguish the contributions from different (i.e. deep or shallow) sources. The initial temperatures of deep-derived fluids can be calculated using the silica, Na—K, K—Mg, or Na—K—Ca geothermometers, although application of these often results in a range of disparate equilibrium temperatures, making interpretation problematical. Such variations in temperature estimates probably result from different rates of adjustment of the equilibrium reactions to the changing physical environments encountered during the ascent of water to the surface. Under most circumstances, the K—Mg reaction responds faster than those which involve sodium and potassium ions, thereby giving a lower calculated temperature (Giggenbach 1988). In order to achieve more meaningful results, Giggenbach (1988) developed a graphical method for evaluating Na—K and K—Mg equilibrium temperatures which relied simply on determining the sodium, potassium, and magnesium contents of thermal waters.

Rock-thermal water interaction can be expected in any hydrothermal system, and the degree of this interaction may be determined by looking at the chlorine concentrations of the water. Different Cl contents are related to different $\delta^{18}O$ values, and can be used to determine different contributions of seawater. The Na/Cl ratio can also be used to recognize a marine source for thermal waters. Stable oxygen and deuterium isotopes are also useful in constraining the origin of waters in volcanic terrains, and can be used to determine the relative contributions of local meteoric water, seawater, steam-heated water, and "magmatic" water.

Toutain et al. (1992) have recently sampled water wells near the foot of Vulcano in order to detect lateral emissions of cold soil gases. Such gases have been shown to diffuse through the flanks of both active and dormant volcanoes and, therefore, through their associated hydrothermal systems. These types of emission, consisting essentially of carbon dioxide and rare gases, were initially recognized at Mount Etna in Sicily (Aubert & Baubron 1988, Allard et al. 1991a). They have also been detected at a number of dormant volcanoes including Vesuvius and Vulcano (Allard et al. 1988, Baubron et al. 1990). Because such emissions are devoid of acid gases such as $SO_2$ and HCl, and as they occur at a safe distance from active craters, they provide a valuable tool for continuous volcano monitoring.

At Vulcano, three gaseous components ($CO_2$, He and $^{222}$Rn) were selected for monitoring and extracted from a water well located at the foot of the La Fossa cone. The gas species were selected on the basis of those geochemical and isotopic characteristics that indicated a genetic link with the centrally located high temperature fumarolic gases emitted at the summit crater. Very strong variations of gas composition were recorded by Toutain et al. (1992), even over a short time scale. $CO_2$ concentrations ranged, for example, from 1% to 94% over a 24-h period. Some variations are clearly diurnal, and can be correlated with changes in atmospheric pressure. Such large variations make the measurement of bulk concentrations unsuitable for monitoring at Vulcano. The authors determined, however, that increasing $^{222}$Rn/$CO_2$ and He/$CO_2$ ratios could well prove useful in recording new inputs of magmatic gases from rising bodies of magma. Monitoring of these ratios therefore might provide precursory evidence for renewed activity at Vulcano and other dormant volcanic centres.

## 12.6 Soil gases

As discussed briefly in the previous section, much attention has been focused, during the last few years, on cold or low-temperature gases which percolate diffusively through volcanic piles at some distance from the active craters. Variations in the hydrogen concentration of soil gases at volcanoes were monitored by Sato & McGee (1980), while the importance of $CO_2$ emanations on an active volcano was recognized by Carbonnelle & Zetwoog (1982) and Carbonnelle et al. (1985). These latter studies demonstrated that huge amount of carbon dioxide, together with helium and radon, escape diffusively from the flanks of Mount Etna. Further investigations have demonstrated both the magmatic origin of these lateral emissions (Allard et al. 1991a), and their potential use as markers for thermal anomalies and active faults (Baldi et al. 1984, Baubron & Sabroux 1984, Baubron et al. 1986, Aubert & Baubron 1988). With this in mind, both He and $CO_2$ mapping has been undertaken on volcanic soils by a number of Italian groups (Bertrami et al. 1984, Lombardi et al. 1984, Lombardi & Nappi 1986).

Soil gas studies have been initiated at several dormant volcanoes, including Vesuvius, Ischia, Vulcano, and Campi Flegrei in Italy, the Soufrière of Guadaloupe in the Lesser Antilles, Kilauea volcano in Hawaii, and Lamongan and Dieng in Indonesia. Diffuse gas release has been recognized at all these volcanoes, the degree of emissions being related to their level of activity (Baubron & Sabroux 1984, Baubron 1987, Allard et al. 1988, Baubron et al. 1989, Baubron 1990, Allard et al. 1991a).

### 12.6.1 Soil gas surveys and sampling

Prior to systematic monitoring, a detailed reconnaissance for volcanic soil gas leaks is required. This initial stage involves numerous field measurements, generally at 10-m intervals along traverses across the main structural features of the volcano under study, such as fracture systems and rift zones which may be identified from geological maps or aerial photographs. Currently it is possible to analyze at 50–100 sites per day, thereby allowing for reasonably rapid location of major gas leaks and determination of their distribution and chemical characteristics. Gas anomalies may be related to either visible or hidden fractures and their origin may be evaluated from their geographical distribution, their chemical nature, and their relationship to fluid sources or fumaroles in the area. Data can be used to produce a geochemical map (e.g. Lombardi et al. 1984) showing the size and type of the gas anomalies and their possible relationship to subsurface reservoirs, either magmatic or hydrothermal. Such maps also help to identify areas of increased risk from potentially lethal gas-release or from a forthcoming eruption.

Volcanic soil gases are normally analyzed or sampled at a depth of 0.7–1.0 m below the ground surface, in order to minimize the effects of metereological changes. Stainless steel probes (1 cm diameter), filled with teflon capillary, are introduced into the ground using a sliding hammer and connected to analyzers or sampling containers. Analytical equipment may be portable (e.g. $CO_2$ infrared spectrometer or radon scintillation counter) or require vehicular transportation (e.g. quadrupole mass spectrometer, an He-specific mass spectrometer, and computerized data collection and processing equipment).

## 12.6.2 Gas flux measurements

Flow rate measurements of volcanic soil gases are important, as a high gas flux is often one of the criteria by which gas anomalies are selected for further monitoring. Soil gas fluxes are measured using stainless steel containers (typical capacities are 60–200 l and surface areas are around 0.5–1.3 m$^2$) which are placed face-down on the ground. The gas flux can be determined by either dynamic (flow-through) or static (accumulation) methods (Baubron 1988, Allard et al. 1991b, Miele 1992), both methods giving comparable results. The dynamic method requires continuous and simultaneous analysis of both the air and a mixture of soil gas plus air, and is laborious but fast. The static method, on the other hand, is easier but more time consuming, involving discontinuous analysis (about every hour) over 6-8 h of gas accumulation. The contribution of biogenic $CO_2$ flux on most volcanoes generally appears to be negligible, between 0.02 and 0.0611 m$^{-2}$h$^{-1}$. This flux is lower than both the accepted values for cultivated fields (0.02–0.311 m$^{-2}$h$^{-1}$, Henin et al. (1969)) and for continental flow areas (0.111 m$^{-2}$h$^{-1}$, Gaudry et al. (1990)).

As demonstrated by Allard et al., (1991b) for the $CO_2$ budget of Etna, it is necessary to quantify the $CO_2$ output from both flank emanations and summit vents in order to determine the total gas flux for a volcano. Only then can precise measurements of both soil gas and summit vent flux be used to evaluate changes in the level of volcanic activity. It should also be emphasized that temporal changes in the soil gas flux are not only related to variable levels of activity within volcanoes, but also appear to follow seasonal variations, as well as changes occuring in response to seismic activity (Tedesco et al. 1988a).

## 12.7 Conclusion

Monitoring fluid and gas compositions at active and dormant volcanoes is becoming increasingly important, both as a tool for recognizing precursory geochemical changes associated with impending activity, as well as for increasing our understanding of magmatic and hydrothermal systems and how they interact. As illustrated at both Vulcano and Campi Flegrei, such studies have already proved vital in constraining behavioural models constructed purely on the basis of geophysical data (e.g. thermal, seismic and ground deformation), primarily by demonstrating minimal, if any, contributions of fresh magma at both locations, and thereby obviating any need for evacuation. As analytical equipment and procedures become further refined, the utility of geochemical monitoring cannot fail to increase, so that it will become a standard surveillance tool alongside the more established seismic and ground deformation techniques.

## References

Allard, P. 1978. Composition isotopique et origine des constituants majeurs des gaz volcaniques (H$_2$O, C, S). Thesis, University of Paris, Paris, France.

Allard, P. 1983. Stable isotope composition of fumarolic gases from Vulcano island, Eolian island arc. In *Proceedings of the IAVCEI symposium on volcanic gases*, 4. Hamburg: General Assembly of the International Union of Geodesy and Geophysics (IUGG).

Allard, P. 1986. Geochimie isotopique et origine de l'eau, du carbone et du soufre dans le gaz volcaniques: zone du rift, marges continentales et arcs insulaires. PhD thesis, University of Paris, Paris, France.

Allard, P., J. C. Baubron, G. Luongo, R. Pece, D. Tedesco 1988. Geochemical survey of soil gas emanations and eruption forecasting: the Vesuvius case, Italy. *Proceedings of the International Conference on Volcanoes*, 12. Kagoshima, Japan: National Institute for Research Advancement.

Allard, P., A. Maiorani, D. Tedesco, B. Cortecci Turi 1991a. Isotopic constraints on the origin of sulphur and carbon in Solfatara fumaroles, Campi Flegrei (Italy). *Journal of Volcanology and Geothermal Research* **48**, 139–59.

Allard, P., J. Carbonnelle, D. Dajlevic, J. Le Bronec, P. Morel, M. C. Robe J. M. Maurenas, R. Faivre Pierret, D. Martin, J-C. Sabroux, P. Zetwoog 1991b. Eruptive and diffuse emissions of CO$_2$ from Mount Etna. *Nature* **351**, 387–91.

Aubert, M. & J. C. Baubron 1988. Indentification of a hidden thermal fissure in a volcanic terrain using the combination of hydrothermal convection indicators and soil-atmosphere analyses. *Journal of Volcanology and Geothermal Research* **35**, 217–25.

Baldi, P., A. Ceccarelli, R. Bertrami, I. Friedman, S. Lombardi 1984. Helium in soil gases in geothermal areas. Seminar on utilization of geothermal energi for electrical power production and space heating. *Proceedings*, 24–28. Firenze, Italy.

Barberi, F., P. Gasparini, F. Innocenti, L. Villari 1973. Volcanism of the Southern Tyrrhenian Sea and its geodynamic implications. *Journal of Volcanology and Geothermal Research* **78**, 5221–32.

Barberi, F., F. Innocenti, G. Ferrara, J. Keller, L. Villari 1974. Evolution of Eolian arc volcanism (Southern Tyrrhenian Sea). *Earth and Planetary Science Letters* **21**, 269–76.

Barberi, F., G. Corrado, F. Innocenti, G. Luongo 1984. Phlegrean Fields 1982-1984: brief chronicle of a volcano emergency in a densely populated area. *Bulletin Volcanologique* **47**, 175–85.

Barin, I. & O. Knacke 1973. *Thermochemical properties of inorganic substances*. Berlin: Springer.

Barin, I., O. Knacke, L. Kobascewsky 1977. *Thermochemical properties of inorganic substances (supplement)*. Berlin: Springer.

Baubron, J. C. 1986-90. Geochemical surveillance of some Italian volcanoes (Vulcano, Vesuvio, Solfatara). *BRGM Reports ANA DT 86, 87, 88, 89, 90* (1986, 1987, 1988, 1989, 1990). Orleans, France: Bureau de Recherche Geologique et Miniere.

Baubron, J. C. & J-C. Sabroux 1984. Analysis des gaz du sol sur les zones de Moscou et Bouillante (Guadeloupe). *BRGM Report No. 84 SGN 400 GTH*. Orleans, France: Bureau de Recherche Geologique et Miniere.

Baubron, J. C., P. Briole, D. Tedesco 1986. In situ analysis of soil gases as an indicator of volcanic risk (abstract). *Periodo di Mineralogia* **55**, 157.

Baubron, J. C., P. Allard, J. P. Toutain 1990. Diffuse volcanic emissions of carbon dioxide from Vulcano island, Italy. *Nature* **344**, 51-3.

Beccaluva, L., G. Gabbianelli, R. Lucchini, P. L. Rossi, C. Savelli 1985. Petrology and K/Ar ages of volcanics dredged from the Eolian seamounts: implications for geodynamic evolution of the southern Tyrrhenian basin. *Earth and Planetary Science Letters* **74**, 187-208.

Bertrami, R., E. Antrodicchia, C. Luzi. Prospezione dei gas del suaolo nell'isola di Vulcano. *Rendiconti della Società Italiana di Mineralogia e Petrologia* **39**, 403-13.

Cannata, S., S. Hauser, F. Parello, M. Valenza 1988. Caratterizzazione isotopica della $CO_2$ presente nelle manifestazioni gassose dell'isola di Vulcano. *Rendiconti Società Italiana di Mineralogia e Pethologia* **43**, 153-61.

Carapezza, M., M. Nuccio, M. Valenza 1981. Genesis and evolution of the fumaroles of Vulcano (Aeolian islands, Italy): a geochemical model. *Bulletin Volcanologique* **44**, 547-63.

Carapezza, M., M. Nuccio, M. Valenza 1984. Geochemical surveillance of the Solfatara of Pozzuoli (Campi Flegrei) during 1983. *Bulletin Volcanologique* **47**, 303-11.

Cioni, R. & E. Corazza 1981. Medium-temperature fumarolic gas sampling. *Bulletin Volcanologique* **41**, 23-29.

Cioni, R. & F. D'Amore 1984. A genetic model for the crater fumaroles of Vulcano island (Sicily, Italy). *Geothermics* **13**, 375-84.

Cioni, R., E. Corazza, L. Marini 1984. The gas/steam ratio as indicator of heat transfer at the Solfatara fumaroles, Phlegraean Fields (Italy). *Bulletin Volcanologique* **47**, 295-302.

Cioni, R., E. Corazza, M. Fratta, G. Magro 1986. Sorveglianza geochimica dei Campi Flegrei. Bollettino della. *Gruppo Nazionale per la Vulcanologia (Italy)* 167-73.

Cheminée, J. L., R. Letolle, P. H. Olive 1969. Premiéres données isotopiques sur des fumerolles de volcans Italiens. *Bulletin Volcanologique* **45**, 173-8.

Degens, E. T. 1969. Biogeochemistry of stable carbon isotopes. In *Organic geochemistry: methods and results*, Eglinton & Murphy (eds), 304-29.

Del Pezzo, E. & M. Martini 1981. Seismic events under Vulcano, Aeolian islands, Italy. *Bulletin Volcanologique* **44**, 521-5.

De Natale, G., F. Pingue, P. Allard, A. Zollo 1991. Geophysical and geochemical modelling of the 1982-1984 bradyseismic phenomena at Campi Flegrei caldera (southern Italy). *Journal of Volcanology and Geothermal Research*.

Falsaperla, S., G. Frazzetta, G. Neri, G. Nunnari, R. Velardita, L. Villari 1989. Volcano monitoring in the Aeolian Islands (Southern Tyrrhenian sea): the Lipari-Vulcano eruptive complex. In *IAVCEI Proceedings in Volcanology vol. 1: Volcanic hazards – assessment and monitoring*, J. H. Latter (ed.), 339-56. Berlin: Springer.

Ferrucci, F., G. Gaudiosi, A. Hirn, G. Luongo, L. Mirabile, N. A. Pino 1991. Seismological exploration of Vulcano (Aeolian Islands, Southern Tyrrhenian sea): case history. *Acta Vulcanologica* **1**, 143-52.

Frazzetta, G., L. La Volpe, M. Sheridan 1984. Hazards at Fossa of Vulcano: data from the last 6,000 years. *Bulletin Volcanologique* **47**, 106-24.

Gaudry, A., G. Polian, B. Ardouin, G. Lambert 1990. Radon-calibrated emission of $CO_2$ from South Africa. *Mellus* **42**, 9-19.

Gerlach, T. M. 1980. Evaluation of volcanic gas analyses from Kilauea Volcano. *Journal of Volcanology and Geothermal Research* **7**, 295-317.

Giggenbach, W. F. 1983. Geothermal gas equilibria. *Geochimica et Cosmochimica Acta* **43**, 2021-32.

Giggenbach, W. F. 1988. Geothermal solute equilibria. Derivation of Na-K-Mg-Ca geoindicators. *Geochimica et Cosmochimica Acta* **52**, 2749–65.

Henin S., R Gras, G. Monnier 1969. *Le profil cultural*. Paris: Masson.

Hermance, J. F. 1983. The Long Valley/Mono Basin volcanic complex in eastern California: Status of present knowledge and future reasearch need. *Reviews of Geophysics and Space Physics* **21**, 1545–65.

Hooker, P. J., R. Bertrami, S. Lombardi, R. K. O'Nions, E. R. Oxburgh 1985. Helium-3 anomalies and crust-mantle interactions in Italy. *Geochimica et Cosmochimica Acta* **49**, 2505–13.

Jaggar, T. A. 1913. Magmatic gases. *American Journal of Science* **238**, 313–53.

Javoy, M., F. Pineau, D. Demaiffe 1984. Nitrogen and carbon isotopic composition in the diamonds of Mbuji Maji (Zaire). *Earth and Planetary Science Letters* **68**, 399–412.

Keller, J. 1980. The island of Vulcano. In *The Eolian Islands, an active volcanic arc in the Mediterranean sea*, L. Villari (ed.). Catania: CNR.

Le Guern, F. & R. Faivre Pierret 1982. Différentiation de l'émanation magmatique: réaction $H_2S + SO_2$ dans le gaz volcaniques. Vulcano (Italie) 1923–1979. *Bulletin Volcanologique* **45**, 179–90.

Lombardi, S. & G. Nappi 1986. Helium in soil gases at Lipari and Stromboli volcanoes. *Periodico di Mineralogia* **55**, 165–76.

Lombardi, S., M. Di Filippo, L. Zantedeschi 1984. Helium in Phlegrean Fields in soil gases in 1983. *Bulletin Volcanologique* **47**, 249–55.

Matsuo, S. 1961. On the chemical nature of fumarolic gases of volcano Showa-shinzan, Hokkaido, Japan. *Journal of Earth Science, Nagoya University* **3**, 80–100.

Martini, M. 1986. Thermal activity and ground deformations at Phlegrean Fields, Italy: precursors of eruptions or fluctuations of quiescent volcanism? A contribution of geochemical study. *Journal of Geophysical Research* **91**, 12255–560.

Martini, M., G. Piccardi, P. Cellini Legittimo 1980. Geochemical surveillance of active volcanoes: data on the fumaroles of Vulcano (Aeolian islands, Italy) *Bulletin Volcanologique* **43**, 255–63.

Martini, M., P. Cellini Legittimo, G. Piccardi, L. Giannini 1984. Composition of hydrothermal fluids during the bradyseismic crisis which commenced at Phlegrean Fields in 1982. *Bulletin Volcanologique* **47**, 267–73.

Mazor, E., R. Cioni, E. Corazza, M. Fratta, G. Magro, S. Matsuo, J. Hirabayashi, H. Shinohara, M. Martini, G. Piccardi, P. Cellini Legittimo 1988. Evolution of fumarolic gases boundary conditions set by measured parameters: case study at Vulcano, Italy. *Bulletin of Volcanology* **50**, 71–85.

McKee, C. O., R. W. Johnson, P. L. Lowenstein, S. J. Riley, R. J. Blong, P. De St Ours, B. Talai 1984. Rabaul caldera, Papua New Guinea: volcanic hazards, surveillance and eruption contingency planning. *Bulletin Volcanologique* **47** 195–327.

Mizutani, Y. & S. Matsuo 1959. Successive observations of chemical components in the condensed water from a fumarole of Showa-shinzan, Kazan. *Bulletin of the Volcanological Society of Japan* **3**, 119–27.

Oskarsson, N. 1984. Monitoring of fumarole discharge during the 1975-1982 rifting in Krafla volcanic center, North Iceland. *Journal of Volcanology and Geothermal Research* **22**, 97–121.

Poliak, B. G. & I. N. Tolstikin 1980. Geotectonics, heat flux and helium isotopes: triple relationship. In *Proceedings of the International Symposium KAPG 1-4, 1979*, 234–9. Leningrad, USSR: Geochemical Society of Russia.

Poreda, R. & H. Craig 1989. Helium isotope ratios in circumpacific volcanism arcs. *Nature* **338**, 473–8.

Rosi M. & A. Sbrana (eds). 1987. Phlegraean Fields. *Quaderni della Ricerca Scientifica de CNR (Italie)* **9**, 1–175.

Sabroux, J-C. 1979. Equilibre thermodynamique en phase gazeuse volcanique. In *Haute temperatures et sciences de la terre*, 37–46. Toulouse: CNRS.

Sabroux, J-C. 1983. Volcano energetics: volcanic gases and vapours as geothermometers and geobarometers. In *Forecasting volcanic events*, H. Tazieff & J-C. Sabroux (eds), 17–25. Amsterdam: Elsevier.

Saint-Claire Deville, C. & F. Leblanc 1858. *Ann. Chim. Phys.* 3rd Ser., **52**, 5.

Saint-Claire Deville, C., F. Leblanc, F. Fouquet 1863. Sur les emanations à gas combustible qui se sont echappé des fissures de la lave de 1794 à Torre del Greco lors de la derniere eruption du Vesuve. *Comptes Reudus de l'Academie des Sciences, Paris* **56**.

Sano, Y. & H. Wakita 1985. Geographical distribution of $^3He/^4He$ ratios in Japan: implications for arc tectonics and incipient magmatism. *Journal of Geophysical Research* **90**, 8729–41.

Sano, Y. H. Wakita 1988. Precise measurement of helium isotopes in terrestrial gases. *Bulletin of the Chemical Society of Japan* **61**, 1153–7.

Sano, Y., H. Wakita, F. Italiano, M. Nuccio 1989. Helium isotopes and tectonics interactions in Italy. *Geophysical Research Letters* **16**, 511–14.

Shepherd, E. S. 1921. Kilauea gases 1919. *Hawaiian Volcano Observatory Bulletin* **9**, 83–8.

Sheridan, M. F. & M. C. Malin 1983. Application of computer-assisted mapping to volcanic hazard evaluation of surge eruptions: Vulcano, Lipari and Vesuvius. *Journal of Volcanology and Geothermal Research* **17**, 182–202.

Shinohara, H. & S. Matsuo 1984. Results and analysis on fumarolic gases from F-1 and F-5 fumaroles of Vulcano, Italy. *Geothermics* **15**, 211–15.

Sicardi, L. 1955. Captazione ed analisi chimica dei gas della esalazione solfidrico-solforosa dei vulcani in fase solfatarica. *Bulletin of Volcanology* **17**, 107–12.

Smithsonian Institution 1988. Vulcan. *Bulletin, Scientific Event Alert Network (SEAN)* **6**.

Tedesco, D. 1987. Significato ed elaborazione termodinanmica dei fluidi di ambienti geotermici (Campi Flegrei, Long Valley) e vulcanici (Hawaii, USA). PhD thesis, University of Napoli, Napoli, Italy.

Tedesco, D. 1994. Chemical and isotopic gas emissions at Campi Flegrei: evidence for an aborted period of unrest. *Journal of Geophysical Research*, 15623–31.

Tedesco, D. & J-C. Sabroux 1987. The determintion of deep temperatures by means of the $CO-CO_2-H_2-H_2O$ geothermometer: an example using fumaroles in Campi Flegrei, Italy. *Bulletin of Volcanology* **49**, 381–77.

Tedesco, D., L. Bottiglieri, R. Pece 1988a. 10th of April 1987 seismic swarm: correlation with geochemical parameters in Campi Flegrei caldera (southern Italy). *Geophysical Research Letters* **15**, 661–4.

Tedesco, D., R. Pece, J-C. Sabroux 1988b. No evidences of a new magmatic gas contribution to the Solfatara volcanic gas, during the bradyseismic crisis at Campi Flegrei caldera (Italy). *Geophysical Research Letters* **15**, 1441–4.

Tedesco, D., P. Allard, Y. Sano, R. Pece, H. Wakita 1990. Helium-3 in subaerial and submarine fumaroles at Campi Flegrei caldera, Italy. *Geochimica et Cosmochimica Acta* **54**, 1105–16.

Tedesco, D., J. P. Toutain, P. Allard, R. Losno 1991. Chemical variations in fumarolic gases at Vulcano island: seasonal and volcanic effects. *Journal of Volcanology and Geothermal Research* **45**, 325–34.

Tedesco, D., G. Miele, Y. Sano, J. P. Toutain 1994. Helium isotope ratios in Vulcano island (southern Italy): temporal variations, superficial source mixing and deep magmatic supply. *Journal of Volcanology and Geothermal Research* **64**.

Tonani, F. 1971. Concepts and techniques for the geochemical of volcanic eruptions. In *The surveillance and prediction of volcanic activity*, 145–66. Paris: UNESCO.

Toutain, J. P., J. C. Baubron, J. Le Bronc, P. Allard, P. Briole, B. Marty, G, Miele, D. Tedesco, G. Luongo 1992. Continuous monitoring of distal gas emanations at Vulacno, southern Italy. *Bulletin of Volcanology*, 147–55.

Villardo, G., M. Castellano, G. Gaudiosi, F. Ferrucci 1991. Seismic surveillance at Vulcano by use of a portable digital array: features of the seismicity and relocation of the events in a 3-D heterogeneous structure. *Acta Vulcanologica* **1**, 171–7.

# 13 Forecasting the behaviour of lava flows

C. R. J. Kilburn, H. Pinkerton and L. Wilson

## 13.1 Introduction

A key objective in monitoring lavas is to forecast their behaviour. Tens of square kilometres of land are buried annually by effusions and, though modest on a global scale, the local impact may be severe and long-lived. Fortunately, human life has seldom been at risk, owing to the generally sluggish rates of flow advance: a tragic exception occurred in 1977 when a fast-moving flow from Nyiragongo (Zaire) killed between 60 and 300 people (Tazieff 1977). Land and property, on the other hand, are extremely vulnerable and approximate costs of recent eruptions include: US\$6 million in 1960 on Kilauea, Hawaii, US\$35 million on Heimaey, Iceland, in 1973, and between US\$2 million and US\$3 million each for the 1983 and 1991–93 activity on Etna, in Southern Italy. In all these instances, economically strategic areas (agricultural land, tourist centres and fishing ports) were threatened or destroyed, causing immense hardship to local communities. More dramatically, on Etna and Vesuvius alone, several towns have been obliterated in historical times.

Forecasting flow behaviour requires an understanding of how lava properties, effusion dynamics and environmental factors (such as groundslope and topography) determine the final dimensions and rate of growth of a flow. Two complementary approaches are available: one focuses on local lava properties and combines these theoretically to obtain a flow's bulk characteristics (e.g. mean velocity and thickness); the other starts by monitoring the bulk characteristics, identifies behaviour patterns empirically, and uses these patterns to deduce controls on lava emplacement. The first approach enjoys mathematical rigour but, since it needs measurements of lava properties (e.g. rheology, density, temperature and velocity) at numerous positions within a flow, it suffers from the lack of necessary field data caused by logistical difficulties (Pinkerton 1993, Tilling & Peterson 1993). The second approach, in contrast, is ideally suited to field observations, but has a weaker theoretical foundation.

This chapter is concerned mainly with practical observations and so emphasizes the second approach. The guiding philosophy is that lavas are self-determining systems which follow systematic and repeatable evolutionary sequences. After a general introduction to subaerial lava features, we concentrate on a'a lava, the flow type most commonly observed in historical times and which has posed the most frequent threat to populated areas. We use field data to identify different regimes of emplacement and present expressions for forecasting maxi-

**Table 13.1** Theoretical and laboratory studies of flow growth.

*1. Thermally unmixed*

Flat terrain, Newtonian rheology
  *Isothermal*                                    *Non-isothermal*
  Didden & Maxworthy (1982)              Fink & Griffiths (1990, 1992)
  Huppert (1982a)                              Whitehead & Helfrich 1991
  Huppert et al. (1982)                       Stasiuk et al. (1993b)

Flat terrain, Bingham rheology
  *Isothermal*
  Blake (1990)

Inclined plane, Newtonian rheology
  *Isothermal*                                    *Non-isothermal*
  Smith (1973)                                  Hallworth et al. (1987)
  Huppert (1982b)
  Schwartz & Michaelides (1988)
  Lister (1992)

Inclined plane, Bingham rheology
  *Isothermal*
  Hulme (1974)
  Hulme & Fielder (1977)
  Dragoni et al. (1986)
  Liu & Mei (1989)

*2. Thermally mixed*
  Danes (1972)
  Harrison & Rooth (1976)
  Dragoni (1989)
  Crisp & Baloga (1990, 1994)
  Dragoni et al. (1992)
  Manley (1992)

*3. Turbulent flow, Newtonian, non-isothermal*
  Huppert et al. (1984)
  Huppert & Sparks (1985)

*4. Effects of variable discharge rate*
  Baloga & Pieri (1986)
  Baloga (1987)

*5. Effects of compressibility*
  Jaupart (1991)

mum flow lengths from easily measured variables.

The key point is that, by focusing on emplacement regime, forecasts can be made from simple models without the need for extensive measurements of, say, lava temperature and viscosity. Such measurements, of course, are essential for refining the simple models and for offering deeper physical insights into lava behaviour, the procedures and problems in obtaining such measurements are beyond the scope of this chapter, but are considered at length in Kilburn & Luongo (1993). Similarly, we attempt neither to review the many recent theoretical and laboratory simulations of flows (a selected reference list is given in Table 13.1), nor to discuss methods for changing the course of a flow, for which details can be found in Bolt et al. (1975), Barberi & Villari (1984), Chester et al. (1985), Barberi et al. (1993) and the references therein.

Finally, general overviews of effusive eruptions and their products are given in Macdonald (1972), Williams & McBirney (1979), Cas & Wright (1987) and in the compilations edited by Fink (1987, 1990) and Kilburn & Luongo (1993). Here we summarize only the gross features of historic lava flows and domes, to provide a context for the discussion of flow emplacement.

## 13.2 Basic features

Subaerial effusions in historic times have involved volumes, during individual eruptions, of less than about $15 \, km^3$ (from Iceland's 1783 Laki eruption (Thordarson & Self 1993)), typical order-of-magnitude values (denoted by "$\sim$") lying between $\sim 10^6$ and $\sim 10^8 \, m^3$; for comparison, extraterrestrial lavas and prehistoric flood basalts on Earth commonly have volumes between $\sim 10^2$ and $\sim 10^3 \, km^3$ (Macdonald 1972, Williams & McBirney 1979, Cas & Wright 1987, Lopes-Gautier 1993).

The final shapes of lava bodies range from almost radially symmetric plugs and domes to elongate flows. Slow eruptions of high viscosity magma (usually of intermediate or silicic composition) allow lava to accumulate above and around vents to form steep-sided hills or domes (Macdonald 1972, Blake 1990). They frequently reach $\sim 10^2 \, m$ in maximum height ($H$) and $\sim 10^2 - 10^3 \, m$ in radius ($R$), final values of $H/R$ tending to the range 0.3–0.9 (Moriya 1978, Blake 1990, Swanson & Holcomb 1990). Common emplacement times are $\sim 10^7 - 10^8 \, s$ (Macdonald 1972, Williams & McBirney 1979, Huppert et al. 1982, Swanson & Holcomb 1990).

Domes grow by accommodating new lava through a mixture of intrusion and extrusion (Fig. 13.1) (Swanson & Holcomb 1990). At one extreme (endogenous growth), new lava spreads into the existing dome interior, causing the whole structure to expand; at the other (exogenous growth), internal lava is forced through the outer crust, often near the dome apex, and growth continues by the piling up of successive extrusions and the induced spreading of the whole mass.

As implied by their name, domes tend to be almost axisymmetric, a result of eruption from conduits with equidimensional cross sections and of spreading over nearly horizontal ground (Blake 1990, Fink 1993). The influence of flat terrain is especially important. When the ground is horizontal, lava spreading is driven by hydrostatic forces due to the outward thinning of the dome. As the ground steepens, the downslope component of the lava's weight becomes increasingly significant and eventually concentrates motion down the steepest gradient.

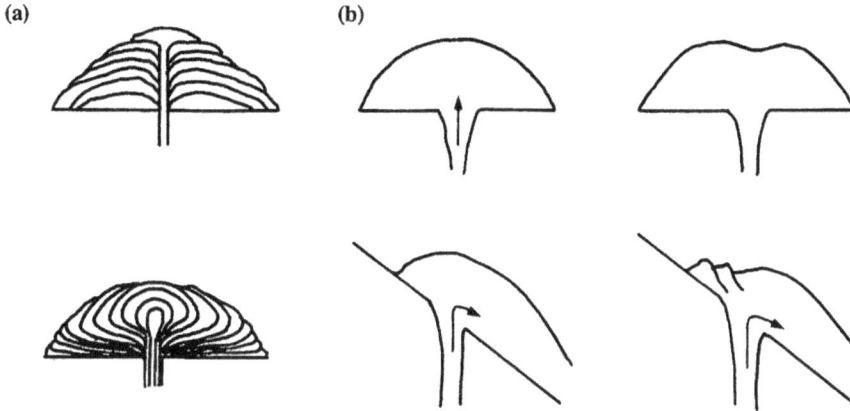

**Figure 13.1**  (a) Idealized exogenous (top) and endogenous (bottom) dome growth (After Fink 1993). (b) Top: domes on horizontal surfaces show radial symmetry. Their tops may sag (right) due to spreading and to drainage of magma back into vent. Bottom: on inclined surfaces, domes elongate downslope to become flows (also called coulées). The uphill side of the dome may move along fault planes (right). (After Macdonald 1972).

Planimetrically, the lava body becomes elongated downslope and the resulting structure classified as a flow (Blake 1990).

In contrast to domes, historic lava flows have intermediate to basic compositions. Their elongation (Fig. 13.2) is clearly illustrated by typical final dimensions: $\sim 1$–$10$ m thick, $\sim 10^2$–$10^3$ m wide and $\sim 10^3$–$10^4$ m long. Common emplacement times are between $\sim 10^5$ and $\sim 10^8$ s.

Characteristic features of flows are early channel formation and downstream flow thickening. Even when a flow is unconfined topographically, widening rapidly slows along its upper and middle reaches. Static lateral margins, or levées, develop and confine motion to a central channel which may later roof-over as a tube. As a result, flows soon develop two distinct zones (Kilburn & Lopes 1991, Kilburn & Guest 1993): a *feeder* zone (channel or tube), the fixed margins of which are older than the lava flowing between them, and a *frontal* zone which, lacking clear marginal structures, is where most flow widening becomes concentrated (Fig. 13.3) (Borgia et al. 1983, Lipman & Banks 1987, Kilburn & Lopes 1991).

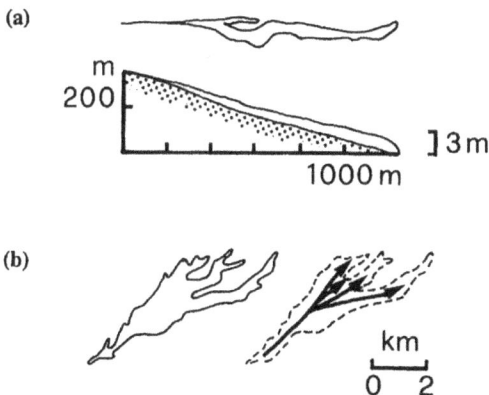

**Figure 13.2**  Shapes of flows and flow fields (a'a lavas). (a) Single flows are elongate and thicken downstream (Etna 1979, Citelli). Scale bar on right refers to lava thickness. (b) Etna's 1983 flow field, whose planimetric outline is controlled by four major flows (after Kilburn & Guest 1993).

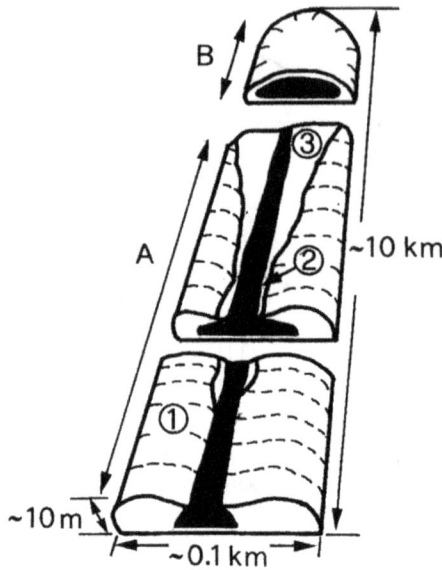

**Figure 13.3**  Feeder (A) and frontal (B) zones in a flow. The example is of an a'a lava, for which feeder zones are usually open channels. (1) fixed levées; (2) main channel; (3) zone of sluggish flow. See also Figure 13.11. (After Lipman & Banks 1987.)

Fronts tend to slow, widen and thicken during emplacement, while channels become narrower because of inward levée growth. Lava accumulates within a channel and may lead to overflow, intrusion or breaching of the flow margins by channel lava (Pinkerton & Sparks 1976, Guest et al. 1987, Lipman & Banks 1987, Kilburn & Lopes 1988a, 1991). Similar responses follow accumulation within a tube, with the obvious requirement for roof rupture before lateral overflow can occur.

Sustained overflow or breaching favours the growth of a new flow. An early lava stream may thus evolve into a collection of adjacent and superposed flows, in a manner reminiscent of exogenous dome growth (Nichols 1936, Walker 1971, Pinkerton & Sparks 1976, Wadge 1978, Kilburn & Lopes 1988a, 1991). When several flows form, the planimetric outline of the resulting flow field is often controlled by the shapes and distributions of a few major, or arterial, flows, smaller streams contributing to local modifications (Fig 13.2). Further details of flow-field growth and classification can be found in Walker (1971), Wadge (1978), Kilburn & Lopes (1988a, 1991) and Pinkerton & Wilson (1994).

## 13.3  Flow surfaces

The most accessible part of a flow is its surface and, indeed, subaerial flows are traditionally classified as pahoehoe, a'a or blocky, according to their dominant surface morphology (Dutton 1884, Macdonald 1953, 1972). At length scales of $\sim 0.1$–$1\,m$, pahoehoe surfaces appear

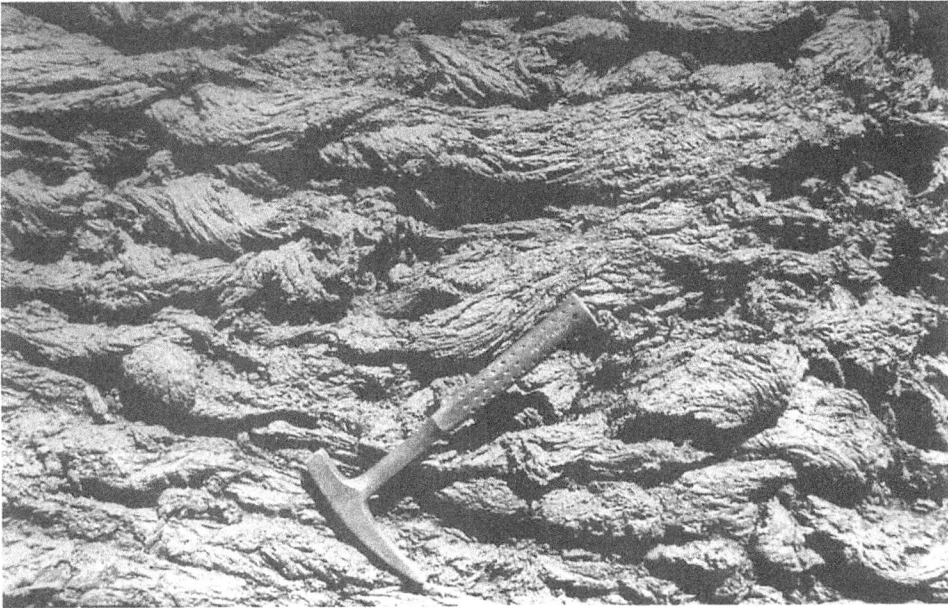

**Figure 13.4** A pahoehoe lava surface on Volcán Fernandina, Galápagos Islands. The flow was erupted sometime between 1980 and 1982. Note the lack of autobreccia. Hammer gives scale. (Photo: D. C. Munro.)

**Figure 13.5** Loose surface material on a'a lava (Etna). The darker fragments (top) have spinose textures, in contrast to the paler, abraded debris of an adjacent flow (bottom). The compass case is 10 cm long. Compare with Figure 13.6.

**Figure 13.6** Loose surface deposits on blocky lava (Nea Kameni, Santorini). Note the planar surfaces of the blocks, which are more angular than a'a debris (see Fig. 13.5).

smooth, while a'a and blocky surfaces are highly uneven and commonly support a layer of broken debris (Figs 13.4, 13.5 & 13.6). A'a and blocky debris are in turn distinguished by size and appearance: a'a fragments are contorted, $\sim 0.1$ m or less across, and have spinose textures on a millimetre scale; blocky fragments may reach metres across, tend to be angular and have planar faces. The detailed structural differences ($< 0.1$ m scales) among lava types reflect differences in surface condition before chilling, while the larger scale variations reflect how surfaces respond to imposed stresses *during* cooling (Macdonald 1953, Kilburn 1990, Kilburn & Guest 1993). Broadly, the pahoehoe-a'a-blocky trend corresponds to chilling of increasingly solidified lava under increasingly concentrated surface stresses (Kilburn 1990, Kilburn & Guest 1993).

Since surface morphology depends on a lava's thermal and dynamic conditions, the morphological classification coincides with fundamental differences in bulk flow structure and mode of advance (Macdonald 1953, 1972, Kilburn 1990, Rowland & Walker 1990, Kilburn 1993). A'a and blocky flows tend to move forward as single units, sometimes subdividing into two or more lobes, and thicken with time, Etnean a'a fronts, for example, can achieve daily thickening rates of $\sim 5$ m. Although tube systems may develop, it is usually along open channels that lava is fed from the vent to the front. Within the channel and at the front, surface disruption is widespread and *mean* advance rates change slowly over intervals of at least $10^4$–$10^5$ s (Section 13.5.2).

Shorter-term advance, however, may be less regular. Thus, the motion of a'a fronts evolves from a simple rolling forward to a puncturing of advancing outer layers by small viscous protrusions or oozing sheets (Krauskopf 1948, Kilburn & Guest 1993). Eventually, an a'a

**Figure 13.7** Portion of the leading edge of a pahoehoe flow, showing accumulation of tongues (after Rowland & Walker 1990).

flow might produce a leading edge of loose autobreccia, although such talus accumulations are more commonly associated with blocky lava fronts (Macdonald 1972).

Pahoehoe flow fronts cover a range of dimensions similar to that for a'a and blocky lavas, but typically advance by local crustal disruption and the simultaneous propagation of numerous small tongues, metres or less across (Fig. 13.7) (Macdonald 1967, Walker 1971). Tongues propagate either from different locations on the main front, or from breaches in the crusts of earlier tongues. The main front lags behind the leading tongues and is normally fed through a tube system, which may extend from the vent; the front maintains its size (thickness $\sim 1$–$10$ m, width $\sim 10^2$–$10^3$ m) by a mixture of endogenous thickening and overlapping of tongues (Rowland & Walker 1990, Hon et al. 1994).

## 13.4 Emplacement regimes

The preceding summaries highlight the tendency of lavas to enclose themselves within carapaces cooler and stronger than their interiors. The style of dome or flow emplacement is then determined by how a lava can disrupt its own carapace. Conceptually, therefore, lavas can be treated as continuously deforming cores beneath colder layers which fail by rupture.

Although both deformation mechanisms contribute to the style of flow, it is convenient to consider emplacement in terms of two limiting regimes: core-dominated and crustal-dominated emplacement (Kilburn & Lopes 1988a, Kilburn 1993). *Core-dominated* emplacement occurs when the rate of mechanical energy supply is fast enough for widespread surface autobrecciation. Open channels form on slopes and fronts advance as single units at rates limited by core resistance. Note that the core need not be isothermal and may have cool outer layers more resistant to flow than the interior; these layers remain part of the core as long as they deform continuously. *Crustal-dominated* emplacement occurs at lower rates of mechanical energy supply, for which surface disruption is less widespread. Tube systems develop and fronts advance by local puncturing of the crust. In this case, the mean advance velocity is limited by the rate of crustal disruption. Implicit here is that the term "crust" refers only to that part of the lava which can fail discontinuously.

All lavas tend from early core-dominated to later crustal-dominated conditions during emplacement. However, their characteristic growth modes depend on how quickly the transition is reached (see the results of laboratory experiments on cooling liquids obtained by Fink & Griffiths (1990, 1992)). If the time to transition ($t'$) is much shorter than the lengthening time ($t_L$), most advance occurs in the crustal regime and this is expected to be appropriate to pahoehoe flows. When $t'/t_L$ is large (approaching or, possibly, exceeding unity), lengthening occurs mainly under core-dominated conditions; such emplacement is expected to be more relevant to a'a and blocky flows. Similarly, exogenous and endogenous dome growth are associated respectively with crustal- and core-dominated regimes.

Most field data on active lavas relate to endogenous domes and to a'a and blocky flows. Of these, a'a flows have posed the most frequent risk to human activity. The first requirements for flow monitoring and hazard forecasting are thus quantifying conditions for core-dominated growth and applying these conditions to a'a flow emplacement.

## 13.5 A'a flow emplacement

### 13.5.1 Empirical relations

Remarkable features of a'a lavas are the first-order correlations obtained empirically between final flow dimensions and other emplacement parameters, such as emplacement time and underlying slope. For example, positive correlations have been found between:
(a)  final length and mean effusion rate (Walker 1973, Wadge 1978, Lopes & Guest 1982)
(b)  final length and volume (Malin 1980, Lopes & Guest 1982)
(c)  final surface area and mean effusion rate (Pieri & Baloga 1986)
(d)  final flow front width and thickness (Wadge & Lopes 1992) and
(e)  final thickness and $1/\sin\alpha$, where $\alpha$ is the ground slope angle (Walker 1967, Kilburn & Lopes 1988b)
Such correlations support the inference that a'a flows are emplaced under physically similar conditions (i.e. core-dominated flow). Importantly, they appear to be independent of a flow's actual size or active lifetime, indicating that a'a formation and emplacement is characterized by specific *ratios* of controlling factors (such as momentum and energy fluxes) rather than the absolute values of those factors. (By analogy, other lava types – e.g. pahoehoe, pillow and blocky lavas – are each expected to have their own specific values for the governing process ratios.) Identifying these ratios and their limiting values may thus lead to realistic forecasts of flow development.

At the beginning of eruption, the most immediate goal for hazard analysis is to estimate the probable length of a flow. The following sections therefore focus on forecasting maximum flow lengths.

### 13.5.2 Rates of advance

At any given moment, lava velocities may vary downstream by more than a factor of ten; for instance, basaltic flows advancing at $\sim 0.01\,\text{m s}^{-1}$ may be fed from vents where mean velocities are $\sim 1\,\text{m s}^{-1}$. Rates of flow advance, however, show surprisingly slow changes. Some examples are given in Figure 13.8. Even though the data cover wide ranges in eruption conditions (Table 13.2), the lengthening trends are similar. Most can be divided into an early

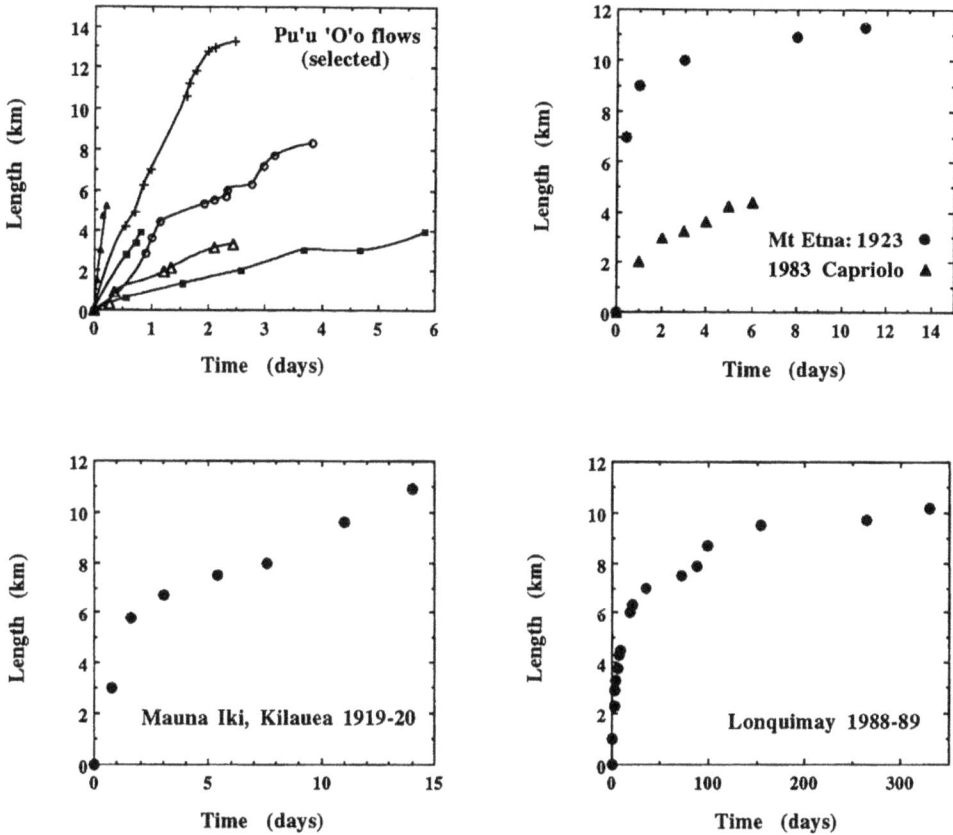

**Figure 13.8** Lengthening patterns in a'a lavas and one blocky flow (Lonquimay). See Table 13.2 for additional data and sources. Note that a'a lengthening times are less than 14 days.

phase of rapid lengthening followed by a longer phase of much slower (by at least a factor of 5) advance; exceptions are the Pu'u 'O'o flows whose mean frontal velocities remain similar throughout. Notably, the flows tend (a) to reach at least half their final length during the phase of rapid lengthening and (b) to approach near-constant mean velocities during initial and final phases. The first feature requires models to focus on early stages of emplacement, the second yields a simplifying condition for analyzing flow-front advance (Section 13.5.6).

The common change from fast to slow advance reflects a sustained change in emplacement conditions. Potential controlling factors are shallowing slopes (reducing the gravitational driving force), waning discharge rates (reducing lava supply to the front) and increasing frontal resistance (due to solidification of the lava core, its crust, or both). Although terrain tends to become shallower downslope on the flanks of volcanoes, sytematic shallowing does not invariably occur along the path of a flow (e.g. Kilburn & Lopes 1988a, Kilburn & Guest 1993) and so either discharge rate or frontal resistance must limit rates of flow lengthening.

Though data are not readily available for opening discharge rates along single a'a flows, some insight is provided by observations of the blocky andesitic flows erupted at Lonquimay in Chile during 1988–89 (Naranjo et al. 1992). Figure 13.9 shows the changes with time of flow length

**Table 13.2** Flow data for Figure 13.8.

| Flow & composition | Volume ($10^6$ m³) | Effusion rate (mean) (m³ s⁻¹) | Slope angle (mean) (°) | Reference |
|---|---|---|---|---|
| 1. Basalt, a'a | | | | |
| Pu'u 'O'o: | | | | |
| Episode 1 | 4 | 308 | 3 | Wolfe et al. (1988) |
| Episode 3 (1123 vent, NE) | 11.7 | 20.8 | 1.2 | Wolfe et al. (1988) |
| Episode 4 | 9.7 | 26.9 | 3 | Wolfe et al. (1988) |
| Episode 14 (SE) | 1.2 | 17.8 | 1.1 | Wolfe et al. (1988) |
| Episode 18 (SE) | 3.4 | 69.2 | 2.6 | Wolfe et al. (1988) |
| | | | | |
| Mauna Iki | 20 | 13 | 1.5 | Rowland & Munro (1993) |
| 2b flow | | | | |
| | | | | |
| 2. Hawaiite, a'a | | | | |
| Etna 1923 | >25 | >30 | 9 | Ponte (1923), Kilburn & Guest (1993) |
| | | | | |
| Etna 1983 | 30 | 48 | 11 | Kilburn & Guest (1993), |
| Capriolo | | | | Frazzetta & Romano (1984) |
| | | | | |
| 3. Andesite, blocky | | | | |
| Lonquimay 1988–89 | 230 | 8 | 3 | Naranjo et al. (1992) |

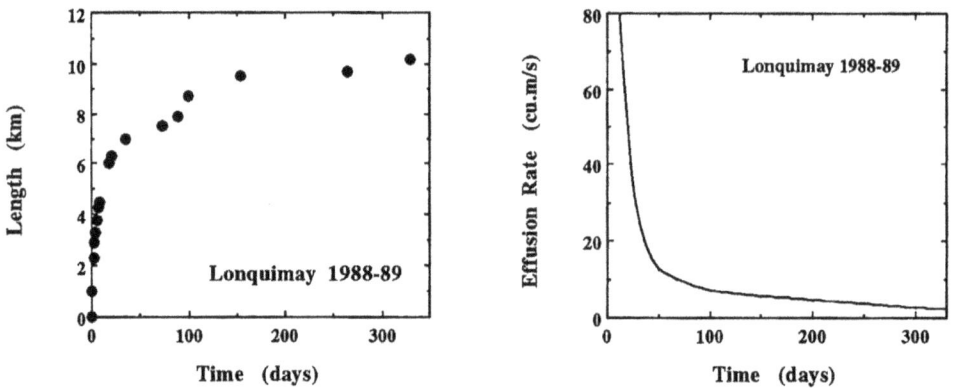

**Figure 13.9** Variations in length and effusion rate during emplacement of the 1988–89 blocky andesitic flow on Lonquimay, Chile (data from Naranjo et al. 1992).

and discharge rate. The transition from rapid to slow near-steady advance (20–100 days after effusion began) coincides with a decrease in discharge rate from about 50 to about 10 m³ s⁻¹, suggesting that waning discharge governed the decrease in frontal velocity. It is tempting to speculate that similar conditions hold generally for a'a and blocky flows. If this proves to be the case, it would allow historic lengthening data to be used to investigate decay rates in effusion rate, thereby constraining models of magma ascent (e.g. Wadge 1981, Stasiuk et al. 1993a).

## 13.5.3 Flow lengthening

Typical emplacement times of a'a lavas are $\sim 10^6$s or less, even if eruption continues for a much longer period (Fig. 13.8) (Wolfe et al. 1988, Kilburn & Guest 1993). For comparison, pahoehoe and blocky lavas may continue lengthening for times of $\sim 10^7$s (Macdonald 1953, Rowland & Walker 1990, Naranjo et al. 1992).

Flows lengthen until the supply rate to the front is unable to overcome frontal resistance. Supply may end because the eruption finishes or because local changes along a flow divert lava in a new direction, beheading the existing front. It is not uncommon, for instance, for broken portions from a channel wall to dam the main channel and initiate a new stream as levées are overflowed (Frazzetta & Romano 1984, Guest et al. 1987, Lipman & Banks 1987). Failure of levées might also result from fatigue at structural weaknesses or from surges in flow rate inducing transient increases in channel pressure.

Once supply has been cut, flow advance may continue at a decreasing rate as channel lava drains into the front (Pinkerton & Sparks 1976, Borgia et al. 1983) The resulting drained channels are easily recognized in the field and are characteristic of lava flows described as *volume limited* (Guest et al. 1987); those whose supply has been stopped by local processes before the end of effusion can then be distinguished as *accidentally breached* flows (Pinkerton & Wilson 1994).

By contrast, a flow may continue to be fed even though its front has come to a halt. As mentioned in Section 13.2, lava accumulates in the channel and may result in overflow, intrusion or breaching of the flow margins. Such *cooling-limited* flows (Guest et al. 1987) achieve the greatest lengths possible for a given set of eruption conditions and so it is their behaviour which is of most interest for hazard assessment.

## 13.5.4 Lava cooling and rheology

Upon eruption, subaerial lavas are typically 1000–1200°C hotter than ambient surface temperatures. Initial cooling is thus concentrated around flow peripheries as the lava exterior seeks thermal equilibrium with the ground and atmosphere. Cooling to the atmosphere is generally considered the more important mode of heat loss. A dark skin rapidly forms as surface temperature drops at a rate limited by radiation. After a critical chilling time, the rate of surface temperature decrease is small and most thermal energy is lost by increasing the thickness of the cooled layer at a rate limited by conduction. Flow motion, however, invariably ruptures the early skin, re-exposing hotter layers beneath. Long-term surface cooling thus depends on slow heat loss through unbroken crust and faster heat loss from areas where underlying lava is exposed (Crisp & Baloga 1990).

The net result is the inward growth of a cool thermal boundary layer as lava moves from the vent to the front. Cooling is important because of the strong temperature dependence of a lava's rheological properties. At temperatures above the liquidus, lavas can be treated as simple newtonian liquids whose rate of deformation increases linearly with applied shear stress (e.g. Shaw 1969, Gauthier 1973, Murase & McBirney 1973). Below the solidus, they can be

357

approximated to elastic bodies. However, within the solidification range, flow resistance tends to increase as the applied shear stress decreases (Pinkerton & Stevenson 1992, Dragoni 1993) and, under conditions typical of field and laboratory measurements (e.g. Shaw et al. 1968, Pinkerton & Sparks 1978, Ryerson et al. 1988), subliquidus lava rheology is most commonly described in terms of a pseudoplastic or Bingham model.

The choice of flow model depends on the specific problem being addressed and, indeed, different models have been used to interpret the same sets of data (Hardee & Dunn 1981, Heslop et al. 1989); further discussion and references can be found in the reviews by Chester et al. (1985), Pinkerton & Stevenson (1992) and Dragoni (1993). The important point here is that rapid increases in resistance occur in the solidification interval, which involves a temperature drop of only about 200°C (e.g. a nominal solidification interval for basalts is from 1200 to 1000°C). Potentially, therefore, resistance in the thermal boundary layer may be several orders of magnitude greater than that in the lava interior. In practice, the boundary layer is weakened by crustal rupture. However, except perhaps at the onset of emplacement, the thermal boundary layer is likely to be thicker than the brittle crust and so boundary-layer resistance is still expected to be significant (Kilburn & Lopes 1991, Stasiuk et al. 1993b).

Beneath the boundary layer, temperatures in the lava interior depend on the dynamical conditions of flow. Extremes are the so-called *thermally unmixed* and *mixed* cases (Crisp & Baloga 1990). The unmixed case corresponds to ground-parallel laminar flow. Heat is transferred through the lava by conduction alone and so temperatures only change across the boundary layers; assuming no internal heat sources (e.g. due to viscous heating or crystallization), the flow interior maintains its eruption temperature.

In the mixed case, complexities of flow motion allow rapid heat transfer through the body of a flow (here complex motion refers to non-ground-parallel flow; it does *not* imply inherent turbulence). The lava interior remains isothermal over every cross section, but decreases in temperature downstream. Mechanisms proposed for encouraging thermal mixing include the overturn of crustal slabs and broken levée walls, convective rolling in a channel, secondary circulations induced by downstream viscosity gradients and channel or tube curvature, and migration of vesicles (e.g. Crisp & Baloga 1990, 1994, Carrigan et al. 1992).

Any thermal mixing accelerates solidification of the flow interior, so reducing the potential for lengthening compared with unmixed conditions. Two other processes which may increase rates of internal solidification are volatile exsolution and lava agitation. Assuming water to be the dominant volatile species, degassing near the surface before eruption will raise the liquidus temperature of the magma. The additional undercooling induced may in turn increase rates of crystal nucleation and growth, which eventually slow as the magma approaches thermochemical equilibrium (Sparks & Pinkerton 1978, Lipman et al. 1985, Pinkerton & Stevenson 1992, Crisp & Baloga 1994). While lava is undercooled, crystallization rates may be increased further by the flow motion itself (Einarsson 1949, Gibb 1974, Corrigan 1982, Kilburn 1983, Kouchi et al. 1986), since greater relative movement in the lava favours the chances of atoms in the melt coming to the arrangement necessary for crystal nucleation (Mullin & Raven 1962, Gibb 1974).

From thermal considerations alone, therefore, the greatest cooling-limited flow lengths are associated with the unmixed regime. The most conservative first-order model thus reduces to the advance of an unmixed flow for which (a) crystallization induced by degassing or flow motion is insignificant, and (b) the boundary-layer thickens at a rate proportional to that for growth beneath unbroken crust. Such conditions for flows unconstrained by topography are considered in the next section.

### 13.5.5 Grätz number model for cooling-limited flows

As a first approximation, we can relate the thickness of the frontal boundary layer to flow front dimensions by considering the proportion of thermal energy lost in boundary layer cooling. Standard conduction theory yields

$$\frac{\text{Initial thermal energy in flow section}}{\text{Energy lost cooling boundary layer}} \approx \left(d_{ef}^2/\kappa t_{tr}\right)^{1/2} \tag{13.1}$$

where the hydraulic (or effective) diameter of the front $d_{ef}$ is defined as $4 \times$ (cross-sectional area/wetted perimeter), $t_{tr}$ is the travel time of the lava from vent to front, $\kappa$ is lava thermal diffusivity and the dimensionless quantity $d_{ef}^2/\kappa t_{tr}$ is known as the "Grätz number" (Gz). The term $(\kappa t_{tr})^{1/2}$ is a measure of the boundary layer thickness after a time $t_{tr}$. Equation 13.1 thus shows that Gz decreases as the boundary layer becomes an increasingly large proportion of the frontal hydraulic diameter.

Assuming conditions in the boundary layer and lava interior are directly related, we may conjecture that flows are halted by cooling when the boundary layer occupies more than a critical fraction of the frontal cross section. In other words, we expect cooling-limited flows to be characterized by critical minimum Grätz numbers (Gz$_c$), volume-limited flows having larger values of Gz. This expectation has been confirmed by empirical studies of a'a flows (Pinkerton & Sparks 1976, Hulme & Fielder 1977, Guest et al. 1987, Pinkerton & Wilson 1994) for which, with $d_{ef}$ as defined above and $\kappa = 4 \times 10^{-7}\,\text{m}^2\text{s}^{-1}$, the critical Grätz number is 500.

The travel time $t_{tr}$ is related to mean flow discharge rate, $Q$, mean cross-sectional area, $a$, of flowing lava (i.e. of the channel and flow front) and flow length, $L$, via

$$t_{tr} = La/Q \tag{13.2}$$

Combining Equations 13.1 & 13.2, and setting Gz to Gz$_c$, we can then express flow length $L$ as

$$L \le (d_{ef}^2/a)(1/\kappa)(1/Gz_c)Q \tag{13.3a}$$

or, denoting the maximum flow length by $L_m$,

$$L_m = (d_{ef}^2/a)(1/\kappa)(1/Gz_c)Q \tag{13.3b}$$

For flows much wider than deep, $d_{ef} = 2h_f$ (where $h_f$ is the mean frontal thickness). Introducing $h$ as the mean flow thickness and $w_f$ and $w$ as the frontal width and mean flow width, Equation 13.3b becomes

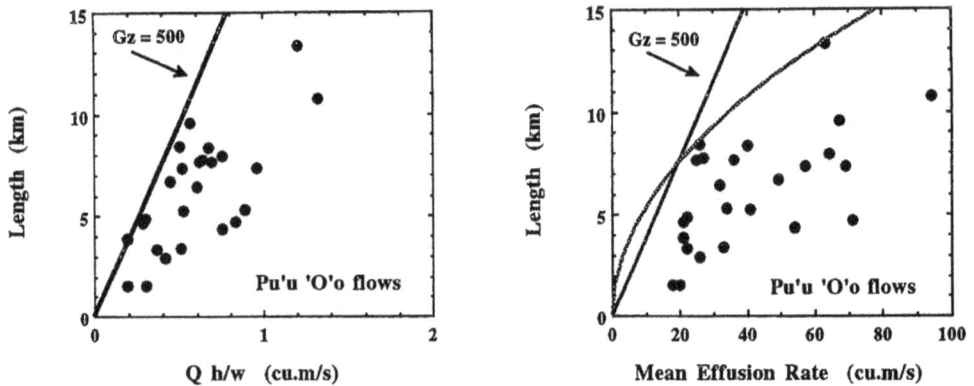

Figure 13.10 Lengths of a'a flows from Pu'u 'O'o, Hawaii. (Left) Variation of length with $Qh/w$. The line shows maximum flow lengths according to the Grätz number model (Eq. 13.11). (Right) Variation of length with mean effusion rate ($Q$). Theoretical maximum lengths are shown for the Grätz number model (straight line; Eq. 13.5) and flow-front model (curve; Eq. 13.11). (Field data from Wolfe et al. 1988.)

$$L_m = 4\beta(1/\kappa)(1/Gz_c)(h/w)Q \qquad (13.4)$$

where the geometrical term $\beta = [(w_f h_f/a)(h_f/h)(w/w_f)]$.

Figure 13.10 compares Equation 13.4 with data from the main channel-fed a'a flows of episodes 1 to 18 of the Pu'u 'O'o eruption in Hawaii (Wolfe et al. 1988); during this analysis, $\beta$ was assigned a nominal value of 1. The agreement is excellent, all the observed flow lengths being less than or equal to the calculated values of $L_m$.

Since initial discharge rates tend to be larger than or comparable to the mean discharge rate $Q$, Equation 13.4 can be used to forecast maximum flow lengths, given early flux measurements and a reasonable value for $h/w$. It happens that suites of a'a lavas from a common vent area often have similar geometries. Thus mean values of $h/w$ from early lavas can be used in Equation 13.4 to estimate the maximum lengths of future flows. For example, typical values of h/w for the 31 principal Pu'u O'o lavas are $\leq 0.02$, a result which could have been inferred from the dimensions of the first half-dozen flows to be erupted. With $h/w = 0.02$, Equation 13.4 becomes

$$L_m = 0.4\,Q \qquad (13.5)$$

for $L_m$ in kilometres and $Q$ in cubic metres per second.

Equation 13.5 is compared with field data in Figure 13.10. Again the expected maximum flow lengths give realistic limits for the observed values.


### 13.5.6 Frontal advance and thickening

The Grätz number model provides a simple means of estimating $L_m$ at the onset of eruption. It also supports the hypothesis that bulk flow behaviour can be characterized in terms of process ratios. However, the model is restricted to the extreme case identified for unmixed, cooling-

360

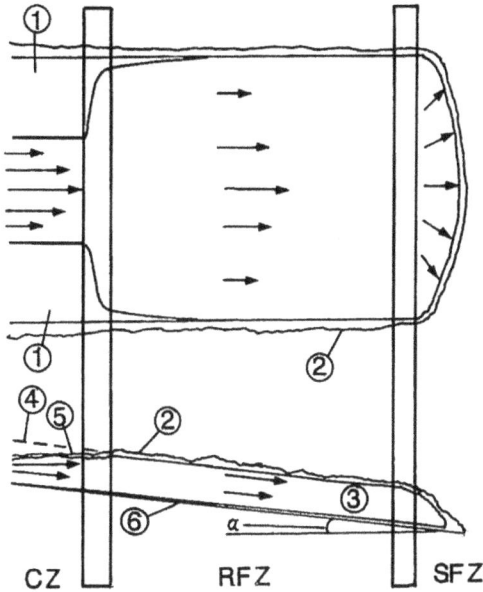

**Figure 13.11** Channel and frontal zones in a'a lava flows (top: plan; bottom, elevation). See also Figure 13.3. The zones are described in the text. CZ, channel (i.e. feeder) zone; RFZ, rear frontal zone; SFZ, snout frontal zone. Arrows show principal flow directions; longer arrows represent faster velocities (schematic). Note relative dimensions of RFZ and SFZ which are approximately to scale. Other features are not to scale. Numbered regions show: (1) fixed levées in CZ; (2) disrupted lava crust; (3) continuous lava core; (4) trace (dashed) of levée top in CZ; (5) surface of lava in active channel (note downstream thickening is more pronounced than in RFZ); (6) basal debris and crust. (After Kilburn & Lopes 1991.)

limited flows. To relax this constraint, we here consider flow front conditions in more detail.

A'a flow fronts tend to have large width/thickness ratios (e.g. Wadge & Lopes 1992) and can be divided into two intergradational zones (Fig. 13.11) (Kilburn & Lopes 1991): the snout frontal zone (SFZ, or snout) and the rear frontal zone (RFZ). The snout forms the leading edge of a flow, its surface dropping steeply downstream (typical dips of between 40° and 50°) as a sloping front. The upflow extent of the snout is $\sim h_f$, the mean thickness of the RFZ. Lava entering the snout spreads laterally, fixing the width of the flow in a time $\sim h_f/u_f$, where $u_f$ is the mean advance velocity. The back of the SFZ grades into the much longer ($\sim 10h_f$) RFZ, which is distinguished by very slow widening, lack of distinct marginal levées and a virtually constant mean thickness throughout (in fact weakly increasing downstream).

Mean advance rates are considered to be limited by the RFZ (Kilburn & Lopes 1991). Motion is treated as uniform (since mean thickness is almost constant) with a constant mean velocity (Section 13.5.2), and one dimensional (since fronts are much wider than deep, and rates of spreading and thickening are much slower than the downstream velocity). Presuming values of $Gz \geq 500$, the thickness of the thermal boundary layer is unlikely to exceed about one-tenth the frontal thickness (from Eq. 13.1); hence, since crustal disruption is also widespread, motion in the RFZ is assumed to be governed by the lava core (the lava interior and continuous thermal boundary layer). Because clear marginal structures are absent, the core is assumed to have a comparable bulk rheology in the downstream and cross-stream directions (Kilburn & Lopes (1991) considered a newtonian rheology for the core, but this restriction is not necessary). The time-scale of downstream deformation ($h_f/u_f$) is similar to that estimated for snout spreading. Together, comparable bulk rheology and deformation timescales ($= 1/$deformation rate) imply that similar mean stresses drive both lava advance and spreading at the curved snout. Importantly, the implied stress condition is not influenced by solidification of the lava interior, provided that solidification affects lengthening and spreading equally.

The preceding arguments suggest that inertial forces are not important and that frontal dynamics depend on the balance between gravitational and rheological forces downstream, and pressure and rheological forces laterally. Assuming laminar flow, these balances then yield (Kilburn & Lopes 1991, Kilburn 1993)

$$u_f \approx (w_f \sin\alpha)/(3\ h_f/u_f) \tag{13.6}$$

where $\alpha$ is the angle of underlying slope.

During advance, fronts continue to thicken slowly and broken crusts attempt to heal themselves by cooling. Two further conditions for steady, core-controlled motion are thus: (a) the time-scale of crustal disruption and healing must be similar to or smaller than the time-scale of flow deformation (Kilburn & Lopes 1991); and (b) the rate of increase in downstream driving force due to flow thickening is balanced by a comparable increase in frontal resistance. In the limit, therefore, steady, core-controlled advance is expected when frontal thickening over periods $\sim h_f/u_f$ can just prevent complete crustal rehealing.

Assuming a constant tensile strength, the resistance of a crust is proportional to its cross-sectional area. During an interval $\sim h_f/u_f$, therefore, the increase in crustal resistance is expected to be proportional to $w_f[\kappa(h_f/u_f)]^{1/2}$, where $[\kappa(h_f/u_f)]^{1/2}$ measures the thickness of newly formed crust.

For gravity driven flow, the downstream force is proportional to the cross-sectional area of the front. Taking a mean rate of frontal thickening $\sim h_f/t$ (where $t$ is the time since emplacement began), the increase in driving force in time $h_f/u_f$ will be proportional to $w_f(h_f/t)(h_f/u_f)$, where $(h_f/t)(h_f/u_f)$ is the increase in the RFZ mean thickness.

Steady core advance is thus expected when the ratio [increase in flow thickness]/[increase in crustal thickness] exceeds a critical value, that is, when

$$\frac{w_f(h_f/t)(h_f/u_f)}{w_f[\kappa(h_f/u_f)]^{1/2}} = \left[(h_f^2/\kappa)(h_f/u_f)(1/t^2)\right]^{1/2} \geq \Omega_c \tag{13.7}$$

where $\Omega_c$ denotes the critical minimum value for steady flow. Further scaling analysis (to be presented elsewhere) suggests that $\Omega_c \approx 1$ and so, from Equation 13.7, limiting steady flow occurs for

$$t^2 \approx (h_f^2/\kappa)(h_f/u_f) \tag{13.8}$$

If limiting conditions prevail for most of a flow's lengthening, Equations 13.6 & 8 can be combined to express flow length $L$ as

$$L = u_f\,t = M\ (w_f\,h_f^2 \sin\alpha)/(3\kappa t) \tag{13.9}$$

where the constant $M \approx 1$.

Equation 13.9 is identical to that derived by Kilburn & Lopes (1991) using a different scaling analysis to estimate the time of flow lengthening. They found a good agreement between the equation and field data for 134 a'a and blocky flow fields, $M$ taking the value $1.34 \pm 0.66$. However, closer inspection of their data (Fig. 3 in Kilburn & Lopes 1991) suggests that $M$

may be weakly dependent on composition, decreasing from about 2 for basalts to about 0.68 for basaltic andesites.

Noting that mean discharge rate $Q = a'L/t$ (where $a'$ is the mean cross-sectional area of the *whole* flow), Equation 13.9 may be re-expressed as

$$L = \psi^{\frac{1}{2}} M^{\frac{1}{2}} [(h_f \sin \alpha)/(3\kappa)]^{\frac{1}{2}} Q^{\frac{1}{2}} \tag{13.10}$$

where $\psi$ is the geometric ratio $w_f h_f/a'$. For flows of constant width and thickness, $\psi = 1$; if width and thickness increase linearly downflow from an effectively point source, $\psi = 3$.

Equation 13.10 can be used to forecast maximum flow lengths for a given discharge rate provided an upper limit for $h_f \sin \alpha$ is known. Conveniently, empirical studies (Walker 1967, Kilburn & Lopes 1988b) indicate that $h_f \sin \alpha$ rarely exceeds 1 m among a'a lavas. For the Pu'u 'O'o flows, $M$ and $\psi$ are both nominally 2, and so, for $h_f \sin \alpha = 1$, Equation 13.10 reduces to

$$L_m = 1.8 \, Q^{\frac{1}{2}} \tag{13.11}$$

for the maximum length $L_m$ in kilometres and $Q$ in cubic metres per second.

As shown in Figure 13.10, Equation 13.11 constrains the Pu'u 'O'o data remarkably well, providing a tighter limit than the corresponding Grätz model (Eq. 13.5). However, the fact that the calculated curve passes almost exactly through the extreme data points in Figure 13.10 is probably fortuitous. First, the choice $\psi = 2$ was a compromise for the conditions given after Equation 13.10; second, we do not know from published data if the limiting flows achieved $h_f \sin \alpha = 1$ m. Nevertheless, a parallel analysis for Etnean flows (Kilburn 1993) also shows very good agreement between theoretical and observed $L_m$–$Q$ trends.

## 13.6 Comparison of lengthening models

The Grätz number model (Eq. 13.4) is derived essentially from thermal criteria. It assumes that flows stop when the thermal boundary layer is a critical fraction of the frontal hydraulic diameter, at which time boundary-layer resistance can inhibit advance. This assumption, however, does not account for possible differences in solidification rates in the lava interior, nor for repeated surface exposure, through crustal disruption, increasing the boundary-layer thickness at a rate no longer proportional to $t_{tr}^{\frac{1}{2}}$ (Eq. 13.1).

The flow-front model (Eq. 13.11), in contrast, utilizes constraints on the stress distribution and bulk rheology of the lava core, as well as on rates of frontal thickening. These constraints specify relations between frontal width, thickness and velocity, while implicitly allowing for any solidification of the lava interior. As a result, advance rates implied by the flow-front model are generally smaller than those associated with the Grätz number analysis.

In the case of the Pu'u O'o lavas, the difference between models is most apparent when effusion rates exceed about $30 \, m^3 s^{-1}$. Agreement is good at lower values of $Q$. The simplest interpretation is that low-$Q$ flows were emplaced under conditions approaching the ideal unmixed regime while, at higher effusion rates, core solidification or increased boundary-layer thickening contributed significantly to limiting maximum flow lengths. Such effects (which may determine the upper limit on $h_f \sin \alpha$) are accounted for by the flow-front model, but not by the Grätz number model.

## 13.7 Hazard assessment and future requirements

As regards hazard analysis, both models can provide estimates of maximum flow length knowing (a) the rate of discharge and (b) a characteristic quantity ($h/w$ or $h_f \sin \alpha$) of flows previously erupted at the volcano in question. The characteristic quantities can be determined easily by flow mapping and should form part of any basic database of the characteristics of a volcano. The need for discharge rate data identifies a primary objective for monitoring the early stages of flow growth.

Both models have been derived on the basis of empirically observed trends. Some refinements and extensions to other flow types have been discussed by Kilburn (1993) and Pinkerton & Wilson (1994). In particular, more data are needed on: rates of crustal disruption and core solidification and their influence on bulk lava rheology; the geometrical evolution of flow fronts and their rates of widening, lengthening and thickening; changes in velocity and channel (or tube) dimensions along a flow; and accurate models of surface topography. Such needs highlight the importance of continuous monitoring of active lavas if flow models are to be improved.

## Acknowledgements

Parts of this work were completed with the assistance of a grant for work on Climatology and Natural Hazards under Contract No. EV5V CT92-0190 of the EC Environment Programme 1991–94 (C.R.J.K. and H.P.) and the Leverhulme Trust (L.W.). Jane Rushton mercifully improved the figures.

## References

Baloga, S. 1987. Lava flows as kinematic waves. *Journal of Geophysical Research* **92**, 271–9.
Baloga, S. & D. Pieri 1986. Time-dependent profiles of lava flows. *Journal of Geophysical Research* **91**, 543–52.
Barberi, F. & L. Villari (eds) 1984. Mt Etna and its 1983 eruption. *Bulletin Volcanologique* **47**, 877–1177.
Barberi, F., M. L. Carapezza, M. Valenza, L. Villari 1993. The control of lava during the 1991–1992 eruption of Mt Etna. *Journal of Volcanology and Geothermal Research* **56**, 1–34.
Blake, S. 1990. Viscoplastic models of lava domes. In *IAVCEI Proceedings in Volcanology, vol. 2: Lava flows and domes – emplacement mechanisms and hazard implications*, J. H. Fink (ed.) 88–126. Berlin: Springer.
Bolt, B. A., W. L. Horn, G. A. Macdonald, R. F. Scott 1975. *Geological hazards*. Berlin: Springer.

Borgia, A., S. Linneman, D. Spencer, L. Morales, L. Andre 1983. Dynamics of the flow fronts, Arenal volcano, Costa Rica. *Journal of Volcanology and Geothermal Research* **19**, 303–29.

Carrigan, C. R., G. Schubert, J. C. Eichelberger 1992. Thermal and dynamical regimes of single and two-phase magmatic flow in dikes. *Journal of Geophysical Research* **97**, 17377–92.

Cas, R. A. F. & J. V. Wright 1987. *Volcanic successions: modern and ancient*. London: Allen & Unwin.

Chester, D. K., A. M. Duncan, J. E. Guest, C. R. J. Kilburn 1985. *Mount Etna. The anatomy of a volcano*. London: Chapman & Hall.

Corrigan, G. M. 1982. Supercooling and the crystallization of plagioclase, olivine, and clinopyroxene from basaltic magmas. *Mineralogical Magazine* **46**, 31–42.

Crisp, J. A. & S. M. Baloga 1990. A model for lava flows with two thermal components. *Journal of Geophysical Research* **95**, 1255–70.

Crisp, J. A. & S. Baloga 1994. Influence of crystallization and entrainment of cooler material on the emplacement of basaltic aa lava flows. *Journal of Geophysical Research* **99**, 11819–31.

Danes, Z. F. 1972. Dynamics of lava flows. *Journal of Geophysical Research* **77**, 1430–2.

Didden, N. & T. Maxworthy 1982. The viscous spreading of plane and axisymmetric gravity currents. *Journal of Fluid Mechanics* **121**, 27–42.

Dragoni, M. 1989. A dynamical model of lava flows cooling by radiation. *Bulletin of Volcanology* **51**, 88–95.

Dragoni, M. 1993. Modelling the rheology and cooling of lava flows. In *Active lavas: monitoring and modelling*, C. R. J. Kilburn & G. Luongo (eds), 235–61. London: UCL Press.

Dragoni, M., M. Bonafede, E. Boschi 1986. Downslope models of a Bingham liquid: implications for lava flows. *Journal of Volcanology and Geothermal Research* **30**, 305–25.

Dragoni, M., S. Pondrelli, A. Tallarico 1992. Longitudinal deformation of a lava flow: the influence of Bingham rheology. *Journal of Volcanology and Geothermal Research* **52**, 247–54.

Dutton, C. E. 1884. Hawaiian volcanoes. *United States Geological Survey 4th Annual Report*, 47–112. Washington, DC: United States Government Printing Office.

Einarsson, T. 1949. The flowing lava. Studies of its main phyusical and chamic properties. In *The eruption of Kekla 1947–1948*, T. Einarsson, G. Kjartansson, S. Thorarinsson (eds), 1–70. Rykjarik: Visindafelag Islandinga and Museum of Natural History.

Fink, J. H. (ed.) 1987. *Geological Society of America special paper, vol. 212: The emplacement of silicic domes and lava flows*. Boulder, Colorado: Geological Society of America.

Fink, J. H. (ed.) 1990. *IAVCEI Proceedings in Volcanology, vol. 2, Lava flows and domes – emplacement mechanisms and hazard implications*. Berlin: Springer.

Fink, J. H. 1993. The emplacement of silicic lava flows and associated hazards. In *Active lavas: monitoring and modelling*, C. R. J. Kilburn & G. Luongo (eds). London: UCL Press.

Fink, J. H. & R. W. Griffiths 1990. Radial spreading of viscous-gravity currents with solidifying crust. *Journal of Fluid Mechanics* **221**, 485–501.

Fink, J. H. & R. W. Griffiths 1992. A laboratory analogue study of the surface morphology of lava flows extruded from point and line sources. *Journal of Volcanology and Geothermal Research* **54** 19–32.

Frazzetta, G. & R. Romano 1984. The 1983 Etna eruption: event chronology and morphological evolution of the lava flow. *Bulletin Volcanologique* **47**, 1079–96.

Gauthier, F. 1973. Field and laboratory studies of the rheology of Mount Etna lava. *Philosophical Transactions of the Royal Society, London, Series A* **274**, 83–98.

Gibb, F. G. F. 1974. Supercooling and the crystallisation of plagioclase from a basaltic magma. *Mineralogical Magazine* **39**, 641–53.

Guest, J. E., C. R. J. Kilburn, H. Pinkerton, A. M. Duncan 1987. The evolution of lava flow fields: observations of the 1981 and 1983 eruptions of Mount Etna, Sicily. *Bulletin of Volcanology* **49**, 527–40.

Hallworth, M. A., H. E. Huppert, R. S. J. Sparks 1987. A laboratory simulation of basaltic lava flows. *Modern Geology* **11**, 93–107.

Hardee, H. C. & J. C. Dunn 1981. Convective heat transfer in magmas near the liquidus. *Journal of Volcanology and Geothermal Research* **10**, 195–207.

Harrison, C. G. A. & C. Rooth 1976. The dynamics of flowing lavas. In *Volcanoes and tectonosphere*, H. Aoki & S. Iizuka (eds), 103–13. Tokai: Tokai University Press.

Heslop, S. E., L. Wilson, H. Pinkerton, J. W. Head III 1989. Dynamics of a confined lava flow on Kilauea volcano, Hawaii. *Bulletin of Volcanology* **51**, 415–32.

Hon, K., J. Kauahikaua, R. Denlinger, K. McKay 1994. Emplacement and inflation of pahoehoe sheet flows: observations and measurements of active lava flows on Kilauea Volcano, Hawaii. *Geological Society of America Bulletin* in press.

Hulme, G. 1974. The interpretation of lava flow morphology. *Geophysical Journal of the Royal Astronomical Society* **39**, 361–83.

Hulme, G. & G. Fielder 1977. Effusion rates and rheology of lunar lavas. *Philosophical transactions of the Royal Society, London, series A* **285**, 227–34.

Huppert, H. E. 1982a. The propagation of two-dimensional and axisymmetric viscous gravity currents over a rigid horizontal surface. *Journal of Fluid Mechanics* **121**, 43–58.

Huppert, H. E. 1982b. Flow and instability of a viscous current down a slope. *Nature* **300**, 427–9.

Huppert, H. E. & R. S. J. Sparks 1985. Komatiites I: eruption and flow. *Journal of Petrology* **26**, 694–725.

Huppert, H. E., J. B. Shepherd, H. Sigurdsson, R. S. J. Sparks 1982. On lava dome growth, with application to the 1979 lava extrusion of the Soufrière of St Vincent. *Journal of Volcanology and Geothermal Research* **14**, 199–222.

Huppert, H. E., R. S. J. Sparks, J. S. Turner, N. T. Arndt 1984. Emplacement and cooling of komatiite lavas. *Nature* **309**, 19–22.

Jaupart, C. 1991. Effects of compressibility on the flow of lava. *Bulletin of Volcanology* **54**, 1–9.

Kilburn, C. R. J. 1983. Studies of lava flow development. In *Forecasting volcanic events*, H. Tazieff & J-C. Sabroux (eds), 83–98. Amsterdam: Elsevier.

Kilburn, C. R. J. 1990. Surfaces of aa flow-fields on Mount Etna, Sicily: morphology, rheology, crystallization and scaling phenomena. In *IAVCEI Proceedings in Volcanology, vol. 2: Lava flows and domes – emplacement mechanisms and hazard implications*, J. H. Fink (ed.) 129–56. Berlin: Springer.

Kilburn, C. R. J. 1993. Lava crusts, aa flow lengthening and the pahoehoe-aa transition. In *Active lavas: monitoring and modelling*, C. R. J. Kilburn & G. Luongo (eds), 263–80. London: UCL Press.

Kilburn, C. R. J. & J. E. Guest 1993. Aa lavas of Mount Etna, Sicily. In *Active lavas: monitoring and modelling*, C. R. J. Kilburn & G. Luongo (eds), 73–106. London: UCL Press.

Kilburn, C. R. J. & R. M. C. Lopes 1988a. The growth of aa lava flow fields on Mount Etna, Sicily. *Journal of Geophysical Research* **93**, 14759–72.

Kilburn, C. R. J. & Lopes R. M. C. 1988b. Lava thicknesses: implications for rheological and crustal development. *Lunar and Planetary Institute Contribution* **660**, 9–11.

Kilburn, C. R. J. & R. M. C. Lopes 1991. General patterns of flow field growth: aa and blocky lavas. *Journal of Geophysical Research* **96**, 19721–32.

Kilburn, C. R. J. & G. Luongo (eds). *Active lavas: monitoring and modelling*. London: UCL Press.

Kouchi, A., A. Tsuchiyama, I. Sunagawa 1986. Effect of stirring on crystallization kinetics of basalt: texture and element partitioning. *Contributions to Mineralogy and Petrology* **93**, 429–38.

Krauskopf, K. B. 1948. Lava movement at Parícutin volcano, Mexico. *Bulletin of the Geological Society of America* **59**, 267–84.

Lipman, P. W. & N. G. Banks 1987. Aa flow dynamics, Mauna Loa 1984. In *United States Geological Survey Professional Paper 1350, Volcanism in Hawaii*, R. W. Decker, T. L. Wright, P. H. Stauffer (eds), 1527–67. Washington DC: United States Government Printing Office.

Lipman, P. W., N. G. Banks, J. M. Rhodes 1985. Gas-release induced crystallization of 1984 Mauna Loa magma, Hawaii, and effects on lava rheology. *Nature* **317**, 604–6.

Lister, J. R. 1992. Viscous flows down an inclined plane from point and line sources. *Journal of Fluid Mechanics* **242**, 631–53.

Liu, K. F. & C. C. Mei 1989. Slow spreading of a sheet of Bingham fluid on an inclined plane. *Journal of Fluid Mechanics* **207**, 505–29.

Lopes, R. M. C. & J. E. Guest 1982. Lava flows on Etna, a morphometric study. In *The comparative study of the planets*, A. Coradini & M. Fulchignoni (eds), 441–58. Hingham: Reidel.

Lopes-Gautier, R. M. C. 1993. Extraterrestrial lava flows. In *Active lavas: monitoring and modelling*, C. R. J. Kilburn & G. Luongo (eds), 107–44. London: UCL Press.

Macdonald, G. A. 1953. Pahoehoe, aa and block lava. *American Journal of Science* **251**, 169–91.

Macdonald, G. A. 1967. Forms and structures of extrusive basaltic rocks. In *The Poldervaart treatise on rocks of basaltic composition, vol. 1: Basalts*, A. A. Poldervaart & H. H. Hess (eds), 1–61. New York: Wiley-Interscience.

Macdonald, G. A. 1972. *Volcanoes*. Englewood Cliffs, New Jersey: Prentice-Hall.

Malin, M. C. 1980. Lengths of Hawaiian lava flows. *Geology* **8**, 306–8.

Manley, C. R. 1992. Extended cooling and viscous flow of large, hot rhyolite lavas: implications of numerical modeling results. *Journal of Volcanology and Geothermal Research* **53**, 27–46.

Moriya, I. 1978. Morphology of lava domes. *Bulletin of the Department of Geography, Kanazawa University, Japan* **14**, 55–69 (in Japanese).

Mullin, J. W. & K. D. Raven 1962. Influence of mechanical agitation on the nucleation of aqueous salt solutions. *Nature* **195**, 35–8.

Murase, T. & A. M. McBirney 1973. Properties of some common igneous rocks and their melts at high temperature. *Bulletin of the Geological Society of America* **84**, 563–92.

Naranjo, J. A., R. S. J. Sparks, M. V. Stasiuk, H. Moreno, G. J. Ablay 1992. Morphological, structural and textural variations in the 1988–1990 andesite lava of Lonquimay Volcano, Chile (38°S). *Geological Magazine* **129**, 17–28.

Nichols, R. L. 1936. Flow units in basalt. *Journal of Geology* **44**, 617–30.

Pieri, D. C. & S. M. Baloga 1986. Eruption rates, areas and length relationships for some Hawaiian lava flows. *Journal of Volcanology and Geothermal Research* **30**, 29–45.

Pinkerton, H. 1993. Measuring the properties of flowing lavas. In *Active lavas: monitoring and modelling*, C. R. J. Kilburn & G. Luongo (eds), 175–91. London: UCL Press.

Pinkerton, H. & R. S. J. Sparks 1976. The 1974 sub-terminal lavas, Mount Etna: a case history of the formation of a compound flow field. *Journal of Volcanology and Geothermal Research* **1**, 167–82.

Pinkerton, H. & R. S. J. Sparks 1978. Field measurements of the rheology of lava. *Nature* **276**, 383–4.

Pinkerton, H. & R. J. Stevenson 1992. Methods of determining the rheological properties of magmas at subliquidus temperatures. *Journal of Volcanology and Geothermal Research* **53**, 47–66.

Pinkerton, H. & L. Wilson 1994. Factors controlling the lengths of channel-fed lava flows. *Bulletin of Volcanology* **56**, 108–120.

Ponte, G. 1923. The recent eruption of Etna. *Nature* **112**, 546–8.

Rowland, S. K. & D. C. Munro 1993. The 1919–1920 eruption of Mauna Iki, Kilauea: chronology, geologic mapping, and magma transport mechanisms. *Bulletin of Volcanology* **55**, 190–202.

Rowland S. K. & Walker G. P. L. 1990. Pahoehoe and a'a in Hawaii: volumetric flow rate controls the lava structure. *Bulletin of Volcanology* **52**, 615–28.

Ryerson, F. J., H. C. Weed, A. J. Piwinskii 1988. Rheology of subliquidus magmas 1. Picritic compositions. *Journal of Geophysical Research* **93**, 421–36.

Schwartz, L. W. & E. E. Michaelides 1988. Gravity flow of a viscous liquid down a slope with injection. *Physics of Fluids* **31**, 739–41.

Shaw, H. R. 1969. Rheology of basalt in the melting range. *Journal of Petrology* **10**, 510–35.

Shaw, H. R., T. L. Wright, D. L. Peck, R. Okamura 1968. The viscosity of basaltic magma: an analysis of field measurements in Makaopuhi lava lake, Hawaii. *American Journal of Science* **261**, 255–64.

Smith, P. C. 1973. A similarity solution for slow viscous flow down an inclined plane. *Journal of Fluid Mechanics* **58**, 275–88.

Sparks, R. S. J. & H. Pinkerton 1978. Effect of degassing on the rheology of basaltic lava. *Nature* **276**, 385–6.

Stasiuk, M. V., C. Jaupart, R. S. J. Sparks 1993a. On the variations of flow rate in non-explosive lava eruptions. *Earth and Planetary Science Letters* **114**, 505–16.

Stasiuk, M. V., C. Jaupart, R. S. J. Sparks 1993b. Influence of cooling on lava-flow dynamics. *Geology* **21**, 335–8.

Swanson, D. A. & R. T. Holcomb 1990. Regularities in growth of the Mount St Helens dacite dome 1980–1986. In *IAVCEI Proceedings in Volcanology, vol. 2: Lava flows and domes – emplacement mechanisms and hazard implications*, J. H. Fink (ed.) 3–24. Berlin: Springer.

Tazieff, H. 1977. An exceptional eruption: Mt Nyiragongo Jan. 10th 1977. *Bulletin Volcanologique* **40**, 189–200.

Thordarson Th. & S. Self 1993. The Laki (Skaftár Fires) and Grímsvotn eruptions in 1783–1785. *Bulletin of Volcanology* **55**, 233–63.

Tilling, R. I. & D. W. Peterson 1993. Field observation of active lava in Hawaii: some practical considerations. In *Active lavas: monitoring and modelling*, C. R. J. Kilburn & G. Luongo (eds). London: UCL Press.

Wadge, G. 1978. Effusion rate and the shape of aa lava flow fields on Mount Etna. *Geology* **6**, 503–6.

367

Wadge, G. 1981. The variation of magma discharge during basaltic eruptions. *Journal of Volcanology and Geothermal Research* **11**, 139–68.

Wadge, G. & R. M. C. Lopes 1992. The lobes of lava flows on Earth and Olympus Mons, Mars. *Bulletin of Volcanology* **54**, 10–24.

Walker, G. P. L. 1967. Thickness and viscosity of Etnean lavas. *Nature* **213**, 484–5.

Walker, G. P. L. 1971. Compound and simple lava flows and flood basalts. *Bulletin Volcanologique* **35**, 579–90.

Walker, G. P. L. 1973. Lengths of lava flows. *Philosophical Transactions of the Royal Society, London, Series A* **274**, 107–18.

Whitehead J. A. & K. R. Helfrich 1991. Instability of a flow with temperature-dependent viscosity: a model of magma dynamics. *Journal of Geophysical Research* **96**, 4145–55.

Williams, H. & A. R. McBirney 1979. *Volcanology*. San Francisco: Freeman, Cooper.

Wolfe, E. W., C. A. Neal, N. G. Banks, T. I. Duggan 1988. Geologic observations and chronology of eruptive events. In *United States Geological Survey Professional Paper 1463, The Puu Oo eruption of Kilauea Volcano, Hawaii: Episodes 1 through 20, January 3 1983, through June 8 1984*, E. W. Wolfe (ed.), 1–97. Washington DC: United States Government Printing Office.

# 14 The role of monitoring in forecasting volcanic events

## R. I. Tilling

*He who does not know how to look back cannot reach where he is going.*
Tagalog proverb (see Newhall & Dzurisin 1988)

## 14.1 Introduction and historical perspective

By decree of Ferdinand II, monarch of the Kingdom of the Two Sicilies, the construction of the Osservatorio Vesuviano was begun in 1841. With construction completed in 1847, this facility on the flank of Vesuvius Volcano became the world's first volcano observatory, marking the beginning of the scientific observations of active volcanoes. Within a decade of the its founding, earthquakes clearly associated with eruptive activity were being recorded instrumentally by Prof. Luigi Palmieri, who directed the observatory from 1855 to 1896. However, another half century was to pass before other volcano observatories were established, in 1911 in Japan by F. Omori at Asama Volcano, and in 1912 in Hawaii by Thomas A. Jaggar, Jr, at Kilauea Volcano. The recognition of the need to study volcanoes on a regular basis was prompted in large measure by the three deadly volcanic disasters in 1902 in the Central America–Caribbean region (Soufrière, St Vincent; Mont Pelée, Martinique; and Santa María, Guatemala), which together caused more than 36000 deaths (Tilling 1989b).

The regular observation and measurement of volcanoes (i.e. *volcano monitoring*), while clearly rooted in the movement to establish volcano observatories, contributed substantially to the overall development of volcanology into the multidisciplinary science that it is today. Modern volcanology is based on two fundamental premises: (1) knowledge of the past *and* present behaviour of a volcano furnishes the best (only?) clues to its possible future activity; and (2) observations of the eruptive processes and products at active volcanoes provide a "key to understanding the origin of volcanic units preserved in the geologic record" (Wright & Swanson 1987, p.231).

The common association of premonitory seismicity and ground deformation with eruptive activity was already becoming well documented by the early 20th century (e.g. Omori 1913, 1914, Wood 1915), but such observations generally were made after the eruption and, thus, not used to anticipate future events. Nonetheless, the instrumental monitoring of magma-induced seismicity and (or) ground deformation obviously led to the recognition of possible precursory patterns and to an awareness of increased potential for eruptive activity. As an example of such awareness, after a description of seismic and ground-tilt behavior at Kilauea, *The Volcano Letter* of the Hawaiian Volcano Observatory in August 1930 (Fiske et al. 1987, No. 294) states: "The conclusion drawn from all this evidence is that lava pressure is increasing under Halemaumau. **It is impossible to say whether or not this will result in an eruption** [bold in original], but it can be said confidently that conditions look more favorable now than at any time in the past several months."

During the 20th century, especially after World War II, tremendous progress was made in electronics and scientific instrumentation, monitoring techniques, observational capabilities, data collection and analysis (computerization), theoretical formulations, and modelling studies (see UNESCO 1972, Civetta et al. 1974, Tazieff & Sabroux 1983, Decker 1986, Tilling 1989a & b, McGuire 1991 and associated papers, Van Ruymbeke & d'Oreye 1991, Ewert & Swanson 1992) (see also other chapters in this volume). Also, from only three volcano observatories in operation early in this century there now exist over 40 institutions in 19 countries that conduct volcano monitoring to some degree (WOVO 1991–92). Only about 150 of the world's 550 or so historically active volcanoes are monitored at present, and of these only about a dozen are being monitored adequately. Consequently, except for a few intensively monitored volcanoes, reliable eruption *forecasting* (see Section 14.2) – at any level of specificity – has not been possible at most volcanoes. However, during the past two decades, some successful eruption forecasts have been achieved with existing monitoring techniques. With wider application of current technology to many more volcanoes, combined with anticipated improvements in the acquisition, analysis, and interpretation of measurement data, volcano monitoring promises to play even a greater role in forecasting volcanic events in the coming decades.

## 14.2 Volcano monitoring and forecasts

Without volcano-monitoring information, there would be no scientific basis to anticipate an impending eruption, or changes in an eruption already in progress. Before examining some selected case studies in which volcano monitoring played – or *could or should* have played – a role in forecasting of volcanic activity, let us review some general considerations.

### 14.2.1 Forecasts and predictions

To date, the term *forecast* has been, and continues to be, used in a vague fashion and is commonly considered synonymous with *prediction*. However, Swanson et al. (1985, p.397) recommended the adoption of the following definitions:

- *factual statement* describes current conditions but does not anticipate future events.
- *forecast* is a comparatively imprecise statement of the time, place, and nature of expected activity.
- *prediction* is a comparatively precise statement of the time, place, and ideally, the nature and size of impending activity. A prediction usually covers a shorter time period than a forecast and is generally based dominantly on interpretations and measurements of ongoing processes and secondarily on a projection of past history.

Forecasts and predictions can be either *long term* or *short term*. Most volcanologists consider *"short -term"* to mean a time scale of months to hours, whereas *"long term"* generally connotes a time scale of years to decades or longer. Long-term forecasts are based primarily on a volcano's eruptive history, including prehistoric activity, and on the nature and distribution of its products and impacts (Scott 1989); short-term forecasts are based exclusively on the volcano's current behaviour, as determined by volcano monitoring, observational and instrumental (Banks et al. 1989).

Depending on degree of specificity, a continuum obviously can exist between a forecast and a prediction, but, for practical purposes, a prediction can be considered as a precise forecast. In this paper, unless otherwise indicated the terms *forecast* and *forecasting* apply only to *short-term* forecasts or predictions. In any case, a prediction or forecast statement must specify a time period ("window"), within which the anticipated phenomenon is expected to occur. Furthermore, strictly speaking, for a statement to constitute a "true" forecast or prediction, it must be issued *publicly before* the onset of the anticipated volcanic event or change in activity. Not surprisingly, the application of the above definitions and stipulation greatly reduces the number of situations wherein "true" forecasts or predictions from monitoring studies were actually made.

## 14.2.2 Monitoring approaches

At the outset, it must be emphasized that, from a global perspective at least, issues pertinent to monitoring approaches, eruption forecasting, and basic studies of volcanoes are inseparable. Significant advances in forecasting will require not only more detailed knowledge of each volcano itself, but also an improved understanding of volcanic phenomena in general. Accordingly, basic studies need to be initiated or greatly accelerated at many heretofore little studied volcanoes.

The preceding chapters give informative summaries of monitoring techniques commonly applied or being tested at volcanoes worldwide. Of the various techniques currently in wide use, seismic monitoring and ground-deformation monitoring to date have proved to be the most reliable instrumental methods in successful forecasts of volcanic activity, in large measure because they have the longest record of use and technology development. Another contributory, and perhaps more important, factor is simply that these two techniques measure the direct responses of the volcanic system – brittle fracture, inflation, and deflation – in adjusting to subsurface magma movement and accompanying stresses and (or) hydrothermal-pressurization effects. Other monitoring techniques measure the associated responses to activation or re-activation of the system, such as variations in composition and emission rates of volcanic gases, changes in gravitational and magnetic field, anomalies in geoelectrical properties, and thermal variations. Commonly, these types of monitoring data tend to exhibit irregular temporal variations less amenable to diagnostic interpretation; while these data complement the characterization of a volcano's behaviour, they alone are usually inadequate to serve as the basis for making a forecast.

In my 20 years experience in volcanology, I have found that when volcanologists are asked the question "If you can use only *one* instrumental monitoring technique at a restless volcano, which one would it be?", the answer almost always is: seismic monitoring. Indeed, participants at the 1981 meeting that established the World Organization of Volcano Observatories (WOVO 1982) unanimously agreed that the core of an "adequate minimum" volcano observatory is a seismic monitoring capability, even if rudimentary or primitive. Recent innovations in seismic monitoring utilize a personal computer (PC) based system for "real-time seismic amplitude measurement" (RSAM), which enables the continuous measurement and registration of total seismic energy release even when the analogue recorders of conventional systems are saturated. The RSAM system (Endo & Murray 1991) has been shown to be highly effective in monitoring dome-building eruptions at Mount St Helens and also explosive eruptions at Redoubt (Alaska) and Mount Pinatubo (Philippines). Data from the RSAM as well as from non-seismic monitoring networks are acquired, processed, and displayed in near-time by means of a software package called BOB (Murray 1990). The integrated PC-based monitoring system (RSAM-BOB), developed mainly by the Cascades Volcano Observatory of the US Geological Survey (USGS), essentially can serve as a "mobile" volcano observatory; it is relatively inexpensive, easily portable, and especially useful in short-term forecasts.

However, just because seismic monitoring happens to be the technique of choice does not mean that other techniques should be discounted. Quite the contrary, to the extent that available scientific and economic resources allow, seismic monitoring should be complemented by, and integrated with, geodetic, geochemical and non-seismic geophysical monitoring techniques, as well as visual and geologic observations. Depending on the local conditions and particular behavior of a volcanic system, non-seismic monitoring techniques sometimes may yield the most diagnostic precursory indicators, for example, some magma movements at Kilauea Volcano were detected *only* by geoelectrical monitoring (Jackson et al. 1985).

Experience worldwide has clearly shown that the optimum monitoring approach is one that employs a combination of techniques rather than reliance on any single one. If the data from such an integrated approach all reveal similar precursory patterns, then much more confidence can be placed on a forecast. As discussed later, not all monitoring techniques need to involve the most advanced technology. I fully share the views eloquently expressed by Stoiber & Williams (1990) and by Swanson (1992) that geological field observations and simple monitoring approaches must be integral components in any comprehensive program of volcano monitoring.

### 14.2.3 Importance of baseline monitoring data

Even though rigorous mathematical analysis has been attempted recently for some high-quality time-series datasets obtained at well-monitored volcanoes (e.g. Voight & Cornelius 1991), to date the forecasting of activity at most volcanoes remains empirical, based almost entirely on pattern recognition. Fundamental to such forecasts is the *reliable* recognition of a recurring characteristic pattern of precursory behaviour, as determined by monitoring, and its time–space relationships to the ensuing activity. If it is known that such a certain pattern has typically preceded previous eruptions, then the recurrence of that pattern suggests the possibility of another eruption. Clearly, the longer the time spanned by the database, the more reliable is the pattern recognition.

An important prerequisite in recognizing possible patterns of precursory behaviour–eruptive activity is the acquisition of baseline monitoring data for an active or potentially active volcano, during periods of inactivity as well as activity. As emphasized by Banks et al. (1989, p.74):

> Only by knowing the range of variations (seasonal or cyclic) in the monitored parameters while a volcano is quiet is it possible to detect unambiguously a departure from *baseline* [italics added] or "normal" behavior that might augur renewed activity. It is important to establish baselines for even long-dormant volcanoes, because they might have eruption frequencies of many centuries and, hence, may not be considered "active" (generally defined as having erupted in historical time).

Ideally, baseline data should be collected for many years, preferably decades, to permit early reliable recognition of departure from normal behaviour. However, it also should be emphasized that the lack of baseline data – in and of itself – does not preclude successful eruption forecasting, provided that intensive monitoring is begun *immediately* after the onset of volcanic unrest (see Mount Pinatubo case study). For long-dormant but potentially hazardous volcanoes, there is no need–and would hardly be cost-effective–to acquire baseline data on a continuous basis. However, it would be prudent to establish simple baseline-monitoring networks on such volcanoes, and to reoccupy these networks on a regular schedule, perhaps initially twice a year for a few years to bracket maximum seasonal variations, and then at progressively longer intervals (e.g. every few years or even decades). If at any time the volcano begins to exhibit signs of unrest, the frequency of remeasurement of the baseline networks can be increased and (or) additional, more detailed monitoring studies can be inaugurated.

## 14.2.4 Usefulness of simple monitoring approaches

Continuing advances in instrumental monitoring techniques, including data acquisition, analysis, and telemetry (Ch. 2) and satellite-based systems (Chs 6, 7 & 11), should yield correspondingly improved time-series data, which in turn should allow more refined and timely forecasts of eruptive activity. Ideally, every potentially dangerous volcano should be densely instrumented and intensively monitored with state-of-the-art geophysical and geochemical surveillance systems, operated by a well-funded volcano observatory staffed with trained, experienced scientists and technicians. Realistically, this ideal scenario has not been, nor is it likely to be, realized fully in any country – industrialized or developing – because of the lack of scientific and economic resources and (or) political will.

Insufficient or unavailable modern instrumental monitoring, however, can be compensated in part by use of on-site observations of visual or manually measurable indicators of volcanic unrest (e.g. ground cracking, appearance of new fumaroles and (or) changed behaviour of existing ones, changes in measured distances, variations in temperature and level of water

(a)

(b)

(c)

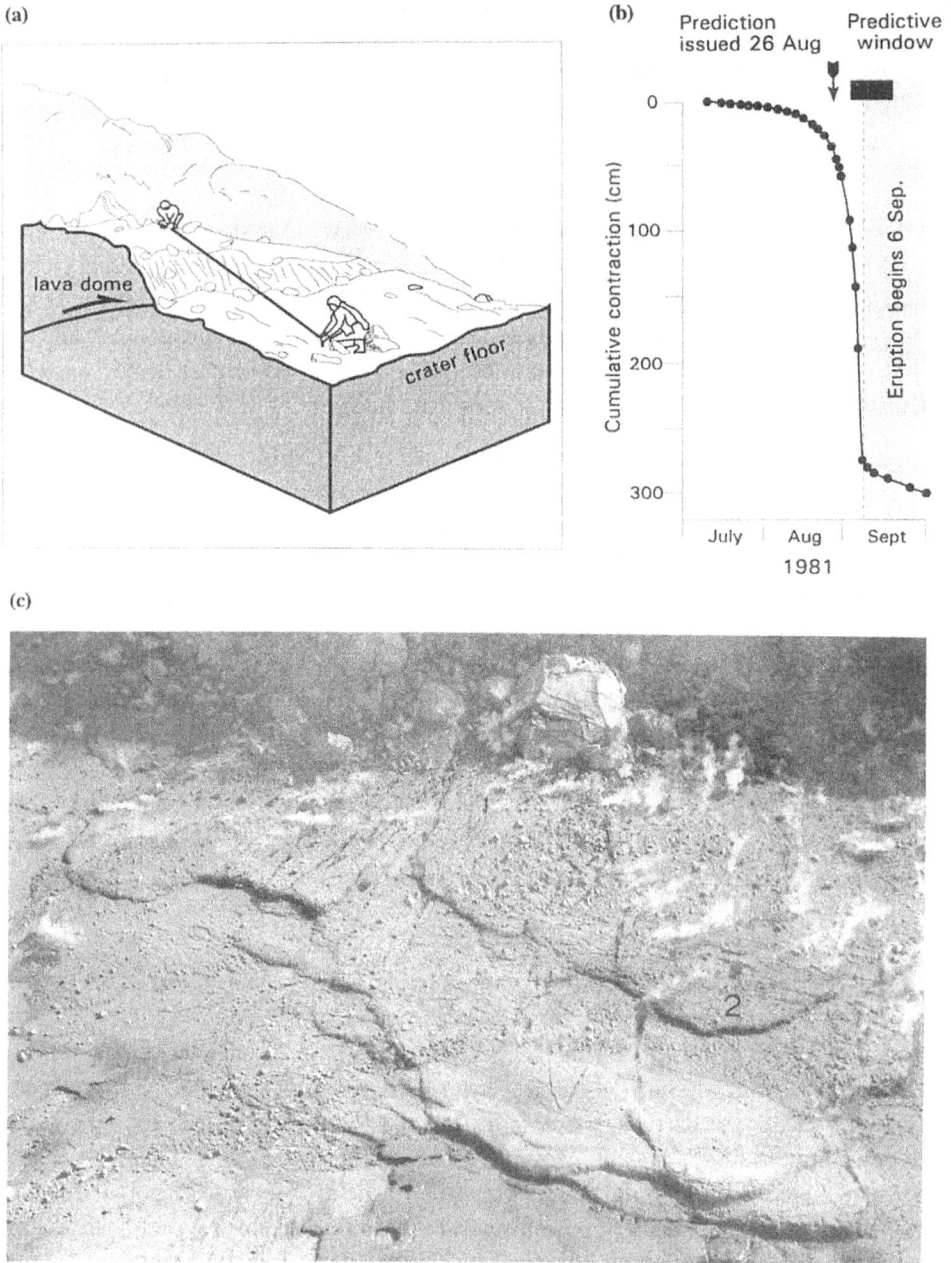

**Figure 14.1** An example of a useful simple method used to monitor horizontal displacements. (a) Distances across the leading edges of thrust faults around the base of the growing lava dome within the crater of Mount St Helens are measured periodically by means of a steel tape (from Brantley & Topinka 1984). (b) Plot of the cumulative contraction of the manually taped distance across the toe of Christina 2 thrust fault (see (c)) showing expansion of dome by magma intrusion. Such data, along with other monitoring data, were used to predict successfully dome building eruptions since June 1980. The arrow marks the issuance dated (26 August 1981) of the prediction, the black rectangle shows the period during which the eruption was predicted to occur ("predictive window"), the dashed vertical line indicates the eruption onset. (Modified from Swanson et al. 1983, Fig. 3.) (c) Southwest part of the crater floor showing several of the thrust faults measured by steel tape (width of view is about 200 m). Site 2 is on the upper plate of Christina 2 thrust, measured distances for which are plotted from (b). (From Swanson et al. 1983, Fig. 2).

bodies, increase or decrease in volcanic-plume size). Such observations and measurements, some of which can be people intensive, are nonetheless very cost effective, because they do not require exacting installation of costly instruments, maintenance of data-telemetry and electronics systems, and substantial computer reduction of data. Instead, a simple approach takes advantage of locally available manpower and inexpensive equipment and materials; another important attribute is that extensive technical training of the local observers or operators is unnecessary.

A highly effective, but simple method used at Mount St Helens was the manual measurement of distances between benchmarks by means of a steel tape to monitor movement across small thrust faults at the base of the growing lava dome (Swanson et al. 1983). Data obtained from this simple, inexpensive, but reliable measurement method, which yielded deformation patterns similar to those determined from more-precise monitoring data using electronic tiltmeters and distance meters, were used in the successful prediction of dome-building eruptions (Fig. 14.1). Descriptions and examples of some other useful simple methods of volcano surveillance are given in Stoiber & Williams (1990) and Swanson (1992).

While simple monitoring methods may not supplant more precise, state-of-the-art instrumental monitoring, the data – if obtained carefully and systematically – can aid eruption forecasting by:

(a) providing the *only* baseline data until such time when more comprehensive monitoring is possible; this especially applies to volcanoes that are little studied;
(b) giving useful ancillary information for parts of the volcano not covered by advanced instrumental monitoring systems;
(c) furnishing on-site verification of data obtained remotely by electronic instrumentation and telemetry (e.g. a continuously recording tiltmeter or extensometer); while the data from more sensitive instrumental monitoring have high *precision*, they may not necessarily have high *accuracy* because of possible installation instability, electronic drift, or other instrument malfunction;
(d) qualitatively indicating the part(s) of the volcano undergoing the most change or no change, thereby providing critical information to guide the optimum deployment of more advanced monitoring networks (Swanson 1992);
(e) under certain circumstances, showing the same pattern of precursory behavior as that defined by more precise, electronic monitoring techniques, thus directly serving as a basis for forecasting volcanic activity (see Mount St Helens case study).

Unfortunately, visual observations, manual measurements, and other on-site simple monitoring techniques often are overlooked as an effective strategy that complements and augments more sophisticated monitoring strategies. In some cases, the simple monitoring approaches are ignored because they are perceived to be "too old fashioned" or "not scientific enough" by emergency-management authorities and even by some scientists. Nothing could be further from the truth!

## 14.2.5 Which volcanoes to monitor?

As noted earlier, only a small handful of the world's active and potentially active volcanoes are being monitored adequately. This problem simply boils down to: too many volcanoes, scarce economic resources, and not enough properly trained scientists and technicians. Worse still, from the standpoint of volcanic-hazards mitigation, most of the world's dangerous, but inadequately monitored or unmonitored, volcanoes are located in densely populated developing countries, mostly in the circum-Pacific region (Tilling 1989b). The high annual population-growth rates of most of these countries are projected to continue into the next century, thus exacerbating an already grave problem of increasing populations exposed to volcanic risk (Tilling 1992).

Given this discouraging demographic outlook, combined with the lack of capability to monitor every volcano, it is necessary to identify for intensive monitoring the volcanoes that pose the greatest threat to populations. A few attempts have been made to compile a list of these high-risk volcanoes using eruption history and population proximity as principal criteria (e.g. Yokoyama et al. 1984), but any such list must be used with extreme caution because eruption histories for most volcanoes cannot be reconstructed with any certainty because of the lack of geological knowledge. Basic mapping and dating studies are essential to determine past eruptive behaviour, including recurrence intervals, and to make long-term assessment of possible renewed activity. In turn, these assessments of eruption potential should be considered in deciding which of the long-dormant, but high-risk volcanoes in populated regions warrant immediate attention.

We must not forget, however, that remote volcanoes in sparsely or even unpopulated regions can also pose a significant threat to life and property (Casadevall 1991, 1993). In recent decades, as commercial air travel has increased dramatically, so has the number of in-flight encounters between jet aircraft and volcanic ash plumes. To date, luckily none of these encounters have resulted in a disastrous crash, but some have caused substantial economic loss. For example, ash damage to a commercial jetliner during the December 1989 eruption of Redoubt Volcano (Alaska) amounted to about US$80 million dollars. Aircraft-ash encounters can potentially be more costly in terms of monetary loss than most volcanic disasters to date. According to industry estimates, a fatal crash of a fully loaded 747–700 jumbo jet with 300 passengers could cost as much as US$500 million dollars, including the financial settlement of post-accident law suits (T. J. Casadevall 1992, personal communication).

The future role that monitoring might play in forecasting volcanic events thus also hinges on which additional volcanoes should be targeted for increased monitoring in the 1990s and beyond. We must avoid deadly surprises in populated areas such as the 1982 El Chichón eruption (see case study), for which a forecast of any quality was impossible because of the total lack of pre-eruption monitoring. How many other El Chichóns or Mount Pinatubos might be waiting to erupt? Unfortunately, with presently available information, we have no idea. We also must identify and monitor the remote volcanoes that pose the greatest threat to civil aviation.

It is feasible to monitor, at relatively reasonable cost, the highest-risk volcanoes with present-day technology by using a combination of ground- and satellite-based observations. The principal obstacle to an effective global volcano-monitoring programme seems to be the lack of political will at national and international levels. Specifically as regards volcanic threat to aviation safety, another impediment is the absence of well-coordinated international system to communicate – accurately and rapidly – eruption information and warnings to the pilots, civil-aviation authorities, and airline operators.

## 14.3 Selected case studies

Since the mid-1970s, volcanologists, public officials, and the general public have faced numerous volcanic disasters and crises. Within the context of the role of monitoring in forecasting, let us briefly examine some case studies of recent eruptions and caldera unrest. Discussion of some of these case studies draws liberally from summaries in several previous review papers (Yokoyama et al. 1984, Punongbayan & Tilling 1989, Tilling 1989b).

### 14.3.1 Volcanic eruptions

Since the mid-1970s, a number of eruptions have caused death and destruction, substantial economic loss, or otherwise adversely affected the daily lives of millions of people. The 1980s, which witnessed the worst volcanic disasters in the recorded histories of the USA, Mexico, and Colombia, registered the most eruption-related deaths in any 10-year period since 1902 (Tilling 1989b). Will the voluminous, devastating eruption of Mount Pinatubo in June 1991 usher in another decade as deadly as the 1980s in terms of volcanic hazards? The case studies selected for review, which include examples from both industrialized and developing countries, span a wide range in terms of eruption size, impact of volcanic hazards, and the role monitoring played – or did not play – in eruption forecasts.

*14.3.1.1 Soufrière, Guadeloupe (Lesser Antilles), France 1976–77*
On 8 July 1976, Soufrière began to erupt after about a year of precursory seismicity, and ash fell on the coastal city of Basse-Terre, the capital of Guadeloupe, and surrounding areas closer to the volcano. Eruptive and seismic activity intensified during the following weeks, peaking in late August, when the emission of ash and steam was most energetic and virtually continuous and the daily count of earthquakes exceeded 400 on some days (Feuillard et al. 1983, Fig. 5). The activity then gradually subsided and ceased by early 1977.

In August 1976 during the height of activity, some scientists reported finding a significant amount of juvenile volcanic glass in the ejecta not identified in the earlier erupted materials. This finding implied that new magma had risen into the volcanic edifice, increasing the concern that a more violent, magmatic phase might ensue. The government officials ordered the immediate evacuation of about 72 000 people, which remained in force for nearly 4 months and caused severe political and socioeconomic problems. However, the anticipated escalation of eruptive activity never materialized, the supposed occurrence of juvenile glass was

later learned to be incorrect (Feuillard et al. 1983). In the end, the 1976–77 Soufrière eruption proved to be comparatively small (about $10^{-3}\,km^3$ solid ejecta) and entirely of phreatic
origin, but it created a volcanic crisis of enormous proportions, exacerbated by what Fiske
(1984) termed euphemistically as the "challenging relationships" between scientists (representing two rival factions), civil authorities, and journalists (see Fiske (1984) for a revealing
and fascinating account).

Could systematic monitoring have played a more beneficial role in the response to the crisis?
Soufrière had been monitored by a modest, semi-permanent seismic network since the early
1960s. The increase in volcanic seismicity, while duly recorded, was inadequate to serve as
a basis for forecasting renewed eruption at the volcano, dormant since 1956. Indeed, because
of the lack of any recognizable precursory patterns, it was felt in April 1976 that there was
". . . no important risk of an eruption of this volcano, but that it is impossible to make a
long-term (several months or years) prognostication . . ." (SEAN 1976, p.4). With the onset
of phreatic activity in July 1976, monitoring was considerably expanded and included electronic
tiltmeters, recording magnetometers, precise levelling surveys, and analysis of volcanic fluids.
Unfortunately, most of these monitoring techniques were not deployed until late July or August,
a year or more after the volcanic unrest had begun. Though comprehensive, these monitoring studies were hampered by the lack of critical *baseline* monitoring data for comparison
with the results obtained after the start of the eruption, thus the monitoring data could not be
used to evaluate the possibility that the phreatic activity was evolving toward magmatic activity,
suggested by the (mistaken) identification of fresh volcanic material in the ejecta.

The Soufrière case study is instructive because it is one of the best examples to illustrate
that, without adequate pre-crisis baseline data – obtained by instrumental and (or) simpler
methods, intensive highly sophisticated studies during the crisis may be insufficient for reliable forecasting. Had more baseline monitoring measurements been available, perhaps it would
have been possible to better forecast the course of the eruption in mid-August 1976 and, in
turn, to assess objectively the need for the evacuation of 72 000 people for 4 months. We will
never know, but it seems clear that the volcano monitoring conducted was "too little, too late"
to serve as a basis for forecasting volcanic events. With the painful lessons learned at Soufrière,
the French government reorganized and modernized its volcano observatories on Guadeloupe
and Martinique, and established a new observatory at Piton de la Fournaise (Réunion Island).

*14.3.1.2 Mount St Helens, Washington, USA. 1980–90*
Following a week of intense precursory seismicity, Mount St Helens reawakened from its
123-year sleep and phreatic explosions began on 27 March 1980 (Lipman & Mullineaux 1981).
During the ensuing intermittent phreatic activity through 17 May, seismic and geodetic monitoring was greatly intensified and confirmed visual and photogrammetric observations that
the volcano's north flank was deforming at an alarmingly high rate (averaging about $1.5\,m\,day^{-1}$
horizontal displacement) because of magma intrusion into the volcanic edifice (Endo et al.
1981, Lipman et al. 1981). Triggered by a magnitude 5.1 earthquake on the morning of 18
May, the climactic eruption occurred, causing 57 human deaths and an economic loss of more
than US$1 billion (Tilling et al. 1990) – the worst volcanic disaster in the history of the USA.
The processes and impacts of the 18 May eruption have been described and analyzed in great

detail in Lipman and Mullineaux (1981) and many hundreds of subsequent scientific studies. The post-May 1980 activity has largely involved the emplacement and growth of a composite dome (Swanson et al. 1987, Chadwick et al. 1988). Other than a few very small phreatic explosions in December 1989, January and April 1990, and winter of 1990–91, the dome at Mount St Helens has been inactive since October 1986 (Swanson 1990).

Prior to 1980, the monitoring of Mount St Helens was minimal, only consisting of a single seismometer on its flank and a few electronic distance measurement (EDM) baselines. Repeated efforts to obtain increased USGS funding for additional monitoring failed, despite the fact that Mount St Helens was recognized to be the most active and explosive of the Cascade Volcanoes. In addition, in one of the few genuine, successful long-term eruption forecasts ever made in the short history of volcanology, USGS scientists, on the basis of detailed mapping and dating studies, concluded in 1975 that Mount St Helens would be the most likely volcano in the conterminous USA to reawaken and erupt, "possibly before the end of this century" (Crandell et al. 1975, p.441). Yet, the USGS still failed to expand monitoring at Mount St Helens; the volcano erupted 5 years later.

Data from the single seismometer on the flank of Mount St Helens operating before 1980, together with those from the regional seismic network for Washington State, alerted scientists to the possible reawakening of Mount St Helens. However, the pre-1980 monitoring was inadequate for making any short-term forecast or prediction of the start of eruptive activity at Mount St Helens on 27 March 1980. Despite the intensive monitoring conducted during the nearly 2 months of phreatic activity (27 March to 17 May), it was also impossible to make a precise short-term forecast of the paroxysmal event on 18 May, even though ground-deformation measurements clearly indicated that the volcano's north flank was becoming dangerously unstable. Some scientists (e.g. D. A. Swanson 1989, personal communication) speculate that the deformation rate probably would have shown a diagnostic inflection, if the magma intrusion-driven deformation process had not been perturbed ("short-circuited") by the magnitude 5.1 earthquake that triggered the climactic eruption. While the monitoring data did not allow a *forecast*, according to the strict definition outlined earlier, they clearly alerted the scientists to the possibility of a major collapse and avalanche triggering a large magmatic eruption. Such a volcanic-hazards scenario was discussed at closed meetings of the monitoring team and explained to emergency-management officials before 1 May, but it was not disclosed publicly (Miller et al. 1981, Decker 1986).

Intensive monitoring at Mount St Helens has produced a remarkably reliable capability to predict dome-forming eruptions. Since 1980, all dome-growth eruptive episodes (except for a very small event in 1984) have been successfully predicted several days to three weeks in advance (Tilling et al. 1990). An example of one of these predictions is illustrated in Figure 14.2 and Table 14.1; other examples are given in detail elsewhere (Swanson et al. 1983, Swanson et al. 1985). Although the successful predictions of dome-building events at Mount St Helens mark one of the most significant advances in modern volcanology, it must be remembered that these predictions apply only to the onset of activity. The Mount St Helens experience also improved the pattern-recognition capability for precursors associated with a complex, rapidly evolving eruption that included cryptodome intrusion, lateral blast, and debris avalanche.

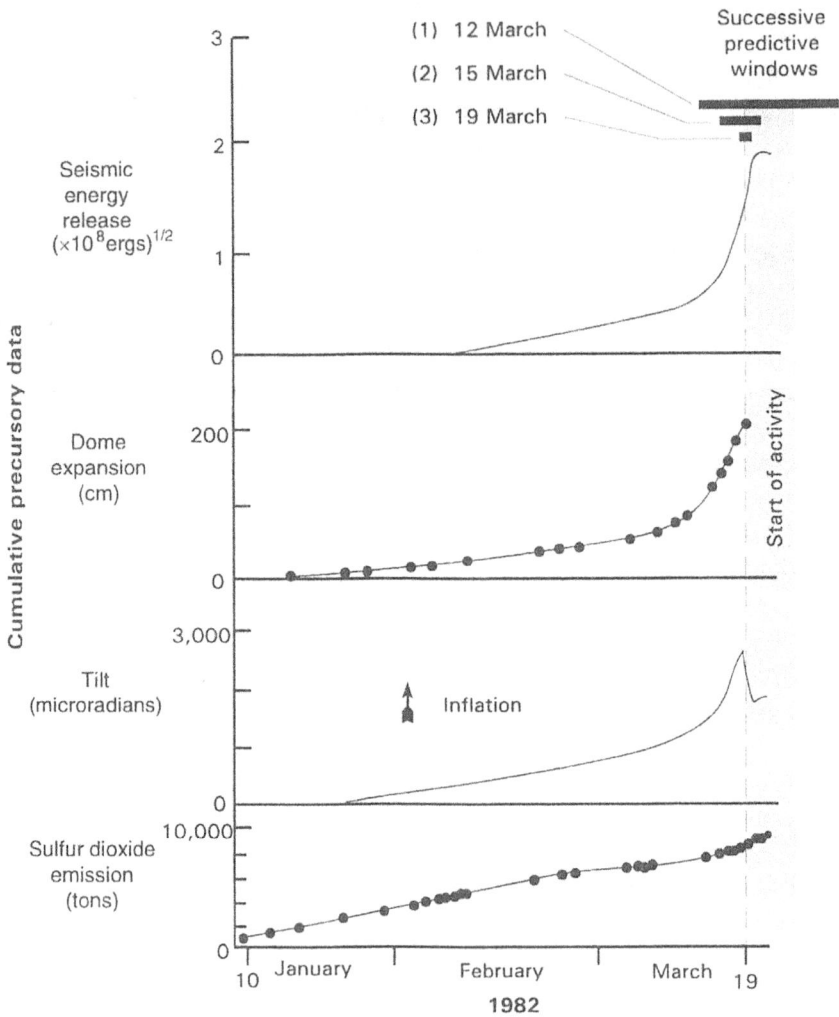

**Figure 14.2** The acceleration of precursory activity before the onset of the March–April 1982 eruptive episode at Mount St Helens (modified from Swanson et al. 1985, Fig. 3.) Dome expansion is measured by electronic distance meter of horizontal displacement (compare with Fig. 14.1) showing similar precursory pattern obtained by a simple manual method), tilt change by electronic tiltmeter, and sulphur dioxide emission by correlation spectrometer (cospec). Excerpts from predictions (1), (2) and (3), issued on 12, 15 and 19 March, respectively, are given in Table 14.1; solid bars indicate the successively narrowing predictive windows.

While the accurate prediction of large explosive events and of the duration or size of an eruption still eludes volcanologists, the Mount St Helens case study attests to the vital role that monitoring can play in short-term eruption forecasting. Moreover, the studies at Mount St Helens also illustrate the importance of basic geologic investigations for long-term forecasts, which in turn help to identify potentially dangerous volcanoes that ideally should receive priority in monitoring studies, if resources are limited.

**Table 14.1** Examples of factual statements and predictions of the onset of the March–April 1982 eruptive episode at Mount St Helens issued to civil authorities and the media. (Excerpted from Swanson et al. 1985, pp 415–6).

| Local time | Date | Factual statement or **prediction** issued |
|---|---|---|
| 09.00h | 5 March | *Factual statement*: "Seismicity . . . increased around 21 February and has remained at a level somewhat above background since that time . . . Measurements made last week (27 February) show only slow ground deformation . . . and no significant increase in gas emissions." |
| (Measurements after 09.00h on 5 March show increased rates of deformation.) | | |
| 08.00h | 12 March | *Factual statement and **prediction** (1)*: "Seismicity . . . continues at elevated levels . . . Rates of ground deformation in the crater have increased during the last two weeks . . . .Based on rates of deformation, an eruption is likely within the next 3 weeks. Deformation is confined to the crater area, suggesting that renewed dome growth will occur." |
| (Measurements on 15 March showed greatly accelerated deformation.) | | |
| 19.00h | 15 March | ***Prediction** (2)*, updated: "An eruption, most likely of the dome-building type, will probably begin within 1 to 5 days." |
| (Rates of deformation and seismic energy release continued to increase rapidly.) | | |
| 09.00h | 19 March | ***Prediction** (3)*, updated: "An eruption will begin soon, probably within 24 hours. The character of both the seismicity and deformation in the crater area indicates that the most likely type of activity is dome growth." |
| 19.27h | 19 March | Eruption begins. |

## 14.3.1.3 El Chichón, Chiapas, Mexico 1982

El Chichón Volcano, located in the state of Chiapas (southeastern Mexico), erupted explosively three times between 28 March and 4 April in 1982 and caused the worst volcanic disaster in the recorded history of Mexico. These eruptions destroyed the volcano's central dome, leaving a 1-km wide, nearly 300-m deep crater in its place. Highly gas-charged pyroclastic flows and surges obliterated all villages within a radius of 7 km of the volcano, killing more than 2000 people (Duffield et al. 1984, Luhr & Varekamp 1984). Although the 1982 El Chichón eruptions produced about the same amount of juvenile silicate ejecta as the 18 May 1980 eruption at Mount St Helens (about 0.4 km³ dense-rock equivalent), they injected far more aerosols into the lower stratosphere and may have affected global climate (Rampino & Self 1984, Galindo et al. 1984).

What is noteworthy about the El Chichón case study is that the first eruption came as *almost* a total surprise, although it should not have. No local monitoring of precursory seismicity was conducted despite notably increased earthquake activity beginning in late 1980, some of which was felt by local inhabitants (SEAN 1982). *Post-eruption* analysis (Havskov et al. 1983) of seismic data from a network in operation to monitor induced seismicity of a nearby reservoir of the Chicoasén hydroelectric project confirmed that the buildup of premonitory earthquakes began months before the eruption, especially evident by early March 1982. Moreover, geologists conducting fieldwork at the volcano during the period December 1980–January 1981 heard loud noises and felt earthquakes that caused them to specifically mention, in a September 1981 report, that "a high volcanic risk" existed at El Chichón (Canul & Rocha 1981). *Post-eruption* geological and radiometric studies also indicated that El Chichón was frequently and violently active during the Holocene, with an average eruption recurrence interval of about $600 \pm 200$ years, and produced deposits substantially more voluminous than the 1982 deposits (Tilling et al. 1984).

In hindsight, the question can be asked: Had the volcano's dangerous prehistoric eruptive behavior been known prior to 1982, might the scientific community and civil-protection authorities have paid more attention to the 1980–81 volcanic unrest at El Chichón, and would monitoring studies been initiated in response to the dramatic increase in precursory seismicity on 1 March 1982? The answer to this rhetorical question will never be known. Monitoring *might* have played an important role in forecasting the 1982 El Chichón eruption, *if* the pre-eruption warning signals had not been overlooked.

### 14.3.1.4 Nevado del Ruiz, Caldas, Colombia 1985

On the afternoon of 13 November 1985, a small-volume (about $0.03 \, km^3$) explosive eruption began at the summit of Nevado del Ruiz, a 5389-m high, glacier-capped mountain that is the northernmost active volcano in the Andes. The hot ejecta melted and mixed with snow and ice to form several mudflows that swept down the steep, narrow valleys draining the volcano. These mudflows destroyed or buried everything in their paths, killing more than 25 000 people, the vast majority in Armero, an agricultural community in the valley of Río Lagunillas. In addition, the mudflows injured another 5000 people, left about 10000 homeless, and caused over US$200 million in economic loss. The 1985 Ruiz catastrophe was the worst volcanic disaster in the recorded history of Colombia and the most deadly in the world since the 1902 eruption of Mont Pelée on the Caribbean Island of Martinique (Herd and the Comité de Estudios Vulcanológicos 1986, CERESIS 1990, Williams 1990a & b).

The 1985 eruption and associated mudflows at Ruiz should not have come as a surprise, because similarly destructive events have occurred in the Lagunillas valley – in AD 1595 and again in 1845 – and affected the same area upon which Armero later developed (Herd and the Comité de Estudios Vulcanológicos 1986, Voight 1990). In addition, in contrast to the 1982 eruption at El Chichón, efforts were made to conduct monitoring studies in response to the precursory signals at Ruiz (felt earthquakes, increased fumarolic activity, phreatic explosions, bursts of volcanic tremor) that were noticed in late November 1984. Due to administrative inertia and logistical problems, it was not until the summer of 1985 that rudimentary seismic monitoring began. However, when a strong phreatic eruption on 11 September produced measurable ash fall at Manizales, the capital of Caldas Province (population 230000), and several sizeable debris flows, governmental concern was heightened and monitoring increased. By mid-October, a monitoring network of five smoke-drum seismometers and four precise levelling ("dry tilt") arrays was established, two electronic tiltmeters operated briefly in early November (Banks et al. 1990). Analysis of fumarolic gases in late October indicated substantial magmatic contributions of $CO_2$ and $SO_2$ and the need "to seriously consider the possibility of an impending magmatic eruption." (Barberi et al. 1990, p.5).

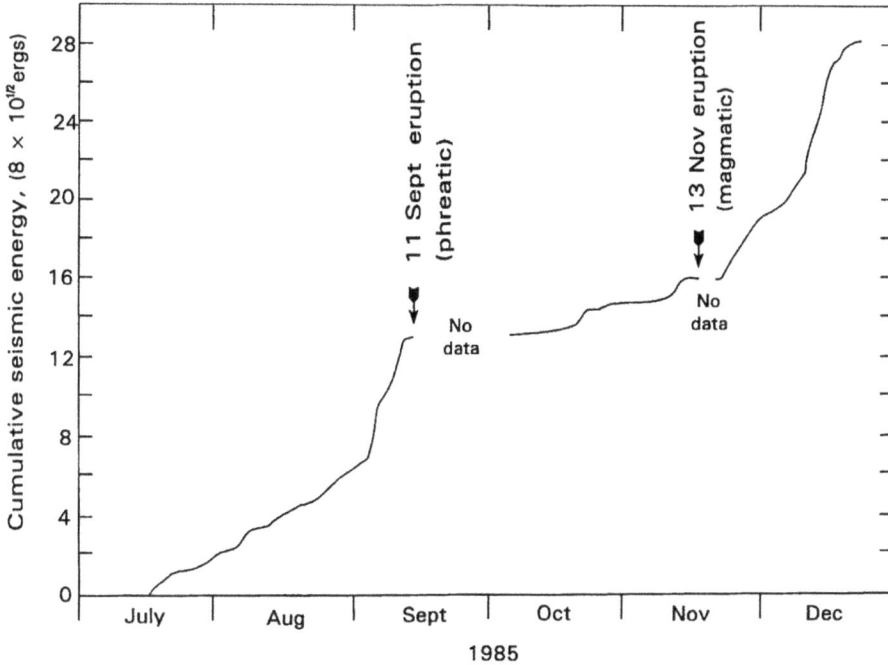

**Figure 14.3** Cumulative seismic energy release at Nevado del Ruiz in the period July–November 1985, with respect to the strongest of the phreatic eruptions on 11 September and the magmatic eruption on 13 November (modified from McClelland et al. 1989, Fig. 15.5). Both the number of earthquakes per day (not shown) and the total seismic energy release were considerably greater before the 11 September phreatic eruption than before the magmatic 13 November eruption. The comparatively weak precursory seismicity precluded forecasting the catastrophic 13 November event and was among the factors that contributed to the lack of decisive action by officials, even after they learned of the beginning of eruptive activity (see Voight 1990).

Although considerably expanded in the month prior to the 13 November 1985 eruption and catastrophe, monitoring studies still were "too little, too late" to allow any precise forecast of impending eruption (Fig. 14.3). As Hall (1990, 114) concluded: "Limited scientific data from a marginal monitoring program, *no baseline data* [italics added], and greatly delayed processing precluded a realistic attempt to understand or predict an eruptive event". In hindsight, however, it is arguable that, even if it had been possible to make a precise forecast based on more extensive monitoring, the Ruiz disaster could have been averted. Well-documented post-mortems of the Ruiz tragedy (e.g. Voight 1990, CERESIS 1990) have clearly shown that the principal contributory factor was "cumulative human error – by misjudgment, indecision, and bureaucratic shortsightedness", rather than "technological ineffectiveness or defectiveness . . ." (Voight 1990, p.349).

### 14.3.1.5 Kelut, Java, Indonesia 1990

The Island of Java contains one of the most dense concentrations of active volcanoes in the world; Kelut, a composite volcano with a crater lake, is in eastern Java, approximately 80 km south-southwest of the city of Surabaya. It has erupted more than 30 times since AD 1000.

Lahars (mudflows) triggered by an eruption in AD 1586 claimed about 10000 lives, and lahars associated with the 1919 eruption destroyed or damaged more than 100 villages and killed 5160 people (Pardyanto 1968, Kusumadinata 1979, Djurmarma & Achmad 1991, Sudradjat 1991). The 1919 eruption was the impetus for establishing Indonesia's first permanent monitoring observatory at Kelut and for the construction of tunnels to lower crater-lake level to minimize the water volume of potential lahars. Kelut remains as one of the most intensively studied and monitored volcanoes in Indonesia, monitoring techniques include seismic (modest network), levelling and electronic-distance-measurement (EDM) surveys, regular measurements of water temperature and chemical composition, acoustical (hydrophone) measurement of volcanic degassing into the crater lake ("gas bubble frequency"), and, since 1986, self-potential variations. In recent years, some of the monitoring data are telemetered via the ARGOS satellite to scientists in Indonesia and France.

Quiet for 23 years following its previous activity in 1967, Kelut exploded violently on 10 February 1990, ejecting about $0.12\,km^3$ tephra and displacing $0.002\,km^3$ of the crater-lake water (Sudradjat 1991). Precursors of the volcano's reawakening were detected beginning in

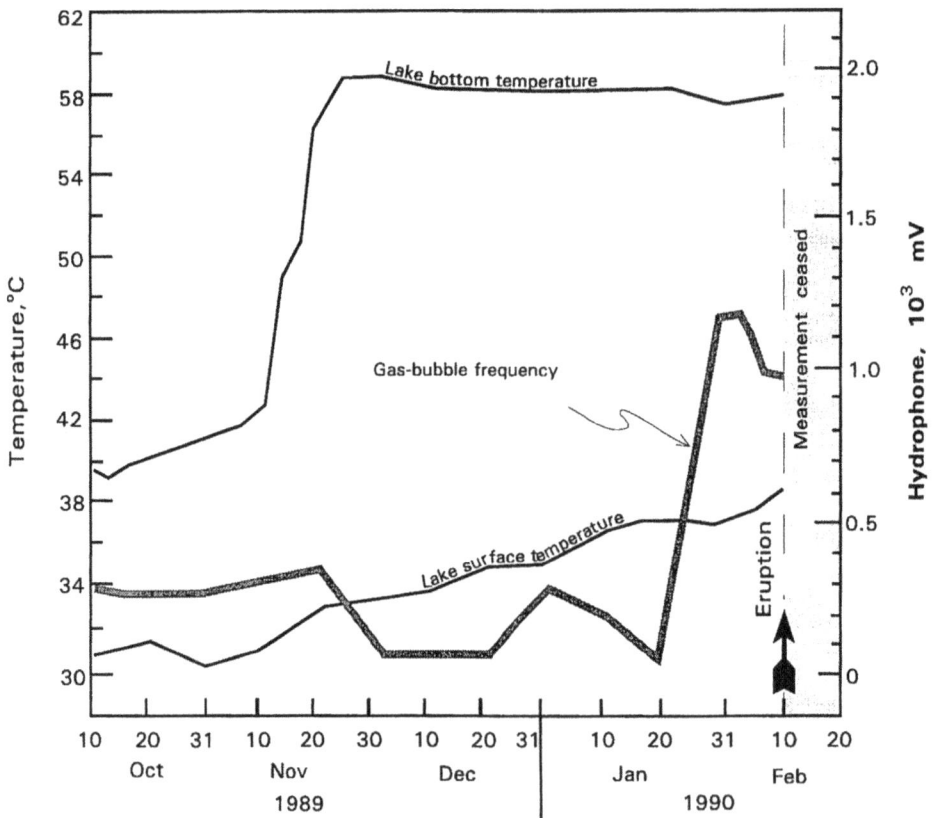

**Figure 14.4** Diagnostic precursory monitoring data before the 10 February 1990 eruption of Kelut Volcano, Java, Indonesia (modified and combined from Sudradjat 1991, Figs 4 & 7). Abrupt increase in the temperature of Kelut's crater lake, especially its bottom water, provides a definitive precursor of the eruptive activity that took about 3 months later.

late October or early November 1989, depending on the parameter monitored. The most diagnostic precursory indicators were the marked warming of both the surface and bottom temperature of the crater, an increase in cumulative seismic-energy release, a change in colour and pH of lake water, and an order-of-magnitude jump in gas-bubble frequency shown by the hydrophone measurements (Fig. 14.4). Less definitive but compatible evidence was provided by the variations in water chemistry and in self-potential.

Although no specific forecasts (in the precise sense) were made in advance, because of the timely detection of precursory indicators, "the alert was issued several months before the eruption. Exercises were carried out in hazard zones" (Sudradjat 1991, p.462). Consequently, because of adequate warning, there were no direct casualties from the eruption, although about 25 people perished because of collapse of roofs laden with wet ash. The Kelut case study is especially pertinent for monitoring active or potentially active volcanoes in developing countries, because most of the techniques used take advantage of relatively simple measurements that can be made by personnel with a minimum of specialized training. The 1990 experience at Kelut again underscores the importance of long-term baseline data in early detection of departures from normal behaviour of the volcano.

The role of monitoring in two other eruptions in Indonesia in the 1980s also deserves mention. Visual and rudimentary seismic monitoring provided important information that made possible the safe evacuation of about 7000 people from Una-Una Island before the largest historical eruption of Colo Volcano (Central Sulawesi) in July 1983 (Katili & Sudradjat 1984). Similarly, a timely eruption alert and orderly evacuation of 1800 people from Gunung Api Island before the May 1988 eruption of Banda Api Volcano (Maluku) were based on monitoring conducted by an observatory post of the Indonesian Volcanological Survey (VSI) located on nearby Neira Island (Casadevall et al. 1989). The VSI I post at Neira is manned by two trained observers and equipped with a single seismometer and a short-wave radio for daily communication with VSI headquarters in Bandung (Java).

Indeed, despite limited scientific and economic resources, Indonesia – with its network of relatively simple, but long-operating volcano observatories – has a remarkably effective volcanic-hazards mitigation program. During the 1980s, Indonesia successfully evacuated more than 140000 persons – 33000 in 1988 alone (Casadevall et al. 1989) – threatened by volcanic activity. In comparison, civil authorities in the USA have had to deal with only a few hundred evacuees in response to volcanic crises in the 20th century. The key to Indonesia's successes in large measure can be attributed to two factors: (a) the high frequency of eruptions, which enhance local populations' awareness of volcanic hazards; and (b) effective utilization of available resources to acquire long-term baseline monitoring data for its high-risk volcanoes, even if this consists only of keeping a round-the-clock log of visual observations and operating a single smoke-drum seismic instrument at some volcanoes.

### 14.3.1.6 Unzen, Kyushu, Japan 1990-present

Unzen is a large volcanic dome complex that occupies much of Shimabara Peninsula (West Kyushu), about 40km east of Nagasaki. Before 1990, the volcano had erupted only five times in history, most recently in 1792, when a volcanogenic tsunami killed about 15000 people – representing one of the world's six deadliest volcanic disasters since AD 1500. Episodes of seismic unrest occur frequently at Unzen, but historically most of these have not culminated in eruption.

Preceded by more than a year of seismic activity in the vicinity of Unzen, a small, short-lived phreatic eruption occurred on 17 November 1990 from two new vents in the summit crater of Fugen-dake, one of several domes making up the Unzen volcanic complex, even though no shallow seismicity was observed beneath the eruptive outbreak (Shimozuru 1991). Seismicity remained high for the next several months, but with no visible eruptive activity. During the period February–May 1991, intermittent phreato-magmatic activity took place along a new line of small vents, and by mid-May a lava dome was seen in the crater, anticipated from the beginning of intense shallow seismicity beneath the vent area 3 days earlier. Electronic distance measurements showed that over the next several days, the growing dome was expanding at 70–80 cm h$^{-1}$ (GVN 1991a). The increased activity produced a number of debris flows, requiring the evacuation of more than 1000 people. Beginning on 24 May, periodic collapses of the expanding dome produced Merapi-type pyroclastic flows and surges.

A large pyroclastic flow on 3 June 1991 took the lives of 43 people, including three volcanologists (Maurice Krafft, Katia Krafft and Harry Glicken), and prompted broader evacuation. Since then, dome extrusion, collapse-induced pyroclastic flows, and debris flows have continued intermittently through the time of this writing (October 1992). By the end of August 1992, the volume of Unzen's composite dome was estimated to be 0.05 km$^3$, which together with the combined volume for the pyroclastic and avalanche deposits yields a total erupted volume (dense rock equivalent) of 0.11 km$^3$ (GVN 1992a).

The ongoing eruption at Unzen (Fig. 14.5) has been, and is being, monitored by state-of-the-art geophysical and geochemical techniques, excellent summaries of the Unzen eruption through August 1992 are given by Nakada (1992), Yanagi et al. (1992) and Suzuki & Furuya (1992). Yet, because no characteristic precursory pattern has been recognized with the advent of modern studies, it was not possible to forecast the start of eruptive activity in 1990 after a dormant period of nearly two centuries following its previous activity in 1792. While the results of seismic and geodetic monitoring did anticipate the start of the magmatic phase of the eruption in May 1991, scientists were unable to specify a precise predictive window. It has been possible to detect and monitor dome growth from monitoring data, but it has been impossible to predict the "time of occurrence and magnitude of discrete pyroclastic flows" triggered by dome collapse (Shimozuru 1991, p.295).

The Unzen case study reemphasizes the sobering reality that, even with the best available monitoring techniques and scientific analysis of the data, it may not be possible to forecast renewed activity at volcanoes that erupt infrequently. As also emphasized by Professor Shimozuru (1991, p.296), chairman of Japan's Coordinating Committee for Prediction of Volcanic Eruptions, the current intensive monitoring to date at Unzen cannot "predict objectively the future of ongoing activity and, consequently, to give suggestions to government officials for the timing of removal of evacuation orders". As of August 1992, 6054 people still remained evacuated (GVN 1992b). The intensive monitoring in progress, however, should be able to detect the cessation of dome growth, thereby providing important data to officials for consideration of modification or cancellation of the evacuation order.

**Figure 14.5** Aerial oblique view of Unzen Volcano on the Shimabara Peninsula, Kyushu, Japan, showing fume rising from the currently active dome complex near its summit (1359-m high Fugendake). The paths of destructuve pyroclastic flows and debris flows down the Mizunashi and Oshigadani valleys can be seen sweeping into Shimabara City. Mount Mayuyama – with its seaward-facing, amphitheatre-like depression (right centre) – marks the site of the major collapse during the 1972 eruption, which generated tsunamis that killed about 15 000 people. (Photograph taken on 20 January 1992 by Asia Air Survey Company.)

*14.3.1.7 Mount Pinatubo, Luzon, Philippines 1991–present*

Prior to 1991, Pinatubo, a dacitic volcanic complex located about 100 km northwest of Manila, had not erupted in historical time. Reconnaissance geologic and radiocarbon dating studies indicated that its summit composite dome was flanked by extensive fans of young (600–8000 years BP) pyroclastic and lahar deposits (Pinatubo Volcano Observatory Team 1991). On 2 April 1991, small phreatic explosions began about 1.5 km northwest of the summit, prompting the evacuation of more than 5000 people within a 10-km radius. No precursory felt earthquakes were reported by local inhabitants. Because Pinatubo was not monitored before 1991, it is not known whether any microearthquakes might have preceded these explosions, which opened a new 1-km-long line of fumaroles.

Within days of the initial explosions, scientists from the Philippine Institute of Volcanology and Seismology (PHIVOLCS) began to monitor the continuing high seismicity with portable seismographs. USGS scientists arrived in late April in response to a request from the Philippine government for assistance. With increasing concern about the possibility of a large magmatic eruption, monitoring was intensified and included the use of correlation spectrometer (COSPEC) measurements of $SO_2$ emission and two electronic tiltmeters. A radio-telemetered network of seven seismic stations was installed by a PHIVOLCS-USGS team, and near-real-time monitoring began with acquisition and processing of the data at nearby Clark Air Base with a PC-based system (Lee 1989). This system made possible the rapid tracking of earthquake foci and total seismic energy output by use of the RSAM techniques of Endo & Murray (1991). Concomitant with the increase in the monitoring programme was the development of a scheme of alert levels (Table 14.2) that could be easily communicated to, and understood by, emergency-management officials.

$SO_2$ flux increased ten fold during the latter part of May, from 500 to 5000 ton day$^{-1}$ and then decreased sharply to only 280 ton day$^{-1}$ by 4 June (GVN 1991b, Pinatubo Volcano Observatory Team 1991). The temporal variation in $SO_2$ emission (Fig. 14.6), together with accompanying increased shallow seismicity and occurrence of volcanic tremor, was interpreted to indicate upward movement of magma and, ultimately, blockage of escaping gas within the conduit. This development led PHILVOCS to declare on 5 June an alert level 3, which states: "If trend of increasing unrest continues, eruption possible within 2 weeks" (Table 14.2). In essence, an alert level 3 constitutes a relatively precise short-term forecast. Intense shallow seismicity and a 32-μrad inflationary tilt during the following 2 days culminated in the extrusion of a lava dome on 7 June (observed on 8 June), marking the beginning of magmatic activity. The appearance and growth of the dome, combined with a shift of seismic foci to directly beneath the dome and an irregular increase in the amplitude and frequency of occurrence of volcanic tremor (Pinatubo Volcano Observatory Team 1991, Harlow et al. 1991), collectively provided the basis for raising the alert level to 4 on 7 June, and then to 5 on 9 June (Table 14.2). The increasing activity and attendant raises in alert levels led to a radius of evacuation to 20 km, involving an additional 20000 evacuees from Zambales, Tarlac and Pampanga Provinces; all aircraft and supporting facilities at Clark Air Base were moved to safer locations elsewhere. On 10 June, more than 14000 military personnel and dependents were evacuated by ground transportation to Subic Bay Naval Station, 30 km to the south-southwest.

**Table 14.2** Pre-eruption alert levels for the 1991 eruption of Mount Pinatubo, Luzon, Philippines (from Pinatubo Volcano Observatory Team 1991, Table 1).

| Alert level | Criteria | Interpretation | Dates(s) declared |
|---|---|---|---|
| No alert | Background, quiet | No eruption in foreseeable future | n/a |
| 1 | Low-level seismic, fumarolic, other unrest | Magmatic, tectonic, or hydrothermal disturbance; no eruption imminent | n/a |
| 2 | Moderate level of seismic, other unrest, with positive evidence for involvement of magma | Probable magmatic intrusion; could eventually lead to an eruption | 13 May 1991 |
| 3 | Relatively high and increasing unrest including numerous b-type earthquakes, accelerating ground deformation, increased vigour of fumaroles, gas emission | If trend of increasing unrest continues, eruption possible within 2 weeks | 5 June 1991 |
| 4 | Intense unrest, including harmonic tremor and/or many "long-period" ( = low-frequency) earthquakes | Eruption possible within 24 hours | 7 June 1991 17.00h |
| 5 | Eruption in progress | Eruption in progress | 9 June 1991 17.15h |

Stand-down procedures: In order to protect against "lull before the storm" phenomena, alert levels will be maintained for the following periods *after* activity decreases to the next lower level. From level **4** to level **3**, wait 1 week; from level **3** to level **2**, wait 72 hours.

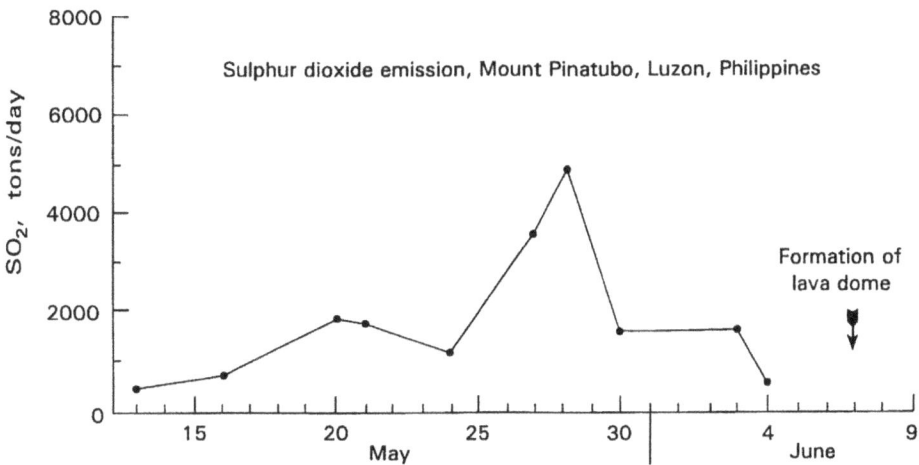

**Figure 14.6** Sulphur dioxide emission at Mount Pinatubo (Luzon, Philippines) during the period May–June 1991, as measured by the correlation spectrometer (COSPEC) (data from gvn 1991b). The sharp increase in $SO_2$ emission in late May was interpreted to indicate rise of magma into the volcano; the subsequent equally sharp decrease was inferred to reflect the plugging of the volcano conduit with continued magma rise (see text).

**Figure 14.7** (a) View from Clark Air Base of Mount Pinatubo's volcanic plume approximately 10 min after the start (08.51 h) of the 12 June 1991 eruption, the first of a series of powerful explosive eruptions culminating in the climactic event on 15 June, as seen from Clark Air Base. The eruption then increased in vigour, becoming strongest by mid-afternoon, when all seismic stations except the one at Clark were destroyed by pyroclastic debris. (US Air Force photograph by R. Lapointe.)

On the morning of 12 June, the first of several vigorous plinian eruptions began, producing a 19-km high ash column. A dozen or so strong explosions took place during the next two days (GVN 1991b, McCormick 1992), culminating in the climactic eruption that began at 05.55h (local time) on 15 June and lasted until early 16 June (Fig. 14.7). The cataclysmic event on 15–16 June – the second or third largest volcanic eruption in this century – expelled 5–7 km$^3$ (dense rock equivalent) of magma (Scott et al. 1991) and released about 20 Mton of SO$_2$ into the stratosphere (Bluth et al. 1992, McCormick 1992). A new 2-km diameter caldera was formed, centered about 1 km north of the former summit, which was lowered about 200 m. Pyroclastic flows and tephra deposits blanketed an area of more than 100 km$^2$ around the volcano, greatly altering local drainage patterns and provided ample starting materials for lahars.

Heavy rains associated with Typhoon Yunya, which passed only 50 km north of Pinatubo on the same day as the climactic eruption, triggered destructive debris flows and also wetted the ash that accumulated on roofs. The increased weight from wetting of the ash, together with intense seismicity accompanying the climactic eruption, resulted in extensive roof collapse, the principal cause of eruption-related deaths. By 16 June, an estimated 250000 people had fled or been evacuated to temporary emergency shelters. The climactic eruptions apparently totally cleared the conduit system, and through the end of July, ash was erupted almost continuously to heights of several kilometers above the volcano (Pinatubo Volcano Observatory Team 1991). Seismicity and tephra emission, however, have gradually diminished with time. During the rainy season (typically June–November), the lahar problem became considerably aggravated and lahar-related erosion of still-hot volcanic debris caused numerous rootless, secondary explosions and large secondary pyroclastic flows that travelled several kilometers from source. Destructive lahar activity again caused much damage and human suffering during the start of 1992 rainy season, and as of June 1992 about 70000 of the 250000 people displaced by the 1991 eruption and associated lahars still remain in the resettlement areas (GVN 1992c). As of this writing (October 1992), sporadic weak eruptive activity, including the growth of another dome, persists within the new caldera of Pinatubo. Good summary accounts of the Pinatubo activity through 1991 are given by Wolfe (1992) and Pierson (1992).

In retrospect, considering the size of the 15 June 1991 Pinatubo eruption and the large number of people affected, the death toll directly related to the climactic eruption was comparatively small, only slightly more than 300; rainfall-induced lahars since 1991 have killed more than 60 people and forced several hundred thousands to evacuate (GVN 1992d). The effectiveness of the scientific response to the reawakening of Pinatubo after a 600-year dormancy can be attributed in large part to:

(a) The rapid deployment of easily portable monitoring techniques, principally seismic and SO$_2$ emission; a major contributory factor in making the rapid response possible was the Volcano Disaster Assistance Program (VDAP), which is sponsored jointly by the USGS and the Office of Foreign Disaster Assistance (OFDA) of the US Department of State. The VDAP maintains a capability to respond quickly to volcanic crises with a scientific team and a cache of volcano-monitoring equipment designed for portability and durability (Miller et al. 1992).

(b) The initiation of intensive monitoring *immediately* after the onset of unrest – soon enough to allow the detection and evaluation of critical shifts in activity, thereby partly compensating for the lack of any previous baseline data.

(c) The near-real-time analysis of geologic and monitoring data that made possible reliable short-term forecasts (i.e. declaration of alert levels), which in turn prompted civil authorities to order timely evacuations of people at risk.

(d) The educational use during the weeks preceding the crisis of a 30-minute videotape produced for the International Association of Volcanology and Chemistry of the Earth's Interior (IAVCEI) by the late Maurice Krafft (killed at Unzen on 3 June 1991) – graphically illustrating the devastating impact of volcanic hazards – was instrumental in convincing the government officials and the populace that Pinatubo posed a very real danger.

In short, the Mount Pinatubo eruption involved a situation where no monitoring baselines existed, yet the mitigation of the hazards was for the most part successful because of the immediate and well-coordinated response by scientists and emergency-management authorities. The encouraging experience at Pinatubo represents a genuine success story for volcano monitoring and short-term forecasts, but volcanologists must not assume that similar successes can be achieved routinely in future volcanic crises. Additionally, several serious aircraft-volcanic ash encounters during the eruption highlighted the growing problem of volcanic-ash hazard to aviation and spurred productive interaction between volcanologists, pilots, aircraft manufacturers, and civil-aviation officials at a first-ever international conference on *Volcanic Ash and Aviation Safety* in August 1991 (Casadevall 1991, 1993) to explore the most effective means of mitigating such hazards.

### 14.3.2 Unrest at calderas

The most powerful but least frequent volcanic events are caldera-forming eruptions from large-volume silicic magmatic systems, such as those that took place at Yellowstone (Wyoming, USA) within the past 2 million years (Christiansen 1984), or at Toba (Sumatra, Indonesia) about 75000 years ago (Zen 1982, Rose & Chesner 1987). These eruptions involved magma volumes 1–2 orders of magnitude larger than that (about $50\,\mathrm{km}^3$ dense rock equivalent) of the largest known historical eruption in 1815 at Tambora (Lesser Sunda Islands, Indonesia). Within recorded history, the world fortunately has not experienced caldera-forming eruptions on the scale of Yellowstone, but there are no geological reasons to believe that such cataclysmic events cannot happen again. During the 1980s, significant unrest occurred at large caldera systems in three widely separated regions of the world: Long Valley (USA), Campi Flegrei (Italy) and Rabaul (Papua New Guinea). While not culminating in eruptive activity, the activity at these calderas spawned major volcanic crises that caused public anxiety and socioeconomic disruption.

The caldera crises at Long Valley and Campi Flegrei have been followed in great detail in many studies and will not be considered here: Long Valley (e.g. Bailey 1982, Hill 1984, Hill et al. 1985, 1990, 1991); and Campi Flegrei (e.g. Barberi et al. 1984a & b, Berrino et al. 1984, Dvorak & Mastrolorenzo 1991, Dvorak & Berrino 1991, Luongo & Scandone 1991). Because it was the only one for which a specific, if implicit, forecast was made, only the caldera crisis at Rabaul is reviewed below; see Newhall & Dzurisin (1988) for a comprehensive survey of historical unrest at large calderas of the world.

### 14.3.2.1 Rabaul, New Britain, Papua New Guinea 1983–85

Rabaul has produced three caldera-forming eruptions within the past 3500 years as well as several small- to moderate-sized explosive eruptions in historical time. Its most recent eruptive activity occurred in 1937–43 and killed about 500 people (Johnson & Threlfall 1985). Today, 70000 to 100000 people could be at risk from renewed activity, depending on eruption size.

Shortly after two magnitude-8.0 earthquakes in 1971 in the nearby New Britain Trench, the number of shallow earthquakes beneath Rabaul doubled to about 100 per month (McKee et al. 1984, 1985, Mori & McKee 1987, Mori et al. 1989). Over the next 10 years, seismic swarms occurred periodically, each becoming successively more energetic, accompanied by an overall uplift of the caldera's central part of about 1 m during the same period. In August 1983, the seismicity and rate of ground deformation increased abruptly. The number of earthquakes increased from a few hundred to a few thousand per month from August to September; by early 1984 the monthly count of earthquakes exceeded 10000 (Fig. 14.8). Correspondingly, tilt rates increased to 50–100 μrad per month, and maximum vertical displacement rates accelerated to 5–10 cm per month. At this point, the Rabaul Volcanological Observatory (RVO) expanded its monitoring to include horizontal deformation measurements, which also indicated horizontal strain in the central part of the caldera of 10–80 ppm, peaking in April 1984 (Archbold et al. 1988).

In late October 1983, after considering socioeconomic factors and the scientific information from RVO on the state of volcanic unrest, Rabaul government officials declared a "stage 2" alert, which implied that an eruption would occur within a *few* months. Such an alert level actually represented a *de facto* short-term forecast of a volcanic event, even though it was not so termed by the scientists or officials. The rate of seismicity and caldera deformation continued to increase for another six months after the declaration of the stage 2 alert, but the expected eruption did not happen. Beginning in June 1984, the level of caldera unrest declined rapidly (Fig. 14.8). On 22 November 1984, civil authorities downgraded the volcanic alert to "stage 1", which implied that the "anticipated eruption is not now expected before *several* months to a *few* years [italics added]" (SEAN 1984, p.9). The decline in activity continued through 1985, effectively terminating the seismo-deformational crisis that had seriously troubled scientists, officials, and the affected public for two anxious years. In the period from 1986 to the present, activity at Rabaul has fluctuated irregularly between low levels, but slightly elevated compared to pre-1983 rates.

The Rabaul case study highlights a well-recognized problem in studies of caldera unrest (e.g. Newhall & Dzurisin 1988): precursory activity does not always culminate in eruptions; the moving magma and (or) pressurized fluids may not breach the surface, resulting only in subsurface intrusions ("aborted eruptions") or hydrothermal pressurization phenomena. The monitoring data at Rabaul caldera clearly indicated that the underlying magmatic system was perturbed during 1983–85, most likely by the remobilization of pre-existing magma and (or) the intrusion of new magma. Yet, in the minds of the civil authorities and the public, the forecast

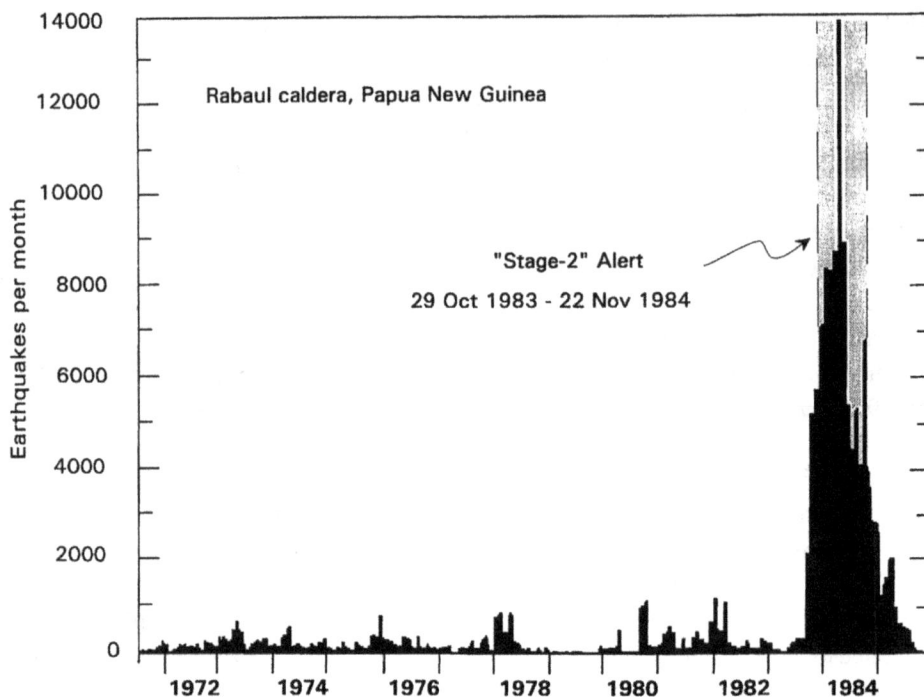

**Figure 14.8** Seismicity beneath Rabaul caldera, Papua New Guinea, from July 1971 through late 1985 (modified from Mori & McKee 1987, Fig. 1). The seismicity and ground deformation peaked in April 1984 and then began to decline sharply, returning to 1982 levels by May 1985. The shaded area indicates the period during which a government imposed "stage-2" alert was in effect (see text).

of an eruption at Rabaul by mid-1984 turned out to be a "false alarm". A major scientific challenge that would advance forecasting capability and materially assist volcanic emergency management would be to distinguish, if possible, "between precursory processes that cease with subsurface intrusion from those that culminate in eruptions" (Banks et al. 1989, p.78). Obviously, such a scientific breakthrough can only be achieved at volcanoes that have been monitored for many decades, coupled with intensive, shorter term monitoring of *many more* volcanoes immediately after they show signs of unrest.

## 14.4 Lessons for the future

As the earlier chapters in this book amply attest, volcano-monitoring techniques have improved tremendously in terms of instrumental sensitivity and precision, computerized data acquisition and telemetry, and near real-time analysis and interpretation compared to the state-of-the-art techniques reviewed two decades ago (UNESCO 1972). Indisputably, the data obtained by these improved instruments and techniques have refined our understanding of how (at least some) volcanoes work. But the case studies summarized above remind us that, except for a few outstanding successes, the state-of-the-art techniques for routine, reliable, and precise forecasting of volcanic events are yet to be fully developed and time tested. Although the case studies may be too few to be representative of all volcanoes being monitored or of all types of volcanic crises, they nonetheless drive home several obvious, but important lessons.

(a) *Too few active or potentially active volcanoes are being monitored adequately, and many are not monitored at all or only minimally*. If indeed the primary motivation to develop reliable eruption forecasting capability is to save human lives and minimize economic loss from volcanic hazards, then we must apply existing technology to many more volcanoes and continue to improve monitoring techniques (Tilling 1989b). Special attention should be paid to high-risk volcanoes in the densely populated developing countries and to those in remote regions underlying heavily travelled air routes. Perhaps the 1982 disaster at El Chichón might have been averted, or at least ameliorated, had the initial precursory signs been followed up by immediate on-site monitoring, as was done at Pinatubo in 1991. In the case of 1985 Ruiz tragedy, which was only marginally monitored for a few months prior to the disastrous eruption, one could ask: "Had there been longer term and better pre-eruption monitoring, would the scientists themselves have been more confident in the results obtained and thereby been more credible and persuasive in urging government officials to take timely action?"

(b) *The importance of baseline monitoring data cannot be over emphasized – the longer the pre-crisis monitoring period, the better*. Because the fundamental basis for forecasting depends on pattern recognition, it is essential to have baseline data to enable early detection of possible deviation from a volcano's normal behavior pattern. The 1975–77 Soufrière case study provides a classic example whereby the lack of baseline data not only precluded forecasts, but also aggravated the volcanic crisis. High-quality data from the sophisticated monitoring obtained well after the onset of the crisis were of little use in forecasting the course of the ongoing activity because they could not be compared with pre-crisis or early crisis monitoring data. To a lesser extent, inadequate baseline data also hampered the monitoring of the 1985 Ruiz eruption and the 1990–91 activity at Unzen. The Kelut case study illustrates that, for forecasting purposes, it is far preferable to have a good pre-crisis baseline than to have advanced electronic monitoring for a short time after the crisis has begun.

(c) *Rapidly deployable PC-based seismic monitoring techniques can be very effective in short-fuse emergency situations*. Although also hampered by the lack of geologic and monitoring baseline data, the pre-eruption monitoring conducted at Mount Pinatubo after the initial phreatic explosions nonetheless was able to obtain near-real-time information on the evolving state of the restless volcano. This would not have been possible had a conventional system of seismic data acquisition and processing been tried, simply because of insufficient time to establish the network and to analyze the data quickly for timely interpretation. The PC-based seismic system (Lee 1989) and the real-time seismic amplitude measurement (RSAM) system (Endo & Murray 1991) truly showed their versatility and utility at Pinatubo, at times under extremely inhospitable field conditions. PC-based seismic monitoring also proved to be highly useful during the 1989–90 eruption of Redoubt Volcano, Alaska (Brantley 1990).

(d) *Monitoring data, no matter how good or complete, do not guarantee "successful" eruption forecasts*. As well illustrated by the Rabaul case study, a dramatic departure from a volcanic system's normal behaviour, including orders-of-magnitude increases in precursory seismicity and ground deformation, may not culminate in eruptive activity. While the declaration of a stage 2 alert (short-term forecast) was certainly justified and reasonable in a scientific context (i.e. "successful") it was perceived publicly to be a failed forecast because the anticipated eruption did not materialize. It now appears that *none* of the seismic swarms at Rabaul in the early 1980s included any definitive long-period events (J. Mori 1992, personal communication) that typically accompany eruptions. Quantitative analysis of long-period seismic activity shows promise in improving eruption forecasting (e.g. Chouet et al. 1994). In hindsight, one wonders if, perhaps knowing such information at the time of the crisis, might have influenced the decision whether or not to declare the stage 2 Alert at Rabaul. In any case, to date available monitoring techniques cannot distinguish between a forthcoming eruption and an intrusion or other subsurface volcanic or hydrothermal processes. Until such distinction can be made analytically (if at all possible), volcanologists can only attempt to recognize possible relationships between actual eruptive or intrusive activity and corresponding precursory patterns. To have any chance of success, such an empirical approach requires many decades of comprehensive monitoring data and detailed eruption chronology, which unfortunately are unavailable for most volcanoes.

(e) *Monitoring data or eruption forecasts, no matter how timely or precise, have little practical consequence unless they can be communicated effectively to, and acted upon in a timely manner by, emergency-management authorities*. In the case of the 1985 Ruiz eruption, the mismanagement of the volcanic emergency by civil authorities was the dominant reason for the disaster, despite the fact that the monitoring data were marginal and no specific forecast was possible. Warnings were given in time by the scientists for the officials to implement protective measures including evacuation, and many lives could have been saved (Williams & Meyer 1988, CERESIS 1989). "In a nutshell, the government on the whole acted responsibly but was not willing to bear the economic or political costs of early evacuation or a false alarm... Armero could have produced no victims, and therein dwells its immense tragedy" (Voight 1990, p.349).

## Acknowledgements

In any brief review of a wide-ranging topic, coverage and emphasis necessarily must be selective and reflect the author's personal bias to some extent. The views summarized in this paper have been influenced by more than two decades of fruitful interaction with many colleagues in the international volcanologic community, who have contributed mightily to the advances in volcano monitoring and eruption forecasting. Many of these colleagues conducted the studies considered in this review. To them, and to many other colleagues whose works were not directly cited, I wish to express my appreciation for generously sharing their knowledge. Special thanks go to several of my USGS colleagues – Thomas J. Casadevell (Denver), John W. Ewert (Vancouver), Donald W. Peterson (Menlo Park), and Donald A. Swanson (Seattle) – for constructive reviews of an earlier draft of this paper.

## Note added in proof

Following the 1983–1985 crisis (Fig. 14.8), the seismic activity and ground deformation at Rabaul caldera, Papua New Guinea, remained at relatively low levels through mid-September 1994. Then, following only 27 hours of premonitory seismic activity, including a magnitude 5.1 event, explosions began at Vulcan and Tavurvur, the two vents on opposite sides of the caldera that have been the sites of previous historical eruptions. The eruptions produced ash clouds ($\geq$18 km high), tephra falls, pyroclastic surges and small lava flows; Rabaul Town suffered extensive damage, principally from tephra accumulation and roof collapse. Over 50 000 people have been displaced by the eruptions, but, fortunately, only 5 deaths have been reported. The eruption at Vulcan ceased on 2 October, but weak activity was continuing at Tavurvur through early November 1994.

## References

Archbold, M. J., C. O. McKee, B. Talai, J. Mori, P. De St Ours 1988. Electronic distance measuring network during the Rabaul seismicity/deformational crisis of 1983–85. *Journal of Geophysical Research* **93**, 123–36.

Bailey, R. A. 1982. Mammoth Lakes earthquakes and ground uplift: Precursors to possible volcanic activity?. In *United States Geological Survey Yearbook Fiscal Year 1982*, 4–13. Reston, Virginia: USGS.

Banks, N. G., R. I. Tilling, D. D. Harlow, J. W. Ewert 1989. Volcano monitoring and short-term forecasts. In *Short courses in geology, vol. 1: Volcanic hazards*, R. I. Tilling (ed.), 51–80. Washington, DC: American Geophysical Union.

Banks, N. G., C. Carvajal, H. Mora, and E. Tryggvason 1990. Deformation monitoring at Nevado del Ruiz, Colombia – October 1985–March 1988. *Journal of Volcanology and Geothermal Research* **41**, 269–95.

Barberi, F., G. Corrado, R. Innocenti, G. Luongo 1984a. Phlegraean Fields 1982–1984: Brief chronicle of a volcano emergency in a densely populated area. *Bulletin Volcanologique* **47**, 175–85.

Barberi, F., D. Hill, F. Innocenti, G. Luongo, M. Treuil (eds) 1984b. The 1982–1984 Bradyseismic crisis at Phlegraean Fields (Italy). *Bulletin Volcanologique* **47**, 173–412.

Barberi, F., M. Martini, M. Rosi 1990. Nevado del Ruiz volcano (Colombia): pre-eruption observations and the November 13 1985 catastrophe. *Journal of Volcanology and Geothermal Research* **42**, 1–12.

Berrino, G., G. Corrado, G. Luongo, B. Toro 1984. Ground deformation and gravity changes accompanying the 1982 Pozzuoli uplift. *Bulletin Volcanologique* **47**, 187–200.

Bluth, G. J. S., S. D. Doiron, C. C. Schnetzler, A. J. Krueger, L. S. Walter 1992. Global tracking of the $SO_2$ clouds from the June 1991 Mount Pinatubo eruptions. *Geophysical Research Letters* **19**, 151–4. [This issue contains a collection of 18 papers on atmospheric studies in a special section of this journal.]

Brantley, S. R. (ed.) 1990. *United States Geological Survey Circular 1061, The eruption of Redoubt Volcano, Alaska, December 14 1989–August 31 1990*. Washington DC: United States Government Printing Office.

Brantley, S. R., L. Topinka 1984, Volcanic studies at the United States Geological Survey's David A. Johnston Cascades Volcano Observatory, Vancouver, Washington. *Earthquake Information Bulletin* **16**, 44–122.

397

Canul, R. F., V. S. Rocha 1981. Informe geológico de la zona geotérmica de "El Chichonal," Chiapas. Unpublished report of the Geothermal Department of the Comisión Federal de Electricidad, Morelia, Michoacán, México. (in Spanish).

Casadevall, T. J. (ed.) 1991. *United States Geological Survey Circular 1065, Program and Abstracts, First International Symposium on Volcanic Ash and Aviation Safety, Seattle Washington, July 8–12 1991*. Washington DC: United States Government Printing Office.

Casadevall, T. J. (ed.) 1993, *United States Geological Survey Bulletin 2047, Volcanic Ash and Aviation Safety, Proceedings Volume, First International Symposium, Seattle, Washington, 8–12 July 1991*. Washington DC: United States Government Printing Office.

Casadevall, T. J., L. Pardyanto, H. Abas, Tulus 1989. The 1988 eruption of Banda Api Volcano, Maluku, Indonesia. *Geologi Indonesia (Journal of the Indonesian Association of Geologists)* 12, 603–35.

CERESIS 1990. *Riesgo volcánico: Evaluación y mitigación en América Latina: Aspectos sociales, institucionales y científicos*. Lima, Perú: Centro Regional de Sismología para América del Sur (CERESIS) (in Spanish).

Chadwick, W. W., Jr, R. J. Archuleta, D. A. Swanson 1988. The mechanics of ground deformation precursory to dome-building extrusions at Mount St Helens 1981–82. *Journal of Geophysical Research* 93, 4351–66.

Chouet, B., R. A. Page, C. D. Stephens, J. C. Lahr 1994. Precursory swarms of long-period events at Redoubt Volcano (1989–1990), Alaska: Their origin and use as a forecasting tool. *Journal of Volcanology and Geothermal Research* 62, 95–135.

Christiansen, R. L. 1984. Yellowstone magmatic evolution: Its bearing on understanding large-volume explosive volcanism. In *Explosive volcanism: inception, evolution and hazards*, Geophysics Study Committee (National Research Council), 84–95. Washington DC: National Academy Press.

Civetta, L., P. Gasparini, G. Luongo, A. Rapolla (eds) 1974. *Developments in solid earth geophysics, vol. 6: Physical volcanology*. Amsterdam: Elsevier.

Crandell, D. R., D. R. Mullineaux, M. Rubin 1975. Mount St Helens: Recent and future behavior. *Science* 187, 438–44.

Decker, R. W. 1986. Forecasting volcanic eruptions. *Annual Reviews Earth and Planetary Sciences* 14, 267–91.

Djurmarma W., Achmad 1991. Some studies of volcanology, petrology and structure of Mt Kelut, East Java, Indonesia. PhD thesis, Victoria University of Wellington, New Zealand.

Duffield, W. A., R. I. Tilling, R. Canul 1984. Geology of El Chichón Volcano, Chiapas, Mexico. *Journal of Volcanology and Geothermal Research* 20, 117–32.

Dvorak, J. J. & G. Berrino 1991. Recent ground movement and seismic activity in Campi Flegrei, southern Italy: Episodic growth of a resurgent dome. *Journal of Geophysical Research* 96, 2309–23.

Dvorak, J. J. & G. Mastrolorenzo 1991. *Special Paper 263, The mechanisms of recent vertical crustal movements in Campi Flegrei caldera, southern Italy*. Boulder, Colorado: Geological Society of America.

Endo, E. T. & T. L. Murray 1991. Real-time seismic amplitude measurement (RSAM): A volcano monitoring and prediction tool. *Bulletin of Volcanology* 53, 533–45.

Endo, E. T., S. D. Malone, L. L. Noson, C. S. Weaver 1981. Locations, magnitudes, and statistics of the March 20–May 18 earthquake sequence. In *United States Geological Survey Professional Paper 1250, The 1980 eruptions of Mount St Helens, Washington*, P. W. Lipman & D. R. Mullineaux (eds), 93–107. Washington DC: United States Government Printing Office.

Ewert, J. W. & D. A. Swanson (eds) 1992. *United States Geological Survey Bulletin 1966, Monitoring volcanoes: Techniques and strategies used by the staff of the Cascades Volcano Observatory 1980–1990*. Washington DC: United States Government Printing Office.

Feuillard, M., C. J. Allègre, G. Brandeis, R. Gaulon, J. L. Le Mouel, J. C., Mercier, J. P. Pozzo, M. P. Semet 1983. The 1975–1977 crisis of la Soufrière de Guadeloupe (FWI): A still-born magmatic eruption. *Journal of Volcanology and Geothermal Research* 16, 317–34.

Fiske, R. S. 1984. Volcanologists, journalists, and the concerned local public: A tale of two crises in the eastern Caribbean. In *Explosive volcanism: inception, evolution and hazards*, Geophysics Study Committee (National Research Council), 170–6. Washington DC: National Academy Press.

Fiske, R. S., T. Simkin, E. A. Nielsen (eds) 1987. *The Volcano Letter 1925–1955*. Washington DC: Smithsonian Institution Press. [Compiled and reprinted; originally published by the Hawaiian Volcano Observatory.]

Galindo, I., D. J. Hofmann, M. P. McCormick (eds) 1984. Proceedings of the Symposium on Atmospheric Effects of the 1982 Eruptions of El Chichón Volcano, XVIII General Assembly of the International Union of Geodesy and Geophysics (IUGG), August 1983, Hamburg, Germany. *Geofísica Internacional* **23**, 113-304.

GVN 1991a. Unzen (Kyushu). *Bulletin, Global Volcanism Network (GVN)* **16**, 8–11. GVN 1991b. Pinatubo (Luzon). *Bulletin, Global Volcanism Network (GVN)* **16**, 2–8.GVN 1992a. Unzen (Kyushu). *Bulletin, Global Volcanism Network (GVN)* **17**, 11–12. GVN 1992b. Unzen (Kyushu). *Bulletin, Global Volcanism Network (GVN)* **17**, 10–11.

GVN 1992c. Pinatubo (Luzon). *Bulletin, Global Volcanism Network (GVN)* **17**, 4–5.

GVN 1992d. Pinatubo (Luzon). *Bulletin, Global Volcanism Network (GVN)* **17**, 12. Hall, M. L. 1990. Chronology of the principal scientific and governmental actions leading up to the November 13 1985 eruption of Nevado del Ruiz. *Journal of Volcanology and Geothermal Research* **42**, 101–15.

Harlow, D. H., R. S. Punongbayan, J. A. Power, C. G. Newhall, R. P. Hoblitt, A. B. Lockhart, T. L. Murray, E. W. Wolfe, J. W. Ewert 1991. Seismic activity and forecasting of the climactic eruption of Pinatubo Volcano, Luzon, Philippines on June 15 1991 (abstract). *Eos, Transactions, American Geophysical Union* **72**, 61.

Havskov, J., S. De la Cruz-Reyna, S. K. Singh, F. Medina, C. Gutiérrez 1983. Seismic activity related to the March–April 1982 eruptions of El Chichón Volcano, Chiapas, Mexico. *Geophysical Research Letters* **10**, 293–6.

Herd, D. G., & the Comité de Estudios Vulcanológicos 1986. The 1985 Ruiz Volcano disaster. *Eos, Transactions, American Geophysical Union* **67**, 457–60.

Hill, D. P. 1984. Monitoring unrest in a large silicic caldera, the Long Valley-Inyo Craters Volcanic Complex in east-central California. *Bulletin Volcanologique* **47**, 371–96.

Hill, D. P., R. E. Wallace, R. S. Cockerham 1985. Review of evidence on the potential for major earthquakes and volcanism in the Long Valley-Mono Craters-White Mountains region of eastern California. *Earthquake Prediction Research* **3**, 571–94.

Hill, D. P., W. L. Ellsworth, M. J. S. Johnston, J. O. Langbein, D. H. Oppenheimer, A. M. Pitt, P. A. Reasenberg, M. L. Sorey, S. R. McNutt 1990. The 1989 earthquake swarm beneath Mammoth Mountain, California: An initial look at the 4 May through 30 September activity. *Bulletin, Seismological Society of America* **80**, 325–39.

Hill, D. P., M. J. S. Johnston, J. O. Langbein, S. R. McNutt, C. D. Miller, C. E. Mortensen, A. M. Pitt, S. Rojstaczer 1991. *United States Geological Survey Open-File Report 91-270, Response plans for volcanic hazards in the Long Valley caldera and Mono Craters area, California*. Washington DC: United States Government Printing Office.

Jackson, D. B., J. Kauahikaua, C. J. Zablocki 1985. Resistivity monitoring of an active volcano using the controlled-source electromagnetic technique, Kilauea, Hawaii. *Journal of Geophysical Research* **90**, 545–55.

Johnson, R. W. & N. A. Threlfall 1985. *Volcano Town: The 1937–43 eruptions at Rabaul*. Bathurst, Australia: Robert Brown.

Katili, J. A. & A. Sudradjat 1984. The devastating 1983 eruption of Colo Volcano, Una-Una Island, Central Sulawesi, Indonesia. *Geologisches Jahrbuch* **A75**, 27–47.

Kusumadinata, K. 1979. *Data Dasar Gunungapi Indonesia (Basic data of Indonesian volcanoes)*. Bandung: Volcanological Survey of Indonesia. (In Indonesian).

Lee, W. H. K. (ed.) 1989. Toolbox for seismic data acquisition, processing, and analysis. International Association of Seismology and Physics of the Earth's Interior (IASPEI) Software Library. *Seismological Society of America* **1**.

Lipman, P. W. & D. R. Mullineaux (eds) 1981. *The 1980 eruptions of Mount St Helens, Washington. United States Geological Survey Professional Paper 1250*. Washington DC: United States Government Printing Office.

Lipman, P. W., J. G. Moore, D. A. Swanson 1981. Bulging of the north flank before the May 18 eruption--Geodetic data. In *United States Geological Survey Professional Paper 1250, The 1980 eruptions of Mount St Helens, Washington*, P. W. Lipman & D. R. Mullineaux (eds), 143–55. Washington DC: United States Government Printing Office.

Luhr, J. F. & J. C. Varekamp (eds) 1984. Special Issue on El Chichón Volcano, Chiapas, Mexico. *Journal of Volcanology and Geothermal Research* **23**, 1–191.

Luongo, G. & R. Scandone (eds) 1991. Special Issue on Campi Flegrei. *Journal of Volcanology and Geothermal Research* **48**, 1–227.

399

McClelland, L., T. Simkin, M. Summers, E. Nielsen, T. C. Stein (eds) 1989. Ruiz. In *Global Volcanism 1975–1985*, 517–23. Englewood Cliffs, New Jersey: Prentice-Hall.

McCormick, M. P. (ed.) 1992. Selected papers on the stratospheric and climatic effects of the 1991 Mt. Pinatubo Eruption: An initial assessment: *Geophysical Research Letters* **19**, 149–218. [A collection of 18 papers on atmospheric studies in a Special Section of this journal.]

McGuire, W. J. 1991, Monitoring active volcanoes. *Journal of the Geological Society* **148**, 516–62 [And associated papers on monitoring, 563–93.]

McKee, C. O., P. L. Lowenstein, P. De St Ours, B. Talai, I. Itikarai, J. J. Mori 1984. Seismic and ground deformation crises at Rabaul caldera: prelude to an eruption? *Bulletin Volcanologique* **47**, 397–411.

McKee, C. O., R. W. Johnson, P. L. Lowenstein, S. J. Riley, R. J. Blong, P. De St Ours, B. Talai 1985. Rabaul Caldera, Papua New Guinea: Volcanic hazards, surveillance, and eruption contingency planning. *Journal of Volcanology and Geothermal Research* **23** 195–237.

Miller, C. D., D. R. Mullineaux, D. R. Crandell 1981. Hazards assessments at Mount St Helens. In *United States Geological Survey Professional Paper 1250, The 1980 eruptions of Mount St Helens, Washington*, P. W. Lipman & D. R. Mullineaux (eds), 789–802. Washington DC: United States Government Printing Office.

Miller, C. D., J. W. Ewert, A. B. Lockhart, B. N. Heyman 1992. The USGS/OFDA Volcano Disaster Assistance Program (abstract). *Eos, Transactions, American Geophysical Union* **73**, 68.

Mori, J. & C. McKee 1987. Outward dipping ring fault structure at Rabaul Caldera as shown by earthquake locations. *Science* **235** 193–5.

Mori, J., C. McKee, I. Itikarai, P. Lowenstein, P. De St Ours, B. Talai 1989. Earthquakes of the Rabaul seismo-deformational crisis September 1983 to July 1985: Seismicity on a caldera ring fault. In *IAVCEI proceedings of volcanology, Volcano hazards: assessment and monitoring*, J. H. Latter (ed.), 429–62. New York: Springer.

Murray, T. L. 1990. *United States Geological Survey Open-File Report 90-56, A user's guide to the PC-based time-series data-management and plotting program BOB*. Washington DC: United States Government Printing Office.

Nakada, S. 1992. Volcanic Hazard at Unzen, Japan: (1) 1990–1992 eruption of Unzen Volcano. *Landslide News* **6**, 2–4.

Newhall, C. G. & D. Dzurisin 1988. *United States Geological Survey Bulletin 1855, vol. 1: Historical unrest at large calderas of the world*. Washington DC: United States Government Printing Office.

Omori, F. 1913. The Usu-san eruption and the elevation phenomena. II. *Bulletin Imperial Earthquake Investigation Committee* **5**, 105–7. [Comparison of bench mark heights in the base district before and after the eruption.]

Omori, F. 1914. The Sakura-jima eruptions and earthquakes. *Bulletin Imperial Earthquake Investigation Committee* **8**, 525.

Pardyanto, L. 1968. Gunung Kelut, Geologi Aktivitas dan Pengawa Sannja. *Report, Dept. Teknik Geologi, Institut Teknologi Bandung (ITB)* **87**. [In Bahasa Indonesia.]

Pierson, T. C. 1992. Rainfall-triggered Lahars at Mt Pinatubo, Philippines, following the June 1991 eruption. *Landslide News* **6**, 6–9.

Pinatubo Volcano Observatory Team 1991. Lessons from a major eruption: Mt Pinatubo, Philippines. *Eos, Transactions, American Geophysical Union* **72**, 545, 552–3, 555.

Punongbayan, R. S. & R. I. Tilling 1989. Recent case histories. In *Short courses in geology, vol. 1: Volcanic hazards*, R. I. Tilling (ed.), 81–101. Washington DC: American Geophysical Union.

Rampino, M. R. & S. Self 1984. The atmospheric effects of El Chichón. *Scientific American* **253**, 48–57.

Rose, W. I. & Chesner, C. A. 1987. Dispersal of ash in the great Toba eruption, 75 ka. *Geology* **15**, 913–7.

Scott, W. E. 1989. Volcanic-hazard zonation and long-term forecasts. In *Short courses in geology, vol 1: Volcanic hazards*, R. I. Tilling (ed.), 25–49. Washington DC: American Geophysical Union.

Scott, W. E., R. P. Hoblitt, J. A. Daligdig, G. Besana, B. S. Tubianosa 1991. 15 June 1991 pyroclastic deposits at Mount Pinatubo, Philippines (abstract). *Eos, Transactions, American Geophysical Union* **72**, 61–2.

SEAN 1976. Soufrière de Guadeloupe. *Bulletin, Scientific Event Alert Network (SEAN)* **1**, 3–4.

SEAN 1982. El Chichón (Mexico). *Bulletin, Scientific Event Alert Network (SEAN)* **7**, 2–6 & 23.

SEAN 1984. Rabaul Caldera (Papua New Guinea). *Bulletin, Scientific Event Alert Network (SEAN)* **9**, 9.

Shimozuru, D. 1991. In the volcano's shadow: *Nature* **353**, 295–6.

Stoiber, R. E. & S. N. Williams 1990. Monitoring active volcanoes and mitigating volcanic hazards: The case for including simple approaches. *Journal of Volcanology and Geothermal Research* **42**, 129–49.

Sudradjat, A. 1991. A preliminary account of the 1990 eruption of the Kelut Volcano. *Geologisches Jahrbuch* **A127**, 447–62.

Suzuki, H. & T. Furuya 1992. Volcaic Hazard at Unzen, Japan: (2) Hazard mapping at Unzen Volcano and the 1792 Mayuyama landslide. *Landslide News* **6**, 5–6.

Swanson, D. A. 1990. A decade of dome growth at Mount St Helens 1980–90. *Geoscience Canada* **17**, 154–7.

Swanson, D. A. 1992. The importance of field observations for monitoring volcanoes, and the approach of "keeping monitoring as simple as practical". In *United States Geological Survey Bulletin 1966, Monitoring volcanoes: Techniques and strategies used by the staff of the Cascades Volcano Observatory 1980–1990*, J. W. Ewert & D. A. Swanson (eds), 219–23. Washington DC: United States Government Printing Office.

Swanson, D. A., T. J. Casadevall, D. Dzurisin, S. D. Malone, C. G. Newhall, C. S. Weaver 1983. Predicting eruptions at Mount St Helens, June 1980 through December 1982. *Science* **221**, 1369–76.

Swanson, D. A., T. J. Casadevall, D. Dzurisin, R. T. Holcomb, C. G. Newhall, S. D. Malone, C. S. Weaver 1985. Forecasts and predictions of eruptive activity at Mount St Helens, USA: 1975-1984. *Journal of Geodynamics* **3**, 397–423.

Swanson, D. A., D. Dzurisin, R. T. Holcomb, E. Y. Iwatsubo, W. W. Chadwick, Jr, T. J. Casadevall, J. W. Ewert, C. C. Heliker 1987. Growth of the lava dome at Mount St Helens, Washington (USA) 1981–1983. *In Special Paper 212, The emplacement of silicic domes and lava flows*, J. H. Fink (ed.), 1–16. Boulder, Colorado: Geological Society of America.

Tazieff, H. & J-C. Sabroux (eds) 1983. *Forecasting volcanic events: Developments in volcanology, vol. 1.* Amsterdam: Elsevier.

Tilling, R. I. (ed.) 1989a. *Short courses in geology, vol 1: Volcanic hazards.* Washington DC: American Geophysical Union.

Tilling, R. I. 1989b. Volcanic hazards and their mitigation: Progress and problems. *Reviews of Geophysics* **27**, 237–69.

Tilling, R. I. 1992. Volcanic-hazards mitigation and population growth: *Volcano Quarterly* **1**, 13–6.

Tilling, R. I., M. Rubin, H. Sigurdsson, S. Carey, W. A. Duffield, W. I. Rose 1984. Holocene eruptive activity of El Chichón Volcano, Chiapas, Mexico. *Science* **224**, 747–9.

Tilling, R. I., L. Topinka, D. A. Swanson 1990. *United States Geological Survey General Interest Publication, Eruptions of Mount St Helens: Past, present, and future* (revised edition). Reston, Virginia: USGS.

Van Ruymbeke, M. & N. d'Oreye (eds) 1991. *Proceedings of the Workshop: Geodynamical Instrumentation applied to Volcanic Areas,* 1–3 October 1991. Walferdange, Grand-Duchy of Luxembourg. [Available through the Observatoire Royal de Belgique, Bruxelles, Belgique.]

UNESCO 1972. *The surveillance and prediction of volcanic activity: A review of methods and techniques.* Paris: UNESCO.

Voight, B. 1990. The 1985 Nevado del Ruiz volcano catastrophe: Anatomy and retrospection. *Journal of Volcanology and Geothermal Research* **44**, 349–86. [Corrected reprinting of article originally in **42**, 151–88.]

Voight, B. & R. R. Cornelius 1991. Prospects for eruption prediction in near real-time. *Nature* **350**, 695–8.

Williams, S. N. (ed.) 1990a. Special Issue on Nevado del Ruiz, Colombia, I. *Journal of Volcanology and Geothermal Research* **41**, 1–377.

Williams, S. N. (ed.) 1990b. Special Issue on Nevado del Ruiz, Colombia, II. *Journal of Volcanology and Geothermal Research* **42**, 1–224.

Williams, S. N. & H. Meyer 1988. A model of Nevado del Ruiz Volcano, Colombia. *Eos, Transactions, American Geophysical Union* **69**, 1554–6.

Wolfe, E. W. 1992. The 1991 eruptions of Mount Pinatubo, Philippines. *Earthquakes and Volcanoes* **23**, 5–35.

Wood, H. O. 1915. The seismic prelude to the 1914 eruption of Mauna Loa. *Bulletin, Seismological Society of America* **5**, 39–50.

WOVO 1982. Announcement: World Organization of Volcano Observatories (WOVO). *Journal of Volcanology and Geothermal Research* **12**, 181–6.

WOVO 1991-92. *Directory of the World Organization of Volcano Observatories (WOVO) 1991–92.* Reykjavik, Iceland: WOVO.

Wright, T. L. & D. A. Swanson 1987. The significance of observations at active volcanoes: A review and annotated bibliography of studies at Kilauea and Mount St Helens. In *The Geochemical Society, Special Publication No. 1, Magmatic Processes: Physicochemical Principles*, B. O. Mysen, (ed.), 231–40. Lancaster, Pennsylvania: Lancaster Press.

Yokoyama, I., R. I. Tilling, R. Scarpa 1984. *Report FP/2106-82-01 (2286), International mobile early-warning system(s) for volcanic eruptions and related seismic activities*. Paris: UNESCO.

Yanagi, T., H. Okada, K. Ohta (eds) 1992. *Unzen volcano: The 1990–1992 eruption*. Fukuoka, Japan: Nishinippon & Kyushu University Press.

Zen, M. T. 1982. Toba resurgent cauldron: Volcano-tectonic evolution and magmatic cycles. *Bulletin Dept. Teknik Geologi, Institut Teknologi Bandung (ITB)* **7**, 27–43.

# 15 Prospects for volcano surveillance

W. J. McGuire

## 15.1 An approach for the 21st century

It is estimated that more than 10% of the world's population currently lives in close proximity to an active volcano (Peterson 1986). These numbers are certain to increase as the populations of developing countries grow. By the year 2000, rapidly increasing urbanization combined with this population rise will result in over 100 cities with populations exceeding 2 million (Bilham 1988). Nearly half of these cities will be located within 200 km of a plate boundary where both volcanic and seismic activity are concentrated (Fig. 15.1). In terms of increased risk from the effects of volcanic eruptions, the Pacific rim regions of south and central America, and southeast Asia will be most susceptible.

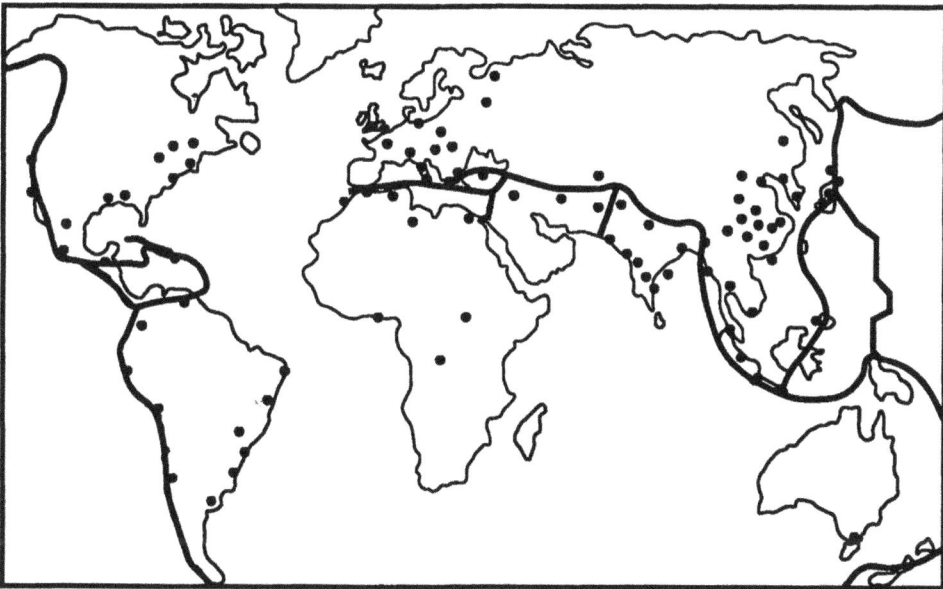

**Figure 15.1**  By the end of the century there will be over 100 cities with populations exceeding 2 million. Forty per cent of these major urban centres (solid circles) will lie within 200 km of a plate boundary or other tectonically active region where volcanic activity is a common hazard. Bold lines represent convergent or strike-slip boundaries (adapted from Bilham 1988).

In the previous chapter, Tilling points out that although volcano monitoring has become more sophisticated and has achieved some success over the past decade, important lessons for the future remain to be learnt. In particular, it is quite clear that volcano surveillance on a global scale is at present inadequate and functioning on an *ad hoc* basis, a situation which must be improved if the future impact of volcanic eruptions is to be minimized. At the same time, every effort must be made to improve communications between scientists and the authorities, and to raise public awareness about volcanic eruptions and their effects (Tilling 1989). Failure to achieve the latter may mean that even the most precise eruption prediction will prove largely worthless.

The report, *Reducing volcanic disasters in the 1990s*, by the International Association of Volcanology and the Chemistry of the Earth's Interior (IAVCEI) UN International Decade for Natural Disaster Reduction (IDNDR) Task Group (1990), provides an excellent blueprint for more effective volcano surveillance and other initiatives designed to mitigate the effects of erupting volcanoes over the remainder of this century and into the next. A recommendation of the report has already been adopted by the United Nations, which has launched, as part of its International Decade for Natural Disaster Reduction initiative, the Decade Volcano Demonstration Projects. At the International Union of Geodesy and Geophysics (IUGG) General Assembly in Vienna in 1991, nine volcanoes were selected as "decade volcanoes", and thereby targeted for intensive, integrated, and multidisciplinary research involving international co-operation, and the number has since increased to twelve. Seven of these volcanoes are located in developing countries, two in the USA, two in Japan, and one in Italy (Fig. 15.2, Table

**Figure 15.2** The main lake-filled crater of Taal volcano in the Philippines. Taal has been selected for special study over the next 10 years, as part of the Decade of Volcano Demonstration Projects.

404

**Table 15.1** Volcanoes selected for special study during the United Nations International Decade for Natural Disaster Reduction (IDNDR).

| Decade volcanoes | |
| --- | --- |
| Colima (Mexico) | Sakurajima (Japan) |
| Galeras (Columbia) | Santa Maria (Guatemala) |
| Mauna Loa (USA) | Ta al (Philippines) |
| Merapi (Indonesia) | Ulawan (Papua New Guinea) |
| Mount Rainier (USA) | Unzen (Japan) |
| Nyirogongo (Zaire) | Vesuvius (Italy) |

| Laboratory volcanoes |
| --- |
| Etna (Sicily, Italy) |
| Furnas (Azores, Portugal) |
| Krafla (Iceland) |
| Piton de la Fournaise (Réunion Island, France) |
| Teide (Teneriffe, Spain) |
| Santorini (Greece) |

15.1). To a large extent, the success of the decade volcano concept will rely upon the level of funding available, and this in turn will be strongly dependent on the effectiveness of lobbying by concerned scientists and other parties, in their own countries and on the international platform. Although no long-term funding had been arranged at the time of writing, the decision of the IDNDR committee of the International Council of Scientific Unions (ICSU) to designate the decade volcano concept as a *Spearhead Project* worthy of govermental and United Nations support provides some cause for optimism.

Operating alongside the Decade Volcano Demonstration Projects is a European initiative designated the European Laboratory Volcano concept. This was proposed by the European Volcanology Network (now the European Volcanology Programme) of the European Science Foundation, with the aim of targeting a number of European volcanoes for special study throughout the 1990s. In 1992 and 1993, under the auspices of its Environment research initiative, the European Commission allocated significant funds to support multidisciplinary studies on five volcanoes (Table 15.1), Mount Etna (Sicily) (Fig. 15.3), Furnas (the Azores), Piton de la Fournaise (Réunion Island), Teide (Teneriffe), and Santorini (Greece), and in 1995 Krafla (Iceland) will be added to this list.

The idea of targeting small numbers of active volcanoes for special study is a sensible one, given the constraints imposed by limited funding and the paucity of available expertise. The overriding rationale being that much of the knowledge gained from the decade and laboratory volcano studies will be directly and immediately applicable to many other potentially hazardous volcanoes around the world.

Within the previously stated framework of rapidly growing populations in areas at risk from volcanic activity, it is unlikely that specialist studies of limited numbers of volcanoes will, on their own, reduce the detrimental impact of eruptions on society. In addition to proposing the decade volcano concept, however, the IAVCEI IDNDR Task Group (1990) also presented a list of further projects which would be required to be undertaken if the risk of future volcanic

**Figure 15.3** Mount Etna in Sicily is one of the most active volcanoes in the world. Largely for this reason, it has been chosen as one of the European "laboratory" volcanoes, and will constitute a major focus of attention for European volcanologists at least until the end of the century.

disasters is to be minimized. Although designed for implementation during the current decade, the proposed projects are likely to be largely open ended, and will continue to form a major part of a strategy for volcanic disaster reduction well into the 21st century. Among other initiatives, the Task Group advocate special attention be paid to the following: hazard and risk mapping of increasing numbers of potentially dangerous volcanoes; establishment of baseline monitoring programmes on currently unmonitored volcanoes which pose a threat to the local population; raising the level of public awareness with regard to volcanoes and their associated hazards; improving communications, both between scientists, and with public officials and authorities; volcanological training for scientists from the developing countries; development of low-cost monitoring instrumentation; and optimum use of satellite monitoring.

## 15.2 Future technological developments

To a large extent, the technology currently available provides the capability to monitor effectively any volcano in the world, at least in terms of being able to provide warning of reactivation, make some forecast or prediction about future eruptive activity, and provide advice on how the authorities and local population should respond. Only a combination of insufficient funding, limited expertise, and a lack of political will, prevents available technologies

406

being utilized now at all the world's potentially destructive volcanoes. Over the next decade, and into the next century, the capabilities of instruments currently used in volcano monitoring programmes are certain to increase. For example, improvements in a future satellite global positioning system should eventually allow small (1–2 cm), vertical displacements in the ground surface to be detected, thereby making satellite precise-levelling a plausible technique. The Global Positioning Systym (GPS) is likely to be increasingly used to regularly and repeatedly monitor both horizontal and vertical ground-surface changes in response to magma intrusion. This will be accomplished using permanently sited receivers, and the results telemetered to an observatory for data interpretation. Ground-based deformation-monitoring techniques are also likely to follow this pattern with, for example, automated electronic distance meters (AEDMs) repeatedly measuring the distances to a number of strategically placed, permanently established reflectors, and relaying the data to a central observatory (see Ch. 4). Monitoring of submarine volcanic eruptions, important because of the potential tsunamis threat they pose, should become more effective with the development and use of more advanced ocean-bottom tiltmeters (Staudigel et al. 1991), and hydrophones (Talandier 1989, McCreery et al. 1993). Improvements in thermal monitoring, particularly by means of satellite-based instrumentation, may allow the total energy budget of a volcano to be more tightly constrained, and enable changes to be recognized (see Ch. 7). This might offer the possibility of recognizing the optimum conditions for reactivation well in advance of the onset of other premonitory behaviour, providing time for early upgrading of the monitoring network. Larger and better seismic arrays, combined with more sophisticated data handling and interpretation techniques, should allow seismic tomography (Evans & Zucca 1988, Foulger & Toomey 1989, Hirn et al. 1991) to become a more powerful tool in constructing increasingly accurate pictures of the density distributions beneath and within active volcanoes, allowing magma bodies to be located, and their form determined.

The role of satellite-based monitoring of active volcanoes is set to expand considerably over the next ten years (Mouginis-Mark & Francis 1992) following the launch of several new instruments and orbital platforms. Satellites such as the European Remote Sensing Satellites (ERS-1 and ERS-2) and Japanese Earth Resources Satellite-1 (JERS-1) carry instruments capable of observing in the radar and infrared regions of the spectrum which will have potential in monitoring, respectively, topographic and thermal changes at active volcanoes. The potential applications to volcano monitoring of the Japanese satellite are noted in Oppenheimer et al. (1993). The radar interferometry technique used by Massonnet et al. (1993) to determine the displacement field associated with the Landers earthquake in California, using ERS-1 imagery, is already being applied to active volcano terrains allowing production of high-resolution volcano deformation maps (Rossi, 1994). Other radar systems are due to be carried on the Canadian Radarsat and by the Space Shuttle. A number of total ozone mapping spectrometer (TOMS) instruments capable of monitoring volcanic-derived $SO_2$ will be launched on Russian, Japanese and US satellites during the 1990s, and sensitivities should be significantly improved with the operation of the global ozone monitoring experiment (GOME) spectrometer on the European ERS-2 satellite from 1995.

Volcanology will undoubtedly benefit enormously from the establishment of the Earth Observing System (EOS), a multisatellite global monitoring system which is currently planned

**Table 16.3** Earth Observing System (EOS) instruments and their application to volcanology (updated from Mouginis-Mark et al. 1991).

| Instrument | Application |
|---|---|
| *Surface phenomena and plumes* | |
| GLAS | Deformation rates (laser ranging) and topography of eruption plumes and determination of plume height (altimeter) |
| ASTER | Energy output from very high temperature features such as lava flows and lava lakes, thermal properties of summit lakes, eruption plumes, and fumaroles; investigation of volcano lithologies; topography of plumes |
| MISR | Stereo imagery of surface topography; determining velocities of rapidly moving flows (e.g. mudflows and pyroclastic flows) |
| MODIS | Eruption detection by means of low resolution (1 km per pixel) thermal-infrared monitoring to search for thermal and $SO_2$ anomalies; eruption plume dispersal rates over days/weeks |
| SAR | Separate orbital platform to other EOS instruments. Morphological and topographical changes during eruptions; rates of lava flow advance at night and under adverse weather conditions; studies of volcano morphology in cloud-prone areas |
| *Volcanic emissions* | |
| AIRS | Temperature and humidity profiles |
| MLS | Vertical profiles of $SO_2$ (15–30 km altitude), HCl (15–60 km), and $H_2O$ (10–90 km) |
| TES | Tropospheric $SO_2$, CO, HCl (daytime only), $H_2S$ (high concentrations only), and qualitative amounts of $H_2O$ and $CO_2$ |
| *Volcanic aerosols* | |
| MISR | Aerosol data and cloud angular reflectance |
| SAGE III | Tropospheric and stratospheric aerosols and $H_2O$ |

to have its first platforms in orbit by 1998. As one of the 28 interdisciplinary investigations selected for the EOS, volcanology will benefit from a greatly increased capability in a number of important areas, including short- and long-term surveillance of selected volcanoes, the detection and analysis of eruption precursive behaviour, and eruption monitoring. The principal instruments carried on EOS satellites, which are applicable to volcanology, are detailed in Table 16.3 (updated from Mouginis-Mark et al. 1991).

Increasing application of satellite-based instrumentation to volcanology is to be welcomed. The technology required to receive and interpret data, and the cost of some kinds of data have hindered scientists and authorities in developing countries from accessing, at least in real-time or near-time, what might be crucial information, although this need not necessarily be so (Rothery 1992). This situation can be alleviated, to some extent by providing low-cost, ground-based monitoring equipment which is more readily affordable. A priority for the next decade, as pointed out by the IAVCEI IDNDR Task Group (1990) should, therefore be development of such instrumentation, particularly seismometers and tiltmeters. Such systems should not only be cheap, they should if possible also be of simple, modular design, durable, and easy to maintain and repair in a hostile environment where high-tech support is often unavailable.

The manipulation of volcanological data is also set to improve in the coming years, notably in how it is used and integrated to produce volcanic hazard and volcanic risk maps. Digital terrain models (DTMs) generated directly from satellite imagery will provide detailed information on topography, enabling purpose-developed algorithms to be used to predict, for example, the paths of lava flows or mudflows. These data will be routinely combined, using geographic information system (GIS) techniques, with a whole range of digitized geographical data such as population centres and communications routes, to define areas at risk, and to plan evacuation routes and emergency centres.

## 15.3 Conclusion

During the 1980s, nearly 30 000 lives were lost as a direct result of erupting volcanoes, a simple and straightforward way of assessing the effectiveness of current volcano monitoring, along with efforts to educate and communicate with the populations at risk, will be to look at the toll for the 1990s and the first decade of the next century. Despite the increasing numbers of people living near potentially dangerous volcanoes, these numbers should come down both in percentage and real terms. At least 22 000 of the 1980s death toll came from one small eruption which should never have had such a catastrophic impact. A volcanic disaster on the scale of Nevada del Ruiz ought not to have occurred in the late 20th century, and must not be allowed to happen again. If it does, then volcanology as a profession must shoulder some of the blame, even if this is shared with poorly informed authorities and an unwitting population. While by no means demeaning the importance of future technological developments in volcano surveillance, it may well be that the best way to avoid such a situation in the future, is for every volcanologist to become as much an educator and communicator as a scientist.

## References

Bilham, R. 1988. Earthquakes and urban growth. *Nature* **336**, 625–6.

Evans, J. R. & J. J. Zucca 1988. Active high-resolution seismic tomography of compositional wave velocity and attenuation structure at Medicine Lake volcano, northern California Cascade range. *Journal of Geophysical Research* **93**, 15016–36.

Foulger, G. R. & D. R. Toomey 1989. Structure and evolution of the Hengill-Grensdalur volcanic complex, Iceland: geology, geophysics, and seismic tomography. *Journal of Geophysical Research* **94**, 17511–22.

Hirn, A., A. Nercessian, M. Sapin, F. Ferrucci, G. Wittlinger 1991. Seismic heterogeneity of Mount Etna: structure and activity. *Geophysical Journal International* **105**, 139–153.

IAVCEI IDNDR Task Group 1990. Reducing volcanic disasters in the 1990s. *Bulletin of the Volcanological Society of Japan* **35**, 80–95.

McCreery, C. S., D. A. Walker, J. Talandier 1993. Hydroacoustics detect submarine volcanism. *Eos* **74**, 85–6.

Massonnet, D., M. Rossi, C. Carmona, F. Adragna, G. Peltzer, K. Feigl, T. Rabaute 1993. The displacement field of the Landers earthquake mapped by radar interferometry. *Nature* **364**, 138–42.

Mouginis-Mark, P. J. & P. Francis 1992. Satellite observations of active volcanoes *Episodes* **15**, 46–55.

Mouginis-Mark, P., S. Rowland, P. Francis, T. Friedman, H. Garbeil, J. Gradie, S. Self, L. Wilson, J. Crisp, L. Glaze, K. Jones, A. Kahle, D. Pieri, H. Zebker, A. Krueger, L. Walter, C. Wood, W. Rose, J. Adams, R. Wolff 1991. Analysis of active volcanoes from the Earth Observing System. *Remote Sensing of the Environment* **36**, 1–12.

Oppenheimer, C., P. W. Francis, D. A. Rothery, R. W. Carlton, L. S. Glaze 1993. Infrared image analysis of volcanic thermal features: Láscar volcano, Chile 1984–1991. *Journal of Geophysical Research* **98**, 4269–86.

Peterson, D. W. 1986. Volcanoes: tectonic setting and impact on society. In *Studies in geophysics: active tectonics*, Panel on Active Tectonics, 231–46. Washington DC: National Academy Press.

Rossi, M. 1994. Potential of SAR interferometry in assessment and prediction of natural hazards. In *Natural hazards and remote sensing*, G. Wadge (ed.), 39–43. London: The Royal Society/Royal Academy of Engineering.

Rothery, D. A. 1992. Monitoring and warning of volcanic eruptions by remote sensing. In *Geohazards: natural and man-made*, G. J. H. McColl, D. J. C. Laming, S. C. Scott (eds), 25–32. London: Chapman & Hall.

Staudigel, H., F. K. Wyatt, J. O. Orcutt 1991. Ocean bottom tiltmeter developed for submarine volcano monitoring. *Eos* **72**, 289, 293, 294.

Talandier, J. 1989. Submarine volcanic activity: detection, monitoring, and interpretation. *Eos* **70**, 567–9.

Tilling, R. I. 1989. Volcanic hazards and their mitigation. *Reviews of Geophysics* **27**, 237–69.

# APPENDIX

## Safety measures for volcanologists

(Recommendations of the International Association of Volcanology and the Chemistry of the Earth's Interior (IAVCEI) sub-committee for reviewing the safety of volcanologists)

## 1. Planning and logistics

1.1 It is recommended that individuals or groups starting research activity on a volcano, should incorporate a comprehensive safety plan into their overall research strategy. This will minimize hazards while working on the volcano and may save lives.

1.2 During the planning stage, it is strongly advisable to contact the local authorities responsible for civil defence, disaster mitigation, and rescue. Ideally, emergency procedures should be discussed with these groups prior to commencing field activities.

1.3 The work schedule of a field party should be lodged with the local authorities, volcano observatory staff, or colleagues stationed outside the hazardous zone.

1.4 A researcher or group of researchers planning to study an unfamiliar volcano are advised to contact local scientists who may be more familiar with potential hazards.

1.5 The size of any research group should be optimized to minimize risk. Working alone should be avoided as far as possible while at the other extreme, large groups, such as those attending field excursions associated with scientific meetings, should be discouraged from visiting hazardous areas.

1.6 The company of members of the public, including tourists and journalists, should be discouraged.

1.7 Research groups or individuals intending to work in potentially hazardous areas may wish to consider both personal insurance and the preparation and signing of appropriate indemnity declarations prior to the commencement of fieldwork.

1.8 Members of research teams should be encouraged to have some experience of first-aid and other safety procedures.

1.9 Safety precautions should take account of both altitude and meteorological conditions, particularly at high altitude during Winter, when experience of mountain survival techniques and procedures would be indispensable.

## 2. Field operations

2.1    Researchers should make full use of available information on precursory phenomena leading to eruption at the particular volcano under study. Consultation with local scientists is essential for this information to be as complete as possible.

2.2    It is strongly advisable for workers in hazardous zones to maintain radio contact with the local volcano observatory or local authorities, who may be able to provide advance warning of impending activity by means of monitoring seismicity and other parameters.

2.3    A work schedule should be lodged with colleagues or with the local authorities every time a field party enters a hazardous zone.

2.4    Researchers in hazardous areas should be constantly alert to danger, and should avoid hasty action. Unless absolutely neccessary for research purposes, active craters, lava flows, debris flows, and pyroclastic flows should not be approached.

2.5    Field parties should minimize the time spent in a hazardous area.

2.6    Wherever possible, field operations should be undertaken upwind of active craters, solfatara, and fumaroles.

2.7    Care should be taken to select routes that minimize fatigue during fieldwork. Similarly, routes should avoid fresh lava surfaces, and topographic lows which may channel debris and pyroclastic flows or trap noxious gases.

2.8    An escape route should always be determined in advance.

## 3. Equipment

3.1    Hand-held two-way radios are highly desirable for communication between members of a field party and with contacts outside the hazardous area.

3.2    A protective helmet (with chin strap) is essential.

3.3    Gas masks are strongly recommended when working in fumes or ash clouds. Care should be taken to obtain the appropriate filters, and spare filters should be carried.

3.4    Clothing should be suitable for harsh weather conditions, and should ideally offer some protection against physical impact and heat. Brightly coloured clothing will increase the visibility of field party members and aid rescue operations in the event of an emergency.

3.5    Heavy-duty boots with good ankle support are recommended.

3.6    Gloves are essential when traversing fresh or recently erupted lavas.

3.7    A basic first-aid kit should always be carried.

3.8    A back-pack together with ample food and (especially) water supplies is essential for all work in active volcanic terrains.

3.9    Essential navigation and safety gear should also be carried, including topographic maps, compass, altimeter, knife, whistle, and signal mirror.

3.10  Identification tags indicating the name of the researcher, blood group, and next of kin, could aid in making rescue operations speedier and more effective.

3.11 Protective eyewear (e.g. goggles) may be required in some circumstances and should be carried.

The wording of the above text has been slightly altered to improve readability, but the recommendations remain essentially those published in *WOVO News* in Autumn 1993: Aramaki, S., F. Barberi, T. Casadevall, S. McNutt, 1993. "Report of the IAVCEI sub-committee for reviewing the safety of volcanologists", *WOVO News*, **4**, 12–14.

# Index

For Product Safety Concerns and Information please contact our EU
representative  GPSR@taylorandfrancis.com
Taylor & Francis Verlag GmbH, Kaufingerstraße 24, 80331 München, Germany

www.ingramcontent.com/pod-product-compliance
Lightning Source LLC
Chambersburg PA
CBHW080138220326
41598CB00032B/5104